# Electric Field Analysis

# Electric Field Analysis

Sivaji Chakravorti

**CRC Press**
Taylor & Francis Group
Boca Raton London New York

CRC Press is an imprint of the
Taylor & Francis Group, an **informa** business

CRC Press
Taylor & Francis Group
6000 Broken Sound Parkway NW, Suite 300
Boca Raton, FL 33487-2742

First issued in paperback 2020

© 2015 by Taylor & Francis Group, LLC
CRC Press is an imprint of Taylor & Francis Group, an Informa business

No claim to original U.S. Government works

ISBN 13: 978-0-367-57585-4 (pbk)
ISBN 13: 978-1-4822-3336-0 (hbk)

# Contents

# Foreword

The knowledge of the physics of electrical discharges and of the electric field distribution are essential design criteria for apparatus applied in physics, medicine and high-voltage engineering. Intensive research work was done particularly in the past, as the support by computers was available. This research work is the subject of numerous articles in scientific journals. For engineers, physicists or students, it is of value that this variety of information is compiled in books by competent scientists experienced in this research field. Dr. Sivaji Chakravorti, professor of electrical engineering at Jadavpur University, Kolkata, India, presents his book *Electric Field Analysis* giving a deep insight into the subject.

The first part of the book (Chapters 1 through 3) introduces the physical and mathematical fundamentals of electrical fields, with explanations to Gauss's law and orthogonal coordinate systems. Chapter 4 considers simple single-dielectric configurations, where an analytical mathematical treatment is possible. The subject of Chapter 5 is dielectric polarization. Electrostatic boundary conditions, multi-dielectric configurations and electrostatic pressures on boundary surfaces are treated in Chapters 6 through 8. These more general subjects are concluded by the description of the classical methods of field determination: method of images (Chapter 9) supplemented by the application to the field calculation of a sphere or a cylinder in uniform external fields (Chapter 10), conformal mapping (Chapter 11) and graphical field plotting (Chapter 12).

The subject of the main part of this book is the numerical computation of electrical fields, after an introduction into these techniques (Chapter 13). The finite difference method (Chapter 14), the finite element method (Chapter 15), the charge simulation method (Chapter 16) and the numerical computation by the surface charge simulation method including the indirect boundary element method (Chapter 17) are described and explained. The possibilities of every method are shown by examples taken from engineering practise and suited as benchmark models for validation. In addition, interesting case studies based on own research works are presented (Chapter 18): electrical field distribution in a cable termination, around a post insulator, in a condenser bushing and around a Gas Insulated Substation (GIS) spacer.

The subject matter of the last chapter (Chapter 19) deals with considerations on the application of numerical field calculation for the electric field optimization, limited to a review of published works; the demonstration of contour correction techniques and the application of artificial neural networks for the optimization of electrode and insulator contours.

The author is a fellow of the Indian National Academy of Engineering, a fellow of the National Academy of Sciences India and is a senior member of the

IEEE (United States). Recently, he was elected as an honorary ambassador of the Technical University of Munich (Germany). He has published numerous papers in refereed international and Indian scientific journals. His research fields are numerical field computation, condition monitoring of transformers, partial discharge analysis, computer-aided design and optimization of insulation systems, application of artificial intelligence in high-voltage systems and lifelong learning techniques.

This book is a valuable tool for engineers and physicists engaged in the design work of high-voltage insulation systems. It opens a view into the research work done in the past and also into the present state of the art. It is a result of the long-time research work of the author in this field.

**Hans Steinbigler**
*Technical University of Munich*

# *Preface*

Since my days as an undergraduate engineering student, I was used to hearing comments from students that field theory is one of the most difficult subjects in the undergraduate curriculum. Later on, as a teacher, I have experienced the fact that not many teachers volunteer to teach field theory for reasons known to them. Moreover, over two decades, I have been told by my students that field theory is taught mostly from a mathematical viewpoint, and the physical understanding of issues related to field theory is normally not discussed in the classes. In this context, my own view is that it is true that nobody has seen electric field. But it exists and many physical phenomena are dependent on the nature of the electric field distribution. Therefore, it is very important to have a physical understanding of the field theory topics, so that one can correlate field theory and the associated physical phenomena. But at the same time, I am not stating that one should not learn the mathematical foundations of field theory, because field-related problem-solving requires a sound knowledge of mathematical aspects of field analysis.

I did my doctoral as well as post-doctoral works on numerical analysis of electric fields and later took up teaching electric field theory for more than one and half decades both at the undergraduate and postgraduate levels at Jadavpur University, Kolkata, India. Over the years, I have had brilliant students at both the undergraduate and postgraduate levels who have asked thought-provoking questions regarding electric field analysis. Quite a few PhD scholars have carried out outstanding works on numerical electric field analysis and electric field optimization under my supervision. They have constantly poked me to write a book on electric field analysis, emphasizing the physical understanding of field theory. Finally, in 2013, I decided to accede to the demands of my students and took up writing of this book on electric field analysis, which would not only present the mathematical aspects but also would strongly deal with the physical understanding of the basic issues of electric field theory.

This book is written with an eye on power engineering, and most of the examples that bring forth the real-life issues related to electric field analysis are from power-engineering applications in general and dielectric engineering in particular. It is to be made clear here that this book is not intended to be a book that helps in solving problems related to electric field, nor is it a book that intends to look everything from a detailed mathematical viewpoint. All the theories have been discussed in such a way that the reader gets an idea of how these theories are useful in real life. The problems given in this book are mainly intended to make the understanding of the theories clear. Wherever possible, I have tried to give objective type questions, which could be used to test whether a student has understood the finer yet practical points of electric field analysis.

This book is clearly divided into two parts. The first part, comprising Chapters 1 through 12, deals with the conventional aspects of electric field theory. These topics are not new and several other books discussing these issues are available. But I have tried to present these fundamental aspects in a way which is easily accepted by my students over the years. The second part of the book, comprising Chapters 13 through 19, concentrates on the numerical analysis and optimization of electric field. Here also, I have tried not to burden the readers with too many theoretical issues of numerical analysis. My target here is to help readers to develop their own codes after going through the chapters. Whether I have succeeded or not is left to the readers.

I thought that I should incorporate some practical case studies of numerical analysis of electric field to give readers a clear idea of how to carry out electric field analysis in a real-life case. All the examples are taken from power-engineering applications, which the students will be able to connect with their acquired knowledge. In the recent past, the optimization of engineering system has been a major area of research and the optimization of high-voltage system is no exception. Therefore, in the end, I have included a chapter on high-voltage field optimization. In this chapter, the aim is not to present optimization techniques, but the goal is to discuss how high-voltage field optimization studies are carried out by researchers. To highlight the evolution of such optimization studies, classical as well as soft computing-based optimization studies have been presented in the concerned chapter.

Over the years, I have developed a software package called *TWIN*, which is based on surface charge simulation method for electric field analysis and is copyrighted in my name. TWIN does not have a user-friendly graphical user interface, but is very easy to use in conjunction with simple help files that are written by me. I proposed to the publisher that this software package may be given to the purchasers of this book as a downloadable code. The publisher agreed to this request and hence I am offering TWIN© in tandem with this book.

I would like to emphasize here that this book is neither intended for the only few who are at the very top of any group, which studies electric field theory, nor is it intended to be a made easy book. I have tried to write in such a way that this book is useful for a majority of students.

I am thankful to many who have helped me to prepare this manuscript in different ways. I cannot name them all, as I will definitely miss some of them. But I will be failing in my duties if I do not mention Dr. Biswendu Chatterjee, because without his support this book could not have been completed. I would also like to mention Dr. Kesab Bhattacharya, Dr. Abhijit Lahiri, Mr. Sandip Saha Chowdhury and Mr. Arijit Baral for their valuable inputs and help.

I am indebted to Prof. Hans Steinbigler of Technical University of Munich in so many ways starting from my post-doctoral days in the university at Munich in the mid-1990s. I thought it to be an honour to request him to write the foreword of this book and express my heartfelt thanks to him for agreeing to write the foreword.

Finally, I would like to express my gratitude to my family members, particularly my wife and daughters who beared with me, when I neglected my family duties to prepare this manuscript. It was also a pleasure to work with the publishing team of Taylor & Francis and I express my sincere thanks to all members of this publishing team.

I am sure that this book is not free from mistakes in spite of my best efforts. The readers are most welcome to send their constructive suggestions to me either on the content of the book or on TWIN. Please write to me at s_chakrav@yahoo.com.

Additional material is available from the CRC Press website: http://www.crcpress.com/product/isbn/9781482233360.

**Sivaji Chakravorti**
*Jadavpur University*

MATLAB® is a registered trademark of The MathWorks, Inc. For product information, please contact:

The MathWorks, Inc.
3 Apple Hill Drive
Natick, MA 01760-2098, USA
Tel: +1 508 647 7000
Fax: +1 508 647 7001
E-mail: info@mathworks.com
Web: www.mathworks.com

# Author

Dr. Sivaji Chakravorti did his bachelor of electrical engineering, master of electrical engineering and PhD from Jadavpur University, Kolkata, India, in 1983, 1985 and 1993, respectively. He has 29 years of teaching experience and is a full professor in electrical engineering at Jadavpur University since 2002. He teaches electric field analysis at UG and PG levels for more than one and half decades. In 1984, he worked at Indian Institute of Science, Bangalore, as Indian National Science Academy Visiting Fellow. He worked at the Technical University of Munich, Germany, as Humboldt Research Fellow during 1995–1996, 1999 and 2007, respectively. In 1998, he served as development engineer in Siemens AG in Berlin, Germany. He has also worked as Humboldt Research Fellow in ABB Corporate Research at Ladenburg, Germany, in 2002. He worked as US-NSF guest scientist at the Advanced Research Institute of Virginia Tech, Alexandria, Virginia, in 2003, and as guest scientist at Technical University Hamburg-Harburg, Germany, in 2005. He is the recipient of the Technical University Munich Ambassador Award in 2013 and the All India Council for Technical Education (AICTE) Technology Day Award for best R&D project in 2003. He has published more than 160 research papers, authored two books, edited three books and developed three online courses. He is a fellow of Indian National Academy of Engineering, fellow of National Academy of Sciences India, senior member of the Institute of Electrical and Electronics Engineers (IEEE) and distinguished lecturer of IEEE Power and Energy Society. He is currently an associate editor of *IEEE Transactions on Dielectrics and Electrical Insulation*. He was the chairman of IEEE Kolkata section during 2011–2012, chairman of its Power Engineering Chapter during 2003–2006 and the founder chairman of its Dielectrics and Electrical Insulation Society Chapter in 2012–2014. In 2001, he was the vice-chairman of IEEE India Council. He was the Asia Pacific West Chapter Representative of IEEE Power and Energy Society (IEEE PES) in 2008 and is currently the Global Chapter secretary of IEEE PES (2013–2014). He is actively involved in several sponsored projects funded by US-NSF, AICTE, Ministry of Human Resource Department (Govt. of India), Department of Science and Technology (Govt. of India) and World Bank. He provides consultancy services in the area of high-voltage engineering to large number of companies and power utilities. He has given many invited talks and has been the key resource person of several workshops in India and abroad.

# 1

## Fundamentals of Electric Field

**ABSTRACT**  Leaving aside nuclear interactions, there are two non-contact forces that act at a distance, namely, gravitational and electric forces. Gravitational force is dominant at large distances whereas electric forces are dominant at shorter distances. The root cause of electric forces is electric charge. The effect of electric charge is spread over the entire space around it, but it falls rapidly with distance. The presence of an electric field is detected by observing the force on a charged body located within the field region. Electric field intensity is obtained by dividing the electric force by the magnitude of the test charge. To have an electric field parameter, which is independent of the charge of the test body, electric potential is introduced in the analysis, such that electric field intensity, which is a vector quantity, is the spatial derivative of electric potential, which is a scalar quantity. For a given material, electric field intensity depends on the electric flux density, which in turn depends on the amount of source charges present in the field region. For the purpose of electric field analysis, several types of charge configurations are considered, such as point, line, ring and disc charges.

## 1.1 Introduction

When I was a student of third semester in engineering degree course, I had an unassuming yet a man of profound knowledge as our teacher who taught us electric field theory. In the very first class, he asked us a simple question: 'Could you name just one thing which is the cause of electric field?' Then he went on to explain the answer to that question, which is 'Electric Charge'. It struck a chord somewhere within me, and I explored further on the effects of electric charge. From that day onwards, I was hooked to electric charge.

To quote the legendary physicist, Richard Feynman, 'Observation, reason, and experiment make up what we call the scientific method'. This is precisely true for electric field theory. Students often ask me how I discussed so many things about electric field when nobody has seen electric field. Yes, it is true that electric field cannot be directly seen. But observations on physical occurrences and related reasoning give theoretical understanding and when such understanding is validated by experimentation, then it establishes the existence of electric field.

## 1.2 Electric Charge

Even in the seventeenth century, it was known that matters exert force on each other, which varies inversely as the square of the distance between them. It is the so-called long-range interaction, commonly known as *gravitation*. But what is the nature of interaction between two things when the distance between them is very small? Is it gravitation? The answer is no. Gravitation is very weak at that dimension. Then it must be some other force. Observation shows that it is analogous to gravitational force in the sense that it also varies inversely as the square of the distance between the two things. But it is also observed that there is a big difference between gravity and this short-range force. In gravitation, any one matter attracts another matter. But this is not the case in this short-range force. Here, dissimilar things attract and similar things repel each other. In other words, there are two different types of things involved in short-range interactions. This thing, which causes the short-range interaction, has been named as 'electric charge'. It has been found that electric charge is a basic property of matter carried by elementary particles. Typically, electric charges are of two types: positive and negative charges. When the atomic structure was properly understood, it was found that the positive charges (primarily protons) are located at the centre of the atom, that is, nucleus, and the negative charges (electrons) revolve around the nucleus.

Experimental evidence shows that all electrons have the same amount of negative charge, which is also equal to the amount of positive charge of each proton. Consequently, it follows that charge exists in quantized unit equal to the charge of an electron or a proton ($e$), which is a fundamental physical constant. Thus electric charge of anything comes in integer multiples of the elementary charge, $e$, except for particles called *quarks*, which have charges that are integer multiples of $e/3$. The unit of electric charge in the SI system is coulomb (C). One C consists of $6.241509324 \times 10^{18}$ natural units of electric charge, such as charge of individual electrons or protons. Conversely, one electron has a negative charge of $1.60217657 \times 10^{-19}$ C and one proton has a positive charge of $1.60217657 \times 10^{-19}$ C. Other particles (e.g. positrons) also carry charge in multiples of the electronic charge magnitude. However, these are not going to be discussed for the sake of simplicity.

Electric charge is also conserved; that is, in any isolated system or in any chemical or nuclear reaction, the net electric charge is constant. The algebraic sum of the elementary charges remains the same. In physical terms, it implies that if a given amount of negative charge appears in one part of an isolated system, then it is always accompanied by the appearance of an equal amount of positive charge in another part of the system. In modern atomic theory, it has been proved that although fundamental particles of matter continually and spontaneously appear, disappear and change into one another, they always obey the constraint that the net quantity of charge is preserved.

## 1.3 Electric Fieldlines

From logical reasoning, it can be stated that if a charge is present in space, then another charge will experience a force when brought into that space. In other words, the effect of the first charge, which may also be called the *source charge*, extends into the space around it. This is known as the electric field caused by the source charge. If there is no charge in space, then there will be no electric field. If the source charge is of positive polarity and the test charge is of negative polarity, then the test charge will experience an attractive force. On the other hand, if the test charge is also of positive polarity, then it will experience a repulsive force. If the test charge is free to move, then it will move in accordance with the direction of the force. The loci of the movement of the test charge within the electric field are known as *electric lines of force* or *electric fieldlines*.

Behaviour of an electric field is conventionally analyzed considering the test charge to be a unit positive charge. Hence, the test charge will experience repulsive force from a positive source charge and attractive force from a negative source charge. Therefore, the test charge will move away from the positive source charge and move towards the negative source charge. Accordingly, the directions of electric fieldlines are such that they originate from a positive charge and terminate on a negative charge, as shown in Figure 1.1. Figure 1.1a shows the electric fieldlines originating from a positive source charge whose magnitude is integer ($N$) multiple of $+e$, whereas Figure 1.1b shows the electric fieldlines terminating on a negative source charge of the same magnitude. The electric fieldlines are the directions of the force experienced by a unit positive charge $+e$, as shown in Figure 1.1.

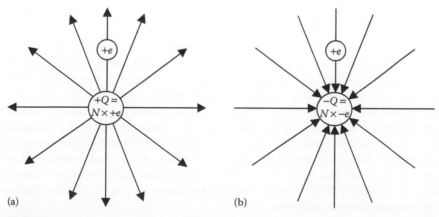

**FIGURE 1.1**
Electric fieldlines due to (a) positive source charge and (b) negative source charge.

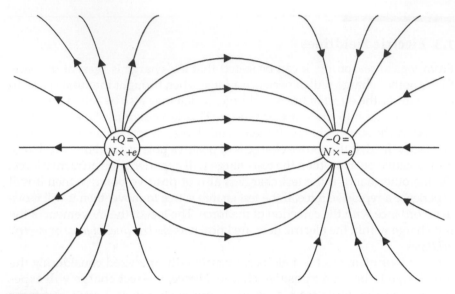

**FIGURE 1.2**
Electric fieldlines due to a pair of positive and negative charges.

Figure 1.2 depicts the electric fieldlines due to a pair of positive and negative charges, showing the fieldlines to be originating from the positive charge and terminating on the negative charge.

According to SI unit system, one C of source charge gives rise to one C of electric fieldlines.

## 1.4 Coulomb's Law

It is the law that describes the electrostatic interaction between electrically charged particles. It was published by the French physicist Charles Augustin de Coulomb in 1785. He determined the magnitude of the electric force between two point charges using a torsion balance to study the attraction and repulsion forces of charged particles. The interaction between charged particles is a non-contact force that acts over some distance of separation. There are always two charges and a separation distance between them as the three critical variables that influence the strength of the electrostatic interaction. The unit of the electrostatic force, like all forces, is Newton. Being a force, the strength of the electrostatic interaction is a vector quantity that has both magnitude and direction.

According to the statement of Coulomb's law (1) the magnitude of the electrostatic force of interaction between two point charges is directly

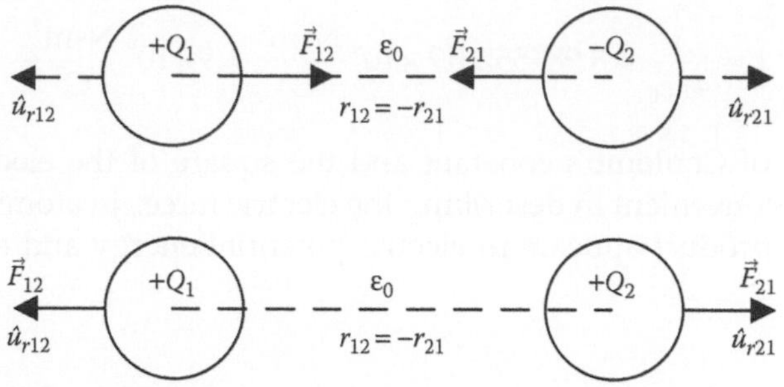

**FIGURE 1.3**
Electrostatic forces of interaction between two point charges.

proportional to the scalar multiplication of the magnitudes of point charges and inversely proportional to the square of the separation distance between the point charges and (2) the electrostatic force of interaction acts along the straight line joining the two point charges. If the two point charges are of same polarity, the electrostatic force between them is repulsive; if they are of opposite polarity, the force between them is attractive.

It should be noted here that two conditions are to be fulfilled for the validity of Coulomb's law: (1) the charges involved must be point charges and (2) the charges should be stationary with respect to each other.

Mathematically, the force between two point charges, as shown in Figure 1.3, could be written as follows:

$$\vec{F}_{21} = \frac{\pm Q_2 \cdot Q_1}{4\pi\varepsilon_0 \left|\vec{r}_{21}\right|^2} \hat{u}_{r21} \text{ and } \vec{F}_{12} = \frac{Q_1 \cdot \pm Q_2}{4\pi\varepsilon_0 \left|\vec{r}_{12}\right|^2} \hat{u}_{r12} \text{ so that } \vec{F}_{21} = -\vec{F}_{12} \tag{1.1}$$

where:
$\varepsilon_0$ is permittivity of free space ($\approx 8.854187 \times 10^{-12}$ F/m)
$\hat{u}_{r21} = \vec{r}_{21} / \left|\vec{r}_{21}\right|$ and $\hat{u}_{r12} = \vec{r}_{12} / \left|\vec{r}_{12}\right|$ are the unit vectors
$\vec{r}_{21} = \vec{r}_2 - \vec{r}_1$ and $\vec{r}_{12} = \vec{r}_1 - \vec{r}_2$ are the distance vectors where $\vec{r}_1$ and $\vec{r}_2$ are the position vectors of the location of the point charges $Q_1$ and $Q_2$, respectively, with respect to a defined origin

When the scalar product of $Q_1$ and $\pm Q_2$ is positive, the force is repulsive, and when the product is negative, the force is attractive.

### 1.4.1 Coulomb's Constant

The constant of proportionality $k$ that appears in Coulomb's law is often called *Coulomb's constant*. In the SI unit system,

$$k = \frac{1}{4\pi\varepsilon_0} = 8.987552617 \times 10^9 \frac{\text{N} \cdot \text{m}^2}{\text{C}^2} \approx 9 \times 10^9 \frac{\text{N} \cdot \text{m}^2}{\text{C}^2} \qquad (1.2)$$

The product of Coulomb's constant and the square of the electron charge ($ke^2$) is often convenient in describing the electric forces in atoms and nuclei, because that product appears in electric potential energy and electric force expressions.

### 1.4.2 Comparison between Electrostatic and Gravitational Forces

The expression for electrostatic force, as obtained from Coulomb's law, bears a strong resemblance to the expression for gravitational force given by Newton's law for universal gravitation.

$$\left| \vec{F}_{\text{electrostatic}} \right| = k \cdot \frac{Q_1 Q_2}{r^2} \text{ and } \left| \vec{F}_{\text{gravitational}} \right| = G \cdot \frac{m_1 m_2}{r^2} \qquad (1.3)$$

where:
$k \approx 9 \times 10^9 \text{ N} \cdot \text{m}^2/\text{C}^2$
$G \approx 6.67 \times 10^{-11} \text{ N} \cdot \text{m}^2/\text{kg}^2$

Both the expressions show that the force is (1) inversely proportional to the square of the separation distance and (2) directly proportional to the scalar product of the quantity that causes the force, that is, electric charge in the case of electrostatic force and mass in the case of gravitational force. But, there are major differences between these two forces. First, gravitational forces are only attractive, whereas electrical forces can be either attractive or repulsive. Second, a comparison of the proportionality constants reveals that the Coulomb's constant ($k$) is significantly greater than Newton's universal gravitational constant ($G$). Consequently, the electrostatic force between two electric charges of unit magnitude is significantly higher than the gravitational force between two masses of unit magnitude.

From Equation 1.3, it is seen that the electrostatic force between two electric charges of magnitude 1 C separated by a distance of 1 m will be a colossal $9 \times 10^9$ *N*! On the other hand, the gravitational force between two masses of magnitude 1 kg separated by a distance of 1 m will be a meagre $6.67 \times 10^{-11}$ *N*! These values clearly show the enormous difference between magnitudes of electrostatic and gravitational forces.

The comparison can also be made between the electrostatic and gravitation forces between two electrons separated by a given distance. Considering the charge of an electron as $e$ ($1.60217657 \times 10^{-19}$ C) and the mass of the electron as $m_e$ ($9.10938291 \times 10^{-31}$ kg), the ratio of electrostatic to gravitational forces between two electrons is given by

$$\frac{\left|\vec{F}_{\text{electrostatic}}\right|}{\left|\vec{F}_{\text{gravitational}}\right|} = \left(\frac{e}{m_e}\right)^2 \frac{k}{G} \approx 4.174 \times 10^{42} \tag{1.4}$$

The above equation shows how strong electrostatic forces are compared to gravitational forces! If this reasoning is applied to the motion of particles in the universe, one may expect the universe to be governed entirely by electrostatic forces. However, this is not the case. The electrostatic force is enormously stronger than the gravitational force, but is usually hidden inside neutral atoms. On astronomical length scales, gravity is the dominant force and electrostatic forces are not relevant. The key to understanding this paradox is that electric charges could be of either positive or negative polarity, whereas the masses that cause gravitational forces are only positive, as there is nothing called *negative mass*. This means that gravitational forces are always cumulative, whereas electrical forces can cancel each other. For the sake of an argument, consider that the universe starts out with randomly distributed electric charges. Initially, electrostatic forces are expected to completely dominate gravity. Because of the dominant electrostatic forces, every positive charge tries to get as far away as possible from the other positive charges, and to get as close as possible to the other negative charges. After some time, the positive and negative charges come near enough ($\approx 10^{-10}$ m) to form close pairs. Exactly how close the charges would come is determined by quantum mechanics. The electrostatic forces due to the charges in each pair effectively cancel one another out on length scales that are much larger than the mutual spacing of the charge pair. If the number of positive charges in the universe is almost equal to the number of negative charges, then only the gravity becomes the dominant long-range force. For effective cancellation of long-range electrostatic forces, the relative difference in the number of positive and negative charges in the universe must be extremely small. In fact, cancellation of the effect of positive and negative charges has to be of such accuracy that most physicists believe that the net charge of the universe is exactly zero. In other words, electric charge is a conserved quantity; that is, the net charge of the universe can neither increase nor decrease. As of today, no elementary particle reaction has been discovered that creates or destroys electric charge.

### 1.4.3 Effect of Departure from Electrical Neutrality

The fine balance of electrostatic forces due to positive and negative electric charges starts to break down on atomic scales. In fact, interatomic and intermolecular forces are all electrical in nature. But, this is electric field on the atomic scale, usually termed as *quantum electromagnetism*. This book is about classical electromagnetism, which is electromagnetism on length scales much larger than the atomic scale. Classical electromagnetism generally describes phenomena in which some sort of disturbance is caused to matter, so that the

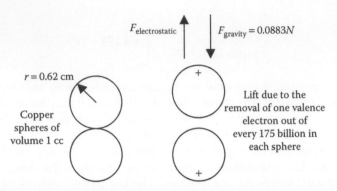

**FIGURE 1.4**
Lifting of a copper sphere due to electrostatic force against gravity.

close pairing of positive and negative charges is disrupted. Such disruption allows electrical forces to manifest themselves on macroscopic length scales. Of course, very little disruption is necessary before gigantic forces are generated, which may be explained with the help of the following example.

Figure 1.4 shows two copper spheres of volume 1 cubic centimetre (cc) lying on one another. Copper is a good electrical conductor and has one valence electron in the outermost shell of its atom, and that electron is fairly free to move about in the volume of solid copper material. The density of metallic copper is approximately 9 g/cc and one mole of copper is 63.55 g. Thus 1 cc of copper contains approximately $8.5 \times 10^{22} [(9/63.55) \times 6.022 \times 10^{23}]$ copper atoms. With one valence electron per atom, and with the electron charge of $1.6 \times 10^{-19}$ C, there are about 13,600 C of potentially mobile charge within a volume of 1 cc of copper. How much electron charge needs to be removed from two spheres of copper, so that there is enough net positive charge on them to suspend the top sphere over the bottom? The force required to lift the top sphere of copper against gravity would be its weight, that is, 0.0883 ($\approx 9 \times 10^{-3} \times 0.807)N$. It is fair to assume that the net charge resides at the points of the spheres most distant from each other because of the charge repulsion. The radius of a sphere of volume 1 cc is 0.62 cm. Therefore, the repulsive force to be considered should be that between two point charges 2.48 cm apart, that is, twice the sphere diameter apart. From Coulomb's law

$$0.0883 = \frac{1}{4\pi\varepsilon_0} \times \frac{Q^2}{0.0248^2} \text{ or } Q \approx 7.75 \times 10^{-8} \text{ C} \tag{1.5}$$

Compared to the total valence charge of approximately 13,600 C, this $7.75 \times 10^{-8}$ C amounts to removing just one valence electron out of every 175 billion copper atoms from each sphere. In summary, the removal of just one out of every 175 billion free electrons from each copper sphere would cause enough electrostatic repulsive force on the top sphere to lift it, overcoming the gravitational pull of the entire Earth!

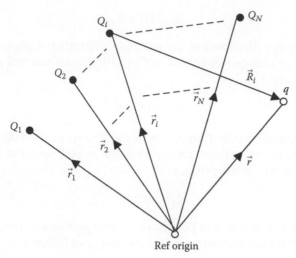

**FIGURE 1.5**
Force due to a system of discrete charges.

### 1.4.4 Force due to a System of Discrete Charges

Consider $N$ charges, $Q_1$ through $Q_N$, which are located at position vectors $\vec{r}_1$ through $\vec{r}_N$, as shown in Figure 1.5. Because the electrostatic forces obey the principle of superposition, the electrostatic force acting on a test charge $q$ at position vector $\vec{r}$ is simply the vector sum of all of the forces from each of the $N$ charges taken in isolation. Thus, the total force acting on the test charge $q$ is given by

$$\vec{F}(r) = q \sum_{i=1}^{N} \frac{q_i}{4\pi\varepsilon_0} \frac{\vec{r} - \vec{r}_i}{\left|\vec{r} - \vec{r}_i\right|^3} \tag{1.6}$$

where:
the distance vector $\vec{R}_i = \vec{r} - \vec{r}_i$ is directed from the $i$th charge $Q_i$ to $q$

### 1.4.5 Force due to Continuous Charge Distribution

Instead of having discrete charges, consider a continuous distribution of charge represented by a charge density, which could be linear, surface or volume charge density, depending on the distribution of charge. For a continuous charge distribution, an integral over the entire region containing the charge is equivalent to a summation for infinite number of discrete charges, where each infinitesimal element of space is treated as a discrete point charge $dq$.

For linear charge distribution, for example, charge in a wire, considering linear charge density as $\lambda(r')$ and the infinitesimal line element $dl'$ at the position $r'$,

$$dq = \lambda(r')dl'$$

For surface charge distribution, for example, charge on a plate or disc, considering surface charge density as $\sigma(r')$ and the infinitesimal area element $dA'$ at the position $r'$,

$$dq = \sigma(r')dA'$$

For volume charge distribution, for example, charge in the volume of a bulk material, considering volume charge density as $\rho(r')$ and the infinitesimal volume element $dV'$ at the position $r'$,

$$dq = \rho(r')dV'$$

The force on a test charge $q$ at position $r$ in free space is given by the integral over the entire continuous distribution of charge as follows:

$$\vec{F}(r) = q \int \frac{dq}{4\pi\varepsilon_0} \frac{\vec{r} - \vec{r'}}{\left|\vec{r} - \vec{r'}\right|^3} \tag{1.7}$$

In the above equation the integration is line, surface or volume integral according to the nature of charge distribution. The integral is over all space, or, at least, over all space for which the charge density is non-zero.

---

## 1.5 Electric Field Intensity

At this juncture, it is useful to define a vector field $\vec{E}(r)$, called the *electric field intensity*, which is the force exerted on a unit test charge of positive polarity located at position vector $\vec{r}$. Then, the force on a test charge could be written as follows:

$$\vec{F}(r) = q\vec{E}(r) \tag{1.8}$$

The electric field intensity could be written from Equation 1.6 as follows:

$$\vec{E}(r) = \sum_{i=1}^{N} \frac{q_i}{4\pi\varepsilon_0} \frac{\vec{r} - \vec{r_i}}{\left|\vec{r} - \vec{r_i}\right|^3} \tag{1.9}$$

or, from Equation 1.7 as follows:

$$\vec{E}(r) = \int \frac{dq}{4\pi\varepsilon_0} \frac{\vec{r} - \vec{r'}}{\left|\vec{r} - \vec{r'}\right|^3} \tag{1.10}$$

The electric fieldlines from a single charge $Q$ located at a given position are purely radial and are directed outwards if the charge is positive or inwards if it is negative, as shown in Figure 1.1. Therefore, the electric field intensity at any point located at a radial distance $r$ from the source charge $Q$ will be the force experience by a unit test charge of positive polarity at that point and is given by

$$\vec{E}(r) = \frac{Q}{4\pi\varepsilon_0 r^2}\hat{u}_r \tag{1.11}$$

The unit of electric field intensity as per the above definition is $N/C$. However, the practical unit of electric field intensity is a different one and will be discussed in Section 1.7.

## 1.6 Electric Flux and Electric Flux Density

Consider the case of air coming in through a window. The amount of air that comes through the window depends on the speed of the air, the direction of the air and the area of the window. The air that comes through the window may be called the *air flux*.

Similarly, the amount of electric fieldlines that pass through an area is the electric flux through that area. Consider the case of a source point charge of positive polarity, as shown in Figure 1.1a. If the source charge magnitude is $Q$ C, then the total amount of electric fieldlines coming out of the source charge will be also $Q$ C. Now, if a fictitious sphere of radius $r$ is considered such that the source charge is located at the centre of the sphere, then the electric flux through the surface of the sphere will be $Q$ C, as the surface of the sphere completely encloses the source charge, and all the electric fieldlines coming out radially from the source point charge passes through the spherical surface. Electric flux is typically denoted by $\psi$.

Electric flux density is then defined as the electric flux per unit area normal to the direction of electric flux. In the case of a point charge the electric fieldlines are directed radially from the source charge and hence the electric fieldlines are always normal to the surface of the sphere having the point charge at its center. Hence, for a point source charge of magnitude $Q$ C, the electric flux that passes through the spherical surface area of magnitude $4\pi r^2$ is $Q$. Then the electric flux density ($\vec{D}$) at a radial distance $r$ from the point charge is given by

$$\vec{D}(r) = \frac{Q}{4\pi r^2}\hat{u}_r \tag{1.12}$$

From Equations 1.11 and 1.12, it may be written that

$$\vec{D}(r) = \varepsilon_0 \vec{E}(r), \text{ when the medium is free space or,}$$

$$\vec{D}(r) = \varepsilon_0 \varepsilon_r \vec{E}(r), \text{ for any particular medium of relative permittivity } \varepsilon_r \tag{1.13}$$

Electric flux density is a vector quantity because it has a direction along the electric fieldlines at the position where electric flux density is being computed.

Equation 1.13 is known as one of the basic equations of electric field theory.

However, it is not necessary that electric flux will always be normal to the area under consideration. In such cases, the component of the area that is normal to electric flux has to be taken for computing electric flux density. Figure 1.6 shows such a case, where an electric flux of magnitude $\psi$ passes through an area of magnitude $A$, which is not normal to the direction of electric flux.

With reference to Figure 1.6, electric flux and electric flux density are related as follows:

$$\psi = \left|\vec{D}\right| A \cos\theta \tag{1.14}$$

Again, as depicted in Figure 1.6, $\hat{u}_r \cdot \hat{u}_n = \cos\theta$.

Therefore, from Equation 1.14

$$\psi = \left|\vec{D}\right| A \hat{u}_r \cdot \hat{u}_n = \left|\vec{D}\right| \hat{u}_r \cdot A \hat{u}_n = \vec{D} \cdot \vec{A} \tag{1.15}$$

Equation 1.15 presents an important idea of introducing an area vector, which is a vector of magnitude equal to the scalar magnitude of the area under

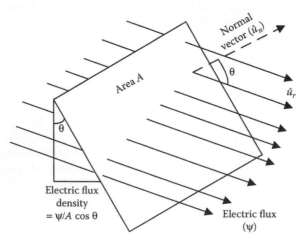

**FIGURE 1.6**
Pertaining to the area related to electric flux density.

consideration, but it has a direction normal to the area under consideration. In the case of closed surfaces, area vector is conventionally taken in the direction of the outward normal. For an open surface, any one normal direction can be taken as positive, whereas the opposite normal direction is to be taken as negative.

## 1.7 Electric Potential

Consider that a test charge of magnitude $q$ is located at a given position within an electric field produced by a system of charges. The test charge will experience a force due to the source charges. If the test charge moves in the direction of the field forces, then the work is done by the field forces in moving the test charge from position 1 to position 2. In other words, energy is spent by the electric field. Hence, the potential energy of test charge at position 2 will be lower than that at position 1. On the other hand, if the charge is moved against the field forces by an external agent, then the work done by the external agent will be stored as potential energy of the test charge. Hence, the potential energy of the test charge at position 2 will be higher than that at position 1. Here, it is to be noted that the force experienced by the test charge within an electric field is dependent on the magnitude of the test charge. Hence, the potential energy of the charge at any position is dependent on its magnitude and the distance by which it moves within the electric field.

The concept of electric potential is introduced to make it a property, which is purely dependent on the location within an electric field and is independent of the test charge. In other words, it is a property of the electric field itself and not related to the test particle. Hence, electric potential ($\phi$) at any point within an electric field is defined as potential energy per unit charge at that point and hence it is a scalar quantity. The unit of electric potential is volt (V), which is equivalent to joules per coulomb (J/C). It is interesting to note that electric field intensity is defined as force per unit charge and electric potential is defined as potential energy per unit charge.

However, it is also practically important to note that absolute values of electric potentials are not physically measurable; only difference in potential energy between two points within an electric field can be physically measured; that is, only the potential difference between two points within an electric field is measurable. The work done in moving a unit positive charge from one point to the other within an electric field is equal to the difference in potential energies and hence difference in electric potentials at the two points.

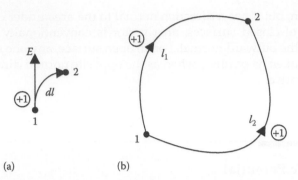

**FIGURE 1.7**
Pertaining to the definition of electric potential.

As shown in Figure 1.7a, consider that a unit positive charge is moved from point 1 to point 2 by a small distance $dl$. The force experienced by the unit positive charge at point 1 is the electric field intensity $\vec{E}$. Then the potential difference between the ending point 2 and starting point 1 is given by

$$\phi_2 - \phi_1 = -\vec{E} \cdot \vec{dl} \tag{1.16}$$

In the above equation $\vec{E} \cdot \vec{dl}$ is the work done by the field forces in moving a unit positive charge from point 1 to point 2. In this case, the potential energy of point 2 will be lower than that of point 1 and hence the potential difference $(\phi_2 - \phi_1)$ will be negative. For this purpose, the minus sign is introduced on the right-hand side (RHS) of Equation 1.16.

As shown in Figure 1.7b, if the unit positive charge moves through a certain distance $l$ within an electric field from point 1 to point 2, then the magnitude as well as direction of $\vec{E}$ may not be same at every location along the path traversed by the unit positive charge. Hence, in such a case, the potential difference between point 2 and point 1 is evaluated by integrating the RHS of Equation 1.16 over the line $l$ from point 1 to point 2 as given in Equation 1.17.

$$\phi_2 - \phi_1 = -\int_1^2 \vec{E} \cdot \vec{dl} \tag{1.17}$$

Equation 1.17 is known as the integral form of relationship between electric field intensity $(\vec{E})$ and electric potential $(\phi)$.

If point 1 is chosen at an infinite distance with respect to the source charges causing the electric field, then the potential energy at point 1 due to the source charges will be zero and hence $\phi_1$ will be zero. Then Equation 1.17 can be rewritten as follows:

$$\phi_2 = -\int_{\infty}^{2} \vec{E} \cdot \vec{dl} \qquad (1.18)$$

Equation 1.18 shows that the electric potential at a point in an electric field can be defined as the work done in moving a unit positive charge from infinity to that point. Because work done is independent of the path traversed between the two end points, the electric potential is a conservative field. It is important to note here that the reference point is arbitrary and is fixed as per nature and convenience of the problem. For most of the problems, taking the reference point at infinity is a sound choice. However, for many others (e.g. a long-charged wire), a different choice may prove to be more useful. From Equation 1.18, the practical unit of electric field intensity is obtained as volt per unit length (e.g. V/m).

It is also evident from Equation 1.18 that the work done in moving a unit positive charge from infinity to a given point within an electric field could be same for several points within that electric field depending upon the distribution of electric field intensity vectors within the field region. Hence, electric potential of all such points will be same. If all these points are joined together then one may get a line or a surface on which every point has the same electric potential. Such a line or surface is called an *equipotential*.

Figure 1.8 shows typical examples of equipotentials in two-dimensional systems, where these will be lines. Figure 1.8a shows equipotential lines and electric fieldlines for an electric field for which $\vec{E}$ is constant everywhere, which is called *uniform field*. Figure 1.8b shows equipotentials and electric fieldlines for a positive polarity point charge. In this case, $\vec{E}$ varies with position and is called *non-uniform field*. In the case of three-dimensional systems, such equipotentials will be surfaces.

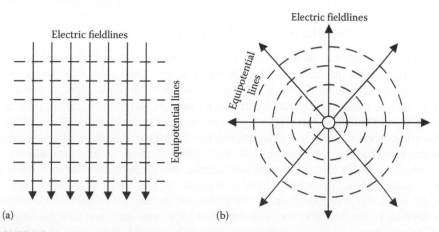

**FIGURE 1.8**
Examples of equipotentials: (a) uniform field and (b) non-uniform field.

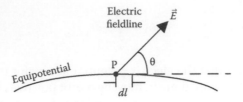

**FIGURE 1.9**
Equipotential vis-à-vis electric fieldline.

### 1.7.1 Equipotential vis-à-vis Electric Fieldline

Consider an equipotential of any electric field, as shown in Figure 1.9. At any point $P$ on this equipotential, consider that the electric fieldline makes an angle $\theta$ with the tangent to the equipotential at that point. If an elementary length $dl$ is considered along the equipotential at $P$, then the potential difference between the two extremities of $dl$ will be given by $|\vec{E}|\,dl\cos\theta$. But if $dl$ lies on the equipotential, then there should not be any potential difference across $dl$. Again, the magnitude of electric field intensity is not zero at $P$ and $dl$ is also a non-zero quantity. Hence, the potential difference across $dl$ could only be zero if $\cos\theta$ is zero (i.e. if $\theta$ is 90°).

Thus, a basic constraint of electric field distribution is that the electric fieldlines are always normal to the equipotential surface. A practical example of this constraint is that the electric fieldlines will always leave or enter conductor surfaces at 90°. This criterion is often used to check the accuracy of electric field computation by numerical techniques. The other properties of equipotential are (1) the tangential component of the electric field along the equipotential is zero and (2) no work is required to move a charged particle along an equipotential.

### 1.7.2 Electric Potential of the Earth Surface

The electric potential of the Earth surface could be determined with the help of the discussion in Section 1.7.1. Consider that a system of source charges has created an electric field over a region located in New York. Now, if one considers a test point on the Earth surface located in New Delhi, India, then the distance of this test point with respect to the source charges is infinite. Hence, electric potential of the test point on the Earth surface in India will be zero due to the stated source charges at New York. Now, Earth is an excellent electrical conductor and in the absence of any conductive current, the Earth surface is an equipotential. Therefore, if the test point on the Earth surface located in India is at zero potential, then all the points on the Earth surface will be at zero potential. Extending the above-mentioned logic, one may see that for any set of source charges located anywhere within this world, there will always be a point on the Earth surface that will be at infinite distance with respect to the source charges. Hence, the Earth surface potential will always be zero due to any set of source charges.

### 1.7.3 Electric Potential Gradient

It is defined as the positive rate of change of electric potential with respect to distance in the direction of greatest change. At any point in a field region, it will be very difficult to comprehend the direction of greatest change. To understand it conveniently, consider that the equipotentials are known within the field region. Figure 1.10 shows three such equipotentials 1, 2 and 3. Then from the point $P$ on the equipotential 2 having an electric potential of $\phi$, if one moves to any point on the equipotential having an electric potential $\phi + \Delta\phi$, the potential difference is $+\Delta\phi$. But the minimum distance between the equipotentials 1 and 2 is the normal distance $\Delta n$. Hence, the greatest rate of change will be along the normal to the equipotential. Moreover, there are two directions of the normal to the equipotential 2 at $P$. Electric potential gradient is defined to be the greatest rate of change of potential in the positive sense. Hence, with reference to Figure 1.10, the electric potential gradient at $P$ will be given by

$$\text{Electric potential gradient (grad } \phi) = (\Delta\phi/\Delta n) \qquad (1.19)$$

where:
   $\Delta\phi$ is the potential difference
   $\Delta n$ is the normal distance between the two equipotentials 1 and 2 of Figure 1.10

Because electric potential gradient has magnitude along with a specific direction, it is a vector quantity and is a spatial derivative of electric potential.

### 1.7.4 Electric Potential Gradient and Electric Field Intensity

As discussed in Section 1.7.1, electric fieldline or $\vec{E}$ will be directed along the normal to the equipotential. But as there are two normal directions to the equipotential, the question is in which direction will $\vec{E}$ be. In this context,

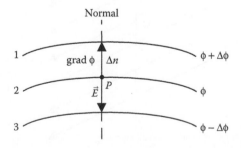

**FIGURE 1.10**
Electric potential gradient and electric field intensity.

recall that if one moves along the direction of electric field, then the potential energy decreases and hence the electric potential also decreases in the direction of electric field or $\vec{E}$. Therefore, with reference to Figure 1.10, $\vec{E}$ will act from equipotential 2 to equipotential 3 at the point $P$; that is, $\vec{E}$ will act along the direction of the decreasing potential. Again, when one moves from equipotential 1 to equipotential 2 along the normal distance $\Delta n$, as shown in Figure 1.10, then the potential drop is $\Delta\phi$ and the work done by the field forces is given by $\left|\vec{E}\right|\Delta n$. Hence,

$$\Delta\phi = \left|\vec{E}\right|\Delta n \text{ or } \left|\vec{E}\right| = \frac{\Delta\phi}{\Delta n} \tag{1.20}$$

Therefore, from Equations 1.19 and 1.20, it is seen that the magnitudes of $\vec{E}$ and grad $\phi$ at $P$ are the same. But, grad $\phi$ acts along the direction of the increasing potential and $\vec{E}$ acts along the direction of the decreasing potential and both grad $\phi$ and $\vec{E}$ act along the normal to the equipotential at $P$. Therefore, it could be concluded that

$$\vec{E} = -\text{grad } \phi \tag{1.21}$$

As shown in Figure 1.11, the potential difference $(\phi_1 - \phi_2)$ between two points 1 and 2 within a field region is given by

$$\Delta\phi = -\vec{E}\cdot\overrightarrow{\Delta l} \text{ or } \Delta\phi = -\left|\vec{E}\right|\left|\overrightarrow{\Delta l}\right|\cos\theta \tag{1.22}$$

where:
  $\overrightarrow{\Delta l}$ is the distance vector from point 1 to point 2
  $\left|\vec{E}\right|\cos\theta$ is the component of $\vec{E}$ along $\overrightarrow{\Delta l}$

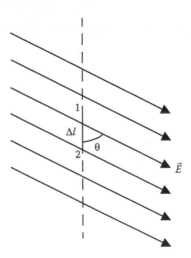

**FIGURE 1.11**
Pertaining to the relationship of electric potential gradient and electric field intensity.

Now, if $\overrightarrow{\Delta l}$ lies along the direction of $x$-axis, then Equation 1.22 can be rewritten in terms of the component of $\vec{E}$ along the $x$-direction, that is, $E_x$, and the distance $\Delta x$, as follows:

$$E_x = -\frac{\Delta\phi}{\Delta x} \tag{1.23}$$

Instead of discrete variation, for continuous variation of electric potential in $x$-direction, Equation 1.23 could be written in partial derivative form as follows:

$$E_x = -\frac{\partial\phi}{\partial x} \tag{1.24}$$

Similarly, in the $y$- and $z$-directions, $E_y = -(\partial\phi/\partial y)$ and $E_z = -(\partial\phi/\partial z)$.

Thus, the electric field intensity vector in terms of the three components could then be written as follows:

$$\vec{E} = E_x\,\hat{i} + E_y\,\hat{j} + E_z\,\hat{k} = -\frac{\partial\phi}{\partial x}\,\hat{i} - \frac{\partial\phi}{\partial y}\,\hat{j} - \frac{\partial\phi}{\partial z}\,\hat{k} = -\left(\frac{\partial}{\partial x}\,\hat{i} + \frac{\partial}{\partial y}\,\hat{j} + \frac{\partial}{\partial z}\,\hat{k}\right)\phi \tag{1.25}$$

On the RHS of Equation 1.25, the vector operator within the parenthesis is the *del* operator, $\vec{\nabla}$. Thus, from Equation 1.25

$$\vec{E} = -\vec{\nabla}\phi \tag{1.26}$$

The $\vec{\nabla}$ operator is an interesting operator. When it acts on a scalar quantity, the result is a vector quantity, and physically it results into the spatial derivative of the scalar quantity, that is, gradient of the scalar quantity. In this case, when $\vec{\nabla}$ acts on scalar electric potential, it results into the vector quantity electric potential gradient.

**PROBLEM 1.1**

The potential field in a medium having relative permittivity of 3.5 is given by $\phi = 4x^3y - 5y^3z + 3xz^3$ V. Find the electric field intensity at the point $(0.1, 0.5, 0.2)$m.

*Solution:*

$$E_x = -\frac{\partial\phi}{\partial x} = -12x^2y - 3z^3$$

Hence,

$$E_x\big|_{(0.1,0.5,0.2)} = -12\times0.1^2\times0.5 - 3\times0.2^3 = -0.084 \text{ V/m}$$

$$E_y = -\frac{\partial\phi}{\partial y} = -4x^3 + 15y^2z$$

Hence,

$$E_y\big|_{(0.1,0.5,0.2)} = -4 \times 0.1^3 + 15 \times 0.5^2 \times 0.2 = 0.746 \text{ V/m}$$

$$E_z = -\frac{\partial \phi}{\partial z} = 5y^3 - 9xz^2$$

Hence,

$$E_z\big|_{(0.1,0.5,0.2)} = 5 \times 0.5^3 - 9 \times 0.1 \times 0.2^2 = 0.589 \text{ V/m}$$

Therefore,

$$\vec{E}\big|_{(0.1,0.5,0.2)} = -0.084\hat{i} + 0.746\hat{j} + 0.589\hat{k} \text{ V/m}$$

Hence,

$$\big|\vec{E}\big|_{(0.1,0.5,0.2)} = 0.954 \text{ V/m}$$

## PROBLEM 1.2

The potential field at any point in a space containing a dielectric medium of $\varepsilon_r = 5$ is given by $\phi = 7x^2y - 3y^2z - 4z^2x$ V, where $x$, $y$ and $z$ are in metres. Calculate the $y$-component of electric flux density at the point (1,4,2)m.

*Solution:*

$$E_y = -\frac{\partial \phi}{\partial y} = -7x^2 + 6yz$$

Hence,

$$E_y\big|_{(1,4,2)} = -7 \times 1^2 + 6 \times 4 \times 2 = 41 \text{ V/m}$$

Therefore,

$$D_y\big|_{(1,4,2)} = 41 \times 5 \times 8.854 \times 10^{-12} = 1.815 \text{ nC/m}^2$$

## PROBLEM 1.3

The potential field in a space containing a dielectric medium of $\varepsilon_{r1}$ is given by $\phi_1 = 7xy - 3yz - 4zx$ V, and another potential field in a space containing a dielectric medium of $\varepsilon_{r2}$ is given by $\phi_2 = -2x - 7yz + 5zx$ V, where $x$, $y$ and $z$ are in metres. If the $x$-component of electric flux density at the point (1,2,2)m is same in both the fields, then find the ratio of $\varepsilon_{r1}$ and $\varepsilon_{r2}$.

*Solution:*

$$E_{x1} = -\frac{\partial \phi_1}{\partial x} = -7y + 4z$$

Hence,

$$E_{x1}|_{(1,2,2)} = -7 \times 2 + 4 \times 2 = -6 \text{ V/m}$$

$$E_{x2} = -\frac{\partial \phi_2}{\partial x} = 2 - 5z$$

Hence,

$$E_{x2}|_{(1,2,2)} = 2 - 5 \times 2 = -8 \text{ V/m}$$

Now, $D_{x1} = \varepsilon_{r1} \times \varepsilon_0 \times E_{x1} = -6\varepsilon_{r1}\varepsilon_0 \text{ C/m}^2$

and $D_{x2} = \varepsilon_{r2} \times \varepsilon_0 \times E_{x2} = -8\varepsilon_{r2}\varepsilon_0 \text{ C/m}^2$

As per the problem statement, $D_{x1} = D_{x2}$, or, $-6\varepsilon_{r1}\varepsilon_0 = -8\varepsilon_{r2}\varepsilon_0$

Hence,

$$\frac{\varepsilon_{r1}}{\varepsilon_{r2}} = 1.333$$

## 1.8 Field due to Point Charge

As shown in Figure 1.12, due to a point charge $+Q_i$, the electric field intensity at any point $P$ at a distance $r$ from the charge is given by

$$\vec{E}_r = \frac{Q_i}{4\pi\varepsilon_0 r^2} \hat{u}_r$$

If $P$ is moved from point $A$ to point $B$, as shown in Figure 1.12, then the potential difference will be as follows:

$$\phi_B - \phi_A = -\int_{r_A}^{r_B} \frac{Q_i}{4\pi\varepsilon_0 r^2} \hat{u}_r \cdot \vec{dl} = \frac{Q_i}{4\pi\varepsilon_0}\left(\frac{1}{r_B} - \frac{1}{r_A}\right) \tag{1.27}$$

If $A$ is located at infinity, then the potential of $B$ will be

$$\phi_B = \frac{Q_i}{4\pi\varepsilon_0 r_B} \tag{1.28}$$

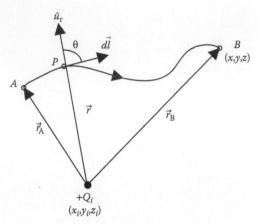

**FIGURE 1.12**
Field due to a point charge.

If $+Q_i$ is located at $(x_i,y_i,z_i)$ and $B$ is located at $(x,y,z)$, then

$$\vec{r_B} = (x - x_i)\hat{i} + (y - y_i)\hat{j} + (z - z_i)\hat{k} \tag{1.29}$$

such that $\left|\vec{r_B}\right| = \sqrt{(x - x_i)^2 + (y - y_i)^2 + (z - z_i)^2}$ and $\hat{u}_{rB} = \dfrac{\vec{r_B}}{\left|\vec{r_B}\right|}$

The electric field intensity at $B$ will then be given by

$$E_B = \frac{Q_i}{4\pi\varepsilon_0\left[(x-x_i)^2 + (y-y_i)^2 + (z-z_i)^2\right]} \frac{(x-x_i)\hat{i} + (y-y_i)\hat{j} + (z-z_i)\hat{k}}{\left[\sqrt{(x-x_i)^2 + (y-y_i)^2 + (z-z_i)^2}\right]} \tag{1.30}$$

Hence, the three components of electric field intensity at $B$ are (dropping the suffix $B$ for the sake of generalization) as follows:

$$E_x = \frac{(x - x_i)Q_i}{4\pi\varepsilon_0\left[\sqrt{(x-x_i)^2 + (y-y_i)^2 + (z-z_i)^2}\right]^3}$$

$$E_y = \frac{(y - y_i)Q_i}{4\pi\varepsilon_0\left[\sqrt{(x-x_i)^2 + (y-y_i)^2 + (z-z_i)^2}\right]^3} \tag{1.31}$$

$$E_z = \frac{(z - z_i)Q_i}{4\pi\varepsilon_0\left[\sqrt{(x-x_i)^2 + (y-y_i)^2 + (z-z_i)^2}\right]^3}$$

In the presence of multiple point charges, at any point electric potential will be the scalar sum of electric potentials and electric field intensity will be the vector sum of electric field intensities at that point due to all the charges. In other words, the effect of all the charges will be superimposed at any point within the field region.

**PROBLEM 1.4**

A point charge $Q_1 = +1.0$ μC is located at $(3,1,1)$m and another point charge $Q_2 = -0.5$ μC is located at $(0.5,2,1.5)$m. Find the magnitude and polarity of the point charge located at $(1,2,2)$m for which the z-component of electric field intensity will be zero at the origin. Medium is air.

*Solution:*

For the point charge 1:

Distance vector to the origin is

$$\vec{r}_{O1} = (0-3)\hat{i} + (0-1)\hat{j} + (0-1)\hat{k} = (-3\hat{i} - \hat{j} - \hat{k})\text{m}$$

Therefore,

$$\left|\vec{r}_{O1}\right| = 3.317 \text{ m}$$

Electric field intensity at the origin:

$$E_{O1} = \frac{10^{-6}}{4\pi \times 1 \times \varepsilon_0} \times \frac{-3\hat{i} - \hat{j} - \hat{k}}{3.317^3} = A(-0.0822\hat{i} - 0.0274\hat{j} - 0.0274\hat{k}) \text{ V/m}$$

where:

$A = 10^{-6} / (4\pi \times 1 \times \varepsilon_0)$

For the point charge 2:

Distance vector to the origin is

$$\vec{r}_{O2} = (0-0.5)\hat{i} + (0-2)\hat{j} + (0-1.5)\hat{k} = (-0.5\hat{i} - 2\hat{j} - 1.5\hat{k})\text{m}$$

Therefore,

$$\left|\vec{r}_{O2}\right| = 2.549 \text{ m}$$

Electric field intensity at the origin:

$$E_{O2} = \frac{-0.5 \times 10^{-6}}{4\pi \times 1 \times \varepsilon_0} \times \frac{-0.5\hat{i} - 2\hat{j} - 1.5\hat{k}}{2.549^3} = A(0.015\hat{i} + 0.06\hat{j} + 0.045\hat{k})\text{V/m}$$

Therefore, the z-component of electric field intensity at the origin due to point charges 1 and 2 is as follows:

$$(-0.0274 + 0.045)A = 0.0176 \, A\text{V/m}$$

For the point charge 3, let the magnitude of the charge be $Q_3$ μC.

Distance vector to the origin is

$$\vec{r}_{O3} = (0-1)\hat{i} + (0-2)\hat{j} + (0-2)\hat{k} = (-\hat{i} - 2\hat{j} - 2\hat{k})\,\text{m}$$

Therefore,

$$\left|\vec{r}_{O3}\right| = 3\,\text{m}$$

Electric field intensity at the origin:

$$E_{O3} = \frac{Q_3 \times 10^{-6}}{4\pi \times 1 \times \varepsilon_0} \times \frac{-\hat{i} - 2\hat{j} - 2\hat{k}}{3^3} = A(-0.037\hat{i} - 0.074\hat{j} - 0.074\hat{k})Q_3 \text{ V/m}$$

If the resultant values of z-component of electric field intensity at the origin is to be zero due to the three point charges, then

$$-0.074AQ_3 + 0.0176A = 0,\ \text{or,}\ Q_3 = 0.238\ \mu\text{C}$$

## PROBLEM 1.5

A right isosceles triangle of side 1 m has charges +1, +2 and –1 nC arranged on its vertices, as shown in Figure 1.13. Find the magnitude and direction

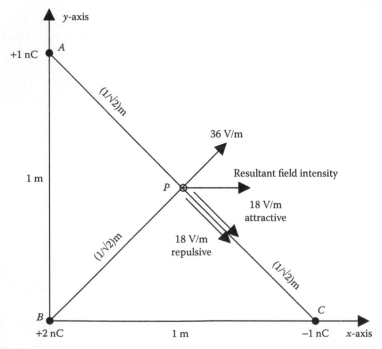

**FIGURE 1.13**
Pertaining to Problem 1.5.

of electric field intensity at the point *P*, which is midway between the line connecting the +1 and –1 nC charges. Medium is air.

*Solution:*

Consider that a unit positive test charge is located at *P*.

The distance of this test charge from +1 nC is $(1/\sqrt{2})$m.

Therefore, the electric field intensity at *P* due to +1 nC is $10^{-9}/[4\pi \times 8.854 \times 10^{-12} \times (1/\sqrt{2})^2] = 18$ V/m and it acts along the line *AC* shown in Figure 1.13, from *P* to *C*, as it is a repulsive force on the test charge located at *P*.

The distance of the test charge from –1 nC is also $(1/\sqrt{2})$m.

Therefore, the electric field intensity at *P* due to –1 nC is also 18 V/m and it also acts along the line *AC* shown in Figure 1.13 from *P* to *C*, as it is an attractive force on the test charge located at *P*.

The distance of this test charge from +2 nC is also $(1/\sqrt{2})$m.

Therefore, electric field intensity at *P* due to +2 nC is $2 \times 10^{-9}/[4\pi \times 8.854 \times 10^{-12} \times (1/\sqrt{2})^2] = 36$ V/m and it acts along the line *BP* shown in Figure 1.13 from *B* to *P*, as it is a repulsive force on the test charge located at *P*. *BP* is perpendicular to *AC*.

Hence, the electric field intensity acting along *AC* from *P* to *C* is 36 V/m and that acting along *BP* from *B* to *P* is also 36 V/m and these two are normal to each other.

Therefore, the resultant electric field intensity at *P* is $\sqrt{36^2 + 36^2} = 50.91$ V/m.

It will make an angle of 45° with respect to *AC*. In other words, it will act along the *x*-axis in the positive sense of the axis.

**PROBLEM 1.6**

An α-particle with a kinetic energy of 1.5 MeV is projected towards a stationary platinum nucleus, which has 78 protons. Determine the distance of closest approach of the α-particle. Neglect the motion of the nucleus.

*Solution:*

The α-particle is positively charged. Hence, it experiences repulsive force from the nucleus and decelerates as it approaches the nucleus. At the closest approach point, it stops before being repulsed back. At this point of closest approach, all its kinetic energy is converted to potential energy.

Initial kinetic energy of the α-particle = 1.5 MeV = $1.5 \times 10^6 \times 1.602 \times 10^{-19}$ J = $2.403 \times 10^{-13}$ J.

Let the distance of closest approach be *d* metres.

An α-particle has a charge of $+3.204 \times 10^{-19}$ C. The platinum nucleus has charge of $(78 \times 1.602 \times 10^{-19})$C = $+124.956 \times 10^{-19}$ C.

The electric potential at a distance of *d* from the platinum nucleus due to all the protons is as follows:

$$\frac{78 \times 1.602 \times 10^{-19}}{4\pi \times 8.854 \times 10^{-12} \times d} = \frac{1.123 \times 10^{-7}}{d} \text{ V}$$

Hence, the potential energy of the α-particle at a distance $d$ from platinum nucleus will be as follows:

$$\frac{1.123 \times 10^{-7}}{d} \times 3.204 \times 10^{-19} \, \text{J} = \frac{3.598 \times 10^{-26}}{d} \, \text{J}$$

Equating this potential energy to the initial kinetic energy of $2.403 \times 10^{-13}$ J, $d = 1.497 \times 10^{-13}$ m.

## 1.9 Field due to a Uniformly Charged Line

Figure 1.14 shows a line charge of uniform charge density $\lambda$. As a practical example, it may be considered to be a non-conducting rod, which is uniformly charged. The length of the charge is $L$. For the sake of simplicity, the field is computed at a point $P$, which is located on the line perpendicular to the line charge and passing through the midpoint of the charge. The normal distance of $P$ from the line charge is $y$. For the purpose of field computation, the origin is considered to be located at the midpoint of the line charge, as shown in Figure 1.14.

Consider an elementary length of charge $dx$ located at distance $x$ from the origin. Then the charge on this elementary length is $dq = \lambda dx$ and it may be considered as a point charge.

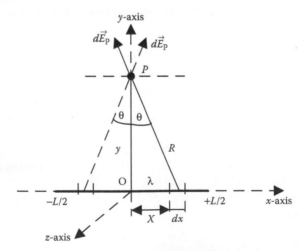

**FIGURE 1.14**
Field due to a uniformly charged line.

Hence, the electric potential at $P$ due the elementary charge $dq$ is given by

$$d\phi_P = \frac{dq}{4\pi\varepsilon_0 R} = \frac{\lambda}{4\pi\varepsilon_0} \frac{dx}{\sqrt{(x^2 + y^2)}}$$

Therefore, the potential at $P$ due to the entire line charge will be as follows:

$$\phi_P = \frac{\lambda}{4\pi\varepsilon_0} \int_{-L/2}^{+L/2} \frac{dx}{\sqrt{(x^2 + y^2)}} = \frac{\lambda}{4\pi\varepsilon_0} \left[ \ln\left(x + \sqrt{x^2 + y^2}\right) \right]_{x=-L/2}^{x=+L/2}$$

$$\text{or, } \phi_P = \frac{\lambda}{4\pi\varepsilon_0} \ln \frac{\sqrt{(L/2)^2 + y^2} + (L/2)}{\sqrt{(L/2)^2 + y^2} - (L/2)} \tag{1.32}$$

As electric potential at $P$ is dependent only on $y$, hence electric field intensity at $P$ will have only the $y$-component, which will be given by

$$E_{yP} = -\frac{\partial \phi_P}{\partial y} = \frac{\lambda}{4\pi\varepsilon_0} \frac{y}{\sqrt{(L/2)^2 + y^2}} \left[ \frac{1}{\sqrt{(L/2)^2 + y^2} - (L/2)} - \frac{1}{\sqrt{(L/2)^2 + y^2} + (L/2)} \right]$$

$$\text{or, } E_{yP} = \frac{\lambda}{4\pi\varepsilon_0} \frac{L}{y\sqrt{(L/2)^2 + y^2}} \tag{1.33}$$

The direction of electric field intensity at $P$ along the $y$-axis can be explained with the help of Figure 1.14. For the elementary length of charge $dx$ on the RHS of the origin $O$, another elementary length of charge can be assumed to be there at the same distance $x$ from the origin on the left-hand side of the origin. Then the electric field intensity at $P$ due to these two elementary charges will be same in magnitude, say $dE_p$. But they will be directed along the distance vectors from the elementary charges to $P$, as shown in Figure 1.14. They will make the same angle $\theta$ with the $y$-axis. Hence, the $x$-components of these two field intensity vectors $dE_p$ at $P$ will cancel each other and only the $y$ component will be present, as the $y$-components of $dE_p$ vectors will be additive at $P$.

## 1.10 Field due to a Uniformly Charged Ring

Consider a uniformly charged ring of radius $a$, as shown in Figure 1.15. The uniform charge density is $\lambda$. As a practical example, it may be considered to be a non-conducting annular strip of very small width, which is uniformly charged. To determine the electric potential at $P$ located on the axis of the ring at a height of $z$ from the plane of the ring, consider an elementary arc

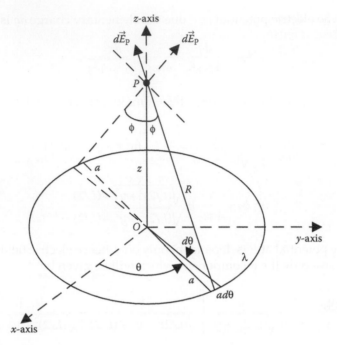

**FIGURE 1.15**
Field due to a uniformly charged ring.

length $ad\theta$, as shown in Figure 1.15. Then this elementary arc length will have a charge $dq = \lambda ad\theta$, and it may be considered as a point charge.

Hence, the electric potential at $P$ due the elementary charge $dq$ is given by

$$d\phi_P = \frac{dq}{4\pi\varepsilon_0 R} = \frac{\lambda}{4\pi\varepsilon_0} \frac{ad\theta}{\sqrt{(a^2 + z^2)}}$$

Therefore, the potential at $P$ due to the entire ring charge will be

$$\phi_P = \frac{\lambda}{4\pi\varepsilon_0} \int_0^{2\pi} \frac{ad\theta}{\sqrt{(a^2 + z^2)}} \tag{1.34}$$

In the integral of Equation 1.34, it may be seen that as $\theta$ is varied from 0 to $2\pi$ along the ring, the distance $R$ to $P$ from any point on the ring always remains the same, that is, $\sqrt{(a^2 + z^2)}$. Hence, this integral is only over $\theta$. Therefore,

$$\phi_P = \frac{\lambda}{4\pi\varepsilon_0} \frac{2\pi a}{\sqrt{(a^2 + z^2)}} = \frac{\lambda a}{2\varepsilon_0 \sqrt{(a^2 + z^2)}} \tag{1.35}$$

Equation 1.35 shows that electric potential at $P$ is dependent only on $z$; hence, electric field intensity at $P$ will have only the $z$-component, which will be given by

$$E_{zP} = -\frac{\partial \phi_P}{\partial z} = \frac{\lambda}{2\varepsilon_0} \frac{az}{(a^2 + z^2)^{3/2}} \tag{1.36}$$

With the help of Figure 1.15, it can be explained why the resultant electric field intensity at the point $P$ is directed along the $z$-axis. Similar to the case of a uniformly charged line, for any elementary arc length $ad\theta$, another elementary arc length $ad\theta$ can be assumed to be located diametrically opposite to the first elementary arc length. Then the electric field intensity at $P$ due to these two elementary arc lengths will be same in magnitude, say $dE_p$. But these two field intensity vectors of magnitude $dE_p$ will act along the distance vectors from the two elementary arc lengths to $P$, as shown in Figure 1.15. It is obvious that these two vectors will make the same angle with the three axes. Hence, considering the directions of these two vectors, it can be seen that the components of electric field intensity in $x$- and $y$-directions will cancel each other, whereas the $z$-components will be additive in nature. Therefore, the net electric field intensity at $P$ due to the ring charge will have only $z$-component.

## 1.11 Field due to a Uniformly Charged Disc

Consider a uniformly charged disc of radius $a$ having a surface charge density $\sigma$, as shown in Figure 1.16. As a practical example, it may be considered to be a non-conducting disc, which is uniformly charged. Point $P$ is located on the axis of the disc at a height $z$ from the plane of the disc. For computing the electric potential at $P$, consider an annular strip of radius $dr$ at a radius $r$, as shown in Figure 1.16. Then take a small arc segment of angular width $d\theta$ of this annular strip. Then this elementary surface of area $dA = rd\theta dr$ will have an elementary charge $dq = \sigma dA = \sigma rd\theta dr$. This elementary charge $dq$ can be considered as a point charge.

Hence, the electric potential at $P$ due the elementary charge $dq$ is given by

$$d\phi_P = \frac{dq}{4\pi\varepsilon_0 R} = \frac{\sigma}{4\pi\varepsilon_0} \frac{rd\theta dr}{\sqrt{(r^2 + z^2)}}$$

Therefore, the potential at $P$ due to the entire annular strip of charge of radius $r$ will be

$$\phi_P' = \frac{\sigma}{4\pi\varepsilon_0} \left[ \int_0^{2\pi} \frac{d\theta}{\sqrt{(r^2 + z^2)}} \right] rdr \tag{1.37}$$

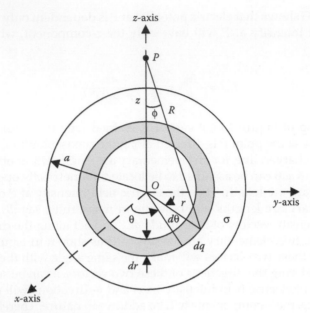

**FIGURE 1.16**
Field due to a uniformly charged disc.

because, when $\theta$ varies from 0 to $2\pi$ along the annular strip of radius $r$, then the distance $R$ to point $P$ from any point on this strip remains the same, that is, $\sqrt{(r^2 + z^2)}$, and the radial distance from the axis of the disc also remains the same, that is, $r$. Hence, the integration of Equation 1.37 is carried out over $\theta$ only. Therefore,

$$\phi_P' = \frac{\sigma}{4\pi\varepsilon_0} \frac{2\pi r dr}{\sqrt{(r^2 + z^2)}} = \frac{\sigma}{2\varepsilon_0} \frac{r dr}{\sqrt{(r^2 + z^2)}}$$

Thus, the potential at $P$ due to the entire disc of radius $a$ will be

$$\phi_P = \frac{\sigma}{2\varepsilon_0} \int_0^a \frac{r dr}{\sqrt{(r^2 + z^2)}} = \frac{\sigma}{2\varepsilon_0} \left[ \sqrt{(r^2 + z^2)} \right]_{r=0}^{r=a} = \frac{\sigma}{2\varepsilon_0} \left[ \sqrt{(a^2 + z^2)} - z \right] \quad (1.38)$$

Equation 1.38 shows that electric potential at $P$ is dependent only on $z$, hence electric field intensity at $P$ will have only the $z$-component, which will be given by

$$E_{zP} = -\frac{\partial \phi_P}{\partial z} = \frac{\sigma}{2\varepsilon_0} \left[ 1 - \frac{z}{\sqrt{(a^2 + z^2)}} \right] \quad (1.39)$$

The charged disc comprises large number of annular strips, as shown in Figure 1.16. Each annular strip of very small radial width $dr$ can be considered as a ring charge of uniform charge density. In Section 1.10, it has been discussed that the electric field intensity due a ring charge at any point located on the axis of the ring has only $z$-component. Hence, the electric field intensity at point $P$ of Figure 1.16 due to large number of co-axial ring charges representing the annular strips of charges will also have only $z$-component as given by Equation 1.39.

From Equation 1.38, it may be seen that the electric potential at the centre of the disc is finite. Putting $z = 0$ in Equation 1.38, $\phi_O = \sigma a/(2\varepsilon_0)$.

The total charge on the entire disc is $Q = \pi a^2 \sigma$. Hence,

$$\phi_O = \frac{Qa}{2\pi a^2 \varepsilon_0} = 2\frac{Q}{4\pi\varepsilon_0 a} \tag{1.40}$$

Equation 1.40 shows that the electric potential at the centre of the disc is equivalent to twice the work done to bring a unit positive charge from infinity to the circumference of the disc ($r = a$) when the entire charge of the disc is assumed to be concentrated as a point charge at the centre.

When $z$ is very large compared to $a$, then

$$\sqrt{z^2 + a^2} = z\left(1 + \frac{a^2}{z^2}\right)^{1/2} = z\left(1 + \frac{a^2}{2z^2} + \cdots\right) \approx z + \frac{a^2}{2z}$$

Then from Equation 1.38,

$$\phi_P\Big|_{z \gg a} = \frac{\sigma}{2\varepsilon_0}\left[z + \frac{a^2}{2z} - z\right] = \frac{\sigma\pi a^2}{4\pi\varepsilon_0 z} = \frac{Q}{4\pi\varepsilon_0 z} \tag{1.41}$$

When $P$ is located at a very large distance from the disc of finite radius $a$, then the disc could be considered to be a point charge of magnitude $Q$. Equation 1.41 is the expression for potential at a point located at distance $z$ from a point charge of magnitude $Q$.

The significance of the above discussion is that the accuracy of field computation for any given charge distribution can be checked by choosing a point $P$, which is located sufficiently far away from the charge distribution under consideration. If the charge distribution is of finite extent, then the electric field should behave as if the charge distribution is like point charge, and decreases with the square of the distance.

## PROBLEM 1.7

A circular disc of charge of radius 1 m having a uniform charge density $\sigma = +1$ nC/m$^2$ lies in the $z = 0$ plane, with centre at the origin. There is also a point charge of $-0.4$ nC at the origin. Find the magnitude and polarity of uniform charge density of a circular ring of charge of radius 1 m lying in the

$z = 0$ plane, with centre at the origin, which would produce the same electric field intensity at the point $(0,0,6)$m as that due to the combined effect of the disc and point charges. Medium is air.

*Solution:*
As stated, the point $P$ is located at $(0,0,6)$m. According to the problem statement, $P$ lies on the axis of the disc as well as on the axis of the ring.

Hence, for the disc charge located in the $x$–$y$ plane having centre at the origin:

$$\sigma = +1\,nC/m^2,\ a = 1\ m\ and\ z = 6\ m$$

$$E_{\text{P-disc}} = \frac{10^{-9}}{2 \times 1 \times 8.854 \times 10^{-12}}\left[1 - \frac{6}{\sqrt{(1^2 + 6^2)}}\right] = 0.768\ V/m$$

Again, for the point charge located at origin: $Q = -0.4\ nC$, $r = 6\ m$

$$E_{\text{P-point}} = \frac{-4 \times 10^{-10}}{4 \times 1 \times 8.854 \times 10^{-12} \times 6^2} = -0.314\ V/m$$

Let the uniform charge density of the ring charge be $\lambda$ C/m.

Then, for the ring charge located in $x$–$y$ plane having centre at the origin, $a = 1$ m and $z = 6$ m.

$$E_{\text{P-ring}} = \frac{\lambda}{2 \times 1 \times 8.854 \times 10^{-12}} \frac{1 \times 6}{(1^2 + 6^2)^{3/2}} = 1.505 \times 10^9 \times \lambda\ V/m$$

As per the statement of the problem, $E_{\text{P-ring}} = E_{\text{P-disc}} + E_{\text{P-point}}$

$$or,\ 1.505 \times 10^9 \times \lambda = 0.768 - 0.314 = 0.454$$

Therefore, $\lambda = 0.302$ nC/m of positive polarity.

## PROBLEM 1.8

Consider a ring charge of radius 10 cm and uniform charge density of $\lambda$ C/m and also a disc charge of radius 15 cm and uniform charge density $\sigma$ C/m$^2$. Both the two charges are placed in the $x$–$y$ plane with their centre at the origin. If the electric field intensity at a point of height 20 cm lying on the z-axis is same due to the ring and disc charges individually, then find the ratio of $\lambda$ and $\sigma$. Relative permittivity of the medium is 2.1.

*Solution:*
The point $P$ is located at $(0,0,0.2)$m. Hence, it is located on the axis of both the ring and disc charges.

For the ring charge, $a = 0.1$ m and $z = 0.2$ m.

$$E_{\text{P-ring}} = \frac{\lambda}{2 \times \varepsilon_r \times \varepsilon_0} \frac{0.1 \times 0.2}{(0.1^2 + 0.2^2)^{3/2}} = 0.894 \frac{\lambda}{\varepsilon_r \varepsilon_0}\ V/m$$

For the disc charge, $a = 0.15$ m and $z = 0.2$ m.

$$E_{P\text{-disc}} = \frac{\sigma}{2 \times \varepsilon_r \times \varepsilon_r}\left[1 - \frac{0.2}{\sqrt{(0.15^2 + 0.2^2)}}\right] = 0.1\frac{\sigma}{\varepsilon_r \varepsilon_r}\, \text{V/m}$$

But as per the problem, $E_{P\text{-ring}} = E_{P\text{-disc}}$

$$\text{or, } 0.894\frac{\lambda}{\varepsilon_r \varepsilon_0} = 0.1\frac{\sigma}{\varepsilon_r \varepsilon_0}$$

Therefore,

$$\frac{\lambda}{\sigma} = 0.112$$

## Objective Type Questions

1. Electric field intensity is defined as
   a. Force per unit charge
   b. Charge per unit force
   c. Potential energy per unit charge
   d. Charge per unit potential energy
2. Electric potential is quantified as
   a   Force per unit charge
   b. Charge per unit force
   c. Potential energy per unit charge
   d. Charge per unit potential energy
3. Electric potential of a point may be defined as
   a. Work done in moving a unit positive charge from that point to infinity
   b. Work done in moving a unit positive charge from infinity to that point
   c. Work done in moving a unit positive charge from the source charge to that point
   d. Work done in moving a unit positive charge from that point to the source charge
4. The angle between the equipotential and electric fieldline is always
   a. 0°
   b. 90°

c.  180°

d.  270°

5.  The conditions to be fulfilled for the validity of Coulomb's law are:

a.  The charges involved must be point charges

b.  The charges should be stationary with respect to each other

c.  Both (a) and (b)

d.  None of the above

6.  When the scalar product of two point charges is positive, then the Coulomb's force between the charges is

a.  Zero

b.  Infinite

c.  Repulsive

d.  Attractive

7.  The electrostatic force between two electric charges of magnitude 1 C separated by a distance of 1 m will be

a.  $\approx 9 \times 10^9$ N

b.  $\approx 9 \times 10^{-9}$ N

c.  $\approx (1/9) \times 10^9$ N

d.  $\approx (1/9) \times 10^{-9}$ N

8.  The ratio of electrostatic to gravitational forces between two electrons separated by a given distance is approximately

a.  $\approx 4 \times 10^{24}$

b.  $\approx (1/4) \times 10^{24}$

c.  $\approx (1/4) \times 10^{42}$

d.  $\approx 4 \times 10^{42}$

9.  If the source charge magnitude is $Q$ coulombs, then the total amount of electric fieldlines coming out of the source charge will be

a.  $Q$ coulombs

b.  $(1/Q)$ coulombs

c.  $Q^2$ coulombs

d.  $1/Q^2$ coulombs

10.  For a given value of electric field intensity, if the permittivity of the medium is doubled, then the electric flux density is

a.  Unchanged

b.  Halved

c.  Doubled

d.  Squared

11. In the case of closed surfaces, area vector is conventionally taken in the direction of
    a. The inward normal
    b. The outward normal
    c. Along the area in clockwise sense
    d. Along the area in anti-clockwise sense

12. The unit of electric field intensity is
    a. J/C
    b. N/C
    c. V/m
    d. Both (b) and (c)

13. Which of the following is true as the properties of an equipotential?
    a. The normal component of the electric field along the equipotential is zero
    b. The tangential component of the electric field along the equipotential is zero
    c. No work is required to move a particle along an equipotential
    d. Both (b) and (c)

14. Which of the following quantity is zero on an equipotential?
    a. Resultant electric field
    b. Normal component of the electric field
    c. Tangential component of the electric field
    d. Electric potential

15. The angle between $\vec{E}$ and grad $\phi$ is
    a. 0°
    b. 90°
    c. 180°
    d. 270°

16. When $\vec{\nabla}$ acts on a scalar quantity, it results into a
    a. Scalar quantity that is the spatial derivative of the scalar quantity
    b. Vector quantity that is the spatial derivative of the scalar quantity
    c. Scalar quantity that is the time derivative of the scalar quantity
    d. Vector quantity that is the time derivative of the scalar quantity

17. In the presence of multiple point charges, electric potential and electric field intensity at any point due to all the charges will be
    a. The scalar sum of electric potentials and the vector sum of electric field intensities, respectively

b.  The scalar sum of electric potentials and the scalar sum of electric field intensities, respectively

c.  The vector sum of electric potentials and the scalar sum of electric field intensities, respectively

d.  The vector sum of electric potentials and the vector sum of electric field intensities, respectively

18. For a uniformly charged ring of positive polarity lying on the $x$–$y$ plane with its axis passing through the origin, the electric field intensity at any point located on the axis is directed along

a.  $x$-axis in the positive sense

b.  $y$-axis in the positive sense

c.  $z$-axis in the positive sense

d.  $z$-axis in the negative sense

19. If the charge distribution is of finite extent, then the electric field at a point that is located sufficiently far away from the charge distribution under consideration should behave as if the charge distribution is like

a.  Point charge

b.  Disc charge

c.  Line charge

d.  Ring charge

20. The electric potential at the centre of a uniformly charged non-conducting disc is

a.  Zero

b.  Finite

c.  Infinite

d.  Undefined

21. A force $\vec{F}$ is conservative if the line integral of the force around a closed path is

a.  Positive

b.  Negative

c.  Zero

d.  Infinite

22. A uniform electric field is parallel to the $z$-axis. In what direction can a charge be displaced in this field without any external work being done on the charge?

a.  $z$-axis

b.  $y$-axis

    c.  *x*-axis

    d.  both (b) and (c)

23. Inside a hollow, uniformly charged conducting sphere

    a.  Both electric potential and electric field intensity are zero

    b.  Both electric potential and electric field intensity are non-zero

    c.  Electric potential is zero but electric field intensity is non-zero

    d.  Electric potential is non-zero but electric field intensity is zero

**Answers:**    1) a; 2) c; 3) b; 4) b; 5) c; 6) c; 7) a; 8) d; 9) a; 10) c; 11) b; 12) d; 13) d; 14) c; 15) c; 16) b; 17) a; 18) c; 19) a; 20) b; 21) c; 22) d; 23) d

---

## Bibliography

1. J.D. Jackson, *Classical Electrodynamics*, John Wiley & Sons Inc, New York, 1975.
2. M. Zahn, *Electromagnetic Field Theory: A Problem Solving Approach*, John Wiley & Sons Inc, New York, 1979.
3. R.K. Wangsness, *Electromagnetic Fields*, John Wiley & Sons Inc, New York, 1979.
4. R.S. Edgar, *Field Analysis and Potential Theory*, Lecture notes in engineering, Vol. 44, Springer-Verlag, USA, 1985.
5. D.K. Cheng, *Field and Wave Electromagnetics*, Addison-Wesley, USA, 1989.
6. P.-B. Zhou, *Numerical Analysis of Electromagnetic Fields*, Springer-Verlag, Berlin, Heidelberg, 1993.
7. D.J. Griffiths, *Introduction to Electrodynamics*, Prentice Hall, New Jersey, 1999.
8. G.L. Pollack and D.R. Stump, *Electromagnetism*, Addison-Wesley, USA, 2002.
9. B.D. Bartolo, *Classical Theory of Electromagnetism*, World Scientific, New Jersey, 2004.
10. W.H. Hayt and J.A. Buck, *Engineering Electromagnetics*, Tata McGraw-Hill Education, New Delhi, India, 2006.
11. G.S.N. Raju, *Electromagnetic Field Theory and Transmission Lines*, Pearson Education, New Delhi, India, 2006.
12. U. Mukherji, *Electromagnetic Field Theory and Wave Propagation*, Alpha Science International, UK, 2006.
13. J. Franklin, *Classical Electromagnetism*, Pearson Education, New Delhi, India, 2007.
14. C.R. Paul, K.W. Whites and S.A. Nasar, *Electromagnetic Fields*, Tata McGraw-Hill Education, New Delhi, India, 2007.
15. G. Lehner, *Electromagnetic Field Theory for Engineers and Physicists*, Springer-Verlag, Berlin, Heidelberg, 2008.
16. S. Sivanagaraju and C.S. Rao, *Electromagnetic Fields*, New Age International, New Delhi, India, 2008.
17. K.R. Meena, *Electromagnetic Fields*, New Age International India, New Delhi, 2008.

18. J.R. Reitz, F.J. Milford and R.W. Christy, *Foundations of Electromagnetic Theory*, Pearson Education, New Delhi, India, 2009.
19. U.A. Bakshi and A.V. Bakshi, *Electromagnetic Field Theory*, Technical Publications, Pune, India, 2009.
20. Y. Singh, *Electro Magnetic Field Theory*, Pearson Education, New Delhi, India, 2011.

# 2

## Gauss's Law and Related Topics

**ABSTRACT** Gauss's law constitutes one of the fundamental laws of electromagnetism. Gauss's law provides an easy means of finding electric field intensity or electric flux density for symmetrical charge distributions such as a point charge, an infinite line charge, an infinite cylindrical surface charge and so forth. The procedure for applying Gauss's law to calculate the electric field involves the introduction of Gaussian surfaces, which are closed surfaces on which electric field is either constant or zero. In most of the cases, electric field is either normal or tangential to the Gaussian surfaces. There are two different ways of stating Gauss's law, namely, integral form and differential form, which are employed to solve field problems depending on suitability. These two forms of Gauss's law in combination lead to *divergence theorem*. The divergence relationship between electric flux density and volume charge density as obtained from Gauss's law and the gradient relationship between electric field intensity and electric potential could be combined in the form of an elliptic partial differential equation known as Poisson's equation. In charge-free field region, Poisson's equation reduces to Laplace's equation. Both Poisson's and Laplace's equations could be solved for field quantities imposing suitable boundary conditions.

## 2.1 Introduction

German mathematician and physicist Karl Friedrich Gauss published his famous flux theorem in 1867, which is now well known in physics as Gauss's law. It is one of the basic laws of classical field theory. Gauss's law is a general law that can be applied to any closed surface. With the help of Gauss's law, the amount of enclosed charge could be assessed by mapping the field on a surface outside the charge distribution. By the application of Gauss's law, the electric field in many practical arrangements could be evaluated by forming a symmetric surface, commonly known as *Gaussian surface*, surrounding a charge distribution and then determining the electric flux through that surface.

In electric field theory, distinction is made between free charge and bound charge. Free charge implies a charge that is free to move over distances large as compared to atomic scale lengths. Free charge typically comes from electrons, for example, in the case of metals, or ions, or in the case of aqueous solutions. Bound charge implies charges of equal magnitude but opposite signs that are held close to each other and are free to move only through atomic scale lengths. Bound charges arise in the context of polarizable dielectric materials. Typical example of bound charge is the positive charge of an atomic nucleus and the negative charge of its associated electron cloud. The microscopic charge displacements in dielectric materials are not as dramatic as the rearrangement of charge in a conductor, but their cumulative effects account for the characteristic behaviour of dielectric materials. Typically, the detailed effect of bound charge is represented through electrical permittivity of dielectric materials, which will be discussed in Chapter 5. The Gauss's law, as discussed in this chapter, is in terms of electric flux and the free charge only.

## 2.2 Useful Definitions and Integrals

### 2.2.1 Electric Flux through a Surface

Consider that the total electric flux through the surface $S$ of an area $A$ needs to be evaluated, as shown in Figure 2.1. The first important point to be noted here is that the electric flux density at different locations on $S$ may not be the same. Therefore, the total flux through $S$ has to be computed by subdividing the entire surface into a large number of smaller surfaces, such that over each small surface area, the electric flux density vector is uniform. As shown in Figure 2.1, for any small surface area $dA$, the area vector $d\vec{A}$ and the electric flux density vector $\vec{D}$ could be along different directions. Then the electric flux through $dA$ is given by $\vec{D} \cdot d\vec{A}$

Then the total flux through $S$ is given by the summation $\vec{D}_1 \cdot d\vec{A}_1 + \vec{D}_2 \cdot d\vec{A}_2 + \cdots$

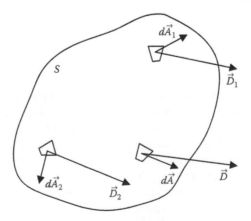

**FIGURE 2.1**
Evaluation of the total electric flux through a surface.

When the sizes of the individual areas become infinitesimally small, then the total flux through $S$ is given by the integral

$$\int_S \vec{D} \cdot d\vec{A} = \iint \vec{D} \cdot d\vec{A} \tag{2.1}$$

## 2.2.2 Charge within a Closed Volume

Consider that the total charge within the volume $V$ enclosed by the surface $S$ needs to be evaluated, as shown in Figure 2.2. Here also, it is to be kept in mind that the distributed charge distribution may not be uniform throughout $V$. Then the entire volume needs to be subdivided into a large number of small volumes such that within each small volume, charge density is constant. Then the total charge within $V$ is given by the summation $\rho_{v1}dV_1 + \rho_{v2}dV_2 + \rho_{v3}dV_3 + \cdots$

When the sizes of the individual volumes become infinitesimally small, then the total charge within $V$ is given by the integral $\int_V \rho_v dV = \iiint \rho_v \, dV$.

## 2.2.3 Solid Angle

The concept of solid angle is a natural extension of two-dimensional plane angle to three dimensions. The solid angle subtended by an area $A$ at the point $O$ is measured by the area $\Omega$ on the surface of the unit sphere centred at $O$, as shown in Figure 2.3. This is the area that would be cut out on the unit sphere surface by lines drawn from $O$ to every point on the periphery of $A$. It is measured in terms of the unit called *steradian*, abbreviated as sr.

Consider an elementary area $dA$, as shown in Figure 2.4. If all the points on the periphery of $dA$ are joined to the point $O$, then these lines cut out an area $d\Omega$ on the surface of the unit sphere. In other words, $dA$ subtends a solid angle $d\Omega$ at $O$. Because the area is infinitesimally small, all points on $dA$

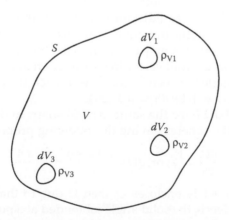

**FIGURE 2.2**
Evaluation of the total charge within a volume.

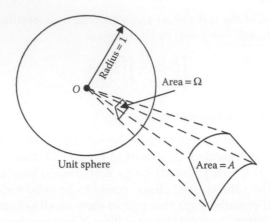

**FIGURE 2.3**
Solid angle subtended by an area at a point.

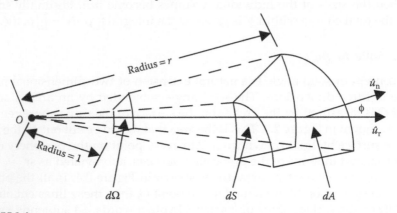

**FIGURE 2.4**
Evaluation of the solid angle subtended by an elementary area.

could be considered to be equidistant from $O$. Let the distance of $dA$ from $O$ be $r$. The unit vector in the direction of the distance vector $\vec{r}$ is $\hat{u}_r$. Because $dA$ is infinitesimally small, it may be taken as flat for practical purposes and hence $\hat{u}_n$ is defined as the unit vector in the direction normal to $dA$, as shown in Figure 2.4. The angle between $\hat{u}_r$ and $\hat{u}_n$ is $\phi$. Then the projection of $dA$ on the sphere of radius $r$ will be $dS = dA \cos\phi$.

The areas $dS$ and $d\Omega$ have the same general shape and are related to the respective radii of the spheres having the following proportionality:

$$\frac{d\Omega}{dS} = \frac{1^2}{r^2}, \text{ or, } d\Omega = \frac{dA \cos\phi}{r^2} = \frac{\hat{u}_r \cdot d\vec{A}}{r^2} \tag{2.2}$$

When an area completely encloses $O$, then $\Omega$ due to that area on the unit sphere will be $4\pi$. Hence, the solid angle subtended at a point by an area that completely encloses the point is $4\pi$ sr, which happens to be maximum possible value of the solid angle.

## 2.3 Integral Form of Gauss's Law

Gauss's law states that the net electric flux through any closed surface enclosing a homogeneous volume of material is equal to the net electric charge enclosed by that closed surface. In other words, the surface integral of electric flux density vector over a closed surface is equal to the volume integral of charge densities within the volume enclosed by that surface.

In integral form,

$$\int_A \vec{D} \cdot d\vec{A} = \int_V \rho_v dV \tag{2.3}$$

Gauss's law is valid for any discrete set of point charges. Nevertheless, this law is also valid when an electric field is produced by charged objects with continuously distributed charges, because any continuously distributed charge may be viewed as a combination of discrete point charges.

Figure 2.5 shows a certain volume $V$ of homogeneous material enclosed by a surface $A$. This volume is continuously charged by distributed charges. Consider these distributed charges to be represented by $N$ number of discrete point charges, as shown in Figure 2.5.

Take a point charge $q_1$ located within $V$. Then consider an elementary area $dA$ on $A$, as shown in Figure 2.5. The distance of $dA$ from $q_1$ is, say, $r$.

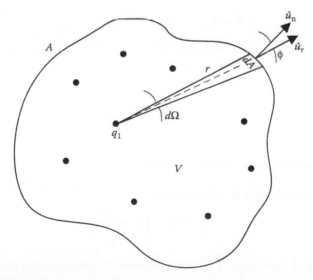

**FIGURE 2.5**
Pertaining to the integral form of Gauss's law.

Then at a distance $r$ from $q_1$,

$$\vec{E}_1 = \frac{q_1}{4\pi\varepsilon_0 r^2}\hat{u}_r \text{ and } \vec{D}_1 = \frac{q_1}{4\pi r^2}\hat{u}_r \tag{2.4}$$

Then, the total flux through $A$ due to $q_1$ could be obtained as follows:

$$\int_A \vec{D}_1 \cdot d\vec{A} = \int_A \frac{q_1}{4\pi}\frac{\hat{u}_r \cdot d\vec{A}}{r^2} = \frac{q_1}{4\pi}\int_A d\Omega, \text{ as } \frac{\hat{u}_r \cdot d\vec{A}}{r^2} = d\Omega \tag{2.5}$$

Because $A$ completely encloses $q_1$, the solid angle subtended by $A$ at the location of $q_1$ is $4\pi$. Thus,

$$\int_A \vec{D}_1 \cdot d\vec{A} = \frac{q_1}{4\pi} \times 4\pi = q_1 \tag{2.6}$$

Considering, $N$ number of point charges within $V$, total electric flux density at the location of $dA$, as shown in Figure 2.5, will be as follows:

$$\vec{D} = \vec{D}_1 + \vec{D}_2 + \vec{D}_3 + \cdots + \vec{D}_N \tag{2.7}$$

Therefore, the total flux through $A$ will be as follows:

$$\int_A \vec{D} \cdot d\vec{A} = \int_A \left(\vec{D}_1 + \vec{D}_2 + \vec{D}_3 + \cdots + \vec{D}_N\right)\cdot d\vec{A}$$

$$= \int_A \vec{D}_1 \cdot d\vec{A} + \int_A \vec{D}_2 \cdot d\vec{A} + \int_A \vec{D}_3 \cdot d\vec{A} + \cdots + \int_A \vec{D}_N \cdot d\vec{A} \tag{2.8}$$

$$= q_1 + q_2 + q_3 + \cdots + q_N$$

The right-hand side (RHS) of Equation 2.8 is equal to the total electric charge enclosed within $V$. Considering continuously distributed charge within $V$,
   The total charge within is

$$V = \int_V \rho_v \, dV$$

where:
$\rho_v$ is the volume charge density within $V$

Hence, Equation 2.8 can be rewritten as

$$\int_A \vec{D} \cdot d\vec{A} = \int_V \rho_v dV$$

From a physical viewpoint, the above expression indicates that the sum of all sources (positive electric charges) and all sinks (negative electric charges) within a volume gives the net flux through the surface enclosing the same volume.

Gauss's law in integral form is only useful for exact solutions when the electric field has symmetry, for example, spherical, cylindrical or planar symmetry. The symmetry of the electric field follows from the symmetry of the charge distribution that is given. Therefore, the electric field of a symmetrically charged sphere will also have the spherical symmetry.

### 2.3.1 Gaussian Surface

A closed surface in a three-dimensional space through which the flux of a vector field is calculated is known as *Gaussian surface*. It is an arbitrary closed surface over which surface integral is performed to evaluate the total amount of enclosed source quantity, for example, electric charge. A Gaussian surface could also be used for calculating the electric field due to a given charge distribution.

Gaussian surfaces are normally carefully chosen in order to exploit symmetries to simplify the evaluation of the surface integral. If the symmetry is such that a surface can be found on which the electric field is constant, then the evaluation of electric flux can be done by multiplying the value of the field with the area of the Gaussian surface.

Two commonly used Gaussian surfaces are as follows:

1. The spherical Gaussian surface, which is chosen to be concentric with the charge distribution. It can be used to determine the electric field or electric flux due to a point charge or a spherical shell of uniform charge density or any other charge distribution with spherical symmetry, as shown in Figure 2.6.

2. The cylindrical Gaussian surface, which is chosen to be co-axial with the charge distribution. It can be used to determine the electric

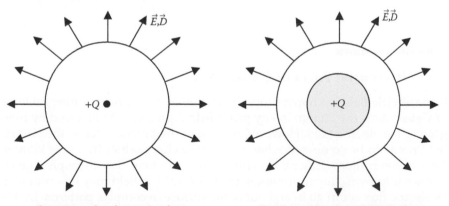

Gaussian surface for a point charge      Gaussian surface for a charged sphere

**FIGURE 2.6**
Spherical Gaussian surface.

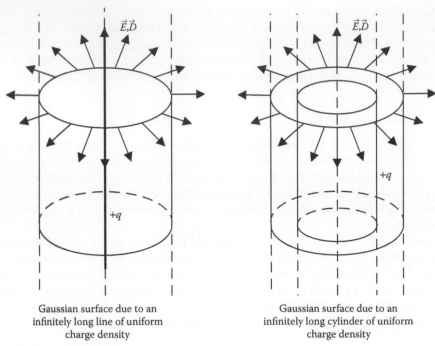

Gaussian surface due to an
infinitely long line of uniform
charge density

Gaussian surface due to an
infinitely long cylinder of uniform
charge density

**FIGURE 2.7**
Cylindrical Gaussian surface.

field or electric flux due to an infinitely long line of uniform charge density or an infinitely long cylinder of uniform charge density, as shown in Figure 2.7.

## 2.4 Differential Form of Gauss's Law

If the electric field is known everywhere, then with the help of integral form of Gauss's law, the charge in any given field region can be deduced by integrating the electric field. However, in most of the practical cases, the electric field needs to be computed when the electric charge distribution is known. This is much more difficult because even if the total flux through a given surface is known, the information about the electric field may be unknown, as electric flux could go in and out of the surface in complex patterns. In this context, the differential form of Gauss's law becomes useful.

Consider that Gauss's law is to be applied for the infinitesimal parallelepiped, as shown in Figure 2.8. The volume of the infinitesimal parallelepiped is

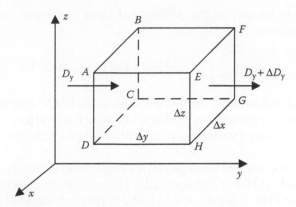

**FIGURE 2.8**
Application of Gauss's law to a differential volume.

$\Delta V = \Delta x \Delta y \Delta z$. Let the component of electric flux density going normally into the surface ABCD be $D_y$ and that coming out of the surface EFGH be $D_y + \Delta D_y$. Surfaces ABCD and EFGH are lying along the $x$–$z$ plane. Then the net electric flux coming out in the $y$-direction, which is normal, to $x$–$z$ plane is given by

$$(D_y + \Delta D_y)\Delta x \Delta z - D_y \Delta x \Delta z = \Delta D_y \Delta x \Delta z = \frac{\Delta D_y}{\Delta y}\Delta x \Delta y \Delta z \approx \frac{\partial D_y}{\partial y}\Delta V \quad (2.9)$$

Similarly, the net electric fluxes coming out in $x$- and $z$-directions are given by $(\partial D_x/\partial x)\Delta V$ and $(\partial D_z/\partial z)\Delta V$, respectively. Hence, net electric flux coming of the volume $\Delta V$

$$= \frac{\partial D_x}{\partial x}\Delta V + \frac{\partial D_y}{\partial y}\Delta V + \frac{\partial D_z}{\partial z}\Delta V = \left(\frac{\partial D_x}{\partial x} + \frac{\partial D_y}{\partial y} + \frac{\partial D_z}{\partial z}\right)\Delta V$$

Let $\rho_v$ be the volume charge density of the small volume $\Delta V$. Then according to Gauss's law, the above expression for the net flux coming out of $\Delta V$ is equal to the total charge within the volume, that is, $\rho_v \Delta V$. Therefore,

$$\left(\frac{\partial D_x}{\partial x} + \frac{\partial D_y}{\partial y} + \frac{\partial D_z}{\partial z}\right)\Delta V = \rho_v \Delta V$$

$$\text{or,} \quad \frac{\partial D_x}{\partial x} + \frac{\partial D_y}{\partial y} + \frac{\partial D_z}{\partial z} = \rho_v$$

(2.10)

Equation 2.10 is known as the differential form of Gauss's law. It can be expanded as follows:

$$\left(\frac{\partial}{\partial x}\hat{i}+\frac{\partial}{\partial y}\hat{j}+\frac{\partial}{\partial z}\hat{k}\right)\left(D_x\hat{i}+D_y\hat{j}+D_z\hat{k}\right)=\rho_v,\ or,\ \vec{\nabla}\cdot\vec{D}=\rho_v \qquad (2.11)$$

In Section 1.7.4, it has been mentioned that when the $\vec{\nabla}$ operator acts on a scalar quantity, it is denoted as gradient. As depicted in Equation 2.11, when the $\vec{\nabla}$ operator acts on a vector quantity with a dot product, then it is denoted as divergence.

The physical meaning of divergence of a vector field can be explained from the left-hand side (LHS) of Equation 2.10. From the above discussion, it may be seen that the LHS of Equation 2.10 is the net flux coming out per unit volume at a given location. Consequently, the divergence of a vector field at any location is the net flux of that vector field coming out per unit volume at that given location. If the divergence of a vector field is positive at any location, then the flux coming out of unit volume is higher than the flux going into the unit volume at that location. On the other hand, if the divergence of a vector field is negative, then the flux going into the unit volume is higher than the flux coming out of the unit volume.

## 2.5 Divergence Theorem

Integral form of Gauss's law is, $\int_A \vec{D}\cdot d\vec{A}=\int_V \rho_v dV$ and differential form of Gauss's law is, $\vec{\nabla}\cdot\vec{D}=\rho_v$.

By substituting $\rho_v$ on the RHS of the integral form by the LHS of the differential form, the above two equations could be combined as follows:

$$\int_A \vec{D}\cdot d\vec{A} = \int_V (\vec{\nabla}\cdot\vec{D})dV \qquad (2.12)$$

where:
The volume $V$ is enclosed by the surface $A$

Equation 2.12 shows that the divergence theorem could be used to convert a volume integral over $V$ to a surface integral over $A$ enclosing $V$.

## 2.6 Poisson's and Laplace's Equations

A useful approach to the calculation of electric potentials is to relate electric potential to electric charge density, which causes electric potential. Because electric potential is a scalar quantity, this approach has advantages over calculation of electric field intensity, which is a vector quantity. Once electric potential has been determined, the electric field intensity can be computed by taking the spatial gradient of electric potential.

The relationship between electric flux density and electric charge density is given by the differential form of Gauss's law, that is, $\vec{\nabla} \cdot \vec{D} = \rho_v$.

For homogeneous medium having uniform dielectric permittivity throughout the volume, the electric flux density and electric field intensity are related as $\vec{D} = \varepsilon \vec{E}$.

The above two equations could be combined as follows:

$$\vec{\nabla} \cdot (\varepsilon \vec{E}) = \rho_v, \text{ or, } \varepsilon \, \vec{\nabla} \cdot \vec{E} = \rho_v, \text{ or, } \vec{\nabla} \cdot \vec{E} = \frac{\rho_v}{\varepsilon} \tag{2.13}$$

Further, electric field intensity and electric potential are related as $\vec{E} = -\vec{\nabla}\phi$. Hence, from Equation 2.13, it may be written that

$$\vec{\nabla} \cdot (-\vec{\nabla}\phi) = \frac{\rho_v}{\varepsilon}, \text{ or, } \vec{\nabla} \cdot \vec{\nabla}\phi = -\frac{\rho_v}{\varepsilon}, \text{ or, } \vec{\nabla}^2 \phi = -\frac{\rho_v}{\varepsilon} \tag{2.14}$$

Equation 2.14 is a partial differential equation of elliptic form and is named after the French mathematician Siméon Denis Poisson. Thus, it is commonly known as *Poisson's equation*.

In Equation 2.14, the mathematical operation, the divergence of gradient of a function ($\vec{\nabla} \cdot \vec{\nabla} = \vec{\nabla}^2$), is called the *Laplacian*, such that

$$\vec{\nabla}^2 = \left( \frac{\partial}{\partial x}\hat{i} + \frac{\partial}{\partial y}\hat{j} + \frac{\partial}{\partial z}\hat{k} \right) \cdot \left( \frac{\partial}{\partial x}\hat{i} + \frac{\partial}{\partial y}\hat{j} + \frac{\partial}{\partial z}\hat{k} \right) = \frac{\partial^2}{\partial x^2} + \frac{\partial^2}{\partial y^2} + \frac{\partial^2}{\partial z^2} \tag{2.15}$$

Therefore, Equation 2.14 can be written as follows:

$$\frac{\partial^2 \phi}{\partial x^2} + \frac{\partial^2 \phi}{\partial y^2} + \frac{\partial^2 \phi}{\partial z^2} = -\frac{\rho_v}{\varepsilon} \tag{2.16}$$

Expressing the Laplacian in different coordinate systems to take advantage of the symmetry of a charge distribution simplifies the solution for the electric potential in many cases.

In electrostatic field problems, the dielectric media may be considered to be ideal insulation. In such case, free charges reside only on the conductor boundaries. Hence, the volume charge density ($\rho_v$) within the field region is zero. Then Equation 2.14 reduces to

$$\vec{\nabla}^2 \phi = 0, \text{ or, } \frac{\partial^2 \phi}{\partial x^2} + \frac{\partial^2 \phi}{\partial y^2} + \frac{\partial^2 \phi}{\partial z^2} = 0, \text{ or, } \frac{\partial^2 \phi}{\partial x^2} + \frac{\partial^2 \phi}{\partial y^2} + \frac{\partial^2 \phi}{\partial z^2} = 0 \tag{2.17}$$

Equation 2.17 is a second-order partial differential equation named after French mathematician Pierre-Simon Laplace and is commonly known as

*Laplace's equation*. There are an infinite number of functions that satisfy Laplace's equation and the proper solution is found by specifying the appropriate boundary conditions, which will be discussed in Chapter 6.

In real-life problems, however, the dielectric media are never ideal dielectrics. Hence, there could be volume leakage as well as surface-leakage currents. Moreover, there could be discharges occurring within a particular zone of the insulation or there may be space charges that have accumulated over time within the field region of interest. In all such cases, volume charge density will be non-zero and hence Poisson's equation needs to be solved to get the correct results.

## 2.7 Field due to a Continuous Distribution of Charge

The electric potential at a point $p$ due to a number of discrete charges could be obtained as simple algebraic superposition of the electric potentials produced at $p$ by each of discrete charges acting in isolation. If $q_1, q_2, q_3, \ldots, q_n$ are discrete charges located at distances $r_1, r_2, r_3, \ldots, r_n$, respectively, from $p$, then the electric potential at $p$ is given by

$$\phi_p = \frac{q_1}{4\pi\varepsilon r_1} + \frac{q_2}{4\pi\varepsilon r_2} + \cdots + \frac{q_N}{4\pi\varepsilon r_N} = \frac{1}{4\pi\varepsilon} \sum_{i=1}^{N} \frac{q_i}{r_i} \tag{2.18}$$

Now, if the charges are distributed continuously throughout the field region, instead of being located at discrete number of points, the field region can be divided into large number of small elements of volume $\Delta V$, such that each element contains a charge $\rho_v \Delta V$, where $\rho_v$ is the volume charge density of the small element $\Delta V$. The potential at a point $p$ will then be given by

$$\phi_p = \frac{1}{4\pi\varepsilon} \sum_{i=1}^{N} \frac{\rho_{vi} \Delta V_i}{r_i} \tag{2.19}$$

where:
$r_i$ is the distance of the $i$th volume element from $p$

As the size of volume element becomes very small, the summation becomes an integration, that is,

$$\phi = \frac{1}{4\pi\varepsilon} \int_V \frac{\rho_v dV}{r} \tag{2.20}$$

The integration is performed over the volume where the volume charge density has finite value. However, it must be noted here that it is not valid for charge distribution that extends to infinity.

Equation 2.20 is often written in the form

$$\phi = \int_V \rho_v G dV \tag{2.21}$$

In Equation 2.21, the function $G = 1/4\pi\varepsilon r$ is the potential due to a unit point charge and is often referred to as the electrostatic Green's function for an unbounded homogeneous region.

## 2.8 Steps to Solve Problems Using Gauss's Law

Typical steps that are often taken while solving electrostatic problems with the help of Gauss's law are as follows:

- An appropriate Gaussian surface with symmetry is selected that matches the symmetry of the charge distribution.
- The Gaussian surface is so chosen that the electric field is either constant or zero at all points on the Gaussian surface.
- The direction of $\vec{E}$ on the Gaussian surface is determined using the symmetry of field.
- The surface integral of electric flux is computed over the selected Gaussian surface.
- The charge enclosed by the Gaussian surface is determined from the surface integral. Solution for $\vec{E}$ is found.

### PROBLEM 2.1

The potential field at any point in a space containing a dielectric material of relative permittivity 3.6 is given by $\phi = (3x^2y - 2y^2z + 5xyz^2)$V, where $x,y,z$ are in metres. Calculate the volume charge density at the point $P(5,3,2)$m.

*Solution:*

$$E_x = -\frac{\partial \phi}{\partial x} = (-6xy - 5yz^2) \text{ V/m, or, } D_x = \varepsilon_r\varepsilon_0 E_x \text{ C/m}^2$$

$$E_y = -\frac{\partial \phi}{\partial y} = (-3x^2 + 4yz - 5xz^2) \text{ V/m, or, } D_y = \varepsilon_r\varepsilon_0 E_y \text{ C/m}^2$$

$$E_z = -\frac{\partial \phi}{\partial z} = (2y^2 - 10xyz) \text{ V/m, or, } D_z = \varepsilon_r\varepsilon_0 E_z \text{ C/m}^2$$

Therefore, $\dfrac{\partial D_x}{\partial x} = -6y\varepsilon_r\varepsilon_0$, $\dfrac{\partial D_y}{\partial y} = 4z\varepsilon_r\varepsilon_0$ and $\dfrac{\partial D_z}{\partial z} = -10xy\varepsilon_r\varepsilon_0$

Therefore, $\rho_v = \dfrac{\partial D_x}{\partial x} + \dfrac{\partial D_y}{\partial y} + \dfrac{\partial D_z}{\partial z} = (-6y + 4z - 10xy)\varepsilon_r\varepsilon_0 \, C/m^3$

Hence, at the point $P(5,3,2)$,

$$\rho_{vP} = -160 \times 3.6 \times 8.854 \times 10^{-12} = -5.1 \, nC/m^3$$

## PROBLEM 2.2
The electric flux density in free space is given by $\vec{D} = e^{-y}(\cos x \, \hat{i} - \sin x \, \hat{j})C/m^2$. Prove that the field region is charge free, that is, no free charge is present in the field region.

*Solution:*
It is required to be proved that $\rho_v$ is zero in the field region. The given expression for electric flux density vector indicates that it is a case of two-dimensional field.

As given,

$$D_x = e^{-y}\cos x \text{ and } D_y = -e^{-y}\sin x$$

Hence,

$$\frac{\partial D_x}{\partial x} = -e^{-y}\sin x \text{ and } \frac{\partial D_y}{\partial y} = e^{-y}\sin x$$

Therefore,

$$\rho_v = \frac{\partial D_x}{\partial x} + \frac{\partial D_y}{\partial y} = -e^{-y}\sin x + e^{-y}\sin x = 0$$

Thus, it is proved that no free charge is present in the field region.

## PROBLEM 2.3
A sphere of radius $R$ carries a volume charge density $\rho_v(r) = kr$, where $r$ is the radial distance from the centre and $k$ is a constant. Determine the electric field intensity inside and outside the sphere.

*Solution:*
Inside the sphere, at any radius $r_i$, the amount of charge enclosed is given by

$$q(r_i) = \int_0^{r_i} kr \frac{4\pi}{3} r^3 dr = \frac{4\pi k}{3} \int_0^{r_i} r^4 dr = \frac{4\pi k r_i^5}{15}$$

Therefore, considering a spherical Gaussian surface of radius $r_i$ ($<R$) within the given sphere, the electric flux is normal to the Gaussian surface and electric field intensity $E_i(r_i)$ is constant over the Gaussian surface. Thus, applying Gauss's law,

$$\varepsilon_0 \int \vec{E}_i(r_i) \cdot d\vec{A} = q(r_i), \text{ or, } \varepsilon_0 \, 4\pi r_i^2 \, E_i(r_i) = \frac{4\pi k r_i^5}{15}, \text{ or, } E_i(r_i) = \frac{k r_i^3}{15\varepsilon_0}$$

Because the electric field is directed radially, its vector form is

$$\vec{E}_i(r_i) = \left( \frac{k r_i^3}{15\varepsilon_0} \right) \hat{u}_r$$

Again, total charge within the sphere of radius $R$ is $(4\pi k R^5)/15$.

Hence, considering a spherical Gaussian surface of radius $r_o$($>R$) outside the given sphere, the electric flux is again normal to the Gaussian surface and electric field intensity $E_o(r_o)$ is constant over the Gaussian surface. Thus, applying Gauss's law,

$$\varepsilon_0 \, 4\pi r_o^2 E_o(r_o) = \frac{4\pi k R^5}{15}, \text{ or, } E_o(r_o) = \frac{k R^5}{15\varepsilon_0 r_o^2}$$

Because the electric field is directed radially, its vector form is

$$\vec{E}_o(r_o) = \left( \frac{k R^5}{15\varepsilon_0 r_o^2} \right) \hat{u}_r$$

**PROBLEM 2.4**

The electric flux density vector in a field region is given by $\vec{D} = (2y+z)\hat{i} + 3xy\hat{j} + 2x\hat{k}$ C/m². Determine the total charge enclosed by the cube defined by $0 \leq x \leq 1, 0 \leq y \leq 1$ and $0 \leq z \leq 1$.

*Solution:*

According to Gauss's law, the total charge enclosed by the cube will be equal to net flux through the cube (Figure 2.9).

For $x = 0$, the flux going into the cube through $y$–$z$ plane, $\int_0^1 \int_0^1 (2y+z)dydz$.

Again, for $x = 1$, the flux coming out of the cube through $y$–$z$ plane, $\int_0^1 \int_0^1 (2y+z)dydz$.

Because these two integrals are same, the net flux coming out of the cube in $x$-direction, that is, through $y$–$z$ plane, is zero.

For $y = 0$, the flux going into the cube through $x$–$z$ plane, $\int_0^1 \int_0^1 3xy\,dxdz = (3y/2)\big|_{y=0} = 0$.

For $y = 1$, the flux coming out of the cube through $x$–$z$ plane, $\int_0^1 \int_0^1 3xy\,dxdz = (3y/2)\big|_{y=1} = 1.5$ C.

Therefore, the net flux coming out of the cube in $y$-direction, that is, through $x$–$z$ plane is 1.5 C.

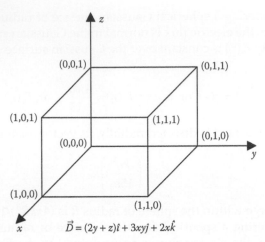

**FIGURE 2.9**
Pertaining to Problem 2.4.

For $z = 0$, the flux going into the cube through $x$–$y$ plane, $\int_0^1 \int_0^1 2x\,dx\,dy$.

Again, for $z = 1$, the flux coming out of the cube through $x$–$y$ plane, $\int_0^1 \int_0^1 2x\,dx\,dy$.

Because these two integrals are same, the net flux coming out of the cube in $z$-direction, that is, through $x$–$y$ plane, is zero.

Therefore, considering all three directions, net flux coming out of the cube is 1.5 C.

Hence, the total charge enclosed by the given cube is also 1.5 C.

### PROBLEM 2.5

A cylinder of unit volume is placed in a uniform field with its axis parallel to the direction of field. Determine the total charge enclosed by the unit cylinder.

*Solution:*
On the curved surface of the cylinder, the direction of the area vector is radially outwards and hence is perpendicular to the electric field intensity, which is directed along the axis of the cylinder. Hence, on the curved surface $\int_A \varepsilon_0 \vec{E}\cdot d\vec{A} = 0$, that is, net flux through the curved surface of the cylinder will be zero.

As shown in Figure 2.10, the two end surfaces of the cylinder S1 and S2 are normal to the direction of electric field intensity. Thus, the electric flux going into the cylinder through the surface S1 on the LHS will be $\varepsilon_0 |\vec{E}| A$, where $A$ is the area of S1. Similarly, the electric flux coming out of the cylinder through the surface S2 on the RHS will also be $\varepsilon_0 |\vec{E}| A$, as the area of S2 is also $A$. Because the field is stated to be uniform, electric field intensity is same everywhere. Therefore, the net flux through the end surfaces of the cylinder is also zero.

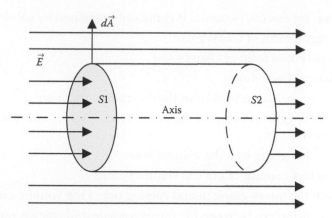

**FIGURE 2.10**   Pertaining to Problem 2.5.

Thus, the net flux through the cylinder considering all the surfaces of the cylinder is zero.

Hence, the total charge enclosed by the cylinder is zero.

## Objective Type Questions

1. The potential field in free space is given by $\phi = (3xy - 4yz + 5zx)$V, where $x,y,z$ are in metres. At the point P(1,1,1)m, which one of the following is true?

    a.  Electric potential is zero

    b.  Electric field intensity is zero

    c.  Electric flux density is zero

    d.  Volume charge density is zero

2. Net electric flux through a closed surface enclosing a volume is equal to the sum of

    a.  Free charges and bound charges within the volume

    b.  Free positive charges within the volume

    c.  Free negative charges within the volume

    d.  Free positive and negative charges within the volume

3. On the Gaussian surface, electric field

    a.  Is constant at all points

    b.  Is zero at all points

    c.  Both (a) and (b)

    d.  Varies linearly

4. Solution for electric potential is commonly obtained by solving
   a. Integral form of Gauss's law
   b. Differential form of Gauss's law
   c. Poisson's equation
   d. The equation derived from the divergence theorem
5. If the divergence of electric flux density through a unit volume is zero, then
   a. The flux going into the volume is zero
   b. The flux coming out of the volume is zero
   c. Both the fluxes going in and coming out of the volume are zero
   d. The difference between the fluxes going in and coming out of the volume is zero
6. Solution for electric potential in charge-free field region is obtained by solving
   a. Laplace's equation
   b. Poisson's equation
   c. Integral form of Gauss's law
   d. Differential form of Gauss's law
7. A volume integral can be converted to a surface integral and vice versa with the help of
   a. Differential form of Gauss's law
   b. Divergence theorem
   c. Poisson's equation
   d. Laplace's equation
8. If the flux coming out of a closed surface enclosing a volume is higher than the flux going into the surface, then within the volume enclosed by the surface
   a. Free positive charges are greater than free negative charges
   b. Free positive charges are lesser than free negative charges
   c. Free positive charges are greater than bound negative charges
   d. Free positive charges are lesser than bound negative charges
9. For a closed surface enclosing a certain volume, Gauss's law relates to
   a. Free charge within the volume and electric potential on the surface
   b. Both free and bound charges within the volume and electric potential on the surface

   c. Free charge within the volume and electric flux through the surface

   d. Both free and bound charges within the volume and electric flux through the surface

10. Which one of the following statements is not true?

   a. Gauss's law is valid for a set of discrete point charges

   b. Gauss's law in not valid for continuously distributed charges

   c. Volume charge density is zero for charge-free field region

   d. Divergence theorem can be obtained from integral and differential forms of Gauss's law

**Answers:** 1) d; 2) d; 3) c; 4) c; 5) d; 6) a; 7) b; 8) a; 9) c; 10) b

# 3

## Orthogonal Coordinate Systems

**ABSTRACT** The purpose of this chapter is to present coordinate systems in such a way that it improves the ability of readers to use them to describe the electric field in a better way. The goal is to put forward the coordinate system as a useful language for describing, visualizing and understanding the concepts that are central to electric field analysis. Many mathematical operators related to electric field analysis have simple forms in Cartesian coordinates and are easy to remember and evaluate. In fact, all the problems could be solved using the Cartesian coordinate system. However, the resulting expressions might be unnecessarily complex. While dealing with problems that have high degree of symmetry, it is helpful to use a coordinate system that exploits the symmetry. As a result, it is useful to have a general method for expressing mathematical operators in non-Cartesian forms. Orthogonal curvilinear coordinate system is a useful approach by which the matching of the symmetry of a given physical configuration could be done and simplified mathematical models could be built to clarify the field concepts clearly.

## 3.1 Basic Concepts

The concept that unifies the different coordinate systems is surfaces of constant coordinates. It means that an equation of the form (coordinate = value) defines a surface. In order to specify a point uniquely in a three-dimensional space, three different types of surfaces are needed. The values associated with the three constant coordinate surfaces that intersect at a specific point are the three coordinates, which uniquely define that point.

Consider that three surfaces are defined by three functions $f_1$, $f_2$ and $f_3$, respectively, in a three-dimensional space. The coordinate, that is, the value, that describes a surface could be defined by the equation $f_i = u_i$. Here, $u$ is introduced to represent a generalized coordinate. In the stated equation, $u_i$ refers to one of the coordinates in a particular system. Thus, the three values $(u_1, u_2, u_3)$ define a point, where the three surfaces defined by Equation 3.1 intersect. These three values $(u_1, u_2, u_3)$ are then called the *coordinates* of that particular point.

$$f_1 = u_1, f_2 = u_2 \text{ and } f_3 = u_3 \tag{3.1}$$

The three functions, however, could not be chosen arbitrarily. The three functions are to be chosen in such a way that for any choice of the three coordinates $(u_1, u_2, u_3)$, the three surfaces generated by Equation 3.1 will intersect only at one point. It will ensure that a given set of three coordinate values $(u_1, u_2, u_3)$ always specifies only one point in space. Moreover, the constant coordinate surfaces are always perpendicular to each other, that is, orthogonal, at the point where they intersect. The intersection of any two constant coordinate surfaces defines a coordinate curve.

### 3.1.1 Unit Vector

An important concept associated with coordinate system is unit vector. It will be easy to understand the concept if it is explained with the help of a specific example involving the Cartesian coordinate system, as shown in Figure 3.1. Each value of $z$, for example, $z_1$, $z_2$ or $z_3$, defines a plane normal to the $z$-axis. The unit vector $\hat{k}$ is defined in such a way that it is normal to all such $z$ = constant planes and points in the direction of the planes with increasing values of $z$, as shown in Figure 3.1. The magnitude of the vector $\hat{k}$ is taken as unity. The other two unit vectors in the Cartesian coordinate system could be visualized in the same manner.

In a generalized manner, the properties of the unit vector associated with a constant coordinate surface defined by the function $f_i$ are as follows: (1) it is perpendicular to the surface $f_i = u_i$ at any point $(u_1, u_2, u_3)$; (2) it points in the direction of the surfaces with increasing value of $u_i$ and (3) it is one unit in magnitude. Orthogonality of the constant coordinate surfaces, as discussed earlier, demands that the dot product of any two different unit vectors will always be zero. Conversely, the dot product of two same unit vectors, for example, $\hat{k} \cdot \hat{k}$, will always be 1.

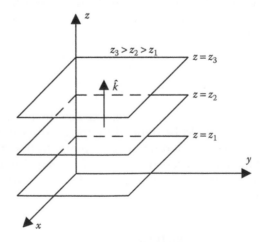

**FIGURE 3.1**
The unit vector $\hat{k}$.

### 3.1.2 Right-Handed Convention

Because three constant coordinate surfaces are involved in determining the coordinates of any point, the question that arises is in which order the three coordinates are to be mentioned while specifying a point. In this context, the choice of the first coordinate $(u_1)$ or surface $(f_1)$ is arbitrary. Once $f_1$ is chosen, then the order in which the other two constant coordinate surfaces $f_2$ and $f_3$ is to be taken is determined by the right-handed system as governed by the definition of vector cross product. In other words, the cross product of the first and second unit vectors must be equal to the third unit vector.

This right-handed property of the coordinate system is an arbitrary convention. But once that choice is made for one coordinate system, it has to be followed for all other coordinate systems to ensure that the expressions for the laws of electric field have the same form in all the coordinate systems.

### 3.1.3 Differential Distance and Metric Coefficient

The constant coordinate surfaces could be of many different types. The simplest of them is constant distance surfaces for which the differential change $du_i$ in the coordinate $u_i$ is the same as the differential distance $dl_i$ between the surfaces $f_i = u_i$ and $f_i = u_i + du_i$. However, there are other types of surfaces such as constant angle surfaces, in which case the differential distance is related to differential coordinate (angle) change as

$$dl_i = h_i du_i = h_i d\theta_i \tag{3.2}$$

where:
The scale factor $h_i$ is the corresponding radius

The scale factors for each of the coordinate system are called the *metric coefficients* of the respective coordinate system. The three metric functions of any coordinate system $(h_1, h_2, h_3)$ constitute a unique set and hence are often known as the *signature of a coordinate system*. Metric coefficients are also very important in writing the vector operators in the general orthogonal curvilinear coordinate system, as will be explained in Section 3.5.

### 3.1.4 Choice of Origin

A basic element of any coordinate system is the choice of origin. If any object or any arrangement is given, then any point within the object or within the whole arrangement could be chosen as origin.

From the earlier discussions, it is clear that the same point will have very different coordinates when defined in different coordinate systems. Therefore, the question that needs to be answered is that whether there is any property of the position of the point that remains the same in different coordinate systems. The answer to this question is that the distance of the point from the origin remains the same in all the coordinate systems if the origin is kept the same in all the cases. As a logical extension, it may be stated that the distance between any two points remains the same no matter how their coordinates are defined in any coordinate system.

## 3.2 Cartesian Coordinate System

The invention of Cartesian coordinates in the seventeenth century by the French mathematician René Descartes revolutionized mathematics by providing the first systematic link between Euclidean geometry and algebra. In this coordinate system, the three constant coordinate surfaces are defined by Equation 3.3

$$f_1 = x, f_2 = y, f_3 = z \tag{3.3}$$

Figure 3.2 shows such constant coordinate surfaces, which are three planes. Cartesian coordinate system has the unique feature that all the three constant coordinate surfaces are constant distance surfaces.

Each point $P$ in space is then assigned a triplet of values $(x,y,z)$, the so-called Cartesian coordinates of that point. The ranges of the values of three coordinates are $-\infty < x < \infty$, $-\infty < y < \infty$ and $-\infty < z < \infty$. Coordinates can be

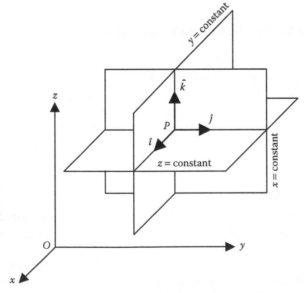

**FIGURE 3.2**
Constant coordinate surfaces in the Cartesian coordinate system.

defined as the positions of the normal projections of the point onto three mutually perpendicular constant coordinate surfaces passing through the origin, expressed as signed distances from the origin.

In the Cartesian coordinate system, three mutually perpendicular axes are commonly defined as follows. The intersection of two mutually perpendicular planes $y$ = constant and $z$ = constant passing through the origin gives a coordinate curve, which is a straight line. This line is then defined as the $x$-axis. Similarly, the intersections of other two pairs of mutually perpendicular constant coordinate planes passing through origin define the $y$- and $z$-axes.

The unit vector $\hat{i}$ is perpendicular to all the surfaces described by $x$ = constant and points in the direction of increasing value of $x$-coordinate. In the same manner, unit vectors $\hat{j}$ and $\hat{k}$ are defined in $y$- and $z$-directions, respectively. Another unique feature of the Cartesian coordinate system is that the directions of the three unit vectors remain the same irrespective of the location of the point in space. The orthogonality of the Cartesian coordinate system is defined by Equation 3.4

$$\hat{i}\cdot\hat{j} = 0,\ \hat{j}\cdot\hat{k} = 0,\ \hat{k}\cdot\hat{i} = 0 \tag{3.4}$$

and the right handedness is defined by Equation 3.5

$$\hat{i}\times\hat{j} = \hat{k},\ \hat{j}\times\hat{k} = \hat{i},\ \hat{k}\times\hat{i} = \hat{j} \tag{3.5}$$

As shown in Figure 3.3, the differential line element in the Cartesian coordinate system is given by Equation 3.6

$$\vec{dl} = dl_1\hat{i} + dl_2\hat{j} + dl_3\hat{k} = dx\hat{i} + dy\hat{j} + dz\bar{k} \tag{3.6}$$

As shown in Figure 3.4, the differential area element corresponding to the surface of a differential cubical volume is given by Equation 3.7

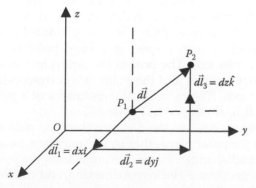

**FIGURE 3.3**
Differential line element in the Cartesian coordinate system.

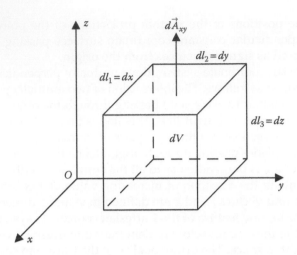

**FIGURE 3.4**
Differential area and volume elements in the Cartesian coordinate system.

$$d\vec{A}_{xy} = dl_1 dl_2 \hat{k} = dxdy\hat{k} \tag{3.7}$$

As shown in Figure 3.4, the differential volume element is given by Equation 3.8

$$dV = dl_1 dl_2 dl_3 = dxdydz \tag{3.8}$$

As depicted in Equation 3.6 the differential distance in $x$-direction is $dl_1 = dx = h_1 dx$, that is, $h_1 = 1$. Similarly, in $y$- and $z$-directions, $h_2 = h_3 = 1$. The metric coefficients of the Cartesian coordinate system are, therefore, $h_1 = 1$, $h_2 = 1$ and $h_3 = 1$.

## 3.3 Cylindrical Coordinate System

Cylindrical coordinates are an alternate way of describing points in a three-dimensional space. In this coordinate system, one of the rectangular coordinate planes, namely, the $x$–$y$ plane, as described by the Cartesian coordinate system, is replaced by a polar plane. In the cylindrical coordinate system, everything is measured with respect to a fixed point called the *pole* and an axis called the *polar axis*. The pole is the equivalent to the origin in the Cartesian coordinate system and the polar axis corresponds to the positive direction of the $x$-axis. The cylindrical coordinates of a point are then the ordered triplet $(r,\theta,z)$, as defined in Figure 3.5.

As shown in Figure 3.5, $r$ is the distance from the pole to the projection of the point $P$ on the polar plane, that is, the $x$–$y$ plane passing through the pole, $\theta$ is the azimuthal angle, that is, the angle from the polar axis spinning around the $z$-axis in counter-clockwise direction, and $z$ is the vertical height from the polar plane. The ranges of the values of the three coordinates are $0 \leq r < \infty$, $0 \leq \theta \leq 2\pi$ and $-\infty < z < \infty$.

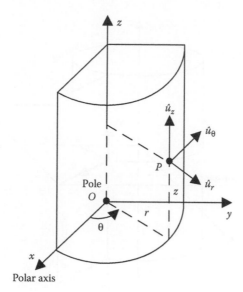

**FIGURE 3.5**
Depiction of cylindrical coordinates of a point.

In the cylindrical coordinate system, the three constant coordinate surfaces are defined by Equation 3.9

$$f_1 = r, f_2 = \theta, f_3 = z \tag{3.9}$$

Figure 3.6 shows the three constant coordinate surfaces in the cylindrical coordinate system. Out of these three surfaces, the first and the third surfaces, namely, $f_1 = r$ and $f_3 = z$, are constant distance surfaces, whereas the second one, that is, $f_2 = \theta$, is a constant angle surface. As shown in Figure 3.6, the surfaces $\theta = $ constant and $z = $ constant are planes, whereas the surface $r = $ constant is a cylindrical surface.

In this coordinate system, two unit vectors are defined on the $x$–$y$ plane. The unit vector $\hat{u}_r$ points in the direction of increasing $r$, that is, radially outwards from the $z$-axis and the unit vector $\hat{u}_\theta$ points in the direction of increasing $\theta$, that is, it points in the direction of the tangent to the circle of radius $r$ in the counter-clockwise sense. The third unit vector $\hat{u}_z$ points in the direction of increasing $z$, that is, vertically upwards from the $x$–$y$ plane. The unit vectors are shown in Figure 3.5. The orthogonality of cylindrical coordinate system is defined by Equation 3.10

$$\hat{u}_r \cdot \hat{u}_\theta = 0, \ \hat{u}_\theta \cdot \hat{u}_z = 0, \ \hat{u}_z \cdot \hat{u}_r = 0 \tag{3.10}$$

and the right handedness is defined by Equation 3.11

$$\hat{u}_r \times \hat{u}_\theta = \hat{u}_z, \ \hat{u}_\theta \times \hat{u}_z = \hat{u}_r, \ \hat{u}_z \times \hat{u}_r = \hat{u}_\theta \tag{3.11}$$

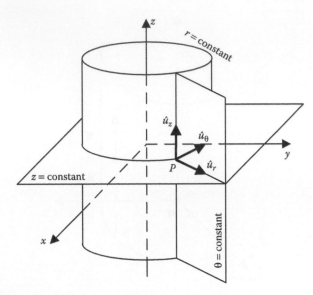

**FIGURE 3.6**
Constant coordinate surfaces in the cylindrical coordinate system.

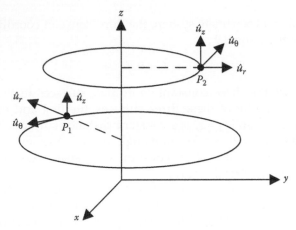

**FIGURE 3.7**
Unit vectors at different points in the cylindrical coordinate system.

Unlike the Cartesian system, in the cylindrical coordinate system, the directions of all the three unit vectors do not remain the same as one moves around the space. As shown in Figure 3.7, the directions of the unit vectors $\hat{u}_r$ and $\hat{u}_\theta$ get changed at different points in space keeping $\hat{u}_z$ unchanged.

As shown in Figure 3.8, the differential line element in the cylindrical coordinate system is given by Equation 3.12

$$d\vec{l} = dl_1\hat{u}_r + dl_2\hat{u}_\theta + dl_3\hat{u}_z = dr\hat{u}_r + rd_\theta\hat{u}_\theta + dz\hat{u}_z \tag{3.12}$$

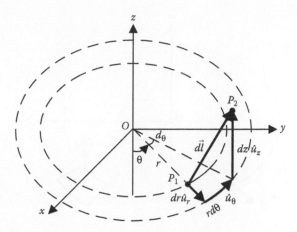

**FIGURE 3.8**
Differential line element in the cylindrical coordinate system.

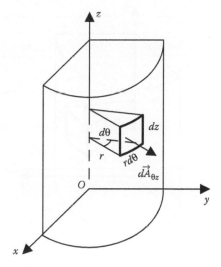

**FIGURE 3.9**
Differential area element on the cylinder surface in the cylindrical coordinate system.

As shown in Figure 3.9, in the cylindrical coordinate system, the differential area element on the surface of a cylinder of radius $r$ is given by Equation 3.13

$$d\vec{A}_{\theta z} = dl_2 dl_3 \hat{u}_r = rd\theta dz \hat{u}_r \qquad (3.13)$$

and the differential area element on the surface of a disc in the $x$–$y$ plane, as shown in Figure 3.10, is given by Equation 3.14

$$d\vec{A}_{r\theta} = dl_1 dl_2 \hat{u}_z = dr r d\theta \hat{u}_z \qquad (3.14)$$

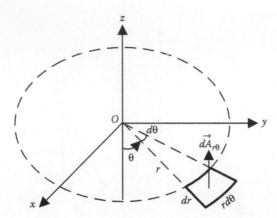

**FIGURE 3.10**
Differential area element on the disc surface in the cylindrical coordinate system.

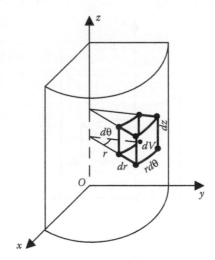

**FIGURE 3.11**
Differential volume element of a cylinder in the cylindrical coordinate system.

As shown in Figure 3.11, the differential volume element is given by Equation 3.15

$$dV = dl_1 dl_2 dl_3 = dr\, rd\theta\, dz \qquad (3.15)$$

As depicted in Equation 3.12, the differential distance in $r$-direction is $dl_1 = dr = h_1 dr$, that is, $h_1 = 1$, in $\theta$-direction, it is $dl_2 = rd\theta = h_2 d\theta$, that is, $h_2 = r$ and in $z$-direction, it is $dl_3 = dz = h_3 dz$, that is, $h_3 = 1$. The metric coefficients of cylindrical coordinate system are, therefore, $h_1 = 1$, $h_2 = r$ and $h_3 = 1$.

## 3.4 Spherical Coordinate System

Spherical coordinates are particularly useful for analyzing fields having spherical symmetry. In the spherical coordinate system, the coordinates of any point in space are the ordered triplet $(r,\theta,\phi)$, as shown in Figure 3.12. The coordinate $r$ measures the radial distance from the origin to the point $P$. The coordinate $\theta$ is the angle that the $r$ vector makes with the positive direction of the $z$-axis, whereas the coordinate $\phi$ is the azimuthal angle with respect to the positive direction of the $x$-axis spinning around the $z$-axis in counterclockwise sense. The angle $\phi$ is defined on the $x$–$y$ plane. The ranges of the values of the three coordinates are $0 \leq r < \infty, 0 \leq \theta \leq \pi$ and $0 \leq \phi \leq 2\pi$.

In the spherical coordinate system, the three constant coordinate surfaces are defined by Equation 3.16

$$f_1 = r, \; f_2 = \theta, \; f_3 = \phi \tag{3.16}$$

Figure 3.13 shows the three constant coordinate surfaces in the spherical coordinate system. Out of these three surfaces, the first surface, namely, $f_1 = r$, is a constant distance surface, whereas the other two surfaces, that is, $f_2 = \theta$ and $f_3 = \phi$, are constant angle surfaces. As shown in Figure 3.13, $r =$ constant is a spherical surface, $\theta =$ constant is a conical surface and $\phi =$ constant is a plane.

In this coordinate system, the unit vector $\hat{u}_r$ points in the direction of increasing $r$, that is, radially outwards from the origin. The unit vector $\hat{u}_\theta$ points in the direction of increasing $\theta$ along the tangent to a circle of radius $r$ in the plane containing the $z$-axis and the $r$ vector. The unit vector $\hat{u}_\phi$ points

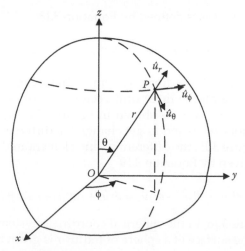

**FIGURE 3.12**
Depiction of spherical coordinates of a point.

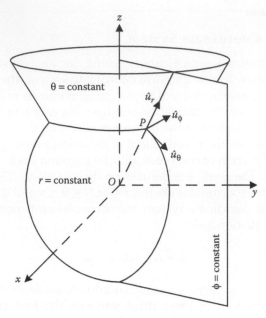

**FIGURE 3.13**
Constant coordinate surfaces in the spherical coordinate system.

in the direction of increasing $\phi$ along the tangent to a circle in the $x$–$y$ plane, which is centred on the $z$-axis. The unit vectors are shown in Figure 3.12. The orthogonality of the spherical coordinate system is defined by Equation 3.17

$$\hat{u}_r \cdot \hat{u}_\theta = 0,\ \hat{u}_\theta \cdot \hat{u}_\phi = 0,\ \hat{u}_\phi \cdot \hat{u}_r = 0 \tag{3.17}$$

and the right handedness is defined by Equation 3.18

$$\hat{u}_r \times \hat{u}_\theta = \hat{u}_\phi,\ \hat{u}_\theta \times \hat{u}_\phi = \hat{u}_r,\ \hat{u}_\phi \times \hat{u}_r = \hat{u}_\theta \tag{3.18}$$

Similar to the cylindrical coordinate system, in the spherical coordinate system, the directions of all the three unit vectors do not remain the same as one move around the space. As shown in Figure 3.14, the directions of all the three unit vectors $\hat{u}_r$, $\hat{u}_\theta$ and $\hat{u}_\phi$ get changed at different points in space.

As shown in Figure 3.15, the differential line element in the spherical coordinate system is given by Equation 3.19

$$d\vec{l} = dl_1 \hat{u}_r + dl_2 \hat{u}_\theta + dl_3 \hat{u}_\phi = dr \hat{u}_r + r d\theta \hat{u}_\theta + r \sin\theta d\phi \hat{u}_\phi \tag{3.19}$$

As shown in Figure 3.16, in the spherical coordinate system, the differential area element on the surface of a sphere of radius $r$ is given by Equation 3.20

$$d\vec{A}_{\theta\phi} = dl_2 dl_3 \hat{u}_r = (rd\theta)(r\sin\theta d\phi)\hat{u}_r = r^2 \sin\theta d\theta d\phi \hat{u}_r \tag{3.20}$$

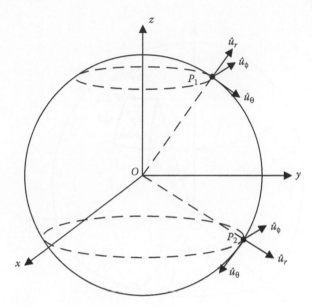

**FIGURE 3.14**
Unit vectors at different points in the spherical coordinate system.

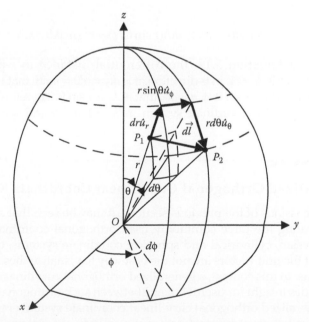

**FIGURE 3.15**
Differential line element in the spherical coordinate system.

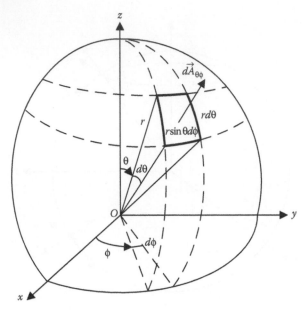

**FIGURE 3.16**
Differential area element on the sphere surface in the spherical coordinate system.

As shown in Figure 3.17, the differential volume element is given by Equation 3.21

$$dV = dl_1dl_2dl_3 = (dr)(rd\theta)(r\sin\theta d\phi) = r^2\sin\theta dr d\theta d\phi \quad (3.21)$$

As depicted in Equation 3.19, the differential distance in $r$-direction is $dl_1 = dr = h_1dr$, that is, $h_1 = 1$, in $\theta$-direction, it is $dl_2 = rd\theta = h_2d\theta$, that is, $h_2 = r$ and in $\phi$-direction, it is $dl_3 = r\sin\theta d\phi = h_3d\phi$, that is, $h_3 = r\sin\theta$. Therefore, the metric coefficients of spherical coordinate system are $h_1 = 1$, $h_2 = r$ and $h_3 = r\sin\theta$.

## 3.5 Generalized Orthogonal Curvilinear Coordinate System

From the discussions of the previous sections, it may be seen that a major hindrance in treating the three commonly used orthogonal coordinate systems, namely, Cartesian, cylindrical and spherical coordinate systems, on an equal footing is that the unit vectors are not the best way to visualize these three coordinate systems. In this context, a generalized orthogonal curvilinear coordinate system provides insight for useful linkage between such orthogonal coordinate systems. Generalized orthogonal curvilinear coordinate system emphasizes on the similarities between different orthogonal coordinate systems rather than highlighting their differences. Thus, the definition and utilization of generalized

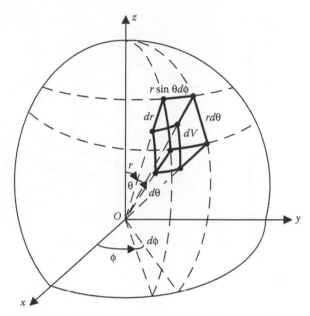

**FIGURE 3.17**
Differential volume element of a sphere in the spherical coordinate system.

orthogonal curvilinear coordinate system is very important for the proper understanding of the formulation and solution of electric field problems.

Consider that $f(x,y,z) = u$ is a function of three independent space variables $x,y,z$ in the Cartesian coordinate system that specify a surface characterized by the constant parameter $u$. Setting such function equal to three different constant parameters $u_1$, $u_2$ and $u_3$ defines three different surfaces as follows:

$$f_1(x,y,z) = u_1, \ f_2(x,y,z) = u_2 \text{ and } f_3(x,y,z) = u_3 \tag{3.22}$$

Considering that these three surfaces are orthogonal, they intersect in space only at one point. In other words, any point in space could be uniquely defined by a set of values of the three parameters $(u_1,u_2,u_3)$, which are then called the *orthogonal curvilinear coordinates* of the point being defined. The constant coordinate surfaces for a generalized orthogonal curvilinear coordinate system are shown in Figure 3.18.

As shown in Figure 3.18, the unit vectors $\hat{a}_1$, $\hat{a}_2$ and $\hat{a}_3$ are normal to the surfaces $u_1 = $ constant, $u_2 = $ constant and $u_3 = $ constant, respectively, and point towards the increasing values of the coordinates $u_1$, $u_2$ and $u_3$, respectively. The orthogonality of the generalized curvilinear coordinate system is defined by Equation 3.23

$$\hat{a}_1 \cdot \hat{a}_2 = 0, \hat{a}_2 \cdot \hat{a}_3 = 0, \hat{a}_3 \cdot \hat{a}_1 = 0 \tag{3.23}$$

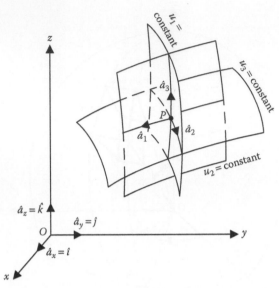

**FIGURE 3.18**
Constant coordinate surfaces in the generalized orthogonal curvilinear coordinate system.

and the right handedness is defined by Equation 3.24

$$\hat{a}_1 \times \hat{a}_2 = \hat{a}_3, \hat{a}_2 \times \hat{a}_3 = \hat{a}_1, \hat{a}_3 \times \hat{a}_1 = \hat{a}_2 \tag{3.24}$$

As shown in Figure 3.19, the surfaces $u = u_1$ and $u = u_1 + du_1$ are separated by a differential length element $dl_1$, which is normal to the surface $u = u_1$, where $dl_1 = h_1(u_1, u_2, u_3)du_1$. The scale factor or the metric coefficient $h_1$ is a function of the three curvilinear coordinates $(u_1, u_2, u_3)$. The other two metric coefficients, that is, $h_2(u_1, u_2, u_3)$ and $h_3(u_1, u_2, u_3)$, are also similarly defined.

As shown in Figure 3.19, the differential line element in the generalized orthogonal curvilinear coordinate system is given by Equation 3.25

$$d\vec{l} = dl_1\hat{a}_1 + dl_2\hat{a}_2 + dl_3\hat{a}_3 = h_1du_1\hat{a}_1 + h_2du_2\hat{a}_2 + h_3du_3\hat{a}_3 \tag{3.25}$$

As shown in Figure 3.19, the differential area elements in the generalized orthogonal curvilinear coordinate system are given by Equation 3.26

$$d\vec{A}_{12} = (h_1du_1)(h_2du_2)\hat{a}_3 = h_1h_2du_1du_2\hat{a}_3$$

$$d\vec{A}_{23} = (h_2du_2)(h_3du_3)\hat{a}_1 = h_2h_3du_2du_3\hat{a}_1 \tag{3.26}$$

$$d\vec{A}_{31} = (h_3du_3)(h_1du_1)\hat{a}_2 = h_3h_1du_3du_1\hat{a}_2$$

and the differential volume element in generalized orthogonal curvilinear coordinate system is given by Equation 3.27

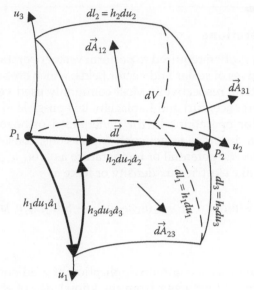

**FIGURE 3.19**
Differential line, area and volume elements in the generalized orthogonal curvilinear coordinate system.

$$dV = (h_1du_1)(h_2du_2)(h_3du_3) = h_1h_2h_3du_1du_2du_3 \tag{3.27}$$

In the light of the generalized orthogonal curvilinear coordinate system, the three commonly used orthogonal coordinate systems could be summarized as in Table 3.1.

Recall that the metric coefficients were earlier stated to be functions of the coordinates $(u_1,u_2,u_3)$. Table 3.1 clearly shows such dependence of the metric coefficients on the respective coordinates in cylindrical and spherical coordinate systems. Cartesian coordinate system is unique in the sense that its metric coefficients are constant.

**TABLE 3.1**

Characterization of Orthogonal Coordinate Systems

| Orthogonal Coordinate System | Coordinates | Unit Vectors | Metric Coefficients |
|---|---|---|---|
| Generalized | $(u_1,u_2,u_3)$ | $(\hat{a}_1,\hat{a}_2,\hat{a}_3)$ | $(h_1,h_2,h_3)$ |
| Cartesian | $(x,y,z)$ | $(\hat{i},\hat{j},\hat{k})$ | $(1,1,1)$ |
| Cylindrical | $(r,\theta,z)$ | $(\hat{u}_r,\hat{u}_\theta,\hat{u}_z)$ | $(1,r,1)$ |
| Spherical | $(r,\theta,\phi)$ | $(\hat{u}_r,\hat{u}_\theta,\hat{u}_\phi)$ | $(1,r,r\sin\theta)$ |

## 3.6 Vector Operations

In field analysis, it is often required to perform vector operations to get vector and scalar derivatives of scalar and vector fields, which are scalar and vector functions of position, respectively. Most commonly used vector operations are gradient, divergence, curl and Laplacian. It is possible to get expressions for all these vector operations in the generalized orthogonal curvilinear coordinate system. In order to get such expressions, consider an arbitrary scalar field like electric potential or temperature as $U(u_1, u_2, u_3)$ and an arbitrary vector field like electric flux density or force as

$$\vec{S} = S_1(u_1, u_2, u_3)\hat{a}_1 + S_2(u_1, u_2, u_3)\hat{a}_2 + S_3(u_1, u_2, u_3)\hat{a}_3 \tag{3.28}$$

### 3.6.1 Gradient

In electric field analysis, a common application of gradient operation is to determine electric field intensity from the knowledge of electric potential. In order to understand this concept clearly, a change $dU$ in the scalar function $U$ is expressed in terms of the changes in differential distances $dl_1$, $dl_2$ and $dl_3$ in the generalized orthogonal curvilinear coordinate system as follows:

$$dU = \frac{\partial U}{\partial l_1}dl_1 + \frac{\partial U}{\partial l_2}dl_2 + \frac{\partial U}{\partial l_3}dl_3 \tag{3.29}$$

Introducing the differential distance as $d\vec{l} = dl_1\hat{a}_1 + dl_2\hat{a}_2 + dl_3\hat{a}_3$, the above equation could be rewritten as follows:

$$dU = \left( \frac{\partial U}{\partial l_1}\hat{a}_1 + \frac{\partial U}{\partial l_2}\hat{a}_2 + \frac{\partial U}{\partial l_3}\hat{a}_3 \right) \cdot \left( dl_1\hat{a}_1 + dl_2\hat{a}_2 + dl_3\hat{a}_3 \right) = \vec{\nabla} U \cdot d\vec{l} \tag{3.30}$$

where:

$$\vec{\nabla} U = \frac{\partial U}{\partial l_1}\hat{a}_1 + \frac{\partial U}{\partial l_2}\hat{a}_2 + \frac{\partial U}{\partial l_3}\hat{a}_3 = \frac{1}{h_1}\frac{\partial U}{\partial u_1}\hat{a}_1 + \frac{1}{h_2}\frac{\partial U}{\partial u_2}\hat{a}_2 + \frac{1}{h_3}\frac{\partial U}{\partial u_3}\hat{a}_3 \tag{3.31}$$

Equation 3.30 indicates that the change $dU$ in the scalar function $U$ over the differential distance $dl$ at any point in space is maximum when $\vec{\nabla} U$ and $d\vec{l}$ are in the same direction. In other words, the magnitude of the vector $\vec{\nabla} U$, which is the spatial derivative of the scalar function $U$ and is called the gradient of $U$, is equal to the maximum value of $dU/dl$ and it is in the direction in which $dU/dl$ is maximum.

With the help of the metric coefficients and other parameters mentioned in Table 3.1, the gradient function can be written in different orthogonal coordinate systems as detailed below:

Cartesian coordinate system:

$$\vec{\nabla}U = \frac{\partial U}{\partial x}\hat{i} + \frac{\partial U}{\partial y}\hat{j} + \frac{\partial U}{\partial z}\hat{k} \tag{3.32}$$

Cylindrical coordinate system:

$$\vec{\nabla}U = \frac{\partial U}{\partial r}\hat{u}_r + \frac{1}{r}\frac{\partial U}{\partial \theta}\hat{u}_\theta + \frac{\partial U}{\partial z}\hat{u}_z \tag{3.33}$$

Spherical coordinate system:

$$\vec{\nabla}U = \frac{\partial U}{\partial r}\hat{u}_r + \frac{1}{r}\frac{\partial U}{\partial \theta}\hat{u}_\theta + \frac{1}{r\sin\theta}\frac{\partial U}{\partial \phi}\hat{u}_\phi \tag{3.34}$$

## 3.6.2 Del Operator

Equation 3.30 introduces a vector differential operator $\vec{\nabla}$ such that when $\vec{\nabla}$ is applied on a scalar function, it results into a vector function. Such vector operation is called *gradient*.

From Equation 3.31

$$\vec{\nabla}U = \frac{1}{h_1}\frac{\partial U}{\partial u_1}\hat{a}_1 + \frac{1}{h_2}\frac{\partial U}{\partial u_2}\hat{a}_2 + \frac{1}{h_3}\frac{\partial U}{\partial u_3}\hat{a}_3$$

$$= \left(\frac{1}{h_1}\frac{\partial}{\partial u_1}\hat{a}_1 + \frac{1}{h_2}\frac{\partial}{\partial u_2}\hat{a}_2 + \frac{1}{h_3}\frac{\partial}{\partial u_3}\hat{a}_3\right)U \tag{3.35}$$

$$\text{so that } \vec{\nabla} = \left(\frac{1}{h_1}\frac{\partial}{\partial u_1}\hat{a}_1 + \frac{1}{h_2}\frac{\partial}{\partial u_2}\hat{a}_2 + \frac{1}{h_3}\frac{\partial}{\partial u_3}\hat{a}_3\right)$$

It has been discussed in Chapter 2 that when $\vec{\nabla}$ is applied on a vector function as scalar or dot product, then it results into a scalar function and such operation is called *divergence*. On the other hand, when $\vec{\nabla}$ is operated on a vector function with vector or cross product, then it results into another vector function and this operation is called *curl*.

## 3.6.3 Divergence

In electric field analysis, divergence is used to relate electric field with the source, that is, the charges, as given by the differential form of Gauss's law.

It has already been discussed in Chapter 2 that the divergence of a vector field $\vec{S}$ could be evaluated by finding the net flux coming out per unit volume. In this context, consider the differential volume element $dV = (dl_1dl_2dl_3)$, as shown in Figure 3.20. Consider that the point $O$ is located at the centre of the volume element $dV$ and also that the vector $\vec{S}$ is known at the point $O$.

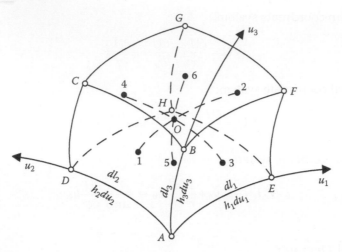

**FIGURE 3.20**
Divergence in the generalized orthogonal curvilinear coordinate system.

The points 1 and 2 are the midpoints of the surfaces $ABCD$ and $EFGH$, respectively, which are normal to $u_1$ direction. Therefore, along the line 1-$O$-2 the derivatives of $S_1$, which is the $u_1$ component of $\vec{S}$, with respect to $u_1$, will be finite, whereas the derivative of $S_1$ with respect to $u_2$ and $u_3$ will be zero, as $S_1$ is orthogonal to $u_2$ and $u_3$. Considering the differential distance $dl_1$ between the two surfaces under reference, namely, $ABCD$ and $EFGH$, $S_1$ at the points 1 and 2 can be evaluated from the knowledge of $S_1$ at $O$ with the help of Taylor series. Noting that the distances between $O$ and both the points 1 and 2 are $dl_1/2$, from the Taylor series expansion, it can be written neglecting higher order terms as follows:

$$S_1\big|_1 = S_1 - \frac{\partial S_1}{\partial l_1}\frac{dl_1}{2}\left(\text{as the distance from } O \text{ to 1 is in the negative sense of } u_1\right)$$

$$S_1\big|_2 = S_1 + \frac{\partial S_1}{\partial l_1}\frac{dl_1}{2}\left(\text{as the distance from } O \text{ to 2 is in the positive sense of } u_1\right)$$

where:
$S_1$ stands for the value of $S_1$ at $O$

Therefore,

$$S_1\big|_1 = S_1 - \frac{\partial S_1}{h_1\partial u_1}\frac{h_1 du_1}{2} = S_1 - \frac{\partial S_1}{\partial u_1}\frac{du_1}{2}$$

$$\text{and } S_1\big|_2 = S_1 + \frac{\partial S_1}{h_1\partial u_1}\frac{h_1 du_1}{2} = S_1 + \frac{\partial S_1}{\partial u_1}\frac{du_1}{2}$$

The area of the surface $ABCD$ is given by $dl_2 dl_3 = h_2 h_3 du_2 du_3$. Therefore, the flux of $\vec{S}$ going into the differential volume through the surface $ABCD$ in the $u_1$ direction is given by

$$d\psi_{ABCD} = \left(S_1 h_2 h_3 du_2 du_3\right) - \frac{\partial\left(S_1 h_2 h_3 du_2 du_3\right)}{\partial u_1}\frac{du_1}{2}$$

$$= \left(S_1 h_2 h_3 du_2 du_3\right) - \frac{\partial\left(S_1 h_2 h_3\right)}{\partial u_1}\frac{du_1 du_2 du_3}{2}$$

as $(h_1, h_2, h_3)$ are functions of $(u_1, u_2, u_3)$, the product $h_2 h_3$ is kept within the partial derivative term.

Similarly, the flux of $\vec{S}$ coming out of the surface $EFGH$ in the $u_1$ direction is given by

$$d\psi_{EFGH} = \left(S_1 h_2 h_3 du_2 du_3\right) + \frac{\partial\left(S_1 h_2 h_3 du_2 du_3\right)}{\partial u_1}\frac{du_1}{2}$$

$$= \left(S_1 h_2 h_3 du_2 du_3\right) + \frac{\partial\left(S_1 h_2 h_3\right)}{\partial u_1}\frac{du_1 du_2 du_3}{2}$$

Hence, the net flux coming out of the differential volume in the $u_1$ direction is given by

$$d\psi_1 = d\psi_{EFGH} - d\psi_{ABCD} = \frac{\partial\left(S_1 h_2 h_3\right)}{\partial u_1} du_1 du_2 du_3 \qquad (3.36)$$

Therefore, considering all the three directions, that is, $u_1$, $u_2$ and $u_3$ directions, the net flux coming out of the volume $dV$ is

$$\psi_{dV} = \left[\frac{\partial\left(S_1 h_2 h_3\right)}{\partial u_1} + \frac{\partial\left(S_2 h_3 h_1\right)}{\partial u_2} + \frac{\partial\left(S_3 h_1 h_2\right)}{\partial u_3}\right] du_1 du_2 du_3 \qquad (3.37)$$

Therefore, the divergence of $\vec{S}$, which is the net flux coming out per unit volume, is given by

$$\text{div}\,\vec{S} = \frac{\psi_{dV}}{dV} = \frac{1}{dl_1 dl_2 dl_3}\left[\frac{\partial\left(S_1 h_2 h_3\right)}{\partial u_1} + \frac{\partial\left(S_2 h_3 h_1\right)}{\partial u_2} + \frac{\partial\left(S_3 h_1 h_2\right)}{\partial u_3}\right] du_1 du_2 du_3$$

$$\text{or, div}\,\vec{S} = \frac{1}{h_1 h_2 h_3 du_1 du_2 du_3}\left[\frac{\partial\left(S_1 h_2 h_3\right)}{\partial u_1} + \frac{\partial\left(S_2 h_3 h_1\right)}{\partial u_2} + \frac{\partial\left(S_3 h_1 h_2\right)}{\partial u_3}\right] du_1 du_2 du_3$$

$$\text{or, } \vec{\nabla}\cdot\vec{S} = \frac{1}{h_1 h_2 h_3}\left[\frac{\partial}{\partial u_1}\left(h_2 h_3 S_1\right) + \frac{\partial}{\partial u_2}\left(h_3 h_1 S_2\right) + \frac{\partial}{\partial u_3}\left(h_1 h_2 S_3\right)\right] \qquad (3.38)$$

The divergence function can be written in different orthogonal coordinate systems as detailed below:

Cartesian coordinate system:

$$\vec{\nabla}\cdot\vec{S} = \left(\frac{\partial S_x}{\partial x} + \frac{\partial S_y}{\partial y} + \frac{\partial S_z}{\partial z}\right) \tag{3.39}$$

Cylindrical coordinate system:

$$\vec{\nabla}\cdot\vec{S} = \frac{1}{r}\left[\frac{\partial}{\partial r}(rS_r) + \frac{\partial S_\theta}{\partial \theta} + \frac{\partial}{\partial z}(rS_z)\right] = \frac{1}{r}\frac{\partial}{\partial r}(rS_r) + \frac{1}{r}\frac{\partial S_\theta}{\partial \theta} + \frac{\partial S_z}{\partial z} \tag{3.40}$$

Spherical coordinate system:

$$\vec{\nabla}\cdot\vec{S} = \frac{1}{r^2 \sin\theta}\left[\frac{\partial}{\partial r}(r^2 \sin\theta S_r) + \frac{\partial}{\partial \theta}(r\sin\theta S_\theta) + \frac{\partial}{\partial \phi}(rS_\phi)\right]$$

$$= \frac{1}{r^2}\frac{\partial}{\partial r}(r^2 S_r) + \frac{1}{r\sin\theta}\frac{\partial}{\partial \theta}(\sin\theta S_\theta) + \frac{1}{r\sin\theta}\frac{\partial S_\phi}{\partial \phi} \tag{3.41}$$

### 3.6.4 Laplacian

$$\text{Let } \vec{S} = \vec{\nabla}U = \frac{1}{h_1}\frac{\partial U}{\partial u_1}\hat{a}_1 + \frac{1}{h_2}\frac{\partial U}{\partial u_2}\hat{a}_2 + \frac{1}{h_3}\frac{\partial U}{\partial u_3}\hat{a}_3 = S_1\hat{a}_1 + S_2\hat{a}_2 + S_3\hat{a}_3$$

so that

$$S_1 = \frac{1}{h_1}\frac{\partial U}{\partial u_1}, \; S_2 = \frac{1}{h_2}\frac{\partial U}{\partial u_2} \text{ and } S_3 = \frac{1}{h_3}\frac{\partial U}{\partial u_3}$$

Then,

$$\vec{\nabla}^2 U = \vec{\nabla}\cdot\vec{\nabla}U = \vec{\nabla}\cdot\vec{S}$$

Therefore, from Equation 3.38

$$\vec{\nabla}^2 U = \vec{\nabla}\cdot\vec{S} = \frac{1}{h_1 h_2 h_3}\left[\frac{\partial}{\partial u_1}(h_2 h_3 S_1) + \frac{\partial}{\partial u_2}(h_3 h_1 S_2) + \frac{\partial}{\partial u_3}(h_1 h_2 S_3)\right]$$

$$= \frac{1}{h_1 h_2 h_3}\left[\frac{\partial}{\partial u_1}\left(\frac{h_2 h_3}{h_1}\frac{\partial U}{\partial u_1}\right) + \frac{\partial}{\partial u_2}\left(\frac{h_3 h_1}{h_2}\frac{\partial U}{\partial u_2}\right) + \frac{\partial}{\partial u_3}\left(\frac{h_1 h_2}{h_3}\frac{\partial U}{\partial u_3}\right)\right] \tag{3.42}$$

Laplacian can be written in different orthogonal coordinate systems as detailed below:

Cartesian coordinate system:

$$\vec{\nabla}^2 U = \left(\frac{\partial^2 U}{\partial x^2} + \frac{\partial^2 U}{\partial y^2} + \frac{\partial^2 U}{\partial z^2}\right) \tag{3.43}$$

Cylindrical coordinate system:

$$\vec{\nabla}^2 U = \frac{1}{r}\left[\frac{\partial}{\partial r}\left(r\frac{\partial U}{\partial r}\right) + \frac{\partial}{\partial \theta}\left(\frac{1}{r}\frac{\partial U}{\partial \theta}\right) + \frac{\partial}{\partial z}\left(r\frac{\partial U}{\partial z}\right)\right]$$

$$= \frac{1}{r}\frac{\partial}{\partial r}\left(r\frac{\partial U}{\partial u_r}\right) + \frac{1}{r^2}\frac{\partial^2 U}{\partial \theta^2} + \frac{\partial^2 U}{\partial z^2}$$

(3.44)

Spherical coordinate system:

$$\vec{\nabla}^2 U = \frac{1}{r^2 \sin\theta}\left[\frac{\partial}{\partial r}\left(r^2 \sin\theta\frac{\partial U}{\partial r}\right) + \frac{\partial}{\partial \theta}\left(\sin\theta\frac{\partial U}{\partial \theta}\right) + \frac{\partial}{\partial \phi}\left(\frac{1}{\sin\theta}\frac{\partial U}{\partial \phi}\right)\right]$$

$$= \frac{1}{r^2}\frac{\partial}{\partial r}\left(r^2\frac{\partial U}{\partial r}\right) + \frac{1}{r^2 \sin\theta}\frac{\partial}{\partial \theta}\left(\sin\theta\frac{\partial U}{\partial \theta}\right) + \frac{1}{r^2 \sin^2\theta}\frac{\partial^2 U}{\partial \phi^2}$$

(3.45)

### 3.6.5 Curl

Curl is evaluated as a closed-line integral per unit area. The closed-line integral of a vector is also known as *circulation*. The circulation of a vector is obtained by multiplying the component of that vector parallel to the specified closed path at each point along it by the differential path length and summing the results of multiplication as the differential lengths approach zero. Because there are three planes that are normal to the three components of a vector, the circulation of a vector is to be computed separately for these three planes, each one of which will give one component of the curl. The curl of any vector thus results into another vector. Any component of the curl is given by the quotient of the closed-line integral of the vector about a small path and the area enclosed by that path, as the path shrinks to zero. The small path is to be chosen in a plane normal to the desired component of the curl.

With reference to Figure 3.21, consider that the values of three components of $\vec{S}$, namely, $S_1$, $S_2$ and $S_3$, are known at the midpoint $O$ of the differential areas as shown.

Considering the differential path length $AB$, the value of $S_1$ at the midpoint of $AB$ is $\left[S_1 - (\partial S_1/\partial l_2)(dl_2/2)\right]$

Therefore,

$$\int_A^B \vec{S}.d\vec{l} = \left[S_1 dl_1 - \frac{\partial(S_1 dl_1)}{\partial l_2}\frac{dl_2}{2}\right]$$

$$= \left[(S_1 h_1 du_1) - \frac{\partial(S_1 h_1 du_1)}{\partial u_2}\frac{du_2}{2}\right]$$

(3.46)

Again, considering the differential path length $CD$, the value of $S_1$ at the midpoint of $CD$ is $\left[S_1 + (\partial S_1/\partial l_2)(dl_2/2)\right]$

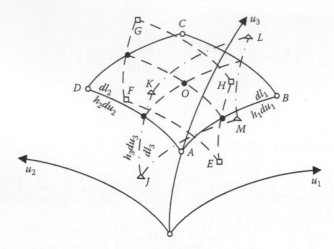

**FIGURE 3.21**
Curl in the generalized orthogonal curvilinear coordinate system.

Therefore,

$$\int_C^D \vec{S} \cdot d\vec{l} = \left[ S_1(-dl_1) + \frac{\partial(-S_1 dl_1)}{\partial l_2} \frac{dl_2}{2} \right]$$

$$= -\left[ (S_1 h_1 du_1) + \frac{\partial(S_1 h_1 du_1)}{\partial u_2} \frac{du_2}{2} \right]$$

(3.47)

Similarly, considering the differential path length $BC$ the value of $S_2$ at the midpoint of $BC$ is $\left[ S_2 + (\partial S_2/\partial l_1)(dl_1/2) \right]$

Therefore,

$$\int_B^C \vec{S} \cdot d\vec{l} = \left[ S_2 dl_2 + \frac{\partial(S_2 dl_2)}{\partial l_1} \frac{dl_1}{2} \right] dl_2$$

$$= \left[ (S_2 h_2 du_2) + \frac{\partial(S_2 h_2 du_2)}{\partial u_1} \frac{du_1}{2} \right]$$

(3.48)

Again, considering the differential path length $DA$ the value of $S_2$ at the midpoint of $DA$ is $\left[ S_2 - (\partial S_2/\partial l_1)(dl_1/2) \right]$

Therefore,

$$\int_D^A \vec{S} \cdot d\vec{l} = \left[ S_2(-dl_2) - \frac{\partial(-S_2 dl_2)}{\partial l_1} \frac{dl_1}{2} \right]$$

$$= \left[ -(S_2 h_2 du_2) + \frac{\partial(S_2 h_2 du_2)}{\partial u_1} \frac{du_1}{2} \right]$$

(3.49)

From Equations 3.46 through 3.49

$$\oint_{ABCDA} \vec{S} \cdot \vec{dl} = \int_A^B \vec{S} \cdot \vec{dl} + \int_B^C \vec{S} \cdot \vec{dl} + \int_C^D \vec{S} \cdot \vec{dl} + \int_D^A \vec{S} \cdot \vec{dl}$$

or, 
$$\oint_{ABCDA} \vec{S} \cdot \vec{dl} = \frac{\partial(h_2 S_2)}{\partial u_1} du_1 du_2 - \frac{\partial(h_1 S_1)}{\partial u_2} du_1 du_2 \tag{3.50}$$

$$= \left[ \frac{\partial(h_2 S_2)}{\partial u_1} - \frac{\partial(h_1 S_1)}{\partial u_2} \right] du_1 du_2$$

Dividing the above equation by the area of the integration, that is, $dl_1 dl_2$,

$$\frac{1}{dl_1 dl_2} \oint_{ABCDA} \vec{S} \cdot \vec{dl} = \frac{1}{h_1 h_2 du_1 du_2} \left[ \frac{\partial(h_2 S_2)}{\partial u_1} - \frac{\partial(h_1 S_1)}{\partial u_2} \right] du_1 du_2$$

$$= \frac{1}{h_1 h_2} \left[ \frac{\partial(h_2 S_2)}{\partial u_1} - \frac{\partial(h_1 S_1)}{\partial u_2} \right] \tag{3.51}$$

Noting that the area enclosed by the path $ABCDA$ is an area normal to $u_3$ direction, Equation 3.51 gives the $u_3$ component of curl $\vec{S}$.

Considering the closed path $EFGHE$ normal to $u_1$ direction, the $u_1$ component of curl $\vec{S}$ can be evaluated as follows:

$$\frac{1}{dl_2 dl_3} \oint_{EFGHE} \vec{S} \cdot \vec{dl} = \frac{1}{h_2 h_3 du_2 du_3} \left[ \frac{\partial(h_3 S_3)}{\partial u_2} - \frac{\partial(h_2 S_2)}{\partial u_3} \right] du_2 du_3$$

$$= \frac{1}{h_2 h_3} \left[ \frac{\partial(h_3 S_3)}{\partial u_2} - \frac{\partial(h_2 S_2)}{\partial u_3} \right] \tag{3.52}$$

and considering the closed path $JKLMJ$ normal to $u_2$ direction, the $u_2$ component of curl $\vec{S}$ can be evaluated as follows:

$$\frac{1}{dl_3 dl_1} \oint_{JKLMJ} \vec{S} \cdot \vec{dl} = \frac{1}{h_3 h_1 du_3 du_1} \left[ \frac{\partial(h_1 S_1)}{\partial u_3} - \frac{\partial(h_3 S_3)}{\partial u_1} \right] du_3 du_1$$

$$= \frac{1}{h_3 h_1} \left[ \frac{\partial(h_1 S_1)}{\partial u_3} - \frac{\partial(h_3 S_3)}{\partial u_1} \right] \tag{3.53}$$

Here, it is to be mentioned that for evaluating the integral the closed paths have to be traversed in a manner that if a right-handed screw is rotated in that direction, then the screw will move in the positive direction of the coordinate to which the plane containing that path is perpendicular, that is

*ABCDA* for the $u_1$ direction, *EFGHE* for the $u_2$ direction and *JKLMJ* for the $u_3$ direction, with reference to Figure 3.21.

From Equations 3.51 through 3.53

$$\text{curl}\,\vec{S} = \vec{\nabla} \times \vec{S} = \frac{1}{h_2 h_3}\left[\frac{\partial(h_3 S_3)}{\partial u_2} - \frac{\partial(h_2 S_2)}{\partial u_3}\right]\hat{a}_1 + \frac{1}{h_3 h_1}\left[\frac{\partial(h_1 S_1)}{\partial u_3} - \frac{\partial(h_3 S_3)}{\partial u_1}\right]\hat{a}_2$$
$$+ \frac{1}{h_1 h_2}\left[\frac{\partial(h_2 S_2)}{\partial u_1} - \frac{\partial(h_1 S_1)}{\partial u_2}\right]\hat{a}_3 \tag{3.54}$$

Equation 3.54 could be written in the following form:

$$\vec{\nabla} \times \vec{S} = \frac{1}{h_1 h_2 h_3}\begin{vmatrix} h_1\hat{a}_1 & h_2\hat{a}_2 & h_3\hat{a}_3 \\ \dfrac{\partial}{\partial u_1} & \dfrac{\partial}{\partial u_2} & \dfrac{\partial}{\partial u_3} \\ h_1 S_1 & h_2 S_2 & h_3 S_3 \end{vmatrix} \tag{3.55}$$

Curl can be written in different orthogonal coordinate systems as detailed below:

Cartesian coordinate system:

$$\vec{\nabla} \times \vec{S} = \begin{vmatrix} \hat{i} & \hat{j} & \hat{k} \\ \dfrac{\partial}{\partial x} & \dfrac{\partial}{\partial y} & \dfrac{\partial}{\partial z} \\ S_x & S_y & S_z \end{vmatrix} \tag{3.56}$$

Cylindrical coordinate system:

$$\vec{\nabla} \times \vec{S} = \frac{1}{r}\begin{vmatrix} \hat{u}_r & r\hat{u}_\theta & \hat{u}_z \\ \dfrac{\partial}{\partial r} & \dfrac{\partial}{\partial \theta} & \dfrac{\partial}{\partial z} \\ S_r & r S_\theta & S_z \end{vmatrix} \tag{3.57}$$

Spherical coordinate system:

$$\vec{\nabla} \times \vec{S} = \frac{1}{r^2 \sin\theta}\begin{vmatrix} \hat{u}_r & r\hat{u}_\theta & r\sin\theta\,\hat{u}_\phi \\ \dfrac{\partial}{\partial r} & \dfrac{\partial}{\partial \theta} & \dfrac{\partial}{\partial \phi} \\ S_r & r S_\theta & r\sin\theta\,S_\phi \end{vmatrix} \tag{3.58}$$

### 3.6.5.1 Curl of Electric Field

With reference to the derivation of Section 3.6.5, consider that the vector quantity be electric field intensity $\vec{E}$. Then the closed line integral $\oint \vec{E}.d\vec{l}$ along any of the three paths as shown in Figure 3.21, namely, *ABCDA*, *EFGHE* and *JKLMJ*, will be zero, as the work done over a closed path is zero. Hence, the results of the integrals as given by Equations 3.51, 3.52 and 3.53 will be zero. In other words, all the three components of curl $\vec{E}$ will be zero. Consequently,

$$\operatorname{curl}\vec{E} = \vec{\nabla} \times \vec{E} = 0, \tag{3.59}$$

which holds good at every point within an electric field. It shows that an electric field is path independent and irrotational.

Alternately, it can be proved by considering $\vec{E} = -\vec{\nabla}U$, where $U$ is electric potential, which is a scalar quantity. Then

$$\vec{\nabla} \times \vec{E} = \vec{\nabla} \times -\vec{\nabla}U = -\left( \frac{\partial}{\partial y}\frac{\partial U}{\partial z} - \frac{\partial}{\partial z}\frac{\partial U}{\partial y} \right)\hat{i} - \left( \frac{\partial}{\partial z}\frac{\partial U}{\partial x} - \frac{\partial}{\partial x}\frac{\partial U}{\partial z} \right)\hat{j}$$

$$- \left( \frac{\partial}{\partial x}\frac{\partial U}{\partial y} - \frac{\partial}{\partial y}\frac{\partial U}{\partial x} \right)\hat{k} = 0$$

Equation 3.59 is commonly used to prove that a valid electric field is conservative.

**PROBLEM 3.1**

Determine whether $\vec{E} = x\hat{i} + y\hat{j} + z\hat{k}$ is a valid form of electric field or not.

*Solution:*

In order to find whether an electric field is valid or not, the curl of the given electric field needs to be evaluated. Therefore, for the given electric field

$$\vec{\nabla} \times \vec{E} = \left( \frac{\partial z}{\partial y} - \frac{\partial y}{\partial z} \right)\hat{i} + \left( \frac{\partial x}{\partial z} - \frac{\partial z}{\partial x} \right)\hat{j} + \left( \frac{\partial y}{\partial x} - \frac{\partial x}{\partial y} \right)\hat{k} = 0$$

Therefore, the given electric field is a valid one.

---

## Objective Type Questions

1. In Cartesian coordinate system
   a. Only one metric coefficient is constant
   b. Two metric coefficients are constant

    c. All three metric coefficients are constant

    d. None of the metric coefficients is constant

2. In cylindrical coordinate system

    a. Only one metric coefficient is constant

    b. Two metric coefficients are constant

    c. All three metric coefficients are constant

    d. None of the metric coefficients is constant

3. In spherical coordinate system

    a. Only one metric coefficient is constant

    b. Two metric coefficients are constant

    c. All three metric coefficients are constant

    d. None of the metric coefficients is constant

4. Three planes as constant coordinate surfaces define

    a. Cartesian coordinate system

    b. Cylindrical coordinate system

    c. Spherical coordinate system

    d. None of the above

5. Two planes as constant coordinate surfaces along with one constant angle surface define

    a. Cartesian coordinate system

    b. Cylindrical coordinate system

    c. Spherical coordinate system

    d. None of the above

6. Only one plane as constant coordinate surface belong to

    a. Cartesian coordinate system

    b. Cylindrical coordinate system

    c. Spherical coordinate system

    d. None of the above

7. A conical surface as constant coordinate surface belong to

    a. Cartesian coordinate system

    b. Cylindrical coordinate system

    c. Spherical coordinate system

    d. None of the above

8. The unit vector associated with a constant coordinate surface

    a. Is perpendicular to the surface coordinate = constant and points in the direction of increasing value of coordinate

b. Is perpendicular to the surface coordinate = constant and points in the direction of decreasing value of coordinate

c. Is parallel to the surface coordinate = constant and points in the direction of increasing value of coordinate

d. Is parallel to the surface coordinate = constant and points in the direction of decreasing value of coordinate

9. The right-handed property of an orthogonal coordinate system is defined by

a. The cross product of the first and second unit vectors equal to the third unit vector

b. The cross product of the second and third unit vectors equal to the first unit vector

c. The cross product of the third and first unit vectors equal to the second unit vector

d. All the above

10. Orthogonality of two constant coordinate surfaces in generalized curvilinear coordinate system is defined by

a. $\hat{a}_1 \cdot \hat{a}_1 = 1$

b. $\hat{a}_1 \cdot \hat{a}_2 = 0$

c. $\hat{a}_1 \times \hat{a}_1 = 1$

d. $\hat{a}_1 \times \hat{a}_2 = 0$

11. The ranges of the values of the three coordinates in cylindrical coordinate system are

a. $0 \leq r < \infty, 0 \leq \theta \leq 2\pi$ and $-\infty < z < \infty$

b. $0 \leq r < \infty, 0 \leq \theta \leq 2\pi$ and $0 < z < \infty$

c. $-\infty \leq r < \infty, 0 \leq \theta \leq 2\pi$ and $0 < z < \infty$

d. $-\infty \leq r < \infty, 0 \leq \theta \leq 2\pi$ and $\infty < z < \infty$

12. The ranges of the values of the three coordinates in spherical coordinate system are

a. $0 \leq r < \infty, 0 \leq \theta \leq 2\pi$ and $0 \leq \phi \leq 2\pi$

b. $0 \leq r < \infty, 0 \leq \theta \leq \pi$ and $0 \leq \phi \leq 2\pi$

c. $-\infty \leq r < \infty, 0 \leq \theta \leq 2\pi$ and $0 \leq \phi \leq 2\pi$

d. $-\infty \leq r < \infty, 0 \leq \theta \leq \pi$ and $0 \leq \phi \leq 2\pi$

13. In spherical coordinate system, the angle $\phi$ is defined on the

a. $x$–$y$ plane

b. $y$–$z$ plane

c. $z$–$x$ plane

d. Plane containing the $r$-vector

14. Curl $\vec{E} = 0$ indicates that
    a. Electric field is conservative
    b. Electric field is irrotational
    c. Electric potential is path independent
    d. All the above
15. Conservative electric field is indicated by
    a. $\vec{E} = -\vec{\nabla}\phi$
    b. $\vec{\nabla}\cdot\vec{E} = 0$
    c. $\vec{\nabla}\times\vec{E} = 0$
    d. $\vec{\nabla}\cdot\vec{D} = \rho_v$

**Answers:**    1) c; 2) b; 3) a; 4) a; 5) b; 6) c; 7) c; 8) a; 9) d; 10) b; 11) a; 12) b; 13) a;
                14) d; 15) c

# 4

## Single-Dielectric Configurations

**ABSTRACT** Although most of the real-life configurations are multi-dielectric in nature, it is required to study some of the single-dielectric arrangements, which are commonly used, for the proper understanding of the application of theory in determining electric field in practical cases. This chapter presents a physical explanation of displacement current and also discusses in details the energy stored in an electric field. Three single-dielectric arrangements are analyzed and how such analysis can be used for most economical use of insulating materials has also been discussed with the help of state-of-the-art examples.

## 4.1 Introduction

Mother Nature has presented the humankind an insulation as a gift, which is available in abundance and is also free. This dielectric is nothing but the atmospheric air. It is not difficult to conceive that no electrical system would have worked had air been not an insulating medium. Because air is present everywhere, most of the real-life arrangements are multi-dielectric configurations where the equipment insulation, whether it is solid, liquid or gaseous or a combination of two or more of these insulating media, is commonly surrounded by atmospheric air. However, there are specific cases where the configuration could be single-dielectric one, for example, a co-axial cable having only one dielectric medium. Analysis of single-dielectric arrangement is necessary to grasp the theoretical aspects of electric field in a meaningful way. Moreover, there are certain aspects such as displacement current and energy stored in an electric field, which need to be understood from a single-dielectric viewpoint to start with.

## 4.2 Displacement Current

Consider a capacitor with a perfect dielectric or, say vacuum, in between its two plates. In such a case, no charge can move through the perfect dielectric or vacuum. In other words, there could be no current flowing between the two plates within the capacitor. Now if a DC voltage is applied between its two plates, then a charging current flows from the source to supply the charges to the plates of the capacitor for a short duration of charging. When the capacitor is fully charged, the charging current ceases to flow. It is to be noted here that this charging current is measurable only in the circuit between the source and the capacitor. If one tries to measure the current between the plates within the capacitor, then the result of the current measurement will be zero. It means that this charging current does not flow through the entire closed loop formed by the source and the capacitor. But why these charges are to be supplied by the source to the capacitor plates? The answer is that the capacitor with a perfect dielectric or vacuum is an open circuit and hence the potential difference between the capacitor plates will be equal to the source voltage. Accordingly, there will be an electric field established between the capacitor plates. In order to establish this electric field, charges have to be supplied from the source to the capacitor plates.

Depending on the area and separation distance between the plates, there is a finite capacitance of the capacitor. Hence, if an AC voltage is applied between the plates of the capacitor, a capacitive current will flow in the circuit, which will be a continuous current varying with time that could be measured by an ammeter placed in between the source and the capacitor. However, due to perfect dielectric or vacuum, as stated earlier, there could be no movement of charge between the plates within the capacitor. Hence, this continuous capacitive current is measurable only in the circuit between the AC source and the capacitor plates. Then the question is how this continuous capacitive current appears in a section of the circuit external to the capacitor plates and not through the whole circuit? The answer to this question lies in the fact that this current is not conduction current. It appears due to the displacement of charges between the source and the capacitor plates, even when there is no charge movement between the plates within the capacitor. Consequently, this current is termed as *displacement current*. The concept of displacement current may be explained as detailed below.

Take the example of a capacitor with vacuum between its plates, as shown in Figure 4.1. If the separation distance between the plates is much smaller than the plate length/diameter and if fringing of flux at the edges is neglected, then the electric field between the capacitor plates will be uniform. If the potential difference between its plates at any time instant is $v$ and the separation distance between the plates is $d$, then the magnitude of electric

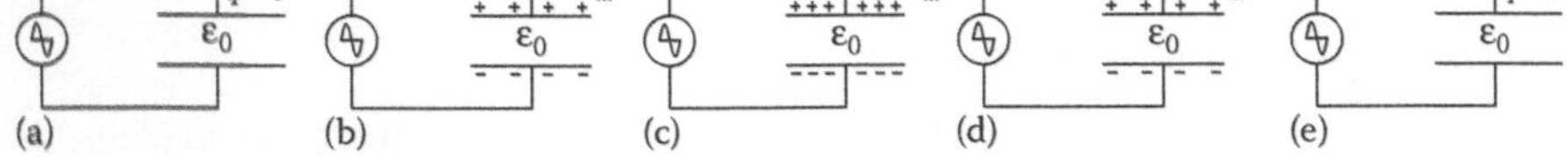

**FIGURE 4.1**
Physical explanation of the displacement current of a capacitor.

field intensity between the capacitor plates at that time instant will be (*v/d*). Thus, a unit positive charge placed in vacuum between the plates within the capacitor will experience this force on the unit change due to the charges present on the capacitor plates. Owing to the application of alternating voltage, the electric field intensity within the capacitor (= *v/d*) will vary in direct proportion to the potential difference between the plates. Consequently, for this variation in electric field intensity, the amount of charge on the capacitor plates will again have to vary in direct proportion to the potential difference between the plates. Let the amount of charge on the capacitor plates corresponding to the peak value ($V_m$) of the sinusoidal applied voltage be $q_m$.

At $\theta = 0°$, the potential difference across the capacitor plates is zero and hence the charge on the plates is also zero, as shown in Figure 4.1a. Then the charge on the plates increases with time as the sinusoidal voltage magnitude increases with time from $\theta = 0°$ to $\theta = 90°$, as shown in Figure 4.1a–c. These charges are supplied by the AC source to the capacitor plates and hence in the circuit between the source and the capacitor plates, there is a change in charge with time (*dq/dt*). Whenever there is such change in charge with time, there is a measurable current. Therefore, if an ammeter is placed in between the source and the capacitor, this current can be measured. The magnitude of this current will be determined by the rate of change of charge, which is the same as the rate of change of voltage with time, as explained earlier in this section. As shown in Figure 4.2, the rate of change of sinusoidal voltage is maximum at the zero crossing of the voltage waveform, that is, at $\theta = 0°$, and is zero at the voltage peak, that is, at $\theta = 90°$. In between this time span, this current will also have a sinusoidal variation with time.

From $\theta = 90°$ to $\theta = 180°$, the negative slope of sinusoidal voltage waveform increases from zero to maximum at the voltage zero crossing at $\theta = 180°$. The magnitude of voltage starts decreasing from $\theta = 90°$ and hence the charges on the capacitor plates also start decreasing from $q_m$ corresponding to $\theta = 90°$. In other words, the charges start to move from the capacitor plate back to the source. Hence, the ammeter between the source and the capacitor will record

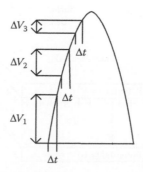

**FIGURE 4.2**
Slope of sinusoidal voltage waveform.

a current in the opposite direction, which will increase in magnitude from $\theta = 90°$ and will become maximum at $\theta = 180°$ in this opposite direction. The same sequential process will be repeated from $\theta = 180°$ to $\theta = 360°$ albeit the direction will be just the opposite of that between $\theta = 0°$ and $\theta = 90°$.

Therefore, it could be seen that although there is no charge movement between the plates within the capacitor itself, there is a continuous displacement of charges between the source and the capacitor plates in the external circuit. As a result, the displacement current flows in the external circuit. From the above discussion, it can be noted that (1) the current is positive maximum when $dq/dt$ and consequently $dv/dt$ is positive maximum at $\theta = 0°$; (2) the current is zero when $dq/dt$ and consequently $dv/dt$ is zero when $\theta = 90°$; (3) the current is at its negative maximum when $dq/dt$ and consequently $dv/dt$ is at its negative maximum at $\theta = 180°$ and so on. Considering the facts from (1) to (3), it can be concluded that there is a phase difference of $90°$ between the potential across the capacitor plates and the displacement current, and this displacement current leads the potential by $90°$.

In real life, the dielectric between the capacitor plates is never a perfect one and hence a small amount of conduction current flows between the capacitor plates. Consequently, the current through a real-life capacitor leads the potential by an angle slightly less than $90°$.

## 4.3 Parallel Plate Capacitor

A capacitor is an arrangement of conductors along with dielectrics that is used primarily to store electric charge. A very simple capacitor is a system consisting of two parallel metallic plates with free space or any dielectric in between, as shown in Figure 4.3.

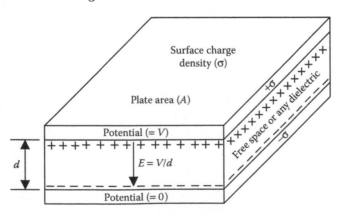

**FIGURE 4.3**
Parallel plate capacitor arrangement.

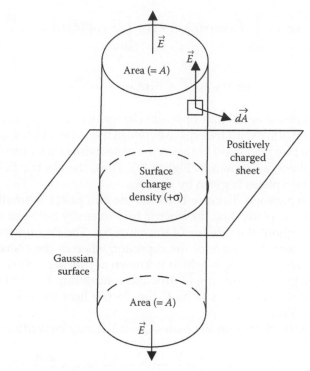

**FIGURE 4.4**
Field due to infinitely long-charged sheet.

In order to understand the electric field within a parallel plate capacitor, it is necessary to know the electric field due to infinitely long-charged planar sheet having uniform surface charge density ($\sigma$). As an infinitely long plane possesses a planar symmetry and the charge is uniformly distributed on the planar surface, the electric field intensity acts in the direction of the normal to the plane. In other words, the magnitude of the electric field intensity is constant on all the planes that are parallel to the charged sheet.

To analyze the field due to such a charged sheet, the best choice for a Gaussian surface is a cylinder whose axis is perpendicular to the plane, as shown in Figure 4.4. As the surface charge density is assumed to be uniform, the charge enclosed by the Gaussian surface is given by ($\sigma A$). As shown in Figure 4.4, the electric field intensity vector is parallel to the area vector at the two end surfaces and is normal to the area vector on the curved wall surfaces. Hence, applying Gauss's law on the cylindrical Gaussian surface

$$\int_{\substack{\text{Cylinder} \\ \text{surface}}} \varepsilon_0 \vec{E} \cdot d\vec{A} = \sigma A$$

$$\text{or, } \varepsilon_0 \int\limits_{\text{Walls}} |\vec{E}| \cos(90°)dA + 2\varepsilon_0 \int\limits_{\text{Ends}} |\vec{E}| \cos(0°)dA = \sigma A$$

$$\text{or, } 0 + 2\varepsilon_0 |\vec{E}| A = \sigma A \text{ or, } |\vec{E}| = \frac{\sigma}{2\varepsilon_0} \tag{4.1}$$

When two plates of equal and opposite charge density are placed near and parallel to each other with free space between them, the electric field intensities due to the two plates add between the plates while they cancel each other outside the plates, as shown in Figure 4.5. Thus, the electric field intensity between the two plates is given by $E = \sigma/\varepsilon_0$.

When the separation distance between the two plates is small compared to the sides of the plate, then the electric field intensity between the plates is constant throughout the interior of the capacitor. The flux lines are parallel to each other near the centre of the capacitor, whereas the concentration of flux lines occurs at the edges, which is known as *fringing of flux*, as shown in Figure 4.6. Neglecting fringing of flux and considering the potential difference between the two plates to be $V$, the electric field intensity can also be written as $E = V/d$.

Noting that the charge on the plates $Q = \sigma A$, it may be written as follows:

$$\frac{\sigma}{\varepsilon_0} = \frac{V}{d} \text{ or, } \frac{QA}{\varepsilon_0} = \frac{V}{d} \text{ or, } \frac{Q}{V} = \frac{\varepsilon_0 A}{d} \text{ or, } C = \frac{\varepsilon_0 A}{d} \tag{4.2}$$

where:
$C$ is the capacitance of the parallel plate capacitor

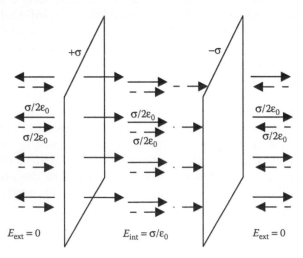

**FIGURE 4.5**
E-field within a parallel plate capacitor.

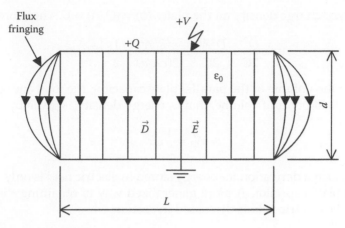

**FIGURE 4.6**
Flux lines in a parallel plate capacitor.

Equation 4.2 shows that the capacitance of a parallel capacitor is independent of the charge on the plates or the potential difference across the capacitor plates.

If any dielectric medium of relative permittivity ($\varepsilon_r$) is inserted between the plates, then the electric field intensity within the capacitor remains unchanged ($= V/d$), but the capacitance ($C$) is changed to ($\varepsilon_r\varepsilon_0 A)/d$.

### 4.3.1 Energy Stored in a Parallel Plate Capacitor

Consider that a parallel plate capacitor has no charge at the beginning. Now, if a voltage source is connected across its plates, so that the potential difference between the two plates becomes $V$, then the capacitor gets charged and these charges are supplied by the source to the capacitor plates. In order to supply these charges to the capacitor plates, some work have to be done by the external source, which is then stored as electrostatic energy in the capacitor.

Let, at any instant, the charge on the capacitor plates be $+q$ and $-q$, respectively, and the potential difference between the plates is $v$. At this instant, consider that an additional amount of charge $dq$ is supplied by the source to the plates. Then the energy spent by the source to deliver this charge is $vdq$ and this energy is stored in the electric field of the capacitor.

Therefore, if the total charge on the plates are $+Q$ and $-Q$, respectively, and the corresponding potential difference between the plates is $V$, then the total energy stored in the electric field is given by

$$W = \int_0^Q vdq = \int_0^Q \frac{q}{C}dq = \frac{Q^2}{C} = \frac{C^2V^2}{2C} = \frac{1}{2}CV^2 \qquad (4.3)$$

Now, surface charge density on the plates ($\sigma$) = ($Q/A$) = $D$. Therefore,

$$W = \frac{Q^2}{2C} = \frac{D^2 A^2}{2C} = \frac{\varepsilon^2 E^2 A^2}{2(\varepsilon A/d)} = \frac{1}{2}\varepsilon E^2 (Ad)$$

($Ad$) being the volume of the parallel plate capacitor, the total energy stored in electric field per unit volume, that is, energy density, is given by

$$W_E = \frac{1}{2}\varepsilon E^2 \tag{4.4}$$

However, such a derivation for energy stored in electric field is only valid for a parallel plate capacitor. A more generalized way of obtaining the energy stored in the electric field is discussed in Section 4.4.

## 4.4 Energy Stored in Electric Field

Consider that an electric field is established by an assembly of charges. To obtain the energy stored in this electric field, the work done to assemble these charges need to be determined. Assume that all the charges are point charges, which are at infinity initially, and an external agent brings these charges one by one and places them at the respective positions, as shown in Figure 4.7. It is obvious that no work is to be done to bring the first charge $q_1$ from infinity to its location at $P_1$, as there is no existing electric field created by another charge. Then the second charge $q_2$ is brought from infinity to $P_2$ within the

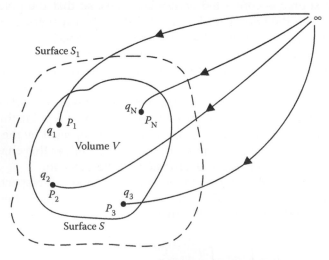

**FIGURE 4.7**
Assembling a set of charges.

electric field created by $q_1$. Therefore, the work done in bringing the charge $q_2$ will be given by

$$W_2 = q_2\phi_{21} = q_2 \frac{q_1}{4\pi\varepsilon_0 r_{21}} \tag{4.5}$$

where:

$\phi_{21}$ is the potential at the location $P_2$ due to the charge $q_1$ located at $P_1$

Then the third charge $q_3$ is brought from infinity to $P_3$ within the electric field created by $q_1$ and $q_2$. Therefore, the work done in bringing the charge $q_3$ will be given by

$$W_3 = q_3\phi_{31} + q_3\phi_{32} = q_3 \frac{q_1}{4\pi\varepsilon_0 r_{31}} + q_3 \frac{q_2}{4\pi\varepsilon_0 r_{32}} = \frac{1}{4\pi\varepsilon_0}\left(\frac{q_1 q_3}{r_{31}} + \frac{q_2 q_3}{r_{32}}\right) \tag{4.6}$$

Therefore, the total work done in bringing $q_1$, $q_2$ and $q_3$ is as follows:

$$0 + W_2 + W_3 = \frac{1}{4\pi\varepsilon_0}\left(\frac{q_1 q_2}{r_{21}} + \frac{q_1 q_3}{r_{31}} + \frac{q_2 q_3}{r_{32}}\right) \tag{4.7}$$

Denoting the charge that is being brought in by the suffix $i$ and the charges that created the field within which this $i$th charge is being brought in by the suffix $j$, the work done in assembling $N$ number of charges $q_1$, $q_2$, ..., $q_N$ can be written as follows:

$$W = \sum_{i=2}^{N} W_i = \frac{1}{4\pi\varepsilon_0} \sum_{i=2}^{N} \sum_{j=1}^{i-1} \frac{q_i q_j}{r_{ij}} \tag{4.8}$$

Figure 4.8 gives a pictorial representation of the numbers over which the summation of Equation 4.8 is being carried out. This summation is clearly over the triangular region marked *I* in Figure 4.8. Because the quantity being summed is symmetric in $i$ and $j$, the same energy $W$ would be obtained by a summation over the triangular region marked *II* in Figure 4.8. The summation over the triangular regions *I* and *II* must then give $2W$. Thus $W$ may also be obtained as follows:

$$W = \frac{1}{8\pi\varepsilon_0} \sum_{i=1}^{N} \sum_{j=1}^{N} \frac{q_i q_j}{r_{ij}}, i \neq j \tag{4.9}$$

Instead of considering discrete point charges, if the charge distribution is taken as continuous throughout the volume $V$ of Figure 4.7, then the potential at any point can be found by summing the contributions from individual differential volume elements of charge. Thus, by writing $\rho(r)dV$ in place of $q_i$ and $\rho(r')dV'$ in place of $q_j$, the summation may be replaced by integrations

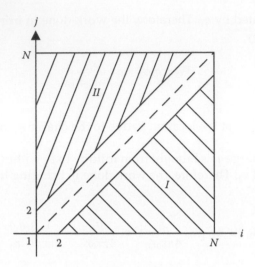

**FIGURE 4.8**
Representation of summation to obtain energy stored.

over volume, which must be large enough to contain all the charges present. Thus, the work done to form a continuous charge distribution is given by

$$W = \frac{1}{8\pi\varepsilon_0} \int\limits_V \int\limits_{V'} \frac{\rho(r)\rho(r')}{R_{rr'}} dV'dV \qquad (4.10)$$

where:

$$R_{rr'} = |\vec{r} - \vec{r}'|$$

The potential at the location $r$ due to all the charges located at the respective positions $r'$ is given by

$$\phi(r) = \int\limits_{V'} \frac{\rho(r')}{4\pi\varepsilon_0 R_{rr'}} dV' \qquad (4.11)$$

Substitution of Equation 4.11 in Equation 4.10 yields

$$W = \frac{1}{2}\int\limits_V \rho(r)\phi(r)dV \qquad (4.12)$$

Because Equation 4.12 deals with quantities related to only one location $r$, the parameter $r$ is omitted hereafter. Noting that $\vec{\nabla}.\vec{D} = \rho$, Equation 4.12 could be rewritten as follows:

$$W = \frac{1}{2}\int\limits_V \phi(\vec{\nabla} \cdot \vec{D})dV \qquad (4.13)$$

For any vector $\vec{D}$ and any scalar $\phi$, the following vector identity could be written:

$$\vec{\nabla}\cdot(\phi\vec{D}) = \vec{D}\cdot\vec{\nabla}\phi + \phi(\vec{\nabla}\cdot\vec{D}) \tag{4.14}$$

$$\text{or, } \phi(\vec{\nabla}\cdot\vec{D}) = \vec{\nabla}\cdot(\phi\vec{D}) - \vec{D}\cdot\vec{\nabla}\phi$$

Therefore, from Equations 4.13 and 4.14,

$$W = \frac{1}{2}\int_V \vec{\nabla}\cdot(\phi\vec{D})dV - \frac{1}{2}\int_V(\vec{D}\cdot\vec{\nabla}\phi)dV \tag{4.15}$$

Applying divergence theorem to the first term on the right-hand side (RHS) of Equation 4.15,

$$W = \frac{1}{2}\oint_S \phi\vec{D}\cdot d\vec{S} - \frac{1}{2}\int_V(\vec{D}\cdot\vec{\nabla}\phi)dV \tag{4.16}$$

As $\vec{E} = -\vec{\nabla}\phi$ and $\vec{D} = \varepsilon_0\vec{E}$, the above equation is rewritten as follows:

$$W = \frac{1}{2}\oint_S \phi\vec{D}\cdot d\vec{S} + \frac{1}{2}\int_V \varepsilon_0(\vec{E}\cdot\vec{E})dV \tag{4.17}$$

In the first term on the RHS of Equation 4.17, $\phi$ varies as $1/r$, whereas $\vec{D}$ varies as $1/r^2$ and surface term varies as $r^2$. Therefore, as a whole the first term varies as $1/r$. If the bounding surface of volume $V$, as shown in Figure 4.7, is expanded from $S$ to $S_1$, then the region between $S$ and $S_1$ does not contribute to the energy integral, as there is no charge located in this region. However, although this region between $S$ and $S_1$ is charge free, the electric field intensity is not zero in this region. Hence, as the bounding surface is expanded, the second term of Equation 4.17 will increase. Then, in order to keep the energy integral unchanged, the first term should decrease. Thus, if the bounding surface is taken to be infinitely large, then the first term on the RHS of Equation 4.17 becomes zero, keeping the total energy same. Therefore, the total work done in forming the continuous charge distribution, that is, the total energy stored in the electric field, is given by

$$W = \frac{1}{2}\int_V \varepsilon_0 E^2 dV \tag{4.18}$$

Hence, energy stored per unit volume, that is, energy density, is given by

$$W_E = \frac{1}{2}\varepsilon_0 E^2 \tag{4.19}$$

Considering the relative permittivity of the dielectric medium to be $\varepsilon_r$, energy density of an electric field is given by

$$W_E = \frac{1}{2} \varepsilon_r \varepsilon_0 E^2 \qquad (4.20)$$

Equation 4.20 is very useful in computing the energy stored in a complex dielectric arrangement where electric field intensity is non-uniform. In such cases, the electric field distribution is first determined using suitable method and then the entire volume under consideration is divided into smaller volume elements, so that each volume element contains only one dielectric medium and the electric field intensity remains constant within the volume element. As a result, the energy stored in each element can be computed using Equation 4.20. The energy stored in all the volume elements could then be summed up to get the total energy stored in any real-life dielectric arrangement having complex geometries and several dielectric media.

## PROBLEM 4.1
Three point charges of magnitude −2, −3 and 1 nC are located in free space at (0,0,0)m, (0,2,0)m and (2,0,0)m, respectively. Find the energy stored in the system of charges.

*Solution:*
Denoting the charges as $q_1 = -2$ nC, $q_2 = -3$ nC and $q_3 = 1$ nC, no work is done to bring in $q_1$.

Electric potential at the location of $q_2$ due to $q_1$ is

$$\phi_{21} = \frac{-2 \times 10^{-9}}{4 \times \pi \times 8.854 \times 10^{-12} \times r_{21}}$$

where:

$$r_{21} = \sqrt{0 + 2^2 + 0} = 2 \text{ m}$$

Hence,

$$\phi_{21} = -8.987 \text{ V}$$

Therefore, work done to bring in $q_2 = -3 \times 10^{-9} \times -8.987 = 26.96$ nJ.

Again, electric potential at the location of $q_3$ due to $q_1$ is

$$\phi_{31} = \frac{-2 \times 10^{-9}}{4 \times \pi \times 8.854 \times 10^{-12} \times r_{31}}$$

where:

$$r_{31} = \sqrt{0 + 0 + 2^2} = 2 \text{ m}$$

Hence,

$$\phi_{31} = -8.987 \text{ V}$$

and electric potential at the location of $q_3$ due to $q_2$ is

$$\phi_{32} = \frac{-3 \times 10^{-9}}{4 \times \pi \times 8.854 \times 10^{-12} \times r_{32}}$$

where:

$$r_{32} = \sqrt{0 + 2^2 + 2^2} = 2\sqrt{2} \text{ m}$$

Hence,

$$\phi_{32} = -6.354 \text{ V}$$

Therefore, work done to bring in $q_3 = 1 \times 10^{-9} \times (-8.987 - 6.354) = -15.34$ nJ.
Hence, the total energy stored in the charge system = (26.96–15.34) = 11.62 nJ.

## 4.5 Two Concentric Spheres with Homogeneous Dielectric

A simple single-dielectric arrangement that have spherical symmetry is two concentric spheres with a homogeneous dielectric medium in between the two spheres, as shown in Figure 4.9. The inner sphere is charged to an electric potential +V, whereas the outer sphere is earthed.

Because this configuration has spherical symmetry, the best choice for a Gaussian surface is a concentric sphere of radius $x$, as shown in Figure 4.9.

Let +Q be total charge on the inner sphere. According to Gauss's law, the total flux leaving the Gaussian surface of radius $x$ is equal to the total charge enclosed, that is, the charge on the inner conductor surface +Q. As the flux lines are symmetrically distributed and are directed radially outwards, the electric flux density at a radial distance of $x$ is given by

$$D_x = \frac{+Q}{4\pi x^2}$$

Hence, electric field intensity at a radial distance of $x$ is $E_x = +Q/(4\pi\varepsilon x^2)$

Then the potential difference between the two spheres of radii $r$ and $R$ could be obtained as follows:

$$V = -\int_R^r E_x dx = -\int_R^r \frac{+Q}{4\pi\varepsilon x^2} dx = \frac{Q}{4\pi\varepsilon_r\varepsilon_0}\left(\frac{1}{r} - \frac{1}{R}\right) \tag{4.21}$$

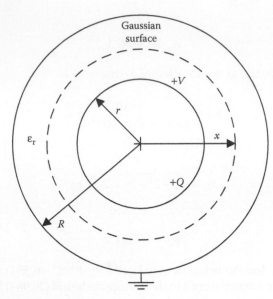

**FIGURE 4.9**
Two concentric spheres with homogeneous dielectric.

Because electric potential is a readily measurable quantity, the charge on the inner sphere could be obtained from the knowledge of electric potential as follows:

$$Q = \frac{4\pi\varepsilon_r\varepsilon_0\, V}{[(1/r)-(1/R)]} \qquad (4.22)$$

The capacitance of the system could be obtained as follows:

$$C = \frac{Q}{V} = \frac{4\pi\varepsilon_r\varepsilon_0}{[(1/r)-(1/R)]} \qquad (4.23)$$

From a practical viewpoint, electric field intensity is a significant quantity that needs to be determined. Hence, electric field intensity at any radius $x$ as expressed in terms of the potential difference between the two spheres is given by

$$E_x = \frac{V}{x^2[(1/r)-(1/R)]} \qquad (4.24)$$

Equation 4.24 shows that electric field intensity varies with radial distance in a non-linear way. Highest value of electric field intensity occurs at a radial distance $r$, which is given by

$$E_{max} = \frac{V}{r^2[(1/r)-(1/R)]} \qquad (4.25)$$

For the above-mentioned concentric spherical system, a value of $r$ for a fixed outer radius $R$ could be obtained that gives the lowest possible value of electric field intensity on the inner conductor. From Equation 4.25, it may be seen that for $E_{max}$ to be minimum, the denominator has to be maximum for a given value of $V$, that is,

$$\frac{d}{dr}\left[r^2\left(\frac{1}{r} - \frac{1}{R}\right)\right] = 0$$

$$\text{or, } r = \frac{R}{2} \tag{4.26}$$

However, it may be noted here that concentric spherical system with a charged inner sphere completely enclosed by an earthed outer sphere is more of theoretical interest, as it is very difficult to realize such a system in practice.

## 4.6 Two Co-Axial Cylinders with Homogeneous Dielectric

A simple single-dielectric arrangement that has cylindrical symmetry is two co-axial cylinders with a homogeneous dielectric medium in between the two cylinders, as shown in Figure 4.10. The inner cylinder is charged to an electric potential $+V$, whereas the outer cylinder is earthed. In real life, a single-core cable having one dielectric medium is a typical example of such a system. In this case, electric field varies with location over the cross-sectional area of the cable. But electric field does not vary along the length of the cable. Hence, the configuration, as shown in Figure 4.10, is represented as a two-dimensional system in the Cartesian coordinates, where the cross-sectional area is taken on $x$–$y$ plane. The field is then independent of $z$-axis, where $z$-direction is along the length of the cylinder.

Because this configuration has cylindrical symmetry, the best choice for a Gaussian surface is a co-axial cylinder of radius $x$, as shown in Figure 4.10.

Let $+q$ be total charge per unit length on the inner cylinder. According to Gauss's law, the total flux leaving the Gaussian surface of radius $x$ is equal to the total charge enclosed, that is, the charge on the inner conductor surface $+q$. As the flux lines are symmetrically distributed and are directed radially outwards, the electric flux density at a radial distance of $x$ is given by

$$D_x = \frac{+q}{2\pi x.1}$$

Hence, electric field intensity at a radial distance of $x$ is $E_x = +q/(2\pi\varepsilon x)$.

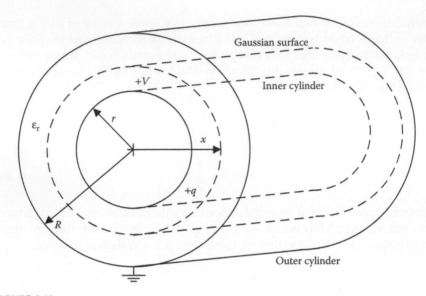

**FIGURE 4.10**
Two co-axial cylinders with homogeneous dielectric.

Then the potential difference between the two cylinders of radii $r$ and $R$ could be obtained as follows:

$$V = -\int_R^r E_x dx = -\int_R^r \frac{+q}{2\pi\varepsilon x} dx = \frac{q}{2\pi\varepsilon_r\varepsilon_0}\ln\frac{R}{r} \qquad (4.27)$$

The charge per unit length on the inner cylinder could be obtained from the knowledge of potential difference between the two cylinders as follows:

$$q = \frac{2\pi\varepsilon_r\varepsilon_0 V}{\ln(R/r)} \qquad (4.28)$$

The capacitance per unit length of the system could be obtained as follows:

$$C = \frac{q}{V} = \frac{2\pi\varepsilon_r\varepsilon_0}{\ln(R/r)} \qquad (4.29)$$

Electric field intensity at any radius $x$ as expressed in terms of the potential difference between the two cylinders is given by

$$E_x = \frac{V}{x\ln(R/r)} \qquad (4.30)$$

From Equation 4.30, it becomes clear that highest value of electric field intensity occurs at a radial distance $r$, which is given by

$$E_{max} = \frac{V}{r \ln(R/r)} \tag{4.31}$$

For the above-mentioned co-axial cylindrical system, a value of $r$ for a fixed outer radius $R$ could be obtained that gives the lowest possible value of electric field intensity on the inner conductor. From Equation 4.31, it may be seen that for $E_{max}$ to be minimum, the denominator has to be maximum for a given value of $V$, that is

$$\frac{d}{dr}(r \ln R - r \ln r) = 0$$

$$\text{or, } \frac{R}{r} = e, \text{ or, } r = \frac{R}{e} \tag{4.32}$$

Then from Equation 4.31, the highest electric field intensity on the inner cylinder becomes

$$E_{max}\big|_{lowest} = \frac{V}{r} \tag{4.33}$$

A practical use of Equation 4.32 in real life is in the finalization of the dimensions of the inner and outer conductors of gas-insulated transmission line (GIL). The typical configuration of a GIL, which is a co-axial cylindrical arrangement where the primary insulation is sulphur hexafluoride ($SF_6$) gas or a mixture of $SF_6$ and $N_2$, is shown in Figure 4.11.

If for a given value of $V$, that is, the potential of the live conductor, values of $R$ and $r$ are chosen as per Equation 4.32 and the value of $r$ is so chosen

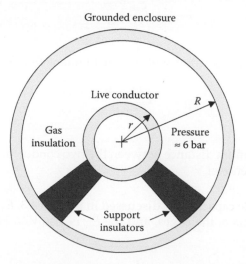

**FIGURE 4.11**
Typical gas-insulated transmission line configuration.

that $E_{max}|_{lowest}$ is equal to the dielectric strength of the gaseous insulation at the designed pressure, then the gaseous insulation is utilized in the most economical way. However, in a practical design, adequate safety margin is taken, so that unwanted breakdown does not take place.

**PROBLEM 4.2**

Find the most economical dimensions of a single-core metal sheathed cable for a working voltage of 76 $kV_{rms}$, if the maximum electric field intensity that can be allowed within the cable insulation is 5 $kV_{rms}$/mm.

*Solution:*

For the most economical design, $r = R/e$.

But, electric field intensity on the conductor surface should not exceed the maximum allowable limit, that is, 5 $kV_{rms}$/mm.

Therefore, electric field intensity on the conductor surface $V/r = 5$, where $V = 76$ $kV_{rms}$.

Hence, the radius of inner conductor $(r) = 76/5 = 15.2$ mm.

Therefore, the radius of the outer conductor $(R) = 15.2 \times 2.718 = 41.3$ mm.

---

## 4.7 Field Factor

For practical configurations, electric field factor $(f)$ is defined as follows:

$$f = \frac{E_{max}}{E_{av}} \tag{4.34}$$

where:

$E_{max}$ is the maximum value of electric field intensity in the system

$$E_{av} = \frac{V}{d_{min}} \tag{4.35}$$

where:

$d_{min}$ is the minimum distance between the two conductors having a potential difference of $V$

For a parallel plate capacitor, as discussed in Section 4.3, $E_{max}$ is $V/d$, neglecting fringing of flux at the edges, and the $E_{av}$ is also $V/d$. Hence, for parallel plate capacitor,

$$f = \frac{E_{max}}{E_{av}} = 1.0 \tag{4.36}$$

For two concentric spheres with one dielectric medium, as discussed in Section 4.5,

$$E_{max} = \frac{V}{r^2[(1/r)-(1/R)]}$$

$$\text{and } E_{av} = \frac{V}{R-r}$$

Hence, for two concentric spheres with single dielectric

$$f = \frac{E_{max}}{E_{av}} = \frac{R}{r} \tag{4.37}$$

For two co-axial cylinders with one dielectric medium, as discussed in Section 4.6,

$$E_{max} = \frac{V}{r\ln(R/r)}$$

$$\text{and } E_{av} = \frac{V}{R-r}$$

Hence, for two concentric spheres with single dielectric

$$f = \frac{E_{max}}{E_{av}} = \frac{(R/r)-1}{\ln(R/r)} \tag{4.38}$$

In fact, for any uniform field system, field factor is unity. The degree of non-uniformity of an electric field is represented by the value of field factor. The higher the field factor, the higher is the non-uniformity of the field distribution. In some references, field utilization factor ($u$) is also used, which is simply defined as the reciprocal of field factor ($f$), such that $u = 1/f = E_{av}/E_{max}$

---

## Objective Type Questions

1. Displacement current due to a capacitor could be measured
   a. Between the source and the capacitor plate
   b. Between the capacitor plates
   c. Both (a) and (b)
   d. None of the above

2. The electric field within a parallel plate capacitor could be best analyzed with the help of the field due to

   a.  Charged point
   b.  Charged sheet
   c.  Charged cylinder
   d.  Charged sphere

3. If the surface charge density on the plates of a parallel capacitor is $\sigma$, then the electric field intensity at the centre of the parallel plate capacitor is given by

   a.  $\varepsilon\sigma$
   b.  $\varepsilon\sigma^2$
   c.  $\sigma/\varepsilon$
   d.  $\sigma^2/\varepsilon$

4. To assemble a set of charge within a body, the energy spent in bringing the first charge from infinity to this body is

   a.  Positive
   b.  Negative
   c.  Infinity
   d.  Zero

5. Energy density of an electric field is given by

   a.  $(CD^2)/2$
   b.  $(CE^2)/2$
   c.  $(\varepsilon E^2)/2$
   d.  $(\varepsilon D^2)/2$

6. For two concentric spheres with a single dielectric between them, the most economical use of the dielectric medium is made when

   a.  $r = R/2$
   b.  $r = R/e$
   c.  $r = \sqrt{R}$
   d.  $r = 1/\sqrt{R}$

7. For two co-axial cylinders with a single dielectric between them, the most economical use of the dielectric medium is made when

   a.  $r = R/2$
   b.  $r = R/e$
   c.  $r = \sqrt{R}$
   d.  $r = 1/\sqrt{R}$

8. In the case of a single-core single-dielectric cable for a given value of the radius of outer sheath ($R$), the value of electric field intensity on the inner conductor of radius $r$ for the most economical use of dielectric is given by

    a. $V/R$
    b. $V/r$
    c. $(r/R)V$
    d. $(R/r)V$

9. In the case of a parallel plate capacitor, energy density of electric field within the capacitor is given by

    a. $CV^2/2$
    b. $\varepsilon E^2/2$
    c. Both (a) and (b)
    d. None of the above

10. For a uniform electric field, field factor

    a. Is equal to 0
    b. Is equal to 1
    c. Lies between 0 and 1
    d. Lies between 0 and $+\infty$

**Answers:** 1) a; 2) b; 3) c; 4) d; 5) c; 6) a; 7) b; 8) b; 9) b; 10) b

# 5

# *Dielectric Polarization*

**ABSTRACT**   Contrary to conductors in which the charges are free to move anywhere within the material, in dielectrics, all the charges are attached to specific atoms or molecules. However, these bound charges could be displaced by a small amount under the action of an external electric field. Many dielectric materials have permanent dipoles that get aligned in the direction of the external electric field more and more as the strength of the external electric field is increased. The cumulative effect of such displacements of charges and/or alignments of dipoles determines the characteristics of dielectric materials. This chapter discusses the mechanisms by which a dielectric material gets polarized, that is, a dielectric material acquires a net dipole moment when influenced by an external electric field. The properties of dielectric materials that are intricately related to polarization are also discussed. Although a lot of dielectric materials are linear, isotropic and homogeneous (LIH), not all dielectrics exhibit such characteristics. Hence, dielectric properties of anisotropic materials are dealt elaborately in this chapter. Frequency dependence of dielectric polarization is also touched on.

## 5.1 Introduction

Inside any material, the electric field varies rapidly with distance in a scale corresponding to the spacing between the atoms or molecules. The local electric field at the site of an atom is significantly different from the macroscopic electric field. It is because of the fact that the local electric field acting on an atom is strongly affected by the nearest atoms, while the macroscopic field is averaged over a large number of atoms or molecules. Therefore, the determination of electric field is extremely complicated, if not impossible, at every *mathematical point* in a given space. As a result, the average value of electric field over a finite volume is commonly determined, particularly in the case of power engineering. This finite volume should be such that it is practically small enough to be considered as a point, but large enough to accommodate enough numbers of atoms or molecules to give smoothly varying average value of electric field.

   Although conductors have large numbers of free charges that can move in response to an external electric field, the materials known as *dielectrics* do not have free charges inside them. Therefore, it may be argued that the dielectrics cannot have any effect of the electrostatic field. But this argument is incorrect, as the mechanism by which dielectric materials affect the electrostatic field is quite different than the mechanism in the case of conductors. Moreover, in reality, there is no electrical equipment or device without conductors as well as dielectrics. Hence, it is important from a practical viewpoint to analyze the behaviour of dielectrics in electrostatic field.

Atoms of all dielectric materials consist of charged constituents such as electrons and nucleus, which could be displaced, albeit through a small distance, by an external electric field. In the process, an electric dipole will be induced by the external electric field in a symmetrical atom or molecule, which originally had zero dipole moment. On the other hand, there are large numbers of dielectric materials containing molecules having permanent dipole moment. In those cases, the permanent dipoles will be aligned by the external electric field in its direction. The degree of alignment of permanent dipoles is higher for stronger external field. The net dipole moment of a dielectric piece is typically zero, when not influenced by external electric field, because the atoms have zero dipole moment and the permanent dipoles are randomly oriented. But due to the induction or alignment of dipoles under the action of external electric field, a dielectric piece may be considered as arrays of oriented electric dipoles. As a result, a dielectric piece acquires a net dipole moment and the dielectric is said to be polarized. The process by which a dielectric material gets polarized is known as *polarization*.

## 5.2 Field due to an Electric Dipole and Polarization Vector

A polarized dielectric can be assumed to be a collection of oriented electric dipoles situated in vacuum. If the charges of the electric dipoles and the distances between them are known, then it is possible to determine the electric potential and electric field intensity at any external location due to the polarized dielectric. But this is practically very difficult due to immensely large number of such dipoles in a polarized dielectric. Because of this reason, a kind of average dipole density is defined in the form of a vector quantity known as *polarization vector* for the ease of analysis.

### 5.2.1 Electric Dipole and Dipole Moment

When two point charges of equal magnitude but of opposite polarities are separated by a small distance, then the arrangement is known as an *electric dipole*, as shown in Figure 5.1. For field analysis, it is required that a single dipole be characterized by a vector quantity. As depicted in Figure 5.1, let the magnitudes of the charges be $+Q$ and $-Q$, respectively, and the distance

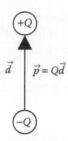

**FIGURE 5.1**
The dipole moment of an electric dipole.

between them is $d$. The distance vector $\vec{d}$ between the two point charges is considered to be directed from the negative charge to the positive charge. Then the dipole moment of the electric dipole is defined as a vector

$$\vec{p} = Q\vec{d} \tag{5.1}$$

The unit of dipole moment is C.m.

## 5.2.2 Field due to an Electric Dipole

The field due to a single electric dipole can be evaluated as the superposition of the field due to two point charges $+Q$ and $-Q$, as shown in Figure 5.2. Then the electric potential at the point $P$ due to the electric dipole is given by

$$V_P = \frac{Q}{4\pi\varepsilon_0 r_1} + \frac{-Q}{4\pi\varepsilon_0 r_2} = \frac{Q}{4\pi\varepsilon_0}\left(\frac{1}{r_1} - \frac{1}{r_2}\right) \tag{5.2}$$

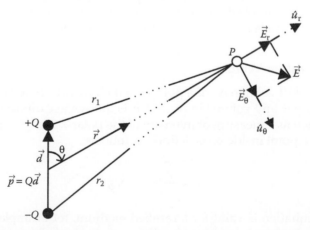

**FIGURE 5.2**
Field due to an electric dipole.

The distance $d$ between the dipole charges is always much smaller than the distance of $P$ from the two charges. Hence, the line segments $r_1$ and $r_2$ will be parallel for all practical purposes. Hence,

$$r_1 = r - \frac{d}{2}\cos\theta \text{ and } r_2 = r + \frac{d}{2}\cos\theta \qquad (5.3)$$

where:
  $r$ = distance from the centre of the electric dipole to $P$
  $\theta$ = the angle between the distance vectors $\vec{d}$ and $\vec{r}$

Thus, assuming $r \gg d$, electric potential at a distance $r$ from the electric dipole may be written as follows:

$$V_P = \frac{Qd\cos\theta}{4\pi\varepsilon_0 r^2} = \frac{\vec{p}\cdot\hat{u}_r}{4\pi\varepsilon_0 r^2} \qquad (5.4)$$

From the above equation, it may be seen that the field due to an electric dipole is two dimensional in nature when represented in spherical coordinate system, as the field depends on $r$ and $\theta$ coordinates and not on $\phi$ coordinate.

Then electric field intensity at $P$ can be expressed as follows:

$$\vec{E}_P = -\vec{\nabla}V_P = -\frac{\partial V_P}{\partial r}\hat{u}_r - \frac{1}{r}\frac{\partial V_P}{\partial\theta}\hat{u}_\theta = \frac{Qd\cos\theta}{2\pi\varepsilon_0 r^3}\hat{u}_r + \frac{Qd\sin\theta}{4\pi\varepsilon_0 r^3}\hat{u}_\theta \qquad (5.5)$$

Thus, the $r$ and $\theta$ components of electric field intensity are

$$E_r = \frac{Qd\cos\theta}{2\pi\varepsilon_0 r^3} \text{ and } E_\theta = \frac{Qd\sin\theta}{4\pi\varepsilon_0 r^3} \qquad (5.6)$$

Equations 5.4 and 5.5 show that electric potential and electric field intensity due to an electric dipole depend on dipole moment of the electric dipole and not on the magnitude of the charges and their separation distance separately.

### 5.2.3 Polarization Vector

Consider a small volume $\Delta v$ of a polarized dielectric. If there are $N$ number of molecules per unit volume in $\Delta v$ and $\vec{p}$ is the average dipole moment per molecule, then as a measure of intensity of the polarization, the polarization vector, $\vec{P}$, at a point inside $\Delta v$ is defined as follows:

$$\vec{P} = \frac{\sum_{\Delta v}\vec{p}}{\Delta v} = N\vec{p} \qquad (5.7)$$

The above equation is valid for a rarefied medium, for example, a gas. The relationship needs to be written in a different way for a dense medium such as a liquid or a solid.

In a generalized manner, the net dipole moment in the small volume $\Delta v$ is given by

$$Q_1\vec{d}_1 + Q_2\vec{d}_2 + \cdots + Q_{N1}\vec{d}_{N1} = \sum_{i=1}^{N1} Q_i\vec{d}_i \qquad (5.8)$$

where:
$N1 = N \times \Delta v$

Then,

$$\vec{P} = \frac{\sum_{i=1}^{N1} Q_i\vec{d}_i}{\Delta v} \qquad (5.9)$$

The unit of dipole moment is C.m and hence the unit of polarization vector is $C/m^2$, which is the same as that of surface charge density.

In other words, if $\vec{P}$ is known at a point, then $\Delta v$, which encloses that point and contains large number of dipoles, can be replaced by a single dipole of moment

$$\vec{p} = \vec{P}\Delta v \qquad (5.10)$$

With the help of Equation 5.10, electric potential and electric field intensity due to a polarized dielectric can be evaluated by an integral.

---

## 5.3 Polarizability

The molecules of dielectric materials, which are the basic building blocks of the material, either have zero dipole moment or have some permanent dipole moments depending on their structure. When an external electric field is applied, then the opposite polarity charges are pulled apart and/or the permanent dipoles get aligned under the action of the external field. In this way, the dielectric material becomes polarized and this property of dielectric materials is known as *polarizability*.

### 5.3.1 Non-Polar and Polar Molecules

The molecules of a dielectric material are classified into two categories, namely, non-polar and polar. Symmetrical molecules such as $CO_2$, as shown in Figure 5.3c, fall in this category. In non-polar molecules the *centres of gravity* of positive and negative charge distribution usually coincide at one point and hence the molecules have zero dipole moment. On the other hand, a polar molecule such as $H_2O$ and CO, as shown in Figure 5.3a and b, respectively, have permanent dipoles even in the absence of any external electric field. However, in the absence of any external field, a macroscopic piece

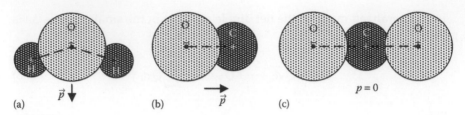

(a)     $\vec{p}$                 (b)    $\vec{p}$          (c)      $p = 0$

**FIGURE 5.3**
Non-polar and polar molecules: (a) polar $H_2O$; (b) polar CO and (c) non-polar $CO_2$.

of polar dielectric is not polarized; that is, it does not contain any dipole moment, because the molecules are randomly oriented due to thermal agitation. When the polar dielectric is subjected to an external electric field, then the individual permanent dipoles within the polar dielectric experiences torques, which tend to align these dipoles in the direction of the external field, and the dielectric gets polarized.

### 5.3.2 Electronic Polarizability of an Atom

A simplified model of an atom is a uniformly charged electron cloud, which is spherical in shape having radius $R$, surrounding the total positive charge located at the point nucleus. The centres of gravity of the total negative charge $(-q)$ of the electron cloud and the total positive charge $(+q)$ coincides at the same point, as shown in Figure 5.4, in the absence of any external electric field.

When an external electric field is applied, the electron cloud is displaced by a small distance $d$ until the mutual attractive force between the negatively charged electron cloud and the positively charged point nucleus balances the force due to the external electric field.

The attractive force between the electron cloud and the point nucleus is given by

$$F_{int} = q \times \frac{\left(d^3/R^3\right)q}{4\pi\varepsilon_0 d^2} \tag{5.11}$$

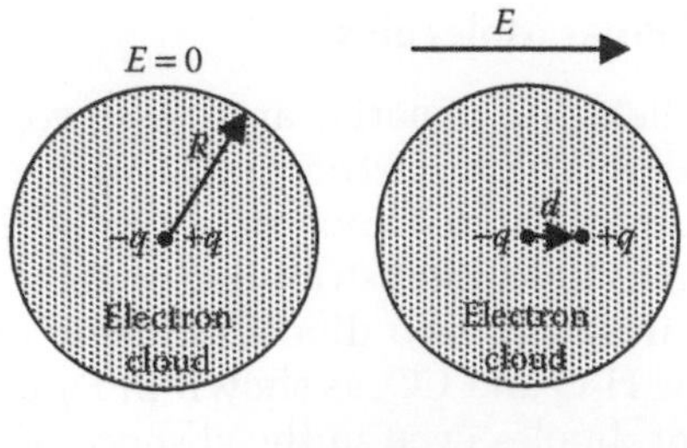

$E = 0$                     $E$

**FIGURE 5.4**
Electronic polarizability of an atom.

The force due to the external electric field is given by

$$F_{ext} = qE \tag{5.12}$$

At equilibrium, $F_{int} = F_{ext}$, or, $\dfrac{dq^2}{4\pi\varepsilon_0 R^3} = qE$

$$\text{or, } d = \frac{4\pi\varepsilon_0 R^3 E}{q} \tag{5.13}$$

Hence, the magnitude of dipole moment induced by the external electric field

$$p = qd = 4\pi\varepsilon_0 R^3 E = \alpha_e E \tag{5.14}$$

where:
$\alpha_e = 4\pi\varepsilon_0 R^3$ = electronic polarizability of the atom

### 5.3.3 Types of Polarizability

The physical processes that give rise to polarizability can be subdivided into four categories: (1) electronic polarizability, (2) ionic polarizability, (3) orientational or dipolar polarizability and (4) interfacial polarizability.

#### 5.3.3.1 Electronic Polarizability

Electronic polarizability arises due to the displacement of negatively charged electron cloud with respect to the positively charged nucleus under the influence of external electric field, as discussed in Section 5.3.2. Electronic polarizability is present in all types of dielectric materials. It is an elastic process without any power loss and is an extremely fast process, which takes place within $10^{-16}$ to $10^{-13}$ s.

#### 5.3.3.2 Ionic Polarizability

In the case of dielectric molecules that contain ionic bonds, the lengths of the bonds get stretched under the influence of the external electric field. Consider the case of NaCl, as shown in Figure 5.5. The external electric field displaces the positive Na+ ion towards right, whereas the negative Cl- ion

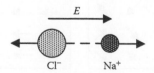

**FIGURE 5.5**
Ionic polarization in NaCl.

is displaced towards left. Thus, the forces due to external field stretch the length of the ionic bond between Na$^+$ and Cl$^-$ ions. As a result of the change in the length of the bond, a net dipole moment appears in the unit cell of NaCl, which does not have any dipole moment in the absence of the external electric field. Because the dipole moment in such cases arises due to the displacement of oppositely charged ions, this process is known as *ionic polarizability*. Ionic polarizability exists in all dielectric materials that contain ionic bonds. Similar to electronic polarizability, ionic polarizability is also an elastic property involving no power loss. But unlike electronic polarizability, ionic polarizability is slower as ions are heavier than electrons. However, it is still a very fast process and occurs within $10^{-13}$ to $10^{-9}$ s.

### 5.3.3.3 Orientational or Dipolar Polarizability

Although the individual molecules of a polar dielectric material have permanent dipoles, the net dipole moment becomes zero in a macroscopic piece of such dielectric material in the absence of external electric field because of random orientation of molecular dipole moments caused by thermal perturbations. Such random orientation of molecular dipoles results in near complete cancellation of dipole moment in any given direction in a macroscopic piece of polar dielectric material. But, when an external electric field is applied, then the molecular dipoles tend to align in the direction of the applied field, as shown in Figure 5.6. It is because of the fact that the energy of a dipole $\vec{p}$ placed in a local electric field $\vec{E}_{loc}$ is $W = -\vec{p} \cdot \vec{E}_{loc}$. This energy is minimum when the dipole is oriented parallel to the applied electric field. As a result of such alignment of molecular dipoles, the net dipole moment in a macroscopic piece of polar dielectric material gets a non-zero value. This mechanism through which a non-zero dipole moment arises in a polar dielectric is known as *orientational* or *dipolar polarizability*. Dipolar polarizability is much slower than electronic or ionic polarizability, as it involves rotation of molecular dipoles that causes molecular friction. It is an inelastic process associated with power loss due to molecular friction and occurs within $10^{-9}$ to $10^{-4}$ s.

Electronic polarizability is present in all dielectrics, but the presence of ionic and dipolar polarizabilities depend on the molecular structure of dielectric materials. The relative magnitudes of the three polarizabilities, as discussed in Sections 5.3.3.1 through 5.3.3.3, are such that in non-polar, ionic

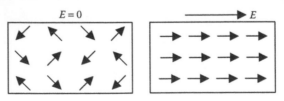

**FIGURE 5.6**
Dipolar polarization in polar dielectric materials.

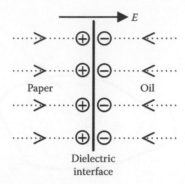

**FIGURE 5.7**
Interfacial polarization at dielectric interface.

dielectric materials, electronic polarizability is of the same order as ionic polarizability. On the other hand, in polar dielectric materials, the dipolar polarizability is much larger than both electronic and ionic polarizabilities.

#### 5.3.3.4 Interfacial Polarizability

It occurs mainly in insulation system composed of different dielectric materials, for example, oil-impregnated paper/pressboard. Under the influence of an external electric field, small numbers of positive and negative charges, which are free to move within the bulk of the dielectric, get trapped at the interfaces of different materials, as shown in Figure 5.7, and thus produces separation of charges at the dielectric interfaces causing polarization. This mechanism known as *interfacial polarizability* is very slow and in general takes hours to complete.

---

## 5.4 Field due to a Polarized Dielectric

Consider a block of polarized dielectric material, as shown in Figure 5.8, containing a polarization vector $\vec{P}$ that varies with position. According to Equation 5.10, the small volume $dV'$ located at $r'(x', y', z')$ can be replaced by a single dipole moment $\vec{p} = \vec{P}dV'$. Then the electric potential at any point $P$ outside the volume of the dielectric and located at $r(x,y,z)$ is given by

$$d\phi_P = \frac{\vec{P} \cdot \hat{u}_R}{4\pi\varepsilon_0 R^2} dV' \tag{5.15}$$

where:
$R = \sqrt{(x-x')^2 + (y-y')^2 + (z-z')^2}$ is the distance between the small volume $dV'$ and the external field point $P$

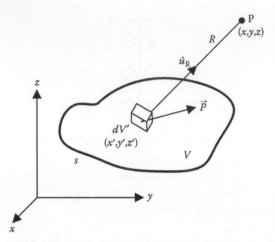

**FIGURE 5.8**
Field at an external point due to a polarized dielectric.

In Equation 5.15,

$$\frac{\vec{P} \cdot \hat{u}_R}{R^2} = \vec{P} \cdot \vec{\nabla}'\left(\frac{1}{R}\right) \tag{5.16}$$

Thus,

$$d\phi_P = \frac{1}{4\pi\varepsilon_0}\left[\vec{P} \cdot \vec{\nabla}'\left(\frac{1}{R}\right)\right]dV' \tag{5.17}$$

where:
$\vec{\nabla}'$ is the gradient with respect to the primed quantities, that is, with respect to the position of the small volume $dV'$ within the dielectric volume

Therefore, the potential at $P$ due to the entire volume $V$ of the polarized dielectric

$$\phi_P = \int_V \frac{1}{4\pi\varepsilon_0}\left[\vec{P} \cdot \vec{\nabla}'\left(\frac{1}{R}\right)\right]dV' \tag{5.18}$$

Consider a scalar quantity $a$, a vector quantity $\vec{A}$ and the vector identity as follows:

$$\vec{\nabla}' \cdot a\vec{A} = a\vec{\nabla}' \cdot \vec{A} + \vec{A} \cdot \vec{\nabla}'a \tag{5.19}$$

Putting $a = 1/R$ and $\vec{A} = \vec{P}$ in the above equation, we get

$$\vec{P} \cdot \vec{\nabla}'\left(\frac{1}{R}\right) = \vec{\nabla}' \cdot \frac{\vec{P}}{R} - \frac{1}{R}\vec{\nabla}' \cdot \vec{P} \tag{5.20}$$

Therefore, from Equations 5.18 and 5.20

$$\phi_P = \int\limits_V \frac{1}{4\pi\varepsilon_0}\left(\vec{\nabla}'\cdot\frac{\vec{P}}{R} - \frac{1}{R}\vec{\nabla}'\cdot\vec{P}\right)dV' \tag{5.21}$$

Applying the divergence theorem to the first term of the above equation,

$$\phi_P = \int\limits_S \frac{\vec{P}\cdot\hat{u}'_n}{4\pi\varepsilon_0 R}dS' + \int\limits_V \frac{-\vec{\nabla}'\cdot\vec{P}}{4\pi\varepsilon_0 R}dV' \tag{5.22}$$

where:
$\hat{u}'_n$ is the outwards unit normal vector to the surface $dS'$ of the small volume $dV'$ of the dielectric

The two terms on the right-hand side of Equation 5.22 can be re-written as follows:

$$\phi_P = \frac{1}{4\pi\varepsilon_0}\left[\int\limits_S \frac{\sigma_{sb}(r')}{R}dS' + \int\limits_V \frac{\rho_{vb}(r')}{R}dV'\right] \tag{5.23}$$

where:

$$\sigma_{sb}(r') = \vec{P}\cdot\hat{u}'_n$$
$$\rho_{vb}(r') = -\vec{\nabla}'\cdot\vec{P} \tag{5.24}$$

and $r'$ denotes the location within the polarized dielectric volume.
Then electric field intensity at $P$ is given by

$$\vec{E}_P = -\vec{\nabla}\phi_P = \frac{1}{4\pi\varepsilon_0}\left[\int\limits_S \frac{\sigma_{sb}(r')\hat{u}_R}{R^2}dS' + \int\limits_V \frac{\rho_{vb}(r')\hat{u}_R}{R^2}dV'\right] \tag{5.25}$$

## 5.4.1 Bound Charge Densities of Polarized Dielectric

Equations 5.23 and 5.25 show that the field at an external point due to a polarized dielectric is superposition of the field due a volume charge density and a surface charge density. In other words, a polarized dielectric can be replaced by an equivalent volume charge density ($\rho_{vb}$) and an equivalent surface charge density ($\sigma_{sb}$). Both volume and surface charge densities can be considered to be in vacuum, as rest of the dielectric does not produce any field at an external point. These charge densities are called *bound volume charge density* and *bound surface charge density*, respectively, as these charges appearing due to polarization are not free to move within the dielectric material. These charges are caused by displacement or rotation occurring in molecular scale during polarization. Such equivalent charge distribution is very useful because the problem of finding the field due to a polarized dielectric is converted to the problem of finding the field due to distribution of charges in vacuum, which is easier to solve. If the polarization vector is known at all the points within the polarized dielectric, then both bound volume and surface charge densities could be found from the polarization vector. Hence, the problem comes down to the determination of polarization vector within the volume of the polarized dielectric, which nowadays is done with the help of numerical techniques.

### 5.4.1.1 Bound Volume Charge Density

From Equation 5.24, bound volume charge density is $\rho_{vb}(r') = -\vec{\nabla}' \cdot \vec{P}$, that is, if the divergence of polarization vector $\vec{P}$ is non-zero, then the bound volume charge density will exist within the volume of the polarized dielectric. For uniform polarization, divergence of $\vec{P}$ is zero and hence there could be no bound volume charge density. But for non-uniform polarization, there can be net increase or decrease of charge within a given volume. For inhomogeneous dielectrics, there will be some net volume charge, because all the molecular dipoles are not identical and hence, their effect does not cancel out on average. In such cases, the divergence of $\vec{P}$ will be non-zero and hence the bound volume charge density will be finite and non-zero. This fact can be understood from Figure 5.9a, where at the centre of the volume the negative ends of the dipoles are concentrated and hence there will be an excess of negative charges at that location giving rise to non-zero polarization vector $\vec{P}$.

### 5.4.1.2 Bound Surface Charge Density

From Equation 5.24, bound surface charge density is $\sigma_{sb}(r') = \vec{P} \cdot \hat{u}_n'$. Such surface charge densities are present for both uniform and non-uniform polarization. As shown in Figure 5.9b, in the case of uniform polarization, for a macroscopic volume of the dielectric there will be equal amount of positive and negative charges and the net charge within the volume will be zero. But if a small volume is considered that includes the upper boundary perpendicular to the

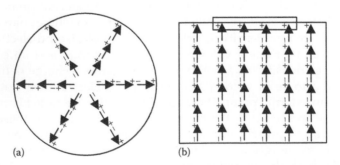

(a)                                              (b)

**FIGURE 5.9**
Equivalent charge distribution of polarized dielectric: (a) volume charge density due to non-uniform polarization, and (b) surface charge density due to uniform polarization.

direction of polarization as shown in Figure 5.9b, there will be net positive charge within the volume, no matter how thin the volume is made. In the limiting case, the thickness tends to zero, but there will still be excess positive charges on the surface. Therefore, bound charge density appears on the surface of the polarized dielectric due to uncompensated charges on the surface.

## 5.4.2 Macroscopic Field

The expressions for electric potential and electric field intensity due to polarized dielectric only are given by Equations 5.23 and 5.25, respectively. But the effects of the external charge distribution that causes the polarization must be added to these to get the resultant field. The effects are simply additive because the bound surface and volume charge densities due to polarization are considered to be in vacuum. Therefore, the complete expressions for electric potential and electric field intensity at a point outside the polarized dielectric due to the external charge distribution and the equivalent bound charge distributions of the polarized dielectric are given by

$$\phi_P = \frac{1}{4\pi\varepsilon_0} \left\{ \int_S \frac{[\sigma_{\text{ext}} + \sigma_{\text{sb}}(r')]}{R} dS' + \int_V \frac{[\rho_{\text{ext}} + \rho_{\text{vb}}(r')]}{R} dV' \right\} \tag{5.26}$$

$$\vec{E}_P = \frac{1}{4\pi\varepsilon_0} \left\{ \int_S \frac{[\sigma_{\text{ext}} + \sigma_{\text{sb}}(r')]\hat{u}_R}{R^2} dS' + \int_V \frac{[\rho_{\text{ext}} + \rho_{\text{vb}}(r')]\hat{u}_R}{R^2} dV' \right\} \tag{5.27}$$

## 5.4.3 Field due to a Narrow Column of Uniformly Polarized Dielectric

Consider a narrow column of polarized dielectric of the cross-sectional area $dS$ with a polarization vector of magnitude $P$ directed along the axis of the column, as shown in Figure 5.10.

Electric potential at the external point $A$ due to the small volume $dSdz$ of the polarized column can be expressed according to Equation 5.15 as follows:

$$d\phi_A = \frac{PdSdz\cos\theta}{4\pi\varepsilon_0 r^2} = \frac{PdSdr}{4\pi\varepsilon_0 r^2} \tag{5.28}$$

Hence, the electric potential at $A$ due to the entire column is

$$\phi_A = \frac{PdS}{4\pi\varepsilon_0} \int_{r_1}^{r_2} \frac{dr}{r^2}$$

$$\text{or, } \phi_A = \frac{PdS}{4\pi\varepsilon_0}\left(\frac{1}{r_1} - \frac{1}{r_2}\right) = \frac{PdS}{4\pi\varepsilon_0 r_1} + \frac{-PdS}{4\pi\varepsilon_0 r_2} \tag{5.29}$$

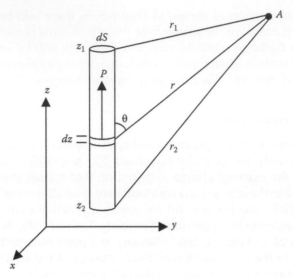

**FIGURE 5.10**
Field due to a narrow column of uniformly polarized dielectric.

For the area $dS$ at $z_1$, $\vec{P} \cdot \hat{u}_n$ is positive and that at $z_2$ is negative. Thus Equation 5.29 shows that the field due to the narrow column of polarized dielectric is the same as that due to positive charges ($+PdS$) on the surface $dS$ at $z_1$ and the negative charges ($-PdS$) on $dS$ at $z_2$. In other words, the bound charges due to polarization appear at the two surfaces as it is a case of uniform polarization.

### 5.4.4 Field within a Sphere Having Uniformly Polarized Dielectric

According to the discussions in Section 5.4, Equations 5.23 and 5.25 are valid if the field observation point is located outside the polarized dielectric. However, these two equations can also be applied when the field observation point is inside the polarized dielectric, provided that average electric field is determined.

Consider a sphere of radius $R$ containing a dielectric medium of uniform polarization $\vec{P}$. Such a uniformly polarized dielectric sphere could be equivalently constructed by superimposing two uniformly charged spheres of opposite polarity with centres displaced by a small distance $d$, as shown in Figure 5.11.

Electric field intensity at a radial distance $r$ within a charged sphere of radius $R$ is given by

$$\vec{E} = \frac{q}{4\pi\varepsilon_0 r^2} \frac{r^3}{R^3} \hat{u}_r = \frac{q\vec{r}}{4\pi\varepsilon_0 R^3} \tag{5.30}$$

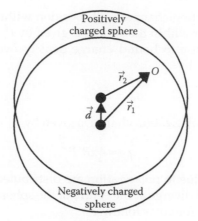

**FIGURE 5.11**
Field within a uniformly polarized dielectric sphere.

where:

$\vec{r}$ points outwards from the centre of the sphere

$q$ is the total charge within the sphere

In the case of two overlapping but slightly displaced spheres, as shown in Figure 5.11, for $r < R$

$$\vec{E} = \frac{q\vec{r}_2}{4\pi\varepsilon_0 R^3} + \frac{-q\vec{r}_1}{4\pi\varepsilon_0 R^3} = \frac{q(\vec{r}_2 - \vec{r}_1)}{4\pi\varepsilon_0 R^3} = \frac{-q\vec{d}}{4\pi\varepsilon_0 R^3} \tag{5.31}$$

Considering uniform polarization, net dipole moment within the dielectric sphere is $\vec{p} = q\vec{d}$.

Therefore, average polarization vector, $\vec{P} = (3\vec{p})/(4\pi R^3) = (3q\vec{d})/(4\pi R^3)$.

Therefore, from Equation 5.31,

$$\vec{E} = -\frac{\vec{P}}{3\varepsilon_0} \tag{5.32}$$

Electric field intensity within the sphere is the average value determined by representing the sphere as a continuum medium with uniform polarization $\vec{P}$.

Outside the uniformly polarized sphere, electric field intensity is equivalent to that due to two point charges of opposite polarity located at the centres of the two slightly displaced spheres.

### 5.4.5 Sphere Having Constant Radial Distribution of Polarization

Consider a sphere of radius $R$ with constant radial distribution of polarization, as shown in Figure 5.9a. Thus the magnitude of polarization ($P$) is constant but the direction changes with position within the sphere. Hence, the polarization vector within the sphere is given by $\vec{P} = P\hat{r}$.

Polarization gives rise to bound charge density over the surface of the sphere, which is given by

$$\sigma_{sb} = \vec{P} \cdot \hat{u}_n = P \tag{5.33}$$

Hence, total surface polarization charge is given by

$$q_{sb} = 4\pi R^2 P \tag{5.34}$$

The volume charges due to polarization are distributed over the entire volume of the sphere and diverge at the centre of the sphere. The bound volume charge density can be found as follows:

$$\rho_{vb} = -\vec{\nabla} \cdot \vec{P} = -P\vec{\nabla} \cdot \vec{r} = -P\frac{1}{r^2}\frac{\partial r^2}{\partial r} = -\frac{2P}{r} \tag{5.35}$$

Total volume charge due to polarization is

$$q_{vb} = \int_V \rho_{vb} dV = \int_0^R \left(-\frac{2P}{r}\right) 4\pi r^2 dr = -4\pi R^2 P \tag{5.36}$$

Hence, the total polarization charge, which is the sum of the bound volume charges ($q_{vb}$) and bound surface charges ($q_{sb}$), is zero.

## PROBLEM 5.1

A thin dielectric rod of cross-sectional area $S$ extends along the z-axis from $z = 0$ to $z = H$. The polarization of the dielectric rod is along the z-axis and is given by $\vec{P} = (a_1 z + a_2)\hat{k}$. Calculate the bound surface charge density at each end, bound volume charge density and the total bound charge within the rod.

*Solution:*
For the surface at $z = H$, $\hat{u}_n = \hat{k}$ and for the surface at $z = 0$, $\hat{u}_n = -\hat{k}$.

Hence, the bound surface charge density at $z = H$,

$$\sigma_{sb}\big|_{z=H} = \vec{P} \cdot \hat{k}\big|_{z=H} = (a_1 z + a_2)\big|_{z=H} = a_1 H + a_2$$

and the bound surface charge density at $z = 0$,

$$\sigma_{sb}\big|_{z=0} = \vec{P} \cdot -\hat{k}\big|_{z=0} = -(a_1 z + a_2)\big|_{z=0} = -a_2$$

Therefore, total bound surface charge $q_{sb} = (a_1 H + a_2)S - a_2 S = a_1 HS$.
The bound volume charge density $\rho_{vb} = -\vec{\nabla} \cdot \vec{P} = -a_1$.
Therefore, total bound volume charge $q_{vb} = -a_1 HS$.
Therefore, total bound charge due to polarization $= q_{sb} + q_{vb} = 0$.

## PROBLEM 5.2

A dielectric cube of side 2 m and centre at origin has a radial polarization given by $\vec{P} = 4\vec{r}$ nC/m² and $\vec{r} = 2x\hat{i} + 2y\hat{j} + 2z\hat{k}$ m. Find bound surface charge density, bound volume charge density and the total bound charge due to polarization.

*Solution:*
Given, $\vec{P} = 8x\hat{i} + 8y\hat{j} + 8z\hat{k}$ nC/m.²
For each of the six faces of the cube, there is a surface charge density $\sigma_{sb}$. For the face at $x = 1$ m,

$$\sigma_{sb} = \vec{P} \cdot \hat{i}\big|_{x=1} = 8x\big|_{x=1} = 8 \text{ nC/m}^2$$

The magnitude of $\sigma_{sb}$ is same for all the six faces. Therefore, considering all the six faces of the cube, total bound surface charge

$$q_{sb} = 6 \times 8 \times 2^2 = 192 \text{ nC}$$

The bound volume charge density is

$$\rho_{vb} = -\vec{\nabla} \cdot \vec{P} = -(8 + 8 + 8) = -24 \text{ nC/m}^3$$

Hence, total bound volume charge $q_{vb} = -24 \times 2^3 = -192$ nC.
Thus, total bound charge within the cube due to polarization = 192 – 192 = 0.

## 5.5 Electric Displacement Vector

According to Gauss's law,

$$\varepsilon_0 \int_S \vec{E} \cdot d\vec{S} = \varepsilon_0 \int_V \vec{\nabla} \cdot \vec{E} dV = q_t \tag{5.37}$$

where:
$q_t$ is the total charge enclosed by the volume $V$, which for dielectric materials include free as well as bound volume charges

Thus, $q_t = \int_V (\rho_f + \rho_{vb}) dV$. Hence, from Equation 5.37

$$\varepsilon_0 \int_V \vec{\nabla} \cdot \vec{E} dV = \int_V (\rho_f + \rho_{vb}) dV$$

$$\text{or, } \vec{\nabla} \cdot \vec{E} = \frac{(\rho_f + \rho_{vb})}{\varepsilon_0} = \frac{\rho_t}{\varepsilon_0} \tag{5.38}$$

But, according to Equation 5.24,

$$\rho_{vb} = -\vec{\nabla} \cdot \vec{P}$$

Therefore,

$$\vec{\nabla} \cdot \vec{E} = \frac{1}{\varepsilon_0}(\rho_f - \vec{\nabla} \cdot \vec{P})$$

$$\text{or, } \vec{\nabla} \cdot (\varepsilon_0 \vec{E} + \vec{P}) = \rho_f \tag{5.39}$$

Equation 5.39 is very important in the sense that the divergence of the vector $(\varepsilon_0 \vec{E} + \vec{P})$ through any volume is equal to the free charge density in that volume. This form of Gauss's law is more convenient because the only charges that can be influenced externally are the free charges.

This vector $(\varepsilon_0 \vec{E} + \vec{P})$ is called *electric displacement vector* $(\vec{D})$, so that

$$\vec{D} = \varepsilon_0 \vec{E} + \vec{P} \tag{5.40}$$

and from Equation 5.39

$$\vec{\nabla} \cdot \vec{D} = \rho_f \tag{5.41}$$

The integral form Equation 5.41 is

$$\int_S \vec{E} \cdot d\vec{S} = \int_V \rho_f dV \tag{5.42}$$

It should be noted here that both $\vec{\nabla} \cdot \vec{D}$ and $\int_S \vec{E} \cdot d\vec{S}$ are related to free charges only and are unaffected by bound charges due to polarization.

From Equation 5.38, it may be seen that both free and bound charges are sources of $\vec{E}$, whereas Equation 5.41 shows that only free charges are sources of $\vec{D}$. In other words, lines of $\vec{D}$ begin and end on free charges only, but the lines of $\vec{E}$ begin and end on either free or bound charges.

From Equation 5.40, it may be written that

$$\vec{E} = \frac{\vec{D}}{\varepsilon_0} - \frac{\vec{P}}{\varepsilon_0} \tag{5.43}$$

It means that the E-field within a dielectric is resultant of two fields, namely, D-field and P-field. D-field is associated with free charges, whereas P-field is associated with bound charges due to polarization. Moreover, P-field acts in opposition to D-field. Thus, in the presence of P-field, that is, in the presence of bound charges due to polarization, the E-field within a dielectric becomes less than the D-field.

### 5.5.1 Electric Susceptibility

For linear dielectric materials, polarization vector $\vec{P}$ varies directly with applied electric field $\vec{E}$. Hence, $\vec{P}$ and $\vec{E}$ are related as follows:

$$\vec{P} = \varepsilon_0 \chi_e \vec{E} \tag{5.44}$$

where:

$\chi_e$ is known as electric susceptibility of the dielectric

It is a dimensionless quantity and is a measure of how susceptible a dielectric material is to applied electric field. In other words, it indicates the relative ease of polarization of the dielectric.

## 5.5.2 Dielectric Permittivity

From Equations 5.40 and 5.44, for linear dielectric materials

$$\vec{D} = \varepsilon_0\left(1+\chi_e\right)\vec{E} = \varepsilon_0\varepsilon_r\vec{E} \tag{5.45}$$

$$\text{or, } \vec{D} = \varepsilon\vec{E} \tag{5.46}$$

where:

$$\varepsilon = \varepsilon_0\varepsilon_r \tag{5.47}$$

$$\varepsilon_r = 1+\chi_e = \frac{\varepsilon}{\varepsilon_0} \tag{5.48}$$

where:

$\varepsilon$ is called the permittivity

$\varepsilon_r$ is called the dielectric constant or relative permittivity of linear dielectric and is also a dimensionless quantity

For any material in which polarization vector is non-zero, electric susceptibility is greater than unity and hence relative permittivity is always greater than unity. For the majority of dielectric materials, $\varepsilon_r$ varies with the frequency of the applied electric field. The value of $\varepsilon_r$ that is relevant to electrostatics is the value in steady electric field or at low frequencies (<1000 Hz).

## 5.5.3 Relationship between Free Charge Density and Bound Volume Charge Density

As discussed above, for linear dielectric media,

$$\vec{P} = \vec{D} - \varepsilon_0\vec{E} = [1-(1/\varepsilon_r)]\vec{D}, \text{ as } \vec{D} = \varepsilon_r\varepsilon_0\vec{E}$$

Therefore,

$$\vec{\nabla}\cdot\vec{P} = \left(1-\frac{1}{\varepsilon_r}\right)\vec{\nabla}\cdot\vec{D} = \left(1-\frac{1}{\varepsilon_r}\right)\rho_f, \text{ as } \vec{\nabla}\cdot\vec{D} = \rho_f$$

$$\text{or, } -\rho_{vb} = \left(1-\frac{1}{\varepsilon_r}\right)\rho_f, \text{ as } \rho_{vb} = -\vec{\nabla}\cdot\vec{P}$$

Hence, total charge density $\rho_t = \rho_f + \rho_{vb} = (\rho_f/\varepsilon_r)$.

Thus for $\varepsilon_r > 1$, $\rho_t < \rho_f$, because $\rho_f$ and $\rho_{vb}$ are of opposite polarities.

**PROBLEM 5.3**

In a dielectric material $E_y = 10$ V/m and polarization vector $\vec{P} = (400\hat{i} + 300\hat{j} - 200\hat{k})$ pC/m$^2$. Calculate (1) electric susceptibility ($\chi_e$), (2) $\vec{E}$ and (3) $\vec{D}$.

*Solution:*

$$P_y = \varepsilon_0\chi_e E_y, \text{ or, } 300\times10^{-12} = 8.854\times10^{-12}\times\chi_e\times10, \text{ or, } \chi_e = 3.39$$

$$E_x = \frac{P_x}{\varepsilon_0\chi_e} = \frac{400\times10^{-12}}{8.854\times10^{-12}\times3.39} = 13.33 \text{ V/m}$$

$$\text{and } E_z = \frac{P_z}{\varepsilon_0\chi_e} = \frac{-200\times10^{-12}}{8.854\times10^{-12}\times3.39} = -6.67 \text{ V/m}$$

Therefore,

$$\vec{E} = (13.33\hat{i} + 10\hat{j} - 6.67\hat{k}) \text{ V/m}$$

Now,

$$\vec{D} = \varepsilon_0\vec{E} + \vec{P}$$

Therefore,

$$\vec{D} = \left[8.854\times(13.33\hat{i} + 10\hat{j} - 6.67\hat{k}) + (400\hat{i} + 300\hat{j} - 200\hat{k})\right]\times10^{-12} \text{ C/m}^2$$

$$\text{or, } \vec{D} = (518.02\hat{i} + 388.54\hat{j} - 259.05\hat{k}) \text{ pC/m}^2$$

---

## 5.6 Classification of Dielectrics

Equations 5.44 through 5.48 are not applicable to dielectric materials in general. These equations are valid for a sub-class of dielectric materials known as LIH materials. LIH dielectrics exhibit the following properties:

1. *Linearity*: A dielectric is said to be linear if $\vec{D}$ varies linearly with $\vec{E}$. For such materials, permittivity is constant and independent of applied electric field. For non-linear dielectrics, $\vec{D}$ and $\vec{E}$ have non-linear relationship.

2. *Isotropy*: A dielectric is said to be isotropic if $\vec{D}$ and $\vec{E}$ are in the same direction. Isotropic dielectrics have same permittivity in all directions. For anisotropic materials, $\vec{D}, \vec{E}$ and $\vec{P}$ are not parallel and hence permittivity varies with direction. Crystalline dielectrics are mostly anisotropic.

3. *Homogeneity*: Dielectric materials for which properties are same at all points within the volume of the material are called *homogeneous*. For inhomogeneous dielectrics, properties such as permittivity vary with space coordinates. A typical example of inhomogeneous dielectric is atmosphere air, as the permittivity of air varies with altitude.

## 5.6.1 Molecular Polarizability of Linear Dielectric

Equation 5.44 relates polarization vector $\vec{P}$ and macroscopic electric field $\vec{E}$ through electric susceptibility ($\chi_e$) in the case of linear dielectrics. The electric field that causes polarization of a molecule of a dielectric is known as molecular field $\vec{E}_{mol}$. Molecular field $\vec{E}_{mol}$ is different from the macroscopic field $\vec{E}$ because the polarization of neighbouring molecules gives rise to an internal field $\vec{E}_{int}$.

Hence,

$$\vec{E}_{mol} = \vec{E} + \vec{E}_{int} \tag{5.49}$$

As shown in Figure 5.12, consider an imaginary sphere that contains the neighbouring molecules. This sphere is much larger in dimension compared to the molecules, but is infinitesimally small in macroscopic scale. The dielectric outside the sphere is replaced by the system of bound charges due to polarization ($\sigma_{sb}$). Then the internal field can be resolved into two components:

$$\vec{E}_{int} = \vec{E}_{near} + \vec{E}_{far} \tag{5.50}$$

where:
  $\vec{E}_{near}$ is the field due to neighbouring molecules, which are located close to the given molecule
  $\vec{E}_{far}$ is the field due to all other molecules, which arises from the bound charge density ($\sigma_{sb}$) on the sphere surface

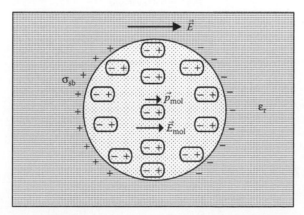

**FIGURE 5.12**
Molecular and macroscopic fields.

$\vec{E}_{far}$ can be expressed in spherical coordinates as follows:

$$\vec{E}_{far} = \frac{1}{4\pi\varepsilon_0} \int_S \sigma_{sb} \frac{-\vec{r}}{r^3} dS = \frac{1}{4\pi\varepsilon_0} \int_0^{2\pi} \int_0^{\pi} (P\cos\theta)\frac{\vec{r}}{r^3}(rd\theta)(r\sin\theta d\phi) \quad (5.51)$$

Considering the Cartesian coordinates, the $x$-component vanishes as it involves the integral $\int_0^{2\pi}\cos\phi d\phi$, which evaluates to zero, and the $y$-component vanishes as it involves the integral $\int_0^{2\pi}\sin\phi d\phi$, which is also zero. Therefore,

$$\vec{E}_{far} = \frac{1}{4\pi\varepsilon_0} \int_0^{2\pi} \int_0^{\pi} P\hat{u}_z \cos^2\theta \sin\theta d\theta d\phi = \frac{\vec{P}}{3\varepsilon_0} \quad (5.52)$$

If the neighbouring molecules are randomly distributed in location, which is the case in most linear dielectrics, then $\vec{E}_{near} = 0$.
   Therefore,

$$\vec{E}_{int} = \vec{P}/(3\varepsilon_0)$$

Then,

$$\vec{E}_{mol} = \vec{E} + \frac{\vec{P}}{3\varepsilon_0} \quad (5.53)$$

Let $N$ be the number of molecules per unit volume and $\vec{p}_{mol}$ be the dipole moment of each molecule, then the polarization vector $\vec{P}$ is given by

$$\vec{P} = N\vec{p}_{mol} \quad (5.54)$$

Molecular dipole moment and molecular field can be related with the help of molecular polarizability ($\alpha_{mol}$) as follows:

$$\vec{p}_{mol} = \varepsilon_0 \alpha_{mol} \vec{E}_{mol} \quad (5.55)$$

Hence, from Equations 5.53 through 5.55,

$$\vec{P} = N\varepsilon_0\alpha_{mol}\vec{E}_{mol} = N\alpha_{mol}\left(\varepsilon_0\vec{E} + \frac{\vec{P}}{3}\right) \quad (5.56)$$

Putting $\vec{P} = \varepsilon_0\chi_e\vec{E}$ in the above equation, we get

$$\chi_e = \frac{3N\alpha_{mol}}{3 - N\alpha_{mol}} \quad (5.57)$$

Using $\varepsilon_r = 1 + \chi_e$ in the above equation, we get

$$\varepsilon_r = 1 + \frac{3N\alpha_{mol}}{3 - N\alpha_{mol}}, \text{ or, } \alpha_{mol} = \frac{3}{N}\frac{(\varepsilon_r - 1)}{(\varepsilon_r + 2)} \quad (5.58)$$

Equation 5.58 is known as Clausius–Mossotti relation.

In this section, we discuss a simple molecular model used to understand the linear behaviour of dielectric, that is, characteristics of a large number of dielectric materials. A detailed treatment, however, will necessitate quantum mechanical consideration.

For a non-polar dielectric, dipole moment induced by the external field is $\vec{p} = \alpha_e \vec{E}$, where, $\alpha_e$ = electronic polarizability of the atom. Then, polarization vector

$$\vec{P} = N\vec{p}$$

where:

$N$ is the number of dipoles per unit volume

But

$$\vec{P} = \varepsilon_0 \chi_e \vec{E}$$

Hence,

$$\varepsilon_0 \chi_e \vec{E} = N\alpha_e \vec{E}, \text{ or, } \chi_e = \frac{N\alpha_e}{\varepsilon_0}, \text{ or, } \varepsilon_r = 1 + \chi_e = 1 + \frac{N\alpha_e}{\varepsilon_0} \qquad (5.59)$$

## 5.6.2 Piezoelectric Materials

The term *piezoelectricity* refers to the fact that when a dielectric is mechanically stressed, then an electric field is produced within the dielectric. As a result of this electric field, measurable quantity of electric potential difference appears across the dielectric sample, which can be measured to find the mechanical strain on the dielectric. This principle is commonly used in piezoelectric strain transducers. The inverse effect also exists, that is, mechanical strain is produced in a dielectric due to the application of electric field. Piezoelectric effect is mostly reversible.

Piezoelectric materials could be natural or synthetic. The most commonly used natural piezoelectric material is quartz ($SiO_2$). But synthetic piezoelectric materials, for example, ceramics and polymers, are more efficient. The piezoelectric materials used in practice are berlinite ($AlPO_4$), gallium orthophosphate ($GaPO_4$), barium titanate ($BaTiO_3$), lead zirconate titanate (PZT: $PbZr_{1-x}Ti_xO_3$), aluminium nitride (AlN) and polyvinylidene fluoride to name a few. In recent years, piezoceramics and piezopolymers are widely used in smart structures. Very recently, breakthrough in single crystal growth technique has enabled the development of high strain and high electric breakdown piezoceramics.

The nature of piezoelectric effect is strongly related to the large number of electric dipoles present in the piezoelectric materials. These dipoles can either be due to ions on crystal lattice sites with asymmetric charge distribution or due to certain molecular groups having asymmetric configurations. When a mechanical stress is applied on a piezoelectric material, the crystalline structure is disturbed and it changes the direction of the polarization vector due to the electric dipoles. If the dipole is due to the ions, then the change in polarization is caused by a re-configuration of ions within the crystalline structure. On the other hand, if the dipole is due to molecular groups, then re-orientation of molecular groups causes the change in the polarization. The electric field developed because of the change in net polarization gives rise to piezoelectric effect.

### 5.6.3 Ferroelectric Materials

A ferroelectric material is a dielectric with at least two discrete stable or metastable states of different non-zero electric polarization under zero applied electric field, referred to as *spontaneous polarization*. For a material to be considered ferroelectric, it must be possible to switch between these states by the application and removal of an applied electric field. In the case of conventional ferroelectrics, the spontaneous polarization is produced by the atomic arrangement of ions in the crystal structure, depending on their positions, and in electronic ferroelectrics the spontaneous polarization is produced by charge ordering of multiple valences. A non-zero spontaneous polarization can be present only in a crystal with a polar space group. Numerical values of spontaneous polarization are customarily given in units of $\mu C/m^2$. All ferroelectric materials are necessarily piezoelectric.

In the ferroelectric state the plot of polarization versus applied electric field shows a hysteresis loop, as shown in Figure 5.13, similar to ferromagnetic materials. A significant feature of ferroelectrics is that the spontaneous polarization can be reversed by an appropriately strong electric field applied in the opposite direction. Dielectric permittivity, which is dependent on the slope of $P–E$ curve, is therefore dependent on the applied field, in the ferroelectric state.

In most ferroelectrics, there is a phase transition from the ferroelectric state, to a non-polar paraelectric phase, with increasing temperature. The phase

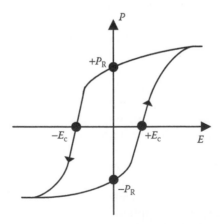

**FIGURE 5.13**
Hysteresis loop in ferroelectric state.

transition temperature is known as *Curie point* ($T_C$) where the ferroelectric material changes from the low temperature polarized state to high temperature unpolarized state. The spontaneous polarization in most ferroelectric crystals is greatest at temperatures well below $T_C$ and decreases to zero at $T_C$. Some ferroelectric materials have no Curie point as these materials melt before the phase transition temperature is reached. The range of phase transition temperature for different ferroelectric materials is very wide varying from very low (50 ~ 100K) to very high (over 1000K) temperature.

While ferroelectricity was discovered in hydrogen-bonded material Rochelle salt ($NaKC_4H_4O_6,4H_2O$), dramatic change in the understanding of this phenomenon came through the discovery of ferroelectricity in the much simpler, non-hydrogen containing, perovskite oxide $BaTiO_3$. $BaTiO_3$ is the typical example of the very large and extensively studied and used perovskite oxide family, which not only includes perovskite compounds but also includes ordered and disordered solid solutions.

Ferroelectrics are used in a variety of applications such as non-volatile memories, capacitors having tunable capacitance, varactors for radio frequency/microwave circuits, electro-optic modulators, high permittivity applications, pyroelectric detectors and so on.

### 5.6.4 Electrets

Electrets are unique, man-made materials that can hold an electrical charge after being polarized in an electric field. It is a piece of dielectric material that has been specially prepared to possess an overall fixed dipolarity. It is the electrical analogy of a permanent magnet. Instead of opposite magnetic poles, the electret has two electrical poles of trapped opposite polarity charges. Therefore, a fixed *static* potential exists between the two opposite poles of the dipolar electret.

Electrets can be prepared from different dielectric materials depending on their structures and properties. The very first electrets were made of carnauba wax (Brazilian palm gum) and its mixtures with rosin, beeswax, ethylcellulose and other components. When a polar dielectric is placed in an electric field, the applied field causes the dipoles to be oriented in such a way that the dipole moments are directed parallel to the applied field. The degree of alignment achieved depends on the freedom with which a dipole can turn around its axis. In the case of a material such as carnauba wax, this freedom is greater when the wax is in molten state, but is almost zero when the wax is in the solid state. Hence, it is possible to melt the wax, and keep it in an electric field so that the dipoles align themselves. Afterwards, while still under the influence of the electric field, the wax is allowed to cool and solidify. Then the molecular dipoles get set in the aligned position, and the wax piece becomes an electret. Electrets prepared in this way are known as *thermoelectrets*. It has been reported in published literature that effective surface charge density of the order of 4 ~ 6 nC/m$^2$ has been preserved practically unchanged in carnauba wax-based electrets for more than 35 years.

However, there are several other methods of preparation of electrets: (1) photoelectrets are produced using light as heat source; (2) radioelectrets are prepared by exposing the dielectric to a beam of charged particles; (3) coronaelectret is produced by placing the dielectric in a corona discharge field and is nowadays preferred for industrial production of electrets and (4) mechanoelectrets are formed by the mechanical compression of dielectric between heated plates. Many modern electrets (such as teflon or polypropylene electrets) have only space or surface charge but no dipole polarization.

Electret materials include ceramics, non-polar/polar semi-crystalline and amorphous synthetic polymers, biopolymers and ferroelectric ceramic/polymer composites. The most common applications of electret are microphones, electroacoustic devices, infrared detection and photocopying machines. Electrets are also being useful as novel devices in biomedical and high-energy charge storage applications.

## 5.7 Frequency Dependence of Polarizabilities

If the behaviour of dielectric polarizability is studied under alternating field, important distinctions between various polarizabilities emerge. Typical dependence of polarizabilities on frequency is depicted in Figure 5.14 over a wide range starting from static field to frequencies above the ultraviolet region.

It may be seen that between $f = 0$ to $f = f_s$, where $f_s$ is typically around 1 kHz, the polarizability gradually increases as frequency is decreased. Such increasing polarizability at lower frequencies arises because of interfacial polarization

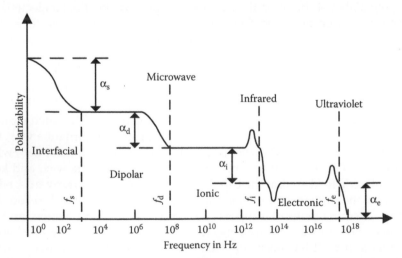

**FIGURE 5.14**
Frequency dependence of polarizabilities.

mechanism. From $f = f_s$ to $f = f_d$, polarizability remains more or less constant, where $f_d$ is typically in microwave region. In this frequency span, dipolar polarization mechanism is predominant. For $f > f_d$ the polarizability decreases by a significant amount and the amount of decrease corresponds to the dipolar polarizability. The reason for disappearance of dipolar polarizability for $f > f_d$ is that the field at these frequencies oscillate at a rapid rate, which the dipoles are unable to follow, and hence the dipolar polarizability vanishes. Further, the polarizability remains nearly constant in the frequency range $f = f_s$ to $f = f_i$, where $f_i$ is in infrared region. The drop in polarizability for $f > f_i$ is due to the absence of ionic polarizability at such higher frequencies, because ions being heavy are unable to follow the very rapidly varying AC field for $f > f_i$. As a result for $f > f_i$ only electronic polarizability is active, because electrons being lighter are still able to follow the oscillating field at these frequencies. At extremely high frequencies for $f > f_e$, where $f_e$ is in ultraviolet region, even the electronic polarizability vanishes, as the electrons are also unable to follow such extremely rapid oscillating field.

Typically, the frequencies $f_e, f_i, f_d$ and $f_s$ characterize electronic, ionic, dipolar and interfacial polarizabilities, respectively. These frequencies depend on the dielectric materials and vary from dielectric to dielectric and also on the condition of the dielectric. Various polarizabilities can be determined by measuring dielectric properties at various frequencies of appropriate value. In fact this principle is the foundation of frequency domain spectroscopy, which is a major technique used for condition monitoring of high-voltage insulation system.

## 5.8 Mass-Spring Model of Fields in Dielectrics

When an electric field is applied to a dielectric material, the nucleus is not much accelerated as it is heavy. But the electrons move in accordance with the applied field. Such a system can be modelled by the classical mass-spring model of an atom, as shown in Figure 5.15, in which the electrons are bound to nucleus like masses on springs. Let the position of the electron be denoted by $x(t)$ and the forcing field in this direction be $\vec{E} = \hat{u}_x E_m e^{j\omega t}$.

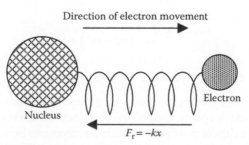

Direction of electron movement

Electron

Nucleus

$F_r = -kx$

**FIGURE 5.15**
Mass-spring model of an atom.

Then the force acting on the electron in relation to its forced movement will be given by

$$F = -c\dot{x} - kx + E_m e^{j\omega t} \tag{5.60}$$

where:

$k$ = spring constant

$c$ = damping coefficient. Spring constant and damping coefficient are dependent on material

From Newton's law of motion,

$$F = m_e \ddot{x} \tag{5.61}$$

where:

$m_e$ = mass of electron

From Equations 5.60 and 5.61 the equation of forced damped harmonic oscillation can be written as follows:

$$m_e \ddot{x} + c\dot{x} + kx = q_e E_m e^{j\omega t} \tag{5.62}$$

where:

$q_e$ = charge of electron

Equation 5.62 can be rewritten as follows:

$$\ddot{x} + 2\delta \dot{x} + \omega_n^2 x = \frac{q_e}{m_e} E_m e^{j\omega t} \tag{5.63}$$

where:

$$\delta = \frac{c}{2m_e} \text{ and } \omega_n^2 = \frac{k}{m_e}$$

To obtain the steady-state solution to Equation 5.63, let $x_p(t) = x_0 e^{j\omega t}$. Then from Equation 5.63

$$\left(-\omega^2 + j2\delta\omega + \omega_n^2\right) x_0 e^{j\omega t} = \frac{q_e}{m_e} E_m e^{j\omega t} \tag{5.64}$$

$$\text{or, } x_0 = \frac{q_e E_m}{m_e} \frac{1}{\left(\omega_n^2 - \omega^2\right) + j2\delta\omega} \tag{5.65}$$

### 5.8.1 Dielectric Permittivity from Mass-Spring Model

Permittivity of a linear dielectric material comprising many electrons can be derived from the mass-spring model. With respect to the rest position ($x_p = 0$) of the electron, the dipole moment of each electron can be written as follows:

$$\vec{p}(t) = \hat{u}_x q_e x_p(t) \tag{5.66}$$

For many electrons undergoing similar motion, the polarization vector can be written as follows:

$$\vec{P}(t) = N\vec{p}(t) = \hat{u}_x N q_e x_p(t) \tag{5.67}$$

where:

$N$ is the number of dipoles per unit volume

But, according to Equation 5.44,

$$\vec{P}(t) = \varepsilon_0 \chi_e \vec{E}(t) \tag{5.68}$$

Then from Equations 5.67 and 5.68, $\hat{u}_x N q_e x_p(t) = \varepsilon_0 \chi_e \vec{E}(t)$

$$\text{or, } \hat{u}_x N q_e \frac{q_e E_m}{m_e} \frac{1}{\left(\omega_n^2 - \omega^2\right) + j2\delta\omega} e^{j\omega t} = \hat{u}_x \varepsilon_0 \chi_e E_m e^{j\omega t}$$

$$\text{or, } \chi_e = \frac{N q_e^2}{m_e \varepsilon_0} \frac{1}{\left(\omega_n^2 - \omega^2\right) + j2\delta\omega} \tag{5.69}$$

Because $\varepsilon_r = 1 + \chi_e$, from Equation 5.69

$$\varepsilon_r = 1 + \frac{N q_e^2}{m_e \varepsilon_0} \frac{1}{\left(\omega_n^2 - \omega^2\right) + j2\delta\omega} \tag{5.70}$$

Although the above equation is obtained from a simple classical model, but it helps to understand several significant dielectric properties as follows:

1. Power loss due to the imaginary part of $\varepsilon_r$
2. Frequency dependence of permittivity
3. Dependence of permittivity on applied electric field
4. Anisotropy of dielectric materials. If the spring constants and damping coefficients are different in different directions, then permittivity will also be different in different directions.

## 5.9 Dielectric Anisotropy

In isotropic dielectric, D-field and E-field are in the same direction and hence $\vec{D}$ and $\vec{E}$ are related by scalar permittivity, that is, $\vec{D} = \varepsilon\vec{E}$. But there are dielectric materials for which the stiffness of the spring, as shown in Figure 5.15, is different in different directions. It happens because of the fact that the wells of the electrostatic potential in which the charges sit are not symmetric and therefore the response of the charges is not necessarily in

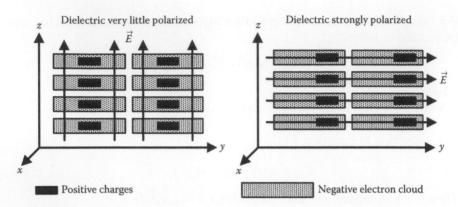

**FIGURE 5.16**
Polarization in anisotropic dielectric.

the direction of the forcing E-field. In other words, the charge distribution within the material responds differently to the three components of the forcing field giving rise to different permittivities in different directions.

To explain this fact consider the material shown in Figure 5.16, which is made up of molecules that can be easily polarized by E-field acting in the $y$-direction but do not respond much to E-fields in the $x$- and $z$-directions. In such a material, $\vec{D}$ and $\vec{E}$ are not necessarily parallel to each other.

### 5.9.1 Tensor of Rank 2

If $\vec{D}$ and $\vec{E}$ are not parallel to each other, then multiplying $\vec{E}$ by a scalar will not yield $\vec{D}$. This is because multiplying any vector by a scalar produces a new vector in the same direction having different magnitude. Multiplying any given vector by another vector with cross product will create a new vector but in a fixed direction orthogonal to both the vectors. If it is needed to change both the magnitude and direction of the given vector other than 90°, then scalar multiplication, vector products such as dot or cross products with another vector will not do. A different kind of mathematical entity has to be used for this purpose. In this case the inner product of the given vector with a *dyad* will be required as explained hereafter.

Consider, two vectors $\vec{A} = A_1\hat{i} + A_2\hat{j} + A_3\hat{k}$ and $\vec{B} = B_1\hat{i} + B_2\hat{j} + B_3\hat{k}$. Then the dyad product of these two vectors is simply $\vec{A}\vec{B}$. It is neither a dot nor a cross product. It may be written as

$$\vec{A}\vec{B} = A_1B_1\hat{\hat{i}\hat{i}} + A_1B_2\hat{\hat{i}\hat{j}} + A_1B_3\hat{\hat{i}\hat{k}} + A_2B_1\hat{\hat{j}\hat{i}} + \cdots \tag{5.71}$$

where:
$\hat{i}, \hat{j}$ and $\hat{k}$ are unit vectors in the usual sense
$\hat{\hat{i}\hat{i}}, \hat{\hat{i}\hat{j}}, \hat{\hat{i}\hat{k}}, \hat{\hat{j}\hat{i}}$ and so on are unit dyads

Note that by setting $A_1B_1 = \alpha_{11}$, $A_1B_2 = \alpha_{12}$ and so on, the dyad $\vec{A}\vec{B}$ can be rewritten as

$$\vec{A}\vec{B} = \alpha_{11}\hat{i}\hat{i} + \alpha_{12}\hat{i}\hat{j} + \alpha_{13}\hat{i}\hat{k} + \alpha_{21}\hat{j}\hat{i} + \cdots \qquad (5.72)$$

and the scalar components $\alpha_{ij}$ can be arranged in the familiar configuration of a $3 \times 3$ matrix:

$$\begin{bmatrix} \alpha_{11} & \alpha_{12} & \alpha_{13} \\ \alpha_{21} & \alpha_{22} & \alpha_{23} \\ \alpha_{31} & \alpha_{32} & \alpha_{33} \end{bmatrix} \qquad (5.73)$$

Inner product of the dyad $\vec{A}\vec{B}$ and a vector $\vec{E}$ yields

$$\vec{A}\vec{B}\cdot\vec{E} = \vec{A}(\vec{B}\cdot\vec{E}) = \vec{A}\lambda = \lambda\vec{A} \qquad (5.74)$$

Equation 5.74 shows that the resultant vector magnitude is determined by the scalar $\lambda$ and its direction is determined by $\vec{A}$, which is not same as that of $\vec{E}$.

In mathematical terminology, a dyad is known as tensor of rank 2. As shown by Equations 5.72 and 5.73, a tensor of rank 2 is defined as a system that has a magnitude and two directions $(\vec{A}, \vec{B})$ associated with it. It has nine components. To generalize the matter more, a scalar is a tensor of rank 0 (magnitude only and one component) and a vector is a tensor of rank 1 (magnitude, one direction and three components).

### 5.9.2 Permittivity Tensor

Consider the polarization characteristics of an anisotropic dielectric, as shown in Figure 5.17. It shows that if a forcing E-field is applied in the

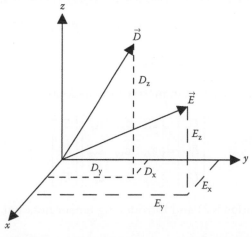

**FIGURE 5.17**
Polarization characteristics of an anisotropic dielectric.

direction shown, then the D-field is produced in a different direction. It may be explained in the following manner. The forcing field $\vec{E}$ has three components $E_x$, $E_y$ and $E_z$, respectively. Let the contribution to $\vec{D}$ from $E_x$ be $\vec{D}_{Ex}$. Then $\vec{D}_{Ex}$ can be written in terms of its three components as follows:

$$\vec{D}_{Ex} = D_{xx}\hat{i} + D_{xy}\hat{j} + D_{xz}\hat{k} \tag{5.75}$$

where:
   $D_{xx}$, $D_{xy}$ and $D_{xz}$ can be obtained by multiplying $E_x$ with three different scalars

This is because of the fact that if an electric field $E_x$ is applied in the $x$-direction, then $\chi_{xz}(=\varepsilon_{xz}-1)$ gives the component of the additional polarization vector in the $z$-direction according to the relationship $\vec{P} = \varepsilon_0 \chi_e \vec{E}$ and similarly for the other components. Hence, Equation 5.75 can be rewritten as

$$\vec{D}_{Ex} = D_{xx}\hat{i} + D_{xy}\hat{j} + D_{xz}\hat{k} = \varepsilon_{xx}E_x\hat{i} + \varepsilon_{xy}E_x\hat{j} + \varepsilon_{xz}E_x\hat{k} \tag{5.76}$$

Similarly, the contribution to $\vec{D}$ from $E_y$ is $\vec{D}_{Ey}$, which can be written as

$$\vec{D}_{Ey} = D_{yx}\hat{i} + D_{yy}\hat{j} + D_{yz}\hat{k} = \varepsilon_{yx}E_y\hat{i} + \varepsilon_{yy}E_y\hat{j} + \varepsilon_{yz}E_y\hat{k} \tag{5.77}$$

and the contribution to $\vec{D}$ from $E_z$ is $\vec{D}_{Ez}$, which can be written as

$$\vec{D}_{Ez} = D_{zx}\hat{i} + D_{zy}\hat{j} + D_{zz}\hat{k} = \varepsilon_{zx}E_z\hat{i} + \varepsilon_{zy}E_z\hat{j} + \varepsilon_{zz}E_z\hat{k} \tag{5.78}$$

Therefore, the net $\vec{D}$ due to $\vec{E}$ will be

$$\vec{D} = \vec{D}_{Ex} + \vec{D}_{Ey} + \vec{D}_{Ez} = D_x\hat{i} + D_y\hat{j} + D_z\hat{k}$$

$$= \left(D_{xx} + D_{xy} + D_{xz}\right)\hat{i} + \left(D_{yx} + D_{yy} + D_{yz}\right)\hat{j} + \left(D_{zx} + D_{zy} + D_{zz}\right)\hat{k}$$

$$= \left(\varepsilon_{xx}E_x + \varepsilon_{xy}E_y + \varepsilon_{xz}E_z\right)\hat{i} + \left(\varepsilon_{yx}E_x + \varepsilon_{yy}E_y + \varepsilon_{yz}E_z\right)\hat{j} \tag{5.79}$$

$$+ \left(\varepsilon_{zx}E_x + \varepsilon_{zy}E_y + \varepsilon_{zz}E_z\right)\hat{k}$$

Equation 5.79 can be written in matrix form as follows:

$$\begin{bmatrix} D_x \\ D_y \\ D_z \end{bmatrix} = \begin{bmatrix} \varepsilon_{xx} & \varepsilon_{xy} & \varepsilon_{xz} \\ \varepsilon_{yx} & \varepsilon_{yy} & \varepsilon_{yz} \\ \varepsilon_{zx} & \varepsilon_{zy} & \varepsilon_{zz} \end{bmatrix} \begin{bmatrix} E_x \\ E_y \\ E_z \end{bmatrix} \tag{5.80}$$

Referring to Equation 5.73 and introducing tensor notation, Equation 5.80 is written as

$$\vec{D} = \varepsilon\vec{E} \tag{5.81}$$

where:

$\underline{\varepsilon}$ is a tensor of rank 2 and is known as permittivity tensor of anisotropic dielectric

From Equations 5.74 and 5.81, it can be seen that the product of the tensor $\underline{\varepsilon}$ and the vector $\vec{E}$ produces a new vector $\vec{D}$ whose direction is not the same as that of E.

For an anisotropic dielectric electric susceptibility is also a tensor. Hence, Equation 5.44 is modified for an anisotropic dielectric as

$$\vec{P} = \varepsilon_0 \underline{\chi_e} \vec{E} \tag{5.82}$$

In matrix form Equation 5.82 is written as

$$\begin{bmatrix} P_x \\ P_y \\ P_z \end{bmatrix} = \varepsilon_0 \begin{bmatrix} \chi_{exx} & \chi_{exy} & \chi_{exz} \\ \chi_{eyx} & \chi_{eyy} & \chi_{eyz} \\ \chi_{ezx} & \chi_{ezy} & \chi_{ezz} \end{bmatrix} \begin{bmatrix} E_x \\ E_y \\ E_z \end{bmatrix} \tag{5.83}$$

For a lossless dielectric, the permittivity matrix, as given in Equation 5.80, is symmetric, that is, $\varepsilon_{ij} = \varepsilon_{ji}$. Then Equation 5.80 can be rewritten as

$$\begin{bmatrix} D_x \\ D_y \\ D_z \end{bmatrix} = \begin{bmatrix} \varepsilon_{xx} & \varepsilon_{xy} & \varepsilon_{xz} \\ \varepsilon_{xy} & \varepsilon_{yy} & \varepsilon_{yz} \\ \varepsilon_{xz} & \varepsilon_{yz} & \varepsilon_{zz} \end{bmatrix} \begin{bmatrix} E_x \\ E_y \\ E_z \end{bmatrix} \tag{5.84}$$

Any symmetric matrix can be diagonalized by a suitable choice for the orientation of the coordinate axes $(x,y,z)$. The choice of coordinate axes that results in a diagonal permittivity matrix is called the principal axes of the material. In the principal axes system, Equation 5.84 can be written as

$$\begin{bmatrix} D_x \\ D_y \\ D_z \end{bmatrix} = \begin{bmatrix} \varepsilon_x & 0 & 0 \\ 0 & \varepsilon_y & 0 \\ 0 & 0 & \varepsilon_z \end{bmatrix} \begin{bmatrix} E_x \\ E_y \\ E_z \end{bmatrix} \tag{5.85}$$

If the diagonal entries in Equation 5.85 are all different, that is, $\varepsilon_x \neq \varepsilon_y \neq \varepsilon_z$, then the material is called *biaxial*. If any two are equal, for example, $\varepsilon_x \neq \varepsilon_y = \varepsilon_z$, then the dielectric medium is called *uniaxial*.

In power engineering applications, in most of the cases LIH dielectric materials are used. For such dielectric media, all formulae derived for free space can be applied simply by replacing $\varepsilon_0$ with $\varepsilon_0 \varepsilon_r$. For example, Coulomb's law of Equation 1.1 becomes

$$\vec{F} = \frac{Q_1 Q_2}{4\pi\varepsilon_0\varepsilon_r |\vec{r}|^2} \hat{u}_r \tag{5.86}$$

## Objective Type Questions

1. At frequencies below 1 kHz, polarizability increases as frequency decreases. This is due to which polarization mechanism?
   a. Electronic
   b. Ionic
   c. Dipolar
   d. Interfacial

2. In the frequency range from 1 kHz to microwave region, which polarization mechanism is predominant?
   a. Electronic
   b. Ionic
   c. Dipolar
   d. Interfacial

3. Above infrared frequency range, which polarization mechanism remains active?
   a. Electronic
   b. Ionic
   c. Dipolar
   d. None of these

4. Above ultraviolet frequency range, which polarization mechanism remains active?
   a. Electronic
   b. Ionic
   c. Dipolar
   d. None of these

5. Which polarization mechanism involves power loss?
   a. Electronic
   b. Ionic
   c. Dipolar
   d. All of these

6. Which polarization occurs due to dipoles induced by applied electric field?
   a. Dipolar
   b. Electronic
   c. Ionic
   d. Both (b) and (c)

7. Which polarization takes place in the case of a non-polar dielectric?
   a. Dipolar
   b. Electronic
   c. Ionic
   d. Both (b) and (c)

8. Which polarization mechanism is slowest?
   a. Interfacial
   b. Dipolar
   c. Ionic
   d. Electronic

9. For a polar dielectric, which type of polarizability contributes maximum to the net polarizability?
   a. Electronic
   b. Ionic
   c. Dipolar
   d. None of these

10. Which type of polarization is present in any dielectric?
   a. Electronic
   b. Ionic
   c. Dipolar
   d. Interfacial

11. Electric field due to a polarized dielectric is equivalent to the field due to
   a. Bound surface charge density
   b. Bound volume charge density
   c. Both (a) and (b)
   d. None of these

12. A uniformly polarized dielectric sphere has divergent polarization directed radially outwards. Then which bound charge density will appear due to polarization?
   a. Bound volume charge density
   b. Bound surface charge density
   c. Both (a) and (b)
   d. None of these

13. Bound volume charge density due to polarization ($\rho_{vb}$) and polarization vector ($\vec{P}$) are related as
   a. $\rho_{vb} = \vec{\nabla} \cdot \vec{P}$
   b. $\rho_{vb} = -\vec{\nabla} \cdot \vec{P}$

c.  $\rho_{vb} = |\vec{P}|$

d.  $\rho_{vb} = -|\vec{P}|$

14. For a linear, isotropic dielectric, relative permittivity ($\varepsilon_r$) and electric susceptibility ($\chi_e$) are related as

a.  $\varepsilon_r = 1 - \chi_e$

b.  $\varepsilon_r = \chi_e - 1$

c.  $\varepsilon_r = 1 + \chi_e$

d.  $\varepsilon_r = \varepsilon_0 \chi_e$

15. For a linear, isotropic dielectric, polarization vector ($\vec{P}$), electric susceptibility ($\chi_e$) and applied electric field ($\vec{E}$) are related as

a.  $\vec{P} = \varepsilon_0 \chi_e \vec{E}$

b.  $\vec{P} = 1 + \varepsilon_0 \chi_e \vec{E}$

c.  $\vec{P} = -\varepsilon_0 \chi_e \vec{E}$

d.  $\vec{P} = 1 - \varepsilon_0 \chi_e \vec{E}$

16. Bound surface charge density due to polarization ($\sigma_{sb}$) and polarization vector ($\vec{P}$) are related as

a.  $\sigma_{sb} = \vec{\nabla} \cdot \vec{P}$

b.  $\sigma_{sb} = -\vec{\nabla} \cdot \vec{P}$

c.  $\sigma_{sb} = \vec{P} \cdot \vec{u}_n$

d.  $\sigma_{sb} = -\vec{P} \cdot \vec{u}_n$

17. Which one of the following is a dimensionless quantity?

a.  Permittivity of free space ($\varepsilon_0$)

b.  Electric susceptibility ($\chi_e$)

c.  Dielectric polarizability ($\alpha$)

d.  All of these

18. Permittivity is a scalar for which type of dielectric?

a.  Homogeneous

b.  Isotropic

c.  Anisotropic

d.  None of these

19. For a linear, isotropic, homogeneous dielectric, which one of the following is true?

a.  $\varepsilon_r$ varies linearly with $\vec{E}$

b.  $\varepsilon_r$ varies linearly with space coordinate

c.  $\varepsilon_r$ varies with direction

d.  $\vec{D}$ varies linearly with $\vec{E}$

20. For an isotropic dielectric, which one of the following is true?
    a. $\vec{D} \parallel \vec{E}$
    b. $\vec{E} \parallel \vec{P}$
    c. $\vec{D} \parallel \vec{P}$
    d. All of these

21. The lines of D-field begin and end on
    a. Free charges
    b. Bound charges
    c. Both (a) and (b)
    d. None of these

22. The lines of E-field begin and end on
    a. Free charges
    b. Bound charges
    c. Both (a) and (b)
    d. None of these

23. For an anisotropic dielectric permittivity is a
    a. Scalar
    b. Vector
    c. Tensor of rank 2
    d. Tensor of rank 3

24. Permittivity of an anisotropic dielectric has how many components?
    a. 1
    b. 3
    c. 6
    d. 9

25. For which type of dielectric electric polarization takes place even without any applied field?
    a. Linear
    b. Anisotropic
    c. Ferroelectric
    d. All of these

**Answers:** 1) d; 2) c; 3) a; 4) d; 5) c; 6) d; 7) d; 8) a; 9) c; 10) a; 11) c; 12) c; 13) b; 14) c; 15) a; 16) c; 17) b; 18) b; 19) d; 20) d; 21) a; 22) c; 23) c; 24) d; 25) c

# 6

## Electrostatic Boundary Conditions

**ABSTRACT**  In electric field analysis, it is important to know the changes in field quantities as one crosses a boundary between a conductor and a dielectric or between two different dielectric media. In practical configurations comprising multi-dielectric media, this is of paramount importance, as quantities defining the field on the two sides of the boundary between two dielectrics often undergo significant changes. Boundary conditions are fundamental equations involving electric field quantities, which describe the changes in electrostatic field specifically in relation to boundary surfaces. In electrostatic field, another major aspect is the presence of charges on the boundaries, either free charges, as in the case of conductor boundaries, or surface charges, as in the case of dielectric–dielectric boundaries. These charges play the most significant role in the changes that the field quantities undergo at the boundary.

### 6.1 Introduction

In real life, any conductor is always surrounded by at least one dielectric. It should be kept in mind that air is also a dielectric, which is present almost everywhere. Therefore, even if there is no solid or liquid or any other gaseous dielectric around a conductor, it will in all probability be surrounded by air. Therefore, there will be boundaries between a conductor and a dielectric in practice. Moreover, except for very few cases like single-core cables having only one dielectric or transmission line conductors surrounded by air at midspan, dielectric materials are arranged either in series or in parallel between two conductors having a particular potential difference. For example, if one takes the case of an outdoor porcelain insulator, it may appear that there is only one dielectric, that is, porcelain, involved. But the porcelain insulator will be surrounded by air and hence it becomes a two dielectric configuration. As a result there will be boundaries between two different dielectric media in practical configurations. Changes in some electric field quantities in direction and/or magnitude occur at such boundaries. The equations that describe such field behaviours by relating electric field quantities on two sides of a boundary surface are known as *boundary conditions*. The transition of electric field from one medium to another medium through a boundary surface is governed by the boundary conditions.

## 6.2 Boundary Conditions between a Perfect Conductor and a Dielectric

A perfect conductor is defined as a material within which the charges are able to move freely. In electrostatics, it is considered that the charges have attained the equilibrium positions and are fixed in space. Theoretically, consider that the charges are initially distributed uniformly throughout the volume of a perfect conductor. Such distributed charges should be of same polarity within a conductor of one particular value of electric potential, because if there are charges of opposite polarity within the volume of the conductor, then such charges will immediately recombine with each other as they are free to move. Hence, the charges of same polarity that are present in the volume of the conductor will exert repulsive forces on each other. Because the charges are able to move without any hindrance, the charges will disperse in such a direction, so that the distance between the charges will increase. In the process all the charges will arrive at the surface of the conductor. But the conductor being surrounded by a dielectric, the charges are unable to move further and the charges will be fixed in space on the surface of the conductor. Consequently, any Gaussian surface within a perfect conductor will enclose zero charge and hence, electric field within a perfect conductor will be zero.

### 6.2.1 Boundary Condition for Normal Component of Electric Flux Density

Consider a coin-like closed volume of cylindrical shape, as shown in Figure 6.1. Such a volume is often termed as a *Gaussian pillbox*. The pillbox has a finite surface area $\Delta A$ and an infinitesimally small height $\delta$, such that half of the pillbox is within the conductor and the other half is within the

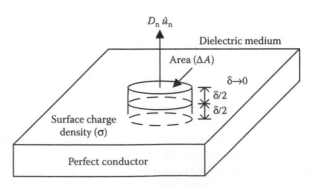

**FIGURE 6.1**
Pertaining to boundary condition for electric flux density at conductor–dielectric boundary.

dielectric medium as shown. The top and bottom surfaces of the pillbox are parallel to the conductor–dielectric interface.

Application of Gauss's law to this pillbox yields

$$\int_{\text{Surface of pillbox}} \vec{D} \cdot d\vec{s} = \text{Charge enclosed by the pillbox} \tag{6.1}$$

For the right-hand side (RHS) of Equation 6.1 the following need to be considered: (1) the volume of the pillbox is infinitesimally small and (2) half of the volume of the pillbox within the perfect conductor does not contain any volume charge density, as the charges reside on the surface of the conductor and the other half of the volume of the pillbox within the dielectric also does not contain any volume charge density, if ideal dielectric is assumed. But the surface charge density on the conductor has a finite value and the area of the pillbox is also finite.

Hence, the RHS of Equation 6.1

$$= \sigma \Delta A \tag{6.2}$$

where:
$\sigma$ is the surface charge density on the conductor

The integral on the left-hand side (LHS) of Equation 6.1 could be expanded as follows:

$$\int_{\text{Surface of pillbox}} \vec{D} \cdot d\vec{s} = \int_{\text{Top surface}} \vec{D} \cdot d\vec{s} + \int_{\text{Bottom surface}} \vec{D} \cdot d\vec{s} + \int_{\text{Wall surface}} \vec{D} \cdot d\vec{s} \tag{6.3}$$

As the height of the pillbox is infinitesimally small, the integral over the wall surfaces is negligible. The integral over the bottom surface is also zero as the field within the perfect conductor is zero. Hence, Equation 6.3 could be rewritten as

$$\int_{\text{Surface of pillbox}} \vec{D} \cdot d\vec{s} = \int_{\text{Top surface}} \vec{D} \cdot d\vec{s} = D_n \Delta A \tag{6.4}$$

where:
$D_n$ is the normal component of electric flux density

Hence, from Equations 6.2 and 6.4,

$$D_n \Delta A = \sigma \Delta A, \text{ or, } D_n = \sigma \tag{6.5}$$

Bringing in the unit normal vector, Equation 6.5 could be written as

$$\hat{u}_n \cdot \vec{D} = \sigma \tag{6.6}$$

## 6.2.2 Boundary Condition for Tangential Component of Electric Field Intensity

Consider an infinitesimally small closed rectangular loop *abcda* as shown in Figure 6.2, of which the length segments *ab* and *cd* are parallel to the conductor–dielectric boundary and the length segments *bc* and *da* are normal to the boundary. The length of the loop parallel to the boundary is $\Delta l$ and is finite, but the length of the loop normal to the boundary is $\delta$, which is negligibly small.

As *E*-field is conservative in nature, the integral of $\vec{E} \cdot d\vec{l}$ over the loop contour *abcda* will be zero, that is,

$$\oint_{abcda} \vec{E} \cdot d\vec{l} = \int_a^b \vec{E} \cdot d\vec{l} + \int_b^c \vec{E} \cdot d\vec{l} + \int_c^d \vec{E} \cdot d\vec{l} + \int_d^a \vec{E} \cdot d\vec{l} = 0 \tag{6.7}$$

As stated, the lengths *bc* and *da* are negligibly small. Hence,

$$\int_b^c \vec{E} \cdot d\vec{l} = \int_d^a \vec{E} \cdot d\vec{l} \approx 0$$

Again, the field within the perfect conductor is zero. Hence,

$$\int_c^d \vec{E} \cdot d\vec{l} = 0$$

Therefore, from Equation 6.7, $\int_a^b \vec{E} \cdot d\vec{l} = 0$

The length *ab* is $\Delta l$, which is small but finite. Then considering $\vec{E}$ to be constant over the small length $\Delta l$, it may be written that

$$E_t \Delta l = 0, \text{ or, } E_t = 0 \tag{6.8}$$

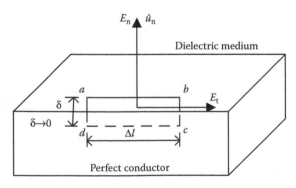

**FIGURE 6.2**
Pertaining to boundary condition for electric field intensity at conductor–dielectric boundary.

where:

$E_t$ is the component of electric field intensity along the length $ab$, which is the tangential component of electric field intensity

Bringing in the unit normal vector, Equation 6.8 could be written as

$$\hat{u}_n \times \vec{E} = 0 \qquad (6.9)$$

### 6.2.3 Field Just Off the Conductor Surface

From Equation 6.8, on the conductor surface $E_t = 0$. In other words, the electric field acts in the direction normal to the conductor surface. Then, from Equation 6.5, the normal component of electric field intensity could be written as

$$E_n = \frac{\sigma}{\varepsilon} = \frac{\sigma}{\varepsilon_r \varepsilon_0} \qquad (6.10)$$

As discussed in Section 6.2.1, the electric field intensity as given by Equation 6.10 is for the top surface of the pillbox. The height of the pillbox is infinitesimally small, but is not zero. Hence, the top surface of the pillbox, as shown in Figure 6.1, is not exactly on the conductor surface. As a result, the value of electric field intensity as obtained from Equation 6.10 is stated to be electric field intensity within the dielectric medium just off the conductor surface.

**PROBLEM 6.1**

A charged conductor is surrounded by air. Calculate the maximum charge density that the conductor can hold at standard temperature and pressure (STP).

*Solution:*

The conductor can hold the maximum charge density for which the electric field intensity just off the surface is equal to the breakdown strength of air at STP. This is due to the fact that any further increase in the value of charge density on the conductor surface will cause breakdown of air and the charges will be drained from the conductor.

Breakdown strength of air at STP = 30 kV/cm = $30 \times 10^3 \times 10^2$ V/m = $3 \times 10^6$ V/m.

Let the maximum charge density that the conductor can hold be $\sigma_m$.

Now, $\varepsilon_r$ for air = 1

Therefore, $\sigma_m / \varepsilon_0 = 3 \times 10^6$

or, $\sigma_m = 3 \times 10^6 \times 8.854 \times 10^{-12} = 26.5$ μC/m².

## 6.3 Boundary Conditions between Two Different Dielectric Media

In the case of practical configurations comprising multiple homogeneous dielectric media, there could be many dielectric–dielectric boundaries. On such dielectric–dielectric boundaries, there could be charges depending on the dissimilarities of the dielectric media that are present on the two sides of the boundaries. These surface charges will serve as source of electric field acting in opposite directions on the two sides of the boundary. Consequently, electric field quantities get changed in direction as well as magnitude on the two sides of the boundary.

### 6.3.1 Boundary Condition for Normal Component of Electric Flux Density

For the boundary between two dielectric media having permittivities $\varepsilon_1$ and $\varepsilon_2$, consider a cylindrical coin-like Gaussian pillbox, as shown in Figure 6.3. The height of the pillbox is infinitesimally small. As in Equation 6.1 application of Gauss's law to this pillbox yields

$$\int_{\text{Surface of pillbox}} \vec{D} \cdot \vec{ds} = \text{Charge enclosed by the pillbox}$$

Subdividing the integral on the LHS into contributions from the top, bottom and wall surfaces of the pillbox and subdividing the wall surface into two halves, one in dielectric 1 and the other in dielectric 2,

$$\int_{\text{Surface of pillbox}} \vec{D} \cdot \vec{ds} = \int_{\text{Top surface dielectric 2}} \vec{D}_2 \cdot \vec{ds} + \int_{\text{Bottom surface dielectric 1}} \vec{D}_1 \cdot \vec{ds}$$

$$+ \int_{\text{Wall surface dielectric 1}} \vec{D}_1 \cdot \vec{ds} + \int_{\text{Wall surface dielectric 2}} \vec{D}_2 \cdot \vec{ds}$$

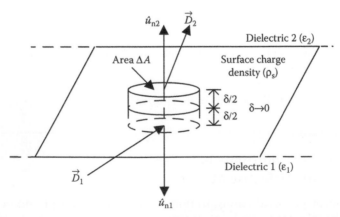

**FIGURE 6.3**
Pertaining to boundary condition for electric flux density at dielectric–dielectric boundary.

As the height of the pillbox is vanishingly small, the contribution of the integral over the wall surfaces will be negligible. Thus

$$\int\limits_{\text{Surface of pillbox}} \vec{D} \cdot d\vec{s} = \int\limits_{\text{Top surface dielectric 2}} \vec{D}_2 \cdot d\vec{s} + \int\limits_{\text{Bottom surface dielectric 1}} \vec{D}_1 \cdot d\vec{s} \quad (6.11)$$

As the top and bottom surfaces are small, $\vec{D}_1$ and $\vec{D}_2$ could be assumed to be constant over the bottom and top surfaces, respectively. Hence,

$$\int\limits_{\text{Top surface dielectric 2}} \vec{D}_2 \cdot d\vec{s} = \vec{D}_2 \cdot \Delta\vec{A} = (\vec{D}_2 \cdot \hat{u}_{n2})\Delta A \quad (6.12)$$

and

$$\int\limits_{\text{Top surface dielectric 1}} \vec{D}_1 \cdot d\vec{s} = \vec{D}_1 \cdot \Delta\vec{A} = (\vec{D}_1 \cdot \hat{u}_{n1})\Delta A \quad (6.13)$$

Now considering the unit normal to be directed into dielectric 2, $\hat{u}_n = \hat{u}_{n2} = -\hat{u}_{n1}$ Therefore, from Equation 6.11,

$$\int\limits_{\text{Surface of pillbox}} \vec{D} \cdot d\vec{s} = \left(\vec{D}_2 \cdot \hat{u}_n - \vec{D}_1 \cdot \hat{u}_n\right)\Delta A \quad (6.14)$$

Let the surface charge density on the boundary be $\rho_s$ and the volume charge densities in the two halves of the pillbox be $\rho_{v1}$ and $\rho_{v2}$. Then the net charge enclosed by the pillbox is given by

$$\text{Charge enclosed by the pillbox} = \rho_s\Delta A + \rho_{v1}\Delta A \frac{\delta}{2} + \rho_{v2}\Delta A \frac{\delta}{2} \quad (6.15)$$

As the height of the pillbox, $\delta$, is vanishingly small, the contribution of the terms involving volume charge densities will be negligible. Hence, Equation 6.15 becomes

$$\text{Charge enclosed by the pillbox} = \rho_s\Delta A \quad (6.16)$$

Therefore, from Equations 6.14 and 6.16, $(\vec{D}_2 \cdot \hat{u}_n - \vec{D}_1 \cdot \hat{u}_n) = \rho_s$

$$\text{or, } D_{2n} - D_{1n} = \rho_s \quad (6.17)$$

as, $\vec{D}_2 \cdot \hat{u}_n = D_{2n}$ and $\vec{D}_1 \cdot \hat{u}_n = D_{1n}$, which are normal components of electric flux density in dielectrics 2 and 1, respectively.

Equation 6.17 is valid in general. For example, consider that the medium 1 is a perfect conductor. Then $D_1$ is zero. Then $D_{2n}$ is equal to the surface charge density, which is in accordance with Equation 6.5.

### 6.3.2 Boundary Condition for Tangential Component of Electric Field Intensity

Considering the infinitesimally small closed rectangular contour *abcdefa*, as shown in Figure 6.4 and applying the principle of conservative E-field along this closed contour

$$\oint_{abcdefa} \vec{E} \cdot d\vec{l} = \int_a^b \vec{E}_2 \cdot d\vec{l} + \int_b^c \vec{E}_2 \cdot d\vec{l} + \int_c^d \vec{E}_1 \cdot d\vec{l} + \int_d^e \vec{E}_1 \cdot d\vec{l} + \int_e^f \vec{E}_1 \cdot d\vec{l} + \int_f^a \vec{E}_2 \cdot d\vec{l} = 0 \quad (6.18)$$

But the length of the segments *bc*, *cd*, *ef* and *fa* are negligibly small and hence the integrals over these length elements contribute insignificantly. Thus from Equation 6.18

$$\int_a^b \vec{E}_2 \cdot d\vec{l} + \int_d^e \vec{E}_1 \cdot d\vec{l} = 0$$

Then considering $\vec{E}_2$ and $\vec{E}_1$ to be constant over the small lengths *ab* and *de* and noting that $d\vec{l}_{ab} = \Delta l = -d\vec{l}_{de}$, it may be written that

$$\int_a^b \vec{E}_2 \cdot d\vec{l} + \int_d^e \vec{E}_1 \cdot d\vec{l} = E_{2t}\Delta l - E_{1t}\Delta l = 0$$

$$\text{or, } E_{2t} = E_{1t} \quad\quad\quad (6.19)$$

where:

$E_{2t}$ and $E_{1t}$ are tangential components of $E_2$ and $E_1$, respectively, along the boundary

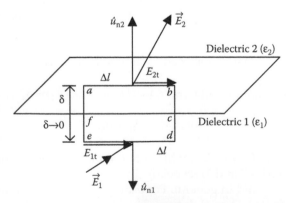

**FIGURE 6.4**
Pertaining to boundary condition for electric field intensity at dielectric–dielectric boundary.

Equation 6.19 is also valid in general. For example, if the medium 1 is considered to be a perfect conductor, then $E_1$ is zero. Then $E_{2t}$ is also zero, which is in accordance with Equation 6.8.

**PROBLEM 6.2**

For a two-dielectric arrangement comprising transformer oil ($\varepsilon_{r1} = 2$) as dielectric 1 and mica ($\varepsilon_{r2} = 6$) as dielectric 2, it is given that $\vec{E}_1 = 6\hat{i} + 4\hat{j} + 9\hat{k}$ kV/cm. Find $\vec{E}_2$: (A) considering the boundary to be charge free and (B) considering a surface charge density of +70 pC/cm$^2$ on the boundary. Given that $x$–$y$ plane is the boundary between the two media.

*Solution:*

Because $x$–$y$ plane is the boundary between transformer oil and mica, $x$- and $y$-components of $\vec{E}$ are the tangential components along the boundary and are equal on both sides of the boundary. The $z$-component of $\vec{D}$ is the normal component of electric flux density on the boundary.

Application of boundary condition for tangential component of electric field intensity yields

$$E_{1x} = E_{2x} = 6 \text{ kV/cm and } E_{1y} = E_{2y} = 4 \text{ kV/cm}$$

The $z$-component of $\vec{E}$ need to be determined from the boundary condition of $D_n$ separately for Case A and Case B.

Case A: Boundary is charge free.

$$\text{As } \rho_s = 0,\ D_{2n} = D_{1n},\ \text{or},\ \varepsilon_{r2}\varepsilon_0 E_{2z} = \varepsilon_{r1}\varepsilon_0 E_{1z}$$

$$\text{As } E_{1z} = 9 \text{ kV/cm},\ E_{2z} = \frac{2}{6} \times 9 = 3 \text{ kV/cm}$$

Therefore,

$$\vec{E}_2 = 6\hat{i} + 4\hat{j} + 3\hat{k} \text{ kV/cm}$$

Case B: Surface charge density ($\rho_s$) on the boundary is +70 pC/cm$^2$.

From the boundary condition of $D_{2n}$,

$$\varepsilon_{r2}\varepsilon_0 E_{2z} - \varepsilon_{r1}\varepsilon_0 E_{1z} = \rho_s$$

$$\text{or, } E_{2z} = \frac{\varepsilon_{r1}\varepsilon_0 E_{1z} + \rho_s}{\varepsilon_{r2}\varepsilon_0} = \frac{\varepsilon_{r1}E_{1z}}{\varepsilon_{r2}} + \frac{\rho_s}{\varepsilon_{r2}\varepsilon_0} = \frac{2 \times 9 \times 10^3 \times 10^2}{6} + \frac{70 \times 10^{-12} \times 10^4}{6 \times 8.854 \times 10^{-12}}$$

$$= 313{,}176 \text{ V/m} = 3.131 \text{ kV/cm}$$

Therefore,

$$\vec{E}_2 = 6\hat{i} + 4\hat{j} + 3.131\hat{k} \text{ kV/cm}$$

### 6.3.3 Boundary Condition for Charge-Free Dielectric–Dielectric Interface

Consider an interface between two different dielectric media having permittivities $\varepsilon_1$ and $\varepsilon_2$, respectively, having no surface charge density on the boundary, as shown in Figure 6.5. Typically, if the dielectric media are ideal in nature, so that there is no volume as well as surface conduction and also if there is no discharge taking place within the dielectric media or on the interface between the dielectric media, then the surface charge density on the interface is zero.

From Equation 6.17,
$D_{2n} - D_{1n} = 0$, as $\rho_s = 0$

$$\text{or, } D_{2n} = D_{1n}, \text{ or, } \varepsilon_{r2}\varepsilon_0 E_{2n} = \varepsilon_{r1}\varepsilon_0 E_{1n}, \text{ or, } \varepsilon_{r2} E_2 \cos\theta_2 = \varepsilon_{r1}E_1 \cos\theta_1 \quad (6.20)$$

Again, from Equation 6.19,

$$E_{2t} = E_{1t}, \text{ or, } E_2 \sin\theta_2 = E_1 \sin\theta_1 \quad (6.21)$$

From Equations 6.20 and 6.21,

$$\frac{\tan\theta_2}{\tan\theta_1} = \frac{\varepsilon_{r2}}{\varepsilon_{r1}} \quad (6.22)$$

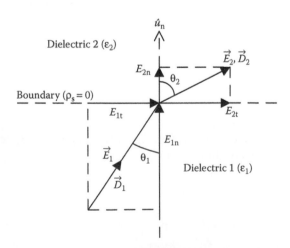

**FIGURE 6.5**
Pertaining to boundary condition for charge-free dielectric–dielectric boundary.

**PROBLEM 6.3**

The flux lines of an electric field pass from air into glass, making an angle 30° with the normal to the plane surface separating air and glass at the air-side of the surface. The relative permittivity of glass is 5.0. The field intensity in air is 200 V/m. Calculate the electric flux density in glass and also the angle, which the flux lines make with the normal on the glass side.

*Solution:*

Given, $\varepsilon_{r1} = 1.0$, $\varepsilon_{r2} = 5.0$ and $\theta_1 = 30°$

$$\text{Therefore,} \quad \frac{\tan\theta_2}{\tan\theta_1} = \frac{\tan\theta_2}{\tan 30°} = \frac{\varepsilon_{r2}}{\varepsilon_{r1}} = \frac{5}{1}, \text{ or, } \tan\theta_2 = 2.887$$

or, $\theta_2 = 70.9°$.

Again, $E_1 = 200$ V/m and $E_1 \sin\theta_1 = E_2 \sin\theta_2$, so that $E_2 \sin 70.9° = 200 \times \sin 30°$ or, $E_2 = 105.8$ V/m.

$$\text{Hence, } D_2 = \varepsilon_{r2}\varepsilon_0 E_2 = 5 \times 8.854 \times 10^{-12} \times 105.8 = 4.68 \text{ nC/m}^2.$$

---

## Objective Type Questions

1. On any conductor–dielectric boundary, the tangential and the normal components of electric field intensity are $E_t$ and $E_n$, respectively. Then

   a. $E_t = E_n/2$

   b. $E_t = 2E_n$

   c. $E_t = \infty$

   d. $E_t = 0$

2. On any dielectric–dielectric boundary, the tangential and the normal components of electric field intensity are $E_t$ and $E_n$, respectively, where the suffixes 1 and 2 denote dielectrics 1 and 2, respectively. Then

   a. $D_n = D_t$

   b. $D_{t1} = D_{t2}$

   c. $E_t = 0$

   d. $E_{t1} = E_{t2}$

3. Surface charge density on a charged conductor is equal to

   a. Tangential component of electric flux density

   b. Normal component of electric flux density

    c. Resultant electric flux density

    d. Both (b) and (c)

4. For any dielectric–dielectric boundary, which one of the following is a valid boundary condition, when the notations have their usual meanings

    a. $(\vec{D}_2 \cdot \hat{u}_n - \vec{D}_1 \cdot \hat{u}_n) = \rho_s$

    b. $(\vec{D}_2 \times \hat{u}_n - \vec{D}_1 \times \hat{u}_n) = \rho_s$

    c. $(\vec{E}_2 \cdot \hat{u}_n - \vec{E}_1 \cdot \hat{u}_n) = \rho_s$

    d. $(\vec{E}_2 \times \hat{u}_n - \vec{E}_1 \times \hat{u}_n) = \rho_s$

5. For any dielectric–dielectric boundary, which one of the following is a valid boundary condition, when the notations have their usual meanings

    a. $D_{2t} = D_{1t}$

    b. $E_{2t} = E_{1t}$

    c. $D_n = \sigma$

    d. $E_t = 0$

6. For any charge-free dielectric–dielectric boundary, which one of the following is a valid boundary condition, when the notations have their usual meanings

    a. $D_{2n} - D_{1n} = \rho_s$

    b. $E_{2t} = E_{1t}$

    c. $D_{2n} = D_{1n}$

    d. Both (b) and (c)

7. On any conductor–dielectric boundary, the tangential and the normal components of electric field intensity are $E_t$ and $E_n$, respectively, and surface charge density is $\sigma$. Then

    a. $D_n = \sigma$

    b. $E_n = \sigma$

    c. $D_t = \sigma$

    d. $E_t = \sigma$

**Answers:** 1) d; 2) d; 3) d; 4) a; 5) b; 6) d; 7) a

# 7

## Multi-Dielectric Configurations

**ABSTRACT**  In multi-dielectric arrangements, electric field distribution is dependent on permittivities of the dielectrics along with the physical dimensions. Because dielectric strengths of the dielectric materials, which are the maximum electric field intensities that the dielectrics can withstand without failure, could be different, it is practically important to analyze electric field distribution in multi-dielectric arrangements that are commonly used in practice to ensure fail-safe operation. Electric field distribution in widely used real-life equipment such as parallel plate capacitor, coaxial cable and high-voltage bushing, which comprise multiple dielectric media, are analyzed in this chapter.

## 7.1 Introduction

Although single-dielectric arrangements are employed in real life, the number of such applications is small. In most of the cases, multiple dielectric media are used in various combinations between the electrodes or conductors. In such multi-dielectric arrangements, contrary to common belief, the location of maximum electric field intensity may not be just off the live conductor in all the cases. Thus, it becomes imperative that not only the magnitude but also the location of maximum electric field intensity should be determined accurately. This is to ensure that the maximum electric field intensity remains well below the dielectric strength of the material within which such maximum electric field intensity occurs. In the case of porous solid dielectric, the small pores are normally filled with air or another gaseous or liquid dielectric. Due to mismatch of the permittivity of the solid dielectric and the dielectric within the pores, electric field intensification may take place in the pores, which can also give rise to unwanted discharges within the equipment if appropriate measures are not taken to eliminate such field intensification. Thus, electric field analysis in multi-dielectric arrangements is not only important from the design viewpoint but also is significant from the viewpoint of life extension of equipment that has such multi-dielectric arrangement.

## 7.2 Parallel Plate Capacitor

Multiple dielectrics within a parallel plate capacitor can be either in series or in parallel between the plates. Such series and parallel dielectric arrangements need to be analyzed separately.

### 7.2.1 Dielectrics in Parallel between the Plates

A parallel plate capacitor with two dielectrics in parallel present between the plates is shown in Figure 7.1, such that the dielectric–dielectric boundary is perpendicular to the plates. If the separation distance between the plates is considered much smaller compared to the length and breadth of the plates, then the flux lines may be considered to be parallel to each other between the plates and also perpendicular to the plates within the dielectrics, neglecting the fringing of flux at the edges of the plates. Then along the paths of integration in both the dielectric media, as shown in Figure 7.1, the flux lines will be tangential. Hence, in dielectric 1,

$$\left|\vec{E}_1\right| l = V, \text{ or, } \left|\vec{E}_1\right| = \frac{V}{l}$$

and in dielectric 2,

$$\left|\vec{E}_2\right| l = V, \text{ or, } \left|\vec{E}_2\right| = V/l = \left|\vec{E}_1\right|$$

where:

$V$ is the potential difference between the plates of the capacitor

Applying Gauss's law considering the Gaussian surface in dielectric 1, as shown in Figure 7.1,

$$\left|\vec{D}_1\right| \times A_1 = Q_1, \text{ or, } \left|\vec{D}_1\right| = \frac{Q_1}{A_1}, \text{ or, } \left|\vec{E}_1\right| = \frac{Q_1}{\varepsilon_1 A_1}, \text{ or, } \frac{V}{l} = \frac{Q_1}{\varepsilon_1 A_1} \tag{7.1}$$

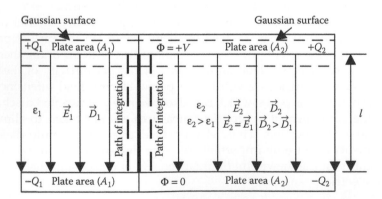

**FIGURE 7.1**
Two dielectrics in parallel within a parallel plate capacitor.

Therefore, the capacitance between the plates comprising dielectric 1,

$$C_1 = \frac{Q_1}{V} = \frac{\varepsilon_1 A_1}{l} \tag{7.2}$$

Again, applying Gauss's law considering the Gaussian surface in dielectric 2, as shown in Figure 7.1,

$$\left|\vec{D}_2\right| \times A_2 = Q_2, \text{ or, } \left|\vec{D}_2\right| = \frac{Q_2}{A_2}, \text{ or, } \left|\vec{E}_2\right| = \frac{Q_2}{\varepsilon_2 A_2}, \text{ or, } \frac{V}{l} = \frac{Q_2}{\varepsilon_2 A_2} \tag{7.3}$$

Therefore, the capacitance between the plates comprising dielectric 2

$$C_2 = \frac{Q_2}{V} = \frac{\varepsilon_2 A_2}{l} \tag{7.4}$$

Hence, the total capacitance of the parallel plate capacitor

$$C = \frac{Q}{V} = \frac{Q_1 + Q_2}{V} = \frac{Q_1}{V} + \frac{Q_2}{V} = C_1 + C_2 = \frac{1}{l}(\varepsilon_1 A_1 + \varepsilon_2 A_2) \tag{7.5}$$

### 7.2.2 Dielectrics in Series between the Plates

A parallel plate capacitor with two dielectrics in series present between the plates is shown in Figure 7.2, such that the dielectric–dielectric boundary is parallel to the plates. As seen in Section 7.2.1, if the separation distance between the plates is considered much smaller compared to the length and breadth of the plates, then the flux lines may be considered to be parallel to each other between the plates and also perpendicular to the plates as well as the dielectric–dielectric boundary within the capacitor, neglecting the fringing of flux at the edges of the plates.

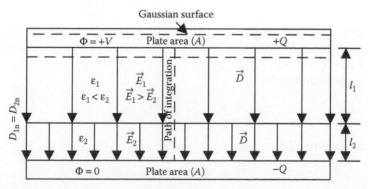

**FIGURE 7.2**
Two dielectrics in series within a parallel plate capacitor.

Applying Gauss's law considering the Gaussian surface, as shown in Figure 7.2,

$$\left|\vec{D}\right| \times A = Q, \text{ or, } \left|\vec{D}\right| = \frac{Q}{A}$$

According to the boundary condition on the dielectric–dielectric boundary of Figure 7.2, $D_{1n} = D_{2n}$. Because the flux lines are perpendicular to the plates and hence to the dielectric–dielectric boundary, $\left|\vec{D}\right| = D_n$. In other words, electric flux density is the same in both the dielectric media.

Therefore, in dielectric 1, $\left|\vec{E}_1\right| = Q/(\varepsilon_1 A)$ and in dielectric 2, $\left|\vec{E}_2\right| = Q/(\varepsilon_2 A)$.

Again, along the path of integration through the two dielectric media, as shown in Figure 7.2, the flux lines will be tangential.

Hence, in dielectric 1,

$$\left|\vec{E}_1\right| l_1 = V_1, \text{ or, } \left|\vec{E}_1\right| = \frac{V_1}{l_1} = \frac{Q}{\varepsilon_1 A} \tag{7.6}$$

where:
$V_1$ = potential difference across the dielectric 1

In dielectric 2,

$$\left|\vec{E}_2\right| l_2 = V_2, \text{ or, } \left|\vec{E}_2\right| = \frac{V_2}{l_2} = \frac{Q}{\varepsilon_2 A} \tag{7.7}$$

where:
$V_2$ = potential difference across the dielectric 2

If the potential difference between the plates of the capacitor is $V$, then

$$V = V_1 + V_2 = \left|\vec{E}_1\right| l_1 + \left|\vec{E}_2\right| l_2 = \frac{Q l_1}{\varepsilon_1 A_1} + \frac{Q l_2}{\varepsilon_2 A_2} \tag{7.8}$$

Let the capacitance of the parallel plate capacitor be $C$, the capacitance of the part comprising dielectric 1 be $C_1$ and the capacitance of the part comprising dielectric 2 be $C_2$.

Then from Equation 7.6,

$$C_1 = \frac{Q}{V_1} = \frac{\varepsilon_1 A}{l_1}$$

and from Equation 7.7,

$$C_2 = \frac{Q}{V_2} = \frac{\varepsilon_2 A}{l_2}$$

Hence, from Equation 7.8,

$$\frac{1}{C} = \frac{V}{Q} = \frac{l_1}{\varepsilon_1 A_1} + \frac{l_2}{\varepsilon_2 A_2} = \frac{1}{C_1} + \frac{1}{C_2} \tag{7.9}$$

Further, from Equations 7.6 and 7.7,

$$\frac{\left|\vec{E}_1\right|}{\left|\vec{E}_2\right|} = \frac{\varepsilon_{r2}}{\varepsilon_{r1}} \tag{7.10}$$

It may be seen from Equation 7.10 that if the dielectric within a capacitor is gas ($\varepsilon_{r1} \approx 1$), then partial filling of the capacitor by a solid or liquid dielectric is actually detrimental to the capacitor. Because in that case, the electric field intensity in the gaseous dielectric will increase by a factor of $\varepsilon_{r2}$ ($\varepsilon_{r2} > \varepsilon_{r1}$) and may even exceed the dielectric strength of the gaseous dielectric resulting in discharge within the capacitor.

### 7.2.3 Void in Insulation

Voids could be present in insulation in many different shapes and due to several reasons. Voids could be in the form of a gas bubble in liquid insulation such as transformer oil, a gas bubble in moulded epoxy resin or a small gap in solid insulation, for example, gap formed due to delamination of pressboard insulation as a result of aging. Such voids are typically filled with air at a pressure slightly less than atmospheric pressure. A void in an insulation is schematically shown in Figure 7.3. Considering the void dimensions to be small in comparison to the insulation surrounding it, it may be reasonably argued that the presence or absence of the void will not alter electric flux density at the location of the void. Then according to the discussions of Section 7.2.2 and Equation 7.10,

$$\frac{\left|\vec{E}_{\text{void}}\right|}{\left|\vec{E}_{\text{insulation}}\right|} = \varepsilon_{r\text{-insulation}} \, (\varepsilon_{r\text{-insulation}} > 1) \tag{7.11}$$

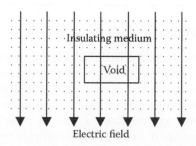

**FIGURE 7.3**
Void in insulating medium.

Thus, from Equation 7.11, it is evident that electric field intensity within the void will be higher than that in the insulating medium. On the other hand, the dielectric strength of air filling the void will be lower than that of the insulating medium. Consequently, there is every possibility that electric field intensity in the void may be in excess of the dielectric strength of air at the pressure and temperature within the void. This could result in discharge within the void. But there will be no discharge within the insulating medium around the void, as electric field intensity there is lower and the dielectric strength of the insulation is higher. Such localized discharge within the void is called *partial discharge* (*PD*) and is highly detrimental to high voltage equipment as PD reduces the life of the equipment significantly.

### 7.2.4 Impregnation of Porous Solid Insulation

Solid insulations that are porous in nature, for example, paper or pressboard, are very commonly used in high-voltage equipment. In such cases, the solid insulation is mainly cellulose, whereas the gas in the pores is air. The relative permittivity of cellulose insulation is of the order of 3 ~ 4 and the dielectric strength of cellulose is about 16 kV/mm. On the other hand, relative permittivity of air is ~1 and the dielectric strength of air is 3 kV/mm. In order that the solid insulation does not fail due to discharges within the pores, electric field intensity within the solid insulation has to be kept around 9 ~ 12 kV/mm, so that electric field intensity in the pores remains below 3 kV/mm. But in that case the capacity of the paper insulation is not utilized to the full extent. Therefore, in practice, such porous solid insulation is always impregnated with a suitable liquid dielectric. Due to impregnation, the pores in the solid insulation will be filled by the liquid insulation, which will have higher dielectric strength compared to air. But here it is important to note that the relative permittivity of the liquid dielectric used for impregnation must be close to the relative permittivity of the solid insulation being impregnated. Otherwise, there will be again a mismatch of the electric field intensity values within the liquid and solid dielectrics and the full benefit of impregnation could not be obtained.

## 7.3 Co-Axial Cylindrical Configurations

Co-axial multi-dielectric arrangements, which are commonly used in power engineering, are cables and bushings. A single-core cable having three different dielectrics in between the core and the earthed metallic screen is shown in Figure 7.4. Bushings are special components used in high-voltage system when a high-voltage conductor has to pass through an earthed barrier, for example, earthed tank of a transformer or a wall and so on. Figure 7.5 shows a solid-type oil-impregnated paper bushing used in transformers typically for voltage ratings below 90 kV. If a cross section is taken

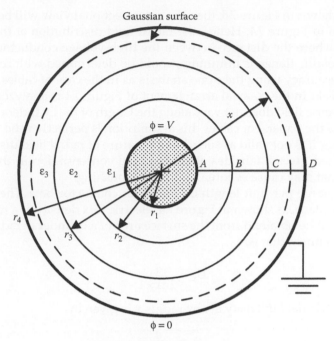

**FIGURE 7.4**
A single-core cable having three different dielectric media.

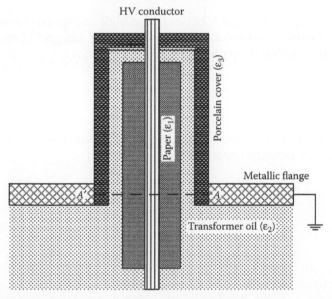

**FIGURE 7.5**
Solid-type oil-impregnated paper bushing used in transformers.

at $AA'$, as shown in Figure 7.5, then the cross-sectional view will be same as that shown in Figure 7.4. Hence, the radial field distribution at the critical zone $AA'$, where the distance between the high-voltage conductor and the earthed metallic flange is minimum, could be determined with reasonable degree of accuracy using the same analysis as in the case of cables.

Electric field in the co-axial arrangement of Figure 7.4 is analyzed assuming that electric field does not vary along the length of the cylinder normal to the plane of the paper. For cables this assumption is perfectly valid, whereas for bushings, it is not valid as such, but the nature of radial field distribution thus obtained gives a fair idea about the field concentration in the critical zone $AA'$ that may cause eventual failure of a bushing.

Let the charge per unit length on the inner conductor be $q$. Then for the Gaussian surface, as shown in Figure 7.4, electric flux density at a radial distance of $x$ can be obtained from the surface area of a cylinder of radius $x$ and axial length unity, that is,

$$D_x = \frac{q}{2\pi x \times 1}$$

Hence, electric field intensity at any radius $x$ is given by

$$E_x = \frac{q}{2\pi \varepsilon_x x} \tag{7.12}$$

where:
$\varepsilon_x$ is the permittivity of the dielectric at the radial distance $x$

Then the potential difference $V$ between the high-voltage conductor and the earthed enclosure could be obtained from the line integral of electric field intensity. Considering the path of integral from $D$ to $A$,

$$V = -\int_D^A E_x dx = -\int_D^A \frac{q}{2\pi \varepsilon_x x} dx \tag{7.13}$$

But the integral of Equation 7.13 needs to be evaluated section-wise where the permittivity remains constant. Thus,

$$V = -\int_B^A \frac{q}{2\pi \varepsilon_1 x} dx - \int_C^B \frac{q}{2\pi \varepsilon_2 x} dx - \int_D^C \frac{q}{2\pi \varepsilon_3 x} dx$$

$$\text{or, } V = -\int_{r_2}^{r_1} \frac{q}{2\pi \varepsilon_1 x} dx - \int_{r_3}^{r_2} \frac{q}{2\pi \varepsilon_2 x} dx - \int_{r_4}^{r_3} \frac{q}{2\pi \varepsilon_3 x} dx$$

$$\tag{7.14}$$

$$= \frac{q}{2\pi} \left( \frac{1}{\varepsilon_1} \ln \frac{r_2}{r_1} + \frac{1}{\varepsilon_2} \ln \frac{r_3}{r_2} + \frac{1}{\varepsilon_3} \ln \frac{r_4}{r_3} \right)$$

In Equation 7.14, it may be seen that the expression within the parenthesis on the right-hand side is a constant for a given multi-dielectric arrangement. Hence, the charge per unit length on the inner conductor can be expressed in terms of the potential difference and the computed constant of Equation 7.14 as follows:

$$q = \frac{2\pi V}{K}, \text{ where } K = \left( \frac{1}{\varepsilon_1} \ln \frac{r_2}{r_1} + \frac{1}{\varepsilon_2} \ln \frac{r_3}{r_2} + \frac{1}{\varepsilon_3} \ln \frac{r_4}{r_3} \right) \qquad (7.15)$$

The capacitance per unit length of the multi-dielectric arrangement is

$$C = \frac{q}{V} = \frac{2\pi}{K} \qquad (7.16)$$

Electric field intensity at any radius $x$ is then given by

$$E_x = \frac{q}{2\pi \varepsilon_x x} = \frac{V}{K \varepsilon_x x} \qquad (7.17)$$

From Equation 7.17, it is clear that electric field intensity varies inversely with radial distance within each section of the arrangement comprising one particular dielectric. Thus, the variation of electric field intensity can be plotted with radial distance, as shown in Figure 7.6. From Figure 7.6, it may be seen that there is a discontinuity in electric field intensity at the boundary

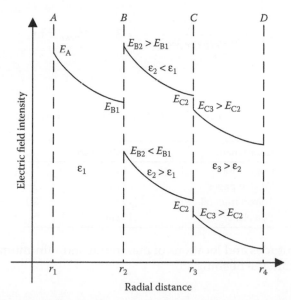

**FIGURE 7.6**
Radial variation of electric field intensity in co-axial multi-dielectric arrangement.

between two different dielectric media. The value of electric field intensity may increase or decrease at the boundary depending on the relative values of the permittivity of the two dielectric media at the boundary. Thus, for a multi-dielectric arrangement, as shown in Figure 7.4, it cannot be stated that the maximum value of electric field intensity will occur just off the surface of the inner conductor in all the cases. On the contrary, it has to be ascertained case by case considering the physical dimensions and the dielectric arrangement.

## PROBLEM 7.1

A single-core, lead sheathed cable joint has a conductor of 10 mm diameter and two layers of different insulating materials, each 10 mm thick. The relative permittivities are 3 and 2.5 for inner and outer dielectrics, respectively. Calculate the potential gradient just off the conductor surface, when the potential difference between the conductor and lead sheath is 33 $kV_{rms}$.

*Solution:*
Given: $\varepsilon_{r1} = 3$ and $\varepsilon_{r2} = 2.5$, $r_1 = (10/2) = 5$ mm, $r_2 = 5 + 10 = 15$ mm and $r_3 = 15 + 10 = 25$ mm.
   Therefore, according to Equation 7.15,

$$K = \frac{1}{\varepsilon_0}\left(\frac{1}{3}\ln\frac{15}{5} + \frac{1}{2.5}\ln\frac{25}{15}\right) = \frac{0.57}{\varepsilon_0}$$

Hence, potential gradient just off the conductor surface

$$E_{r1} = \frac{33}{(0.57/\varepsilon_0)\times 3\times\varepsilon_0\times 5} = 3.85\ kV_{rms}/mm$$

## PROBLEM 7.2

A transformer bushing for 36 $kV_{rms}$ consists of the following:

| Component | Outside Diameter | $\varepsilon_r$ |
|---|---|---|
| Copper rod | 4 cm | |
| Treated paper | 5 cm | 3 |
| Transformer oil | 10 cm | 2.1 |
| Porcelain | 15 cm | 5 |

Find the magnitudes and locations of maximum and minimum electric field intensities within the bushing.

*Solution:*
This problem is solved in accordance with Figure 7.4 and Equations 7.15 and 7.17.

Given: $\varepsilon_{r1} = 3$ and $\varepsilon_{r2} = 2.1$ and $\varepsilon_{r3} = 5$, $r_1 = 2$ cm, $r_2 = 2.5$ cm, $r_3 = 5$ cm and $r_4 = 7.5$ cm.

Therefore, according to Equation 7.15,

$$K = \frac{1}{\varepsilon_0}\left(\frac{1}{3}\ln\frac{2.5}{2} + \frac{1}{2.1}\ln\frac{5}{2.5} + \frac{1}{5}\ln\frac{7.5}{5}\right) = \frac{0.4855}{\varepsilon_0}$$

Therefore, according to Figure 7.4, within paper insulation just off the conductor surface

$$E_A = \frac{36}{(0.4855/\varepsilon_0)\times 3 \times \varepsilon_0 \times 2} = 12.36\,\text{kV}_{\text{rms}}/\text{cm}$$

At the paper–oil boundary, on the paper side

$$E_B\big|_{\text{paper}} = \frac{36}{(0.4855/\varepsilon_0)\times 3 \times \varepsilon_0 \times 2.5} = 9.88\,\text{kV}_{\text{rms}}/\text{cm}$$

At the paper–oil boundary, on the oil side

$$E_B\big|_{\text{oil}} = \frac{36}{(0.4855/\varepsilon_0)\times 2.1 \times \varepsilon_0 \times 2.5} = 14.12\,\text{kV}_{\text{rms}}/\text{cm}$$

At the oil–porcelain boundary, on the oil side

$$E_C\big|_{\text{oil}} = \frac{36}{(0.4855/\varepsilon_0)\times 2.1 \times \varepsilon_0 \times 5} = 7.06\,\text{kV}_{\text{rms}}/\text{cm}$$

At the oil–porcelain boundary, on porcelain side

$$E_C\big|_{\text{porcelain}} = \frac{36}{(0.4855/\varepsilon_0)\times 5 \times \varepsilon_0 \times 5} = 2.97\,\text{kV}_{\text{rms}}/\text{cm}$$

Within porcelain just off the Earth surface

$$E_D = \frac{36}{(0.4855/\varepsilon_0)\times 5 \times \varepsilon_0 \times 7.5} = 1.98\,\text{kV}_{\text{rms}}/\text{cm}$$

Therefore, the maximum electric field intensity is 14.12 kV$_{\text{rms}}$/cm at the oil side of the oil–paper boundary, and the minimum electric field intensity is 1.98 kV$_{\text{rms}}$/cm within porcelain just off the Earth surface.

## PROBLEM 7.3

A conductor 2.8 cm in diameter is passed centrally through a porcelain bushing ($\varepsilon_r = 5$) having internal and external diameters of 3 and 9 cm, respectively. The potential difference between the conductor and the earthed metallic flange around the porcelain is 12 kV$_{rms}$. Determine whether or not partial discharges will be present in the airspace around the conductor. Also, find the electric stress just off the conductor surface, if the airspace is filled with transformer oil ($\varepsilon_r = 2.1$).

*Solution:*

The arrangement as per the statement of the problem is shown in Figure 7.7. With reference to Figure 7.7, the given quantities are as follows:

$$\varepsilon_{r2} = 5, \, r_1 = 1.4 \text{ cm}, \, r_2 = 1.5 \text{ cm and } r_3 = 4.5 \text{ cm}$$

Case I: When the small space between the copper rod and porcelain cover is filled with air. Then $\varepsilon_{r1} = 1$. Therefore, according to Equation 7.15, in this case

$$K1 = \frac{1}{\varepsilon_0}\left(\frac{1}{1}\ln\frac{1.5}{1.4} + \frac{1}{5}\ln\frac{4.5}{1.5}\right) = \frac{0.2887}{\varepsilon_0}$$

Within this airspace, the maximum electric field intensity will occur just off the conductor surface and will be

$$E_{cond}\big|_{air} = \frac{12}{(0.2887/\varepsilon_0)\times\varepsilon_0\times1.4} = 29.69 \text{ kV}_{rms}/\text{cm}$$

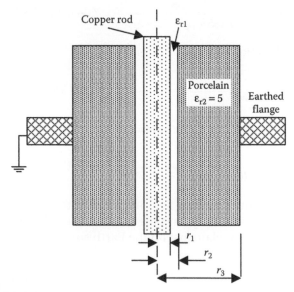

**FIGURE 7.7**
Pertaining to Problem 7.3.

Breakdown strength of air is 30 kV$_p$/cm or 21.21 kV$_{rms}$/cm at standard temperature and pressure (STP). Therefore, partial discharges will be present in the airspace.

Case II: When the small space between the copper rod and porcelain cover is filled with transformer oil instead of air. Then $\varepsilon_{r1} = 2.1$. Therefore, according to Equation 7.15, in this case

$$K2 = \frac{1}{\varepsilon_0}\left(\frac{1}{2.1}\ln\frac{1.5}{1.4} + \frac{1}{5}\ln\frac{4.5}{1.5}\right) = \frac{0.2525}{\varepsilon_0}$$

Then the maximum electric field intensity just off the conductor surface is

$$E_{cond}\big|_{oil} = \frac{12}{(0.2525/\varepsilon_0)\times 2.1\times \varepsilon_0\times 1.4} = 16.16 \text{ kV}_{rms}/\text{cm}$$

It may be seen that filling up the small space by transformer oil brings down the electric field intensity considerably. Moreover, transformer oil has higher breakdown strength than air. Therefore, partial discharges will not occur in the small gap.

This problem brings out an important aspect of transformer bushing design. Because copper and porcelain are both solid, it is very difficult to get a perfect contact between them. Therefore, intentionally a small gap is kept between these two for easy operation. But if this small space is kept filled with air then partial discharge is inevitable, which is undesirable. Hence, the solid-type bushings are always so designed that this small space is filled with transformer oil instead of air.

## Objective Type Questions

1. An air-filled parallel plate capacitor is charged and then the voltage source is disconnected from the plates. Now half of the airspace is filled with transformer oil, so that the oil-filled half is in series with air-filled half. Then
   a. The potential difference between the plates will increase
   b. The potential difference between the plates will decrease
   c. The energy stored in the capacitor will increase
   d. The energy stored in the capacitor will decrease
2. The potential difference between the plates of an air-filled parallel plate capacitor is kept constant by a DC voltage source. Now half of the airspace is filled with pressboard ($\varepsilon_r$), so that the pressboard-filled

half is in series with air-filled half. Then the electric field intensity in air will be

a. $\varepsilon_r^2$ times the electric field intensity in pressboard
b. $\varepsilon_r$ times the electric field intensity in pressboard
c. $1/\varepsilon_r$ times the electric field intensity in pressboard
d. $1/\varepsilon_r^2$ times the electric field intensity in pressboard

3. An air-filled parallel plate capacitor is charged and then the voltage source is disconnected from the plates. Now a solid dielectric slab ($\varepsilon_r = 4$) of thickness equal to the separation distance between the plates is inserted half way into the capacitor such that the air-filled part is in parallel with the solid-filled part. Then

a. The potential difference between the plates will increase
b. The potential difference between the plates will decrease
c. The energy stored in the capacitor will increase
d. The energy stored in the capacitor will decrease

4. If a small void is present in a solid insulation, then the electric field intensity within void is

a. Higher than that in the solid insulation
b. Lower than that in the solid insulation
c. Equal to that in the solid insulation
d. Zero

5. If a small void is present in a solid insulation, then there is very high probability that

a. Complete breakdown of the insulation system will take place
b. Breakdown of solid insulation will take place around the void
c. Breakdown of gaseous insulation will take place within the void
d. No breakdown will take place

6. Impregnation of porous solid insulation is done using a liquid dielectric for which

a. Dielectric strength should be same as that of the solid insulation
b. Volume resistivity should be same as that of the solid insulation
c. Dielectric dissipation factor should be same as that of the solid insulation
d. Permittivity should be same as that of the solid insulation

7. For a single-core screened cable with three layers of different dielectric materials arranged in co-axial form, electric field intensity within the second layer of dielectric is maximum
   a. At the boundary between dielectric 1 and dielectric 2
   b. At the boundary between dielectric 2 and dielectric 3
   c. In the middle of dielectric 2
   d. None of the above

**Answers:**   1) b; 2) b; 3) b; 4) a; 5) c; 6) d; 7) a

# 8

## Electrostatic Pressures on Boundary Surfaces

**ABSTRACT**  In electrostatic field, due to the presence of surface charges on the boundary, there is always a mechanical pressure that acts on any boundary. However, the magnitude of such mechanical pressure due to electrostatic field is usually small compared to the other types of mechanical pressure that one encounters in everyday life. But the mechanical pressures due to the electrostatic field have many interesting engineering applications. Thus, it is important to derive the general expressions for mechanical pressure acting on the boundary between a perfect conductor and a dielectric as well as the boundary between two different dielectric media.

## 8.1 Introduction

There could be two different approaches towards the calculation of electrostatic force acting on any boundary. In the first approach, the macroscopic resultant force may be calculated by summing up the elementary electrostatic forces as obtained from Coulomb's law. On the one hand, this approach should always give correct result, but on the other hand, in most of the practical cases, it is very difficult, if not impossible, to perform such calculation as the actual distributions of charges are mostly unknown.

The second approach is based on the principle of energy conservation and the electrostatic forces are derived indirectly from energy relationship. This approach is advantageous in the sense that the forces can be calculated conveniently even by analytical methods. But the accuracy of the results obtained from this approach depends on the validity of the principle of energy conservation for the specific case under consideration. Some argue that this condition is not always satisfied and consequently this method can lead to wrong results in some cases. Although majority of the scientific community firmly believe that the principle of energy conservation has a general validity, and therefore this approach should always provide correct results. On the whole, it is better to suggest that one should use it with due circumspection and reservation by always verifying the results.

## 8.2 Mechanical Pressure on a Conductor–Dielectric Boundary

As discussed in Section 6.2, only same polarity charges reside on the boundary between a perfect conductor and a dielectric medium. These same polarity charges, residing on the surface of the conductor, exert repulsive forces on each other. These forces will be of such nature that the distance between the charges should increase. In other words, the surface area of the conductor–dielectric boundary will try to increase under the influence of these repulsive forces. Hence, the electrostatic force on the conductor–dielectric boundary always acts along the normal to the boundary directed from the conductor to the dielectric.

The force on a conductor–dielectric boundary could be calculated using the expression for energy density. If an elemental area $\Delta A$ on a conductor–dielectric boundary is depressed by a distance $\Delta l$, the increase in stored energy ($\Delta W$) is equal to the work done against the electrostatic force acting on the conductor–dielectric boundary trying to swell the surface. Hence, considering the energy density of electric field within the dielectric to be $W_E$,

$$\Delta W = W_E \Delta A \Delta l$$

Here, it has been assumed that the charge on the conductor surface remain unchanged even when the geometry is changed slightly. This assumption is valid when the conductor is an isolated one, that is, it is not connected to a source that could alter its charge, for example, a voltage source. Hence, the work done in depressing the boundary could be related directly to the energy content of electric field according to the law of conservation of energy.

If the electrostatic force acting against the depression of the surface is given by $F$, then the work done against this force is

$$F\Delta l = W_E \Delta A \Delta l, \text{ or, } F = W_E \Delta A \tag{8.1}$$

Electrostatic force per unit area is the mechanical pressure due to electrostatic field acting on the conductor–dielectric boundary, which is, therefore, given by

$$P_{mech}\big|_{cond} = \frac{F}{\Delta A} = W_E \tag{8.2}$$

Considering the surface charge density of the conductor–dielectric boundary to be $\sigma$,

$$P_{mech}\big|_{cond} = W_E = \frac{1}{2}\varepsilon E^2 = \frac{1}{2\varepsilon}D^2 = \frac{\sigma^2}{2\varepsilon} \tag{8.3}$$

as $D = D_n = \sigma$, because $D_t = \varepsilon E_t$ is zero on the conductor surface, as $E_t = 0$.

### 8.2.1 Electric Field Intensity Exactly on the Conductor Surface

Dimensionally, the mechanical pressure acting on the charged conductor surface is equal to the product of surface charge density ($\sigma$) and the electric field intensity exactly on the conductor surface ($E_{surface}$), that is, from Equation 8.3

$$P_{mech}\big|_{cond} = \sigma E_{surface} = \frac{\sigma^2}{2\varepsilon}, \text{ or, } E_{surface} = \frac{\sigma}{2\varepsilon} = \frac{\sigma}{2\varepsilon_r\varepsilon_0} \tag{8.4}$$

From Equations 6.8 and 6.10,

$$E_{just\ off\ the\ surface} = \frac{\sigma}{\varepsilon} = \frac{\sigma}{\varepsilon_r\varepsilon_0} \tag{8.5}$$

Therefore,

$$E_{just\ off\ the\ surface} = 2 \times E_{surface} \tag{8.6}$$

### PROBLEM 8.1

A metallic sphere of 20 cm radius is charged with 1 µC, spread uniformly over the surface and is surrounded by a dielectric medium having a relative permittivity of 5. Find the electric field intensity just off the sphere and also on the sphere. Find also the mechanical pressure acting on the sphere.

*Solution:*
Sphere radius = 20 cm = 0.2 m
  Sphere surface area = $4\pi \times (0.2)^2 = 0.5026\ \text{m}^2$
  Therefore, surface charge density ($\sigma$) = $(1 \times 10^{-6})/0.5026 = 1.989 \times 10^{-6}\ \text{C/m}^2$.
  Electric field intensity just off the sphere surface = $(1.989 \times 10^{-6})/(5 \times 8.854 \times 10^{-12}) = 44.94 \times 10^3\ \text{V/m}$.
  Electric field intensity exactly on the sphere surface = $(44.94 \times 10^3)/2 = 22.47 \times 10^3\ \text{V/m}$.
  Therefore, mechanical pressure acting on the sphere surface = $1.989 \times 10^{-6} \times 22.47 \times 10^3 = 0.0447\ \text{N/m}^2$.

### 8.2.2 Electrostatic Forces on the Plates of a Parallel Plate Capacitor

In order to get a simple analytical solution, consider that the distance between the plates ($l$) of the capacitor is much smaller than the area of the plates ($A$). In that case, it may be assumed that the E-field between the plates is homogeneous, and the effect of the inhomogeneity of E-field at the edges, commonly known as *fringing*, may be neglected. The charge may also be assumed to be distributed uniformly over the plates, that is, the charge density may be assumed to be known for the application of Coulomb's law.

Capacitance of the parallel plate capacitor $(C) = \varepsilon A/l$.

Let the potential difference between the two plates of the capacitor be $V$.

Therefore, the uniformly distributed charge on the plates of the capacitor $(Q_{plate}) = C \times V = \varepsilon A V/l$ and the electric field intensity within the dielectric of the capacitor $(E) = V/l$.

As discussed in Section 8.2.1, electric field intensity exactly on the plate surface is given by

$$E_{plate} = \frac{\sigma_{plate}}{2\varepsilon} = \frac{Q_{plate}}{2\varepsilon A} = \frac{1}{2\varepsilon A} \times \frac{\varepsilon A V}{l} = \frac{V}{2l} \tag{8.7}$$

Therefore, from Coulomb's law, the electrostatic force acting on the capacitor plates is given by

$$F_{plate} = Q_{plate} E_{plate} = \frac{\varepsilon A V}{l} \times \frac{V}{2l} = \frac{1}{2} \varepsilon A \left( \frac{V}{l} \right)^2 \tag{8.8}$$

Again, according to Equation 8.2 mechanical pressure acting on the plates

$$P_{mech}\big|_{plate} = W_E = \frac{1}{2} \varepsilon E^2 = \frac{1}{2} \varepsilon \left( \frac{V}{l} \right)^2 \tag{8.9}$$

Hence, the electrostatic force acting on the capacitor plates is

$$F_{plate} = P_{mech}\big|_{plate} \times A = \frac{1}{2} \varepsilon A \left( \frac{V}{l} \right)^2$$

which is the same as that of Equation 8.8.

Equation 8.8 is very useful because if the force could be measured when the voltage is unknown, then the voltage can be calculated from the above formula from the knowledge of the capacitor dimensions. In fact, the measurement of voltage by electrostatic voltmeter is based on this principle.

**PROBLEM 8.2**

For an air-filled parallel plate capacitor, the area of the plates is 50 cm$^2$ and the separation distance between the plates is 5 mm. What will be the maximum electrostatic force on the capacitor plates at standard temperature and pressure?

*Solution:*

Breakdown strength of air at standard temperature and pressure (STP) is 30 kV/cm or $3 \times 10^6$ V/m. That will be the maximum possible electric field intensity within the parallel plate capacitor.

Therefore, the maximum electrostatic pressure on the capacitor plates is

$$P_{\text{plate}}\big|_{\text{max}} = \frac{1}{2} \times 8.854 \times 10^{-12} \times (3 \times 10^6)^2 = 39.84 \text{ N/m}^2$$

$$F_{\text{plate}}\big|_{\text{max}} = 39.84 \times 50 \times 10^{-4} = 0.199 \text{ N}.$$

## 8.3 Mechanical Pressure on a Dielectric–Dielectric Boundary

Mechanical pressure due to electrostatic field that act on a dielectric–dielectric boundary arise due to two reasons. The first one is the polarization of atoms and/or dipoles in the volume of the dielectric media under the action of electric field, while the second one is the change in the polarization vector that takes place at the boundary between two different dielectric media. These two mechanical pressures need to be discussed separately.

### 8.3.1 Mechanical Pressure due to Dielectric Polarization

Consider a homogeneous isotropic dielectric piece of small volume placed in vacuum within an electric field. The atoms and/or dipoles within this dielectric volume will be polarized under the action of the electric field. Because each atom and/or dipole consists of positive and negative charge, it may be considered that under the action of the electric field a uniformly distributed cloud of negative charges is shifted through a small distance from a uniformly distributed cloud of positive charges. The same polarity charges by mutual repulsion develop an outward force, which tends to swell the small dielectric volume into the surrounding vacuum.

The dielectric volume, being small enough, will not alter the electric field intensity due to the external field within which it is placed. But due to dielectric polarization, the electric flux density within the dielectric volume will be higher due to higher permittivity of the dielectric compared to vacuum for the same electric field intensity.

In the absence of dielectric medium, energy density of electric field will be given by

$$W_E\big|_{\text{vacuum}} = \frac{1}{2}\varepsilon_0 E^2 \qquad (8.10)$$

and in the presence of dielectric medium, energy density of electric field for the same electric field intensity within the same volume will be given by

$$W_E\big|_{\text{dielectric}} = \frac{1}{2}\varepsilon E^2 \tag{8.11}$$

Due to electrostatic repulsive force, if unit area of the surface of the small dielectric volume expands through a small distance $dl$ in the normal direction, then the difference in stored energy is equal to the mechanical work done by the repulsive forces, if no heat energy is lost. Let the force on the dielectric boundary due to dielectric polarization be $F_{\text{pol}}$. Then the mechanical work done by $F_{\text{pol}}$ is given by

$$\text{Work done} = F_{\text{pol}} \times dl = \frac{1}{2}(\varepsilon - \varepsilon_0)E^2 \times dl \times 1$$

$$\text{or, } \frac{F_{\text{pol}}}{1} = \frac{1}{2}(\varepsilon - \varepsilon_0)E^2 \tag{8.12}$$

This force per unit area, that is, mechanical pressure, acts normally on the boundary directed from the dielectric to vacuum.

When two different dielectric meet at a boundary, then the mechanical pressure by which each dielectric tends to push the boundary normally outwards could be computed using Equation 8.12. The difference in these two pressures is the net pressure acting on the dielectric–dielectric boundary due to dielectric polarization.

The mechanical pressure on the boundary due to dielectric 1 alone, that is, when dielectric 2 is replaced by vacuum, is given by

$$P_{\text{mech}}\big|_{\text{dielctric1-vacuum}} = \frac{1}{2}(\varepsilon_1 - \varepsilon_0)E_1^2 \tag{8.13}$$

This pressure given by Equation 8.13 acts normally from dielectric 1 to vacuum on the boundary.

Again, the mechanical pressure on the boundary due to dielectric 2 alone, that is, when dielectric 1 is replaced by vacuum, is given by

$$P_{\text{mech}}\big|_{\text{dielctric2-vacuum}} = \frac{1}{2}(\varepsilon_2 - \varepsilon_0)E_2^2 \tag{8.14}$$

This pressure given by Equation 8.14 acts normally from dielectric 2 to vacuum on the boundary.

The difference in these two pressures as given by Equations 8.13 and 8.14 is the net pressure acting on the boundary due to dielectric polarization.

Considering the normal on the dielectric–dielectric boundary to be from dielectric 1 to dielectric 2,

$$P_{mech}\big|_{pol} = \frac{1}{2}\Big[(\varepsilon_1 - \varepsilon_0)\,E_1^2 - (\varepsilon_2 - \varepsilon_0)\,E_2^2\Big] \qquad (8.15)$$

$$\text{or, } P_{mech}\big|_{pol} = \frac{1}{2}\Big[(\varepsilon_1 - \varepsilon_0)(E_{1t}^2 + E_{1n}^2) - (\varepsilon_2 - \varepsilon_0)(E_{2t}^2 + E_{2n}^2)\Big]$$

$$= \frac{1}{2}\Big[(\varepsilon_1 E_{1t}^2 - \varepsilon_2 E_{2t}^2) + (\varepsilon_1 E_{1n}^2 - \varepsilon_2 E_{2n}^2) - \varepsilon_0(E_{1n}^2 - E_{2n}^2)\Big]$$

$$= \frac{1}{2}\left[(\varepsilon_1 - \varepsilon_2)E_t^2 + \left(\frac{D_{1n}^2}{\varepsilon_1} - \frac{D_{2n}^2}{\varepsilon_2}\right) - \varepsilon_0\left(\frac{D_{1n}^2}{\varepsilon_1^2} - \frac{D_{2n}^2}{\varepsilon_2^2}\right)\right] \qquad (8.16)$$

$$= \frac{1}{2}\left[(\varepsilon_1 - \varepsilon_2)E_t^2 + \left(\frac{1}{\varepsilon_1} - \frac{1}{\varepsilon_2}\right)D_n^2 - \varepsilon_0\left(\frac{1}{\varepsilon_1^2} - \frac{1}{\varepsilon_2^2}\right)D_n^2\right]$$

as $E_{1t} = E_{2t} = E_t$ and $D_{1n} = D_{2n} = D_n$.

### 8.3.2 Mechanical Pressure on Surface Film at the Dielectric–Dielectric Boundary

Consider a boundary between a homogeneous isotropic dielectric medium and vacuum, as shown in Figure 8.1. For a boundary having no free charge, the boundary conditions are $E_{1t} = E_{2t}$ and $D_{1n} = D_{2n}$.

Within the dielectric medium $D_1 = \varepsilon_1 E_1 = \varepsilon_0 E_1 + P_1$, where $P_1$ is the polarization vector in the dielectric medium. Hence, $P_1 = D_1 - \varepsilon_0 E_1$.

On the other hand, within vacuum $D_2 = \varepsilon_0 E_2$ and hence, $P_2 = 0$.

Thus, as one crosses the boundary the polarization vector drops from a finite value within the dielectric medium to zero in vacuum. But this change

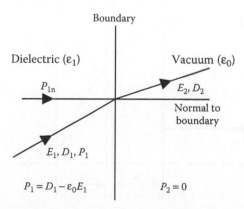

**FIGURE 8.1**
Change of polarization vector at dielectric–vacuum boundary.

in polarization vector cannot occur at one line of infinitesimal thickness representing the boundary. On the contrary, such change in polarization vector occurs within a thin film of finite but very small thickness within the dielectric medium just off the boundary.

Consequently, this boundary film needs to be studied closely. Because only a normal displacement of the boundary film will expand the boundary, only the normal components of electric flux density and polarization vector are considered. The normal component of electric flux density remains constant inside the boundary film. But, the normal component of electric field intensity as well as the normal component of polarization vector vary gradually and finally assume the values equal to those just outside the film on the vacuum side of the boundary. Such variation is schematically shown in Figure 8.2.

Within the boundary film, the polarization vector is $P_n = D_n - \varepsilon_0 E_n$.

Therefore, $dP_n = -\varepsilon_0 dE_n$ as $D_n$ is constant within the film.

Now, consider a thin strip of thickness $ds$ within the film such that the distance of the strip from the beginning of the film is $S$, as shown in Figure 8.2. Also consider that $ds$ be the separation distance between the positive and negative charges of the atoms and/or dipoles within this strip. At a distance $S$, the negative charges of the atoms lie in a field of intensity $E_n$ and at a distance $S + ds$, the positive charges lie in a field of $E_n + dE_n$. Hence, each dipole charge is subjected to a force per unit of $dE_n$ towards right according to Figure 8.2. But the polarization vector being $P_n$, there are $P_n$ charges per unit area. Thus, the force per unit area, that is, mechanical pressure, acting towards right on the thin strip is

$$d\,P_{\text{mech}}\big|_{\text{film}} = P_n dE_n = -\frac{1}{\varepsilon_0} P_n dP_n$$

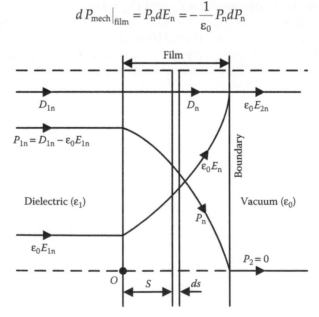

**FIGURE 8.2**
Variation of electric field quantities within the boundary film.

Therefore, total mechanical pressure acting on the entire film is

$$P_{\text{mech}}\Big|_{\text{film-dielectric-vacuum}} = -\int_{P_n=P_{1n}}^{P_n=0} \frac{1}{\varepsilon_0} P_n dP_n = \frac{P_{1n}^2}{2\varepsilon_0} = \frac{(D_{1n} - \varepsilon_0 E_{1n})^2}{2\varepsilon_0}$$

As $D_{1n} = \varepsilon_1 E_{1n}$, $P_{\text{mech}}\Big|_{\text{film-dielectric-vacuum}} = \frac{(\varepsilon_1 - \varepsilon_0)^2}{2\varepsilon_0} E_{1n}^2$  (8.17)

Thus, when two dielectric media meet at a boundary, the normal pressure that acts on the boundary film from dielectric 1 to dielectric 2 may be computed following the same logic as in the case of Equations 8.13 through 8.15. Therefore,

$$P_{\text{mech}}\Big|_{\text{film}} = \frac{1}{2\varepsilon_0}\Big[(\varepsilon_1 - \varepsilon_0)^2 E_{1n}^2 - (\varepsilon_2 - \varepsilon_0)^2 E_{2n}^2\Big]$$  (8.18)

### 8.3.3 Total Mechanical Pressure on the Dielectric–Dielectric Boundary

Total mechanical pressure on the boundary between two dielectric media pushing the boundary from dielectric 1 to dielectric 2 is the sum of Equations 8.15 and 8.18. Therefore,

$$P_{\text{mech}}\Big|_{\text{dielectric}} = \frac{1}{2}\Big[(\varepsilon_1 - \varepsilon_0)E_1^2 - (\varepsilon_2 - \varepsilon_0)E_2^2 + \frac{(\varepsilon_1 - \varepsilon_0)^2 E_{1n}^2}{\varepsilon_0} - \frac{(\varepsilon_2 - \varepsilon_0)^2 E_{2n}^2}{\varepsilon_0}\Big]$$  (8.19)

Noting that $E_1^2 = E_{1t}^2 + E_{1n}^2$, $E_2^2 = E_{2t}^2 + E_{2n}^2$, $E_{1t} = E_{2t}$ and $D_{1n} = D_{2n}$, the above equation may be rewritten as

$$
\begin{aligned}
P_{\text{mech}}\Big|_{\text{dielectric}} &= \frac{1}{2}\left[\begin{array}{l} (\varepsilon_1 - \varepsilon_0)(E_{1t}^2 + E_{1n}^2) - (\varepsilon_2 - \varepsilon_0)(E_{2t}^2 + E_{2n}^2) \\[2mm] + \dfrac{(\varepsilon_1^2 - 2\varepsilon_1\varepsilon_0 + \varepsilon_0^2)E_{1n}^2}{\varepsilon_0} - \dfrac{(\varepsilon_2^2 - 2\varepsilon_2\varepsilon_0 + \varepsilon_0^2)E_{2n}^2}{\varepsilon_0} \end{array}\right] \\[4mm]
&= \frac{1}{2}\left[\varepsilon_1 E_{1t}^2 - \varepsilon_2 E_{2t}^2 - \varepsilon_1 E_{1n}^2 + \varepsilon_2 E_{2n}^2 + \frac{\varepsilon_1^2 E_{1n}^2}{\varepsilon_0} - \frac{\varepsilon_2^2 E_{2n}^2}{\varepsilon_0}\right] \\[4mm]
&= \frac{1}{2}\left[\frac{(\varepsilon_1 E_{1t})^2}{\varepsilon_1} - \frac{(\varepsilon_2 E_{2t})^2}{\varepsilon_2} - \frac{(\varepsilon_1 E_{1n})^2}{\varepsilon_1} + \frac{(\varepsilon_2 E_{2n})^2}{\varepsilon_2}\right] \\[4mm]
&= \frac{1}{2}\left[\frac{D_{1t}^2}{\varepsilon_1} - \frac{D_{2t}^2}{\varepsilon_2} - \frac{D_{1n}^2}{\varepsilon_1} + \frac{D_{2n}^2}{\varepsilon_2}\right]
\end{aligned}
$$  (8.20)

Equation 8.20 may also be written as

$$P_{mech}\big|_{dielectric} = \frac{1}{2}\left[(\varepsilon_1 - \varepsilon_2)E_t^2 + \left(\frac{1}{\varepsilon_2} - \frac{1}{\varepsilon_1}\right)D_n^2\right] \tag{8.21}$$

as $E_{1t} = E_{2t} = E_t$ and $D_{1n} = D_{2n} = D_n$

Equation 8.21 can be used to find the mechanical pressure acting on the conductor–dielectric boundary, by considering $\varepsilon_1 = \infty$ for the conductor and noting the boundary conditions on the conductor–dielectric boundary as $E_t = 0$ and hence $D = D_n$. Then putting $\varepsilon_2 = \varepsilon$ for the dielectric surrounding the conductor in Equation 8.21

$$P_{mech}\big|_{cond} = \frac{D^2}{2\varepsilon}$$

which is the same as Equation 8.3.

## PROBLEM 8.3

There is a paper-insulated transformer coil immersed in oil. $\varepsilon_r$ for paper $= 3$ and $\varepsilon_r$ for transformer oil $= 2.1$. There is a normal electric stress of 25 kV/cm and a tangential electric stress of 10 kV/cm just within the paper at the paper–oil boundary. Calculate the total mechanical pressure acting on the paper–oil boundary.

*Solution:*
Given: $E_{1n} = 25$ kV/cm $= 25 \times 10^5$ V/m and $E_{1t} = 10$ kV/cm $= 10^6$ V/m
     Therefore,

$$D_{1n} = \varepsilon_{r1}\varepsilon_0 E_{1n} = 3 \times \varepsilon_0 \times 25 \times 10^5 = 75 \times \varepsilon_0 \times 10^5 \text{ V/m}$$

Total mechanical pressure acting on the paper–oil boundary

$$= \frac{1}{2}\left[(3 - 2.1) \times \varepsilon_0 \times (10^6)^2 + \frac{1}{\varepsilon_0}\left(\frac{1}{2.1} - \frac{1}{3}\right)(75 \times \varepsilon_0 \times 10^5)^2\right] = 39.56 \text{ N/m}^2$$

## PROBLEM 8.4

A rectangular slab of porcelain ($\varepsilon_r = 5$) is placed in air within an electric field such that the surface of the porcelain slab is perpendicular to the electric fieldlines. Find the maximum possible mechanical pressure acting on the porcelain–air boundary at standard temperature and pressure.

*Solution:*
Because the boundary is perpendicular to the electric fieldlines in air, the air side of the boundary $E_t = 0$. Again the maximum value of electric field

intensity in air is 30 kV/cm at standard temperature and pressure. Therefore, on the air side of the boundary $E_{n-max} = 30$ kV/cm or $3 \times 10^6$ V/m.

Therefore, the maximum possible mechanical pressure acting on the porcelain–air boundary is

$$= \frac{1}{2\varepsilon_0}\left(\frac{1}{1} - \frac{1}{5}\right) \times (\varepsilon_0)^2 \times (3 \times 10^6)^2 = 31.9 \text{ N/m}^2$$

## 8.4 Two Dielectric Media in Series between a Parallel Plate Capacitor

Consider a parallel plate capacitor having two dielectric media in series between the plates such that the boundary between the dielectric media is parallel to the plates, as shown in Figure 8.3. In this case, electric fieldlines will be perpendicular to the boundary. According to the boundary condition, normal component of electric flux density will be same in both the dielectric media, but the magnitude of electric field intensity will be different in the two dielectrics. Further, because the tangential component of electric field is zero, as the electric flux lines are perpendicular to the boundary, hence $D = D_n$.

Let the potential difference across the dielectric 1 ($\varepsilon_1$) is $V_1$ and that across the dielectric 2 ($\varepsilon_2$) is $V_2$. Then

$$V = V_1 + V_2 = E_1 l_1 + E_2 l_2 = \frac{D}{\varepsilon_1} l_1 + \frac{D}{\varepsilon_2} l_2$$

$$\text{or, } D = \frac{\varepsilon_1 \varepsilon_2}{\varepsilon_2 l_1 + \varepsilon_1 l_2} V$$

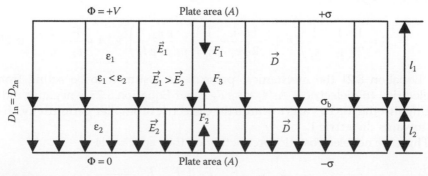

**FIGURE 8.3**
Two dielectric media in series between a parallel plate capacitor.

Hence,

$$E_1 = \frac{D}{\varepsilon_1} = \frac{\varepsilon_2}{\varepsilon_2 l_1 + \varepsilon_1 l_2} V \text{ and } E_2 = \frac{D}{\varepsilon_2} = \frac{\varepsilon_1}{\varepsilon_2 l_1 + \varepsilon_1 l_2} V \tag{8.22}$$

Mechanical pressure acting on the plates will be equal to energy density of electric field just off the plates, that is, $(1/2)\varepsilon E^2$. Considering the area of the plates to be $A$,

Mechanical force acting on the top plate

$$(F_1) = \frac{1}{2}\varepsilon_1 E_1^2 A = \frac{1}{2}\varepsilon_1 \left(\frac{\varepsilon_2}{\varepsilon_2 l_1 + \varepsilon_1 l_2}\right)^2 V^2 A$$

$$\text{or, } F_1 = \frac{1}{2\varepsilon_1}\left(\frac{\varepsilon_1 \varepsilon_2}{\varepsilon_2 l_1 + \varepsilon_1 l_2}\right)^2 V^2 A = \frac{F}{\varepsilon_1} \tag{8.23}$$

where:

$$F = \frac{1}{2}\left(\frac{\varepsilon_1 \varepsilon_2}{\varepsilon_2 l_1 + \varepsilon_1 l_2}\right)^2 V^2 A$$

Similarly, mechanical force acting on the bottom plate

$$(F_2) = \frac{1}{2}\varepsilon_2 E_2^2 A = \frac{F}{\varepsilon_2} \tag{8.24}$$

The force $F_1$ acts on the top plate directed towards dielectric 1, whereas the force $F_2$ acts on the bottom plate directed towards dielectric 2. Thus, $F_1$ and $F_2$ are in opposite direction and $F_1 > F_2$ as $\varepsilon_1 < \varepsilon_2$.

Therefore, it appears that there is a resultant unidirectional reaction-less force acting on the capacitor. But this is not true, because the force acting on the boundary surface between the two dielectric media has to be taken into account, too.

From Equation 8.21 the mechanical force on the dielectric–dielectric boundary is given by

$$F_3 = \frac{1}{2}\left(\frac{1}{\varepsilon_2} - \frac{1}{\varepsilon_1}\right)D^2 A = \frac{1}{2}\left(\frac{1}{\varepsilon_2} - \frac{1}{\varepsilon_1}\right)\left(\frac{\varepsilon_1 \varepsilon_2}{\varepsilon_2 l_1 + \varepsilon_1 l_2}\right)^2 V^2 A = \frac{F}{\varepsilon_2} - \frac{F}{\varepsilon_1} \tag{8.25}$$

In Equation 8.21 the mechanical pressure is assumed to be acting from dielectric 1 to dielectric 2. As $\varepsilon_1 < \varepsilon_2$, $F_3$ as per Equation 8.25 is negative indicating that the force $F_3$ acts towards the dielectric of smaller permittivity, that is, towards dielectric 1.

Thus, the net electrostatic force acting on the capacitor is

$$F_1 - F_2 + F_3 = \frac{F}{\varepsilon_1} - \frac{F}{\varepsilon_2} + \left(\frac{F}{\varepsilon_2} - \frac{F}{\varepsilon_1}\right) = 0$$

## 8.5 Two Dielectric Media in Parallel between a Parallel Plate Capacitor

Consider a parallel plate capacitor having two dielectric media in parallel between the plates such that the boundary between the dielectric media is perpendicular to the plates, as shown in Figure 8.4. In this case, the electric fieldlines will be tangential to the boundary. According to the boundary condition, tangential component of electric field intensity will be the same in both the dielectric media, but the magnitude of electric flux density will be different in the two dielectrics.

In this case,

$$E_1 = E_2 = \frac{V}{l}, D_1 = \varepsilon_1 E_1 \text{ and } D_2 = \varepsilon_2 E_2$$

The mechanical force acting on the part of the plate in contact with dielectric 1 is given by

$$F_1 = \frac{1}{2}\varepsilon_1 E_1^2 A_1 = \frac{1}{2}\varepsilon_1 \left(\frac{V}{l}\right)^2 A_1 \tag{8.26}$$

The forces acting on the two plates within the section of the capacitor containing dielectric 1 will be equal to $F_1$ but will act in opposite directions.

The mechanical force acting on the part of the plate in contact with dielectric 2 is given by

$$F_2 = \frac{1}{2}\varepsilon_2 E_2^2 A_2 = \frac{1}{2}\varepsilon_2 \left(\frac{V}{l}\right)^2 A_2 \tag{8.27}$$

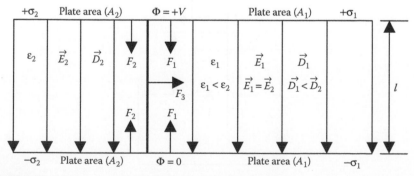

**FIGURE 8.4**
Two dielectric media in parallel between a parallel plate capacitor.

The forces acting on the two plates within the section of the capacitor containing dielectric 2 will be equal to $F_2$ but will act in opposite directions. If $A_1 = A_2$, then $F_2 > F_1$ as $\varepsilon_2 > \varepsilon_1$. But the net force on the capacitor plates will always be zero due to the presence of equal and opposite forces on the plates.

But there will be mechanical force acting on the boundary between the two dielectric media, which according to Equation 8.21 is given by

$$F_3 = \frac{1}{2}(\varepsilon_1 - \varepsilon_2)E^2 A_b = \frac{1}{2}(\varepsilon_1 - \varepsilon_2)\frac{V^2}{l^2}A_b \tag{8.28}$$

where:
$A_b$ = area of the dielectric–dielectric boundary

As stated earlier, in Equation 8.21, the mechanical pressure is considered to be acting from dielectric 1 to dielectric 2. As $\varepsilon_1 < \varepsilon_2$, $F_3$, as per Equation 8.28, is negative, which indicates that the force $F_3$ acts towards the dielectric of smaller permittivity, that is, from dielectric 2 to dielectric 1.

### 8.5.1 Electrostatic Pump

The mechanical force acting on the dielectric–dielectric boundary, as discussed in Section 8.5, can be demonstrated practically with the help of the arrangement shown in Figure 8.5, which is also known as *electrostatic pump*.

If a charged parallel plate capacitor with air between the plates is partially submerged in a liquid dielectric, as shown in Figure 8.5, then the mechanical force on liquid–air boundary within the capacitor will act from the liquid dielectric to air, as the permittivity of air is smaller than the permittivity of liquid dielectric. Hence, this mechanical force will push the liquid up between the plates against the force of gravity. When the electrical force on

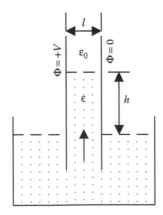

**FIGURE 8.5**
Demonstration of electrostatic pump.

the boundary and the weight of the liquid column between the capacitor plates become equal, then the upward movement of the liquid dielectric between the plates stops, and the surface of the liquid dielectric between the capacitor plates remains at a higher level than the surface of the liquid dielectric in the container.

As per Equation 8.28, the mechanical force on the liquid–air boundary acting towards air is

$$F_{\text{dielectric boundary}} = \frac{1}{2}(\varepsilon - \varepsilon_0)\frac{V^2}{l^2}A_b = \frac{1}{2}(\varepsilon - \varepsilon_0)\frac{V^2}{l^2}l \times l_w = \frac{1}{2}(\varepsilon - \varepsilon_0)\frac{V^2}{l}l_w \quad (8.29)$$

where:
$l_w$ = width of the plates normal to the plane of the paper
$l$ = separation distance between the capacitor plates
$V$ = potential difference between the capacitor plates

The gravitational force on the liquid column between the capacitor plates is

$$F_{\text{gravity}} = \rho h l_w l g \quad (8.30)$$

where:
$h$ = height of the liquid column between the capacitor plates
$\rho$ = density of the liquid dielectric
$g$ = acceleration due to gravity

At equilibrium $F_{\text{gravity}} = F_{\text{dielectric boundary}}$, that is,

$$\rho h l_w l g = \frac{1}{2}(\varepsilon - \varepsilon_0)\frac{V^2}{l}l_w$$

$$\text{or, } h = \frac{1}{2\rho g}(\varepsilon - \varepsilon_0)\left(\frac{V}{l}\right)^2 = \frac{1}{2\rho g}(\varepsilon - \varepsilon_0)E^2 \quad (8.31)$$

In the case of electrostatic pump, for a given liquid dielectric and for fixed dimensions of the capacitor, the height of the liquid column between the capacitor plates can be controlled by controlling the voltage applied across the plates.

## PROBLEM 8.5
A parallel plate capacitor with air between the plates is submerged into trans-former oil in a container in such a way that the top surface of transformer oil in the container is perpendicular to the capacitor plates. The potential difference between the capacitor plates is 15 kV and the separation distance between the plates is 6 mm. Density of transformer oil is 860 kg/m³. Calculate the height of

the transformer oil column between the capacitor plates if $\varepsilon_r$ for transformer oil = 2.1. What will be the height of the liquid column if transformer oil is replaced by water for which density is 1000 kg/m³ and $\varepsilon_r$ is 80?

*Solution:*
Given: $\rho_{oil}$ = 860 kg/m³, $\varepsilon_r$ of oil = 2.1, $V$ = 15 kV = 15 × 10³ V and $l$ = 6 mm = 6 × 10⁻³ m
Acceleration due to gravity $(g)$ = 9.81 m/s²
Therefore,

$$h_{oil} = \frac{1}{2 \times 860 \times 9.81} \times (2.1 - 1) \times 8.854 \times 10^{-12} \times \left(\frac{15 \times 10^3}{6 \times 10^{-3}}\right)^2 = 3.6 \text{ mm}$$

If transformer is replaced by water, then $\rho_{water}$ = 1000 kg/m³, $\varepsilon_r$ of water = 80.

$$\text{Therefore, } h_{water} = \frac{1}{2 \times 1000 \times 9.81} \times (80 - 1) \times 8.854 \times 10^{-12} \times \left(\frac{15 \times 10^3}{6 \times 10^{-3}}\right)^2$$

$$= 222.3 \text{ mm}$$

## Objective Type Questions

1. Mechanical pressure acting on a charged conductor surface due to electrostatic field is equal to
   a. Energy density of electrostatic field
   b. Surface charge density on the conductor
   c. Square of the surface charge density on the conductor
   d. Square of electric flux density on the conductor

2. Mechanical pressure acting on a charged conductor surface due to electrostatic field is proportional to
   a. Square of the electric potential of the conductor
   b. Square of the electric field intensity on the conductor
   c. Square of the surface charge density on the conductor
   d. Square of the permittivity of the dielectric surrounding the conductor

3. Electric field intensity exactly on the conductor surface is equal to
   a. Electric field intensity just off the conductor surface
   b. Half of the electric field intensity just off the conductor surface
   c. Twice the electric field intensity just off the conductor surface
   d. Square of the electric field intensity just off the conductor surface

4. Mechanical pressure acting on the boundary between a dielectric and vacuum due to dielectric polarization is proportional to
   a. The electric field intensity within the dielectric
   b. The electric flux density within the dielectric
   c. Square of the electric field intensity within the dielectric
   d. Square of the electric flux density within the dielectric

5. Mechanical pressure on the boundary between two dielectric media due to electrostatic field acts on the boundary
   a. Normally towards the dielectric having lower permittivity
   b. Normally towards the dielectric having higher permittivity
   c. Tangentially in the clockwise sense
   d. Tangentially in the anti-clockwise sense

6. Within the surface film at the boundary between a dielectric and vacuum, the polarization vector
   a. Remains constant
   b. Varies linearly with distance
   c. Varies from the finite value within the dielectric just off the boundary to infinity on the vacuum side of the boundary
   d. Varies from the finite value within the dielectric just off the boundary to zero on the vacuum side of the boundary

7. From the known values of the dimensions of a parallel plate capacitor, the potential difference between the plates can be conveniently obtained from the measurement of
   a. Surface charge density on the capacitor plates
   b. Electric flux density between the capacitor plates
   c. Polarization vector between the capacitor plates
   d. Mechanical force acting on the capacitor plates

8. For a parallel plate capacitor having two dielectric media in series between the plates, mechanical pressure due to electric field is highest
   a. On the plate having the dielectric of higher permittivity next to it
   b. On the plate having the dielectric of lesser permittivity next to it
   c. On the dielectric–dielectric boundary acting towards the dielectric of higher permittivity
   d. On the dielectric–dielectric boundary acting towards the dielectric of lesser permittivity

9. A parallel plate capacitor with air between the plates is submerged into a liquid dielectric in a container in such a way that the top surface of the liquid in the container is perpendicular to the capacitor

plates. Then the surface of the liquid dielectric column between the capacitor plates will be

a. At a higher level than the surface of the liquid within the container

b. At a lower level than the surface of the liquid within the container

c. At the same level as that of the surface of the liquid within the container

d. Parallel to the capacitor plates

**Answers:**    1) a; 2) c; 3) b; 4) c; 5) a; 6) d; 7) d; 8) b; 9) a

# 9

# Method of Images

**ABSTRACT**   There are many electric field problems involving one or more types of charges in the presence of conducting surfaces with specific boundary conditions, which are difficult to satisfy, if the governing field equation such as Poisson's or Laplace's equations is to be solved directly. Very often these problems do not exhibit any symmetry. However, sometimes the placement of additional fictitious charges of appropriate magnitude at suitable locations does provide a solution to such problems depending on the geometry of the equipotential conducting surfaces. These fictitious charges are called *image charges* and the electric field distribution can then be determined in the region excluding that occupied by the image charges in a straightforward manner. Thus, without solving Poisson's or Laplace's equations, this method can give solutions to problems, which are otherwise difficult to obtain. This method of replacing the equipotential conducting surfaces by appropriate image charges in place of direct solution of Poisson's or Laplace's equations is called the *method of images*. It is possible to use this method when the boundary involves linear or curved surface.

## 9.1 Introduction

To explain the idea behind the method of images, consider two distinctly different electrostatic problems. The real problem is the one in which a charge density is given in a finite domain $V$ bounded by its surface $S$ with specific boundary conditions on $S$. The other is a fictitious problem in which the charge density within the finite domain $V$ is the same as that for the real problem, but the boundary surfaces are replaced by suitable fictitious charge distribution located outside the domain $V$. If the fictitious charge distribution is so chosen that the solution to the fictitious problem satisfies the boundary conditions specified in the real problem, then the solution to the fictitious problem is also the solution to the real problem. The fictitious charge distribution so determined is called the *image of the true charge distribution* for the real problem.

In other words, the idea is to convert an electrostatic problem involving conducting objects, which are spatially extended, in such a way that conducting surfaces having given boundary conditions are replaced by a finite number of appropriately chosen and suitably placed discrete charges known as image charges. Image charges are always located outside the region where the field is to be determined. While doing this replacement of boundary surfaces by image charges, the original boundary conditions of the real problem are retained. Thus, the more complicated original problem could then be solved as a relatively simpler problem having known charge configuration.

## 9.2 Image of a Point Charge with Respect to an Infinitely Long Conducting Plane

Consider a point charge of positive polarity and magnitude $Q$ located at a height $z_1$ from an infinitely long conducting plane present at $z = 0$, as shown in Figure 9.1, such that the location of the point charge is given by $(x_1,y_1,z_1)$. As discussed in Section 1.7.2, an infinitely long conducting plane, which is also an equipotential surface, will have zero electric potential. The practical example of such a plane is the earth surface. In fact, in any electric field distribution the earth surface having zero potential plays a significant role and hence taking the image of a point charge with respect to an infinitely long conducting plane takes into account the effect of the grounded earth surface of zero potential in electric field calculation.

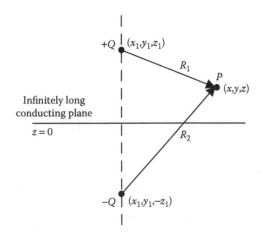

**FIGURE 9.1**
Point charge near an infinitely long conducting plane.

Considering the conducting plane to be the $x$–$y$ plane, the problem is to find a solution for electric field in the region $z > 0$. This solution can be obtained by solving Laplace's equation with the following conditions:

1. The solution $\phi(x,y,z)$ will be valid at every point for $z > 0$ except the location of the point charge.
2. The potential at all the points on the conducting plane is zero, that is, $\phi(x,y,0) = 0$.
3. The potential approaches zero as the distance from the charge approaches infinity.

It is not easy to find a solution to Laplace's equation that satisfies all the above conditions.

The problem can also be viewed from another angle. The positive polarity point charge at $z = z_1$ will induce negative polarity charges on the surface of the conducting plane. Say, the induced surface charge density is $-\sigma_s$. Then the electric field at any point in the region $z > 0$ will be due to the point charge and the induced surface charges. But the difficulty is that $\sigma_s$ needs to be determined using the boundary condition $\phi(x,y,0) = 0$, before the solution to electric field in the region $z > 0$ could be obtained. Furthermore, the evaluation of the surface integral to obtain the field due to the induced charges is not an easy task. From all these considerations, taking the image of the point charge with respect to the infinitely long conducting plane to replace the plane by the image charge is an easy and practical methodology.

The electric potential at a distance $r$ from a point charge is given by $Q/(4\pi\varepsilon_0 r)$. Considering the spherical symmetry of the field due to a point charge, it may be seen that the electric field due to a point charge is dependent only on $r$-coordinate in spherical coordinate system and is independent of $\theta$ and $\phi$ coordinates. Hence, this electric potential satisfies Laplace's equation in spherical coordinates as follows:

$$\vec{\nabla}^2\phi = \frac{1}{r^2}\frac{\partial}{\partial r}\left(r^2\frac{\partial\phi}{\partial r}\right) = \frac{1}{r^2}\frac{\partial}{\partial r}\left(r^2 \times -\frac{Q}{4\pi\varepsilon_0 r^2}\right) = \frac{1}{r^2}\frac{\partial}{\partial r}\left(-\frac{Q}{4\pi\varepsilon_0}\right) = 0$$

Consider that the infinitely long conducting plane is replaced by a fictitious discrete point charge of magnitude $Q_1$ located at $(x_1,y_1,-z_2)$. Because the field due to a point charge satisfies Laplace's equation, the field due to the real point charge and the image point charge will also satisfy Laplace's equation at all points for $z > 0$ except at the location of the real point charge.

Therefore, the electric potential at any point $P(x,y,z)$ will be given by

$$\phi_P = \frac{Q}{4\pi\varepsilon_0 R_1} + \frac{Q_1}{4\pi\varepsilon_0 R_2} \tag{9.1}$$

where:

$$R_1 = \left[ (x - x_1)^2 + (y - y_1)^2 + (z - z_1)^2 \right]^{1/2}$$

$$\text{and } R_2 = \left[ (x - x_1)^2 + (y - y_1)^2 + (z + z_2)^2 \right]^{1/2}$$

From Equation 9.1, it is evident that the potential at a very large distance from the point charge $Q$ will be zero.

Therefore, two of the above-mentioned three conditions are satisfied by Equation 9.1. In order to satisfy the third condition, $\phi_P$ at any point on the conducting plane has to be zero. From Equation 9.1, it is obvious that the polarity of $Q_1$ has to be negative to satisfy this condition. Furthermore, the following has to be satisfied at all the points on the conducting plane

$$\frac{Q}{R_1} = \frac{Q_1}{R_2} \tag{9.2}$$

The best solution for Equation 9.2 is obtained when the magnitude of the image charge ($Q_1$) is equal to the magnitude of the real charge ($Q$) and location of the image charge is such that the magnitudes of $R_1$ and $R_2$ are equal at all the points on the conducting plane. From the expressions of $R_1$ and $R_2$, it is clear that for $z = 0$, $R_1$ will be equal to $R_2$ if $z_1 = z_2$. Therefore, the third condition will also be satisfied if $Q_1 = -Q$ and the location of the image charge is at $(x_1, y_1, -z_1)$.

Thus, the combination of the real charge $Q$ located at $(x_1, y_1, z_1)$ and the image charge $-Q$ located at $(x_1, y_1, -z_1)$ satisfies all the three conditions stated above. Hence, the electric field in the region $z > 0$ due to the real point charge $+Q$ and the infinitely long conducting plane will be the same as that due to $+Q$ and $-Q$ located as mentioned above.

Therefore, the electric potential at any point $P(x,y,z)$ for $z > 0$ will be given by

$$\phi_P = \frac{Q}{4\pi\varepsilon_0} \left( \frac{1}{R_1} - \frac{1}{R_2} \right) \tag{9.3}$$

where:

$$R_1 = \left[ (x - x_1)^2 + (y - y_1)^2 + (z - z_1)^2 \right]^{1/2}$$

$$\text{and } R_2 = \left[ (x - x_1)^2 + (y - y_1)^2 + (z + z_1)^2 \right]^{1/2}$$

In Cartesian coordinates, electric field intensity at any point $P(x,y,z)$ for $z > 0$ will be given by

$$\vec{E}_P = -\frac{\partial \phi_P}{\partial x}\hat{i} - \frac{\partial \phi_P}{\partial y}\hat{j} - \frac{\partial \phi_P}{\partial z}\hat{k} \tag{9.4}$$

where:

$$E_{xP} = -\frac{\partial \phi_P}{\partial x} = \frac{Q}{4\pi\varepsilon_0}\left(\frac{x-x_1}{R_1^3} - \frac{x-x_1}{R_2^3}\right)$$

$$E_{yP} = -\frac{\partial \phi_P}{\partial y} = \frac{Q}{4\pi\varepsilon_0}\left(\frac{y-y_1}{R_1^3} - \frac{y-y_1}{R_2^3}\right)$$

$$E_{zP} = -\frac{\partial \phi_P}{\partial z} = \frac{Q}{4\pi\varepsilon_0}\left(\frac{z-z_1}{R_1^3} - \frac{z+z_1}{R_2^3}\right)$$

For any point very near to the conducting plane, that is, $z \to 0$, $R_1 \approx R_2$. Hence, $E_{xP}$ and $E_{yP}$ will be zero. Thus,

$$\lim_{z \to 0}\vec{E}_{zP} = \frac{\sigma_s}{\varepsilon_0}\hat{k} = \frac{-Qz_1}{2\pi\varepsilon_0\left[(x-x_1)^2 + (y-y_1)^2 + z_1^2\right]^{3/2}}\hat{k}$$

$$\text{or, } \sigma_s = \frac{-Qz_1}{2\pi\left[(x-x_1)^2 + (y-y_1)^2 + z_1^2\right]^{3/2}} \tag{9.5}$$

Equation 9.5 gives the negative surface charge density induced on the conducting plane by the positive polarity real point charge.

This problem can also be viewed from a different perspective. The field due to the real point charge and the image point charge is the same as that due to a spatially extended electric dipole. Considering spherical coordinates and assuming that the electric dipole to be oriented along the axis of symmetry, as shown in Figure 9.2, electric potential and electric field intensity can be expressed in spherical coordinates.

With reference to Figure 9.2,

$$r_1 = (r^2 + h^2 - 2rh\cos\theta)^{1/2} \text{ and}$$

$$r_2 = \left[r^2 + h^2 - 2rh\cos(\pi-\theta)\right]^{1/2} = (r^2 + h^2 + 2rh\cos\theta)^{1/2} \tag{9.6}$$

Then,

$$\phi_P = \frac{Q}{4\pi\varepsilon_0}\left[\frac{1}{(r^2 + h^2 - 2rh\cos\theta)^{1/2}} - \frac{1}{(r^2 + h^2 + 2rh\cos\theta)^{1/2}}\right] \tag{9.7}$$

$$\vec{E}_P = -\frac{\partial\phi_P}{\partial r}\hat{u}_r - \frac{1}{r}\frac{\partial\phi_P}{\partial\theta}\hat{u}_\theta - \frac{1}{r\sin\theta}\frac{\partial\phi_P}{\partial\phi}\hat{u}_\phi \tag{9.8}$$

where:

$$E_{rP} = -\frac{\partial\phi_P}{\partial r} = -\frac{Q}{4\pi\varepsilon_0}\left(\frac{r-h\cos\theta}{r_1^3} - \frac{r+h\cos\theta}{r_2^3}\right)$$

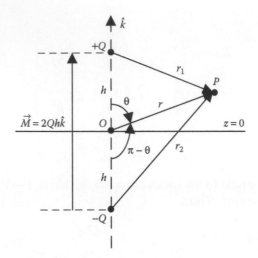

**FIGURE 9.2**
Electric dipole formed by the real and image charges.

$$E_{\theta P} = -\frac{1}{r}\frac{\partial \phi_P}{\partial \theta} = \frac{Qh\sin\theta}{4\pi\varepsilon_0}\left(\frac{1}{r_1^3} - \frac{1}{r_2^3}\right)$$

$$\text{and } \vec{E}_{\phi P} = -\frac{1}{r\sin\theta}\frac{\partial \phi_P}{\partial \phi} = 0.$$

Equations 9.7 and 9.8 are the expressions for the electric field due to an electric dipole of dipole moment $\vec{M} = 2Qh\hat{k}$, as shown in Figure 9.2.

However, it is to be noted here that the solution obtained with the help of image charge is valid only for the region $z > 0$ as for $z < 0$ the region is below an infinitely long conducting plane whose potential is zero and hence for $z < 0$, the electric field is zero. The image charge does not exist in reality as it is a fictitious charge.

### 9.2.1 Point Charge between Two Conducting Planes

Consider two conducting planes making an angle $\theta$ between them, as shown in Figure 9.3a, such that $2\pi$ is integer multiple of $\theta$. Consider also that the potential of the conducting planes is zero and there is a point charge $+Q$ in the wedge-shaped space between the two conducting planes. A direct solution of Laplace's equation subject to boundary condition of zero electric potential on the conducting planes would be quite difficult. However, this problem could be solved by introducing image charges considering the conducting planes as mirrors. Then $n$ number of image charges are introduced at each location where the conducting planes acting as mirrors would form images of the real point charge, where $n = (2\pi/\theta) - 1$. Such images are shown in Figure 9.3a,

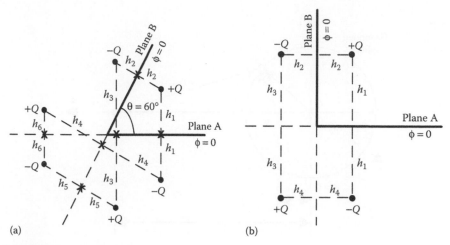

(a)                                              (b)

**FIGURE 9.3**
Point charge between two conducting planes: (a) between two planes making an angle θ and
(b) between two perpendicular planes.

where $\theta = 60°$. The pair of charges $+Q$ and $-Q$ separated by $2h_1$, $2h_2$, $2h_3$, $2h_4$, $2h_5$ and $2h_6$ will make the potential zero for Plane A, Plane B, Plane A, Plane B, Plane B and Plane A, respectively. The solution then will be unique as the boundary conditions are satisfied. But this solution will be valid only in the wedge-shaped space between the two conducting planes. Another case of two conducting planes making an angle $\theta = 90°$ is shown in Figure 9.3b.

## 9.3 Image of a Point Charge with Respect to a Grounded Conducting Sphere

Image of a charge is not necessarily to be taken with respect to infinitely long plane only. It can also be taken with respect to curved surfaces such as sphere, cylinder and so on. To elaborate this issue, consider a point charge $+Q$ located at distance $d$ from the centre ($O$) of the sphere of radius $a$ ($a < d$), as shown in Figure 9.4. Consider also that the electric potential of the sphere is zero. The field due to the point charge and the grounded sphere in the region outside the sphere could be determined by replacing the grounded sphere by an image point charge. From the symmetry of the system, it is evident that the image charge $q$ will be of negative polarity and will be located inside the sphere on the line joining the centre of the sphere and the point charge, as shown in Figure 9.4. However, in this case, the magnitude of $q$ will not be equal to $Q$ because such a pair of charges will not result into a zero-potential spherical surface of radius $a$ as required by the boundary condition.

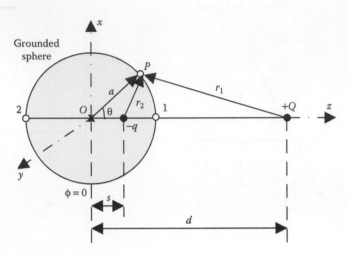

**FIGURE 9.4**
Point charge near a grounded sphere.

Consider that the image charge is located at a distance $s$ from the centre of the sphere, as shown in Figure 9.4. Now the problem is to determine the magnitude as well as the location of the image charge that satisfies the zero-potential boundary condition for the spherical surface. With reference to Figure 9.4, the potential at the point $P$ due to the point charge and its image is given by

$$\phi_P = \frac{1}{4\pi\varepsilon_0}\left(\frac{Q}{r_1} - \frac{q}{r_2}\right)$$                                      (9.9)

Imposition of the boundary condition $\phi_P = 0$ leads to

$$\frac{q}{Q} = \frac{r_2}{r_1} = \alpha$$                                      (9.10)

If $\alpha$ is kept constant, then $r_2/r_1 = $ constant is the equation of a sphere. Hence, the problem now is to find the constant $\alpha$.

For the point 1, as shown in Figure 9.4, $\alpha = r_2/r_1 = (a-s)/(d-a)$ and for the point 2, as shown in Figure 9.4, $\alpha = r_2/r_1 = (a+s)/(d+a)$.

Therefore, $\alpha = \dfrac{r_2}{r_1} = \dfrac{a-s}{d-a} = \dfrac{a+s}{d+a} = \dfrac{a}{d} = \dfrac{s}{a}$ (by componendo–dividendo)  (9.11)

Because the radius of the sphere and the location of the point charge are known, the constant $\alpha$ can be computed from the ratio of $a$ and $d$, as given by Equation 9.11.

Therefore, the magnitude of the image charge is given by

$$q = \frac{a}{d}Q \tag{9.12}$$

and the location of the image charge is given by

$$s = \frac{a^2}{d} \tag{9.13}$$

Considering the line joining the point charge and its image passing through the centre of the sphere to be along the z-axis and the centre of the sphere to be the origin, as shown in Figure 9.4, and also taking into account the spherical symmetry of the configuration, the field can be expressed in spherical coordinates as follows.

With reference to the point *P* of Figure 9.4,

$$r_1 = |\vec{r}_P - \vec{r}_{+Q}| = (a^2 + d^2 - 2ad\cos\theta)^{1/2}$$

where:

$\theta$ is the angle between *a* and *d* at *P*

$$r_2 = |\vec{r}_P - \vec{r}_{-q}| = (a^2 + s^2 - 2as\cos\theta)^{1/2}$$

Similarly, for any point in the field region for which *r* is the distance of the point from the origin, that is, the centre of the sphere, and $\theta$ is the angle between *r* and *d*

$$r_1 = |\vec{r} - \vec{r}_{+Q}| = (r^2 + d^2 - 2rd\cos\theta)^{1/2} \text{ and}$$

$$r_2 = |\vec{r} - \vec{r}_{-q}| = (r^2 + s^2 - 2rs\cos\theta)^{1/2} = \left( r^2 + \frac{a^4}{d^2} - 2r\frac{a^2}{d}\cos\theta \right)^{1/2}$$

$$= \left( \frac{r^2d^2 + a^4 - 2rda^2\cos\theta}{d^2} \right)^{1/2}$$

Therefore, the electric potential at any point due to the point charge and its image is given by

$$\phi(r,\theta) = \frac{Q}{4\pi\varepsilon_0}\left[ \frac{1}{(r^2 + d^2 - 2rd\cos\theta)^{1/2}} - \frac{a}{(r^2d^2 + a^4 - 2rda^2\cos\theta)^{1/2}} \right] \tag{9.14}$$

Therefore,

$$\vec{E}_r(a,\theta) = -\frac{\partial\phi(r,\theta)}{\partial r}\bigg|_{r=a} = -\frac{Q}{4\pi\varepsilon_0 a}\frac{d^2 - a^2}{(a^2 + d^2 - 2ad\cos\theta)^{3/2}} \tag{9.15}$$

Now, $r$-component of electric field intensity is the normal component on the sphere surface. Therefore, assuming the induced surface charge density on the sphere surface to be $\sigma_s$, the normal component of electric field intensity is equal to $\sigma_s/\varepsilon_0$ just off the sphere surface. Equating this expression with the one given by Equation 9.15, the induced surface charge density on the grounded sphere surface is given by

$$\sigma_s = -\frac{Q}{4\pi a}\frac{d^2 - a^2}{(a^2 + d^2 - 2ad\cos\theta)^{3/2}} \tag{9.16}$$

### 9.3.1 Method of Successive Images

Sphere-gap arrangements are very commonly used in high-voltage system for voltage measurement. As shown in Figure 9.5, in this arrangement, two spheres of identical radius ($a$) are separated by a specific distance $s$, where one sphere is charged while the other is earthed. The field within the sphere gap due to the two spheres could be analyzed with the help of image charges, as described in Section 9.3. The live sphere of potential $V$ is at first replaced by a charge of magnitude $Q_A = 4\pi\varepsilon_0 aV$ located at the centre of the live sphere. Then to keep the potential of the grounded sphere at zero, $-q_1$ is introduced within the grounded sphere, which is the image of $Q_A$, as shown in Figure 9.5. The magnitude and location of $q_1$ are given by

$$q_1 = \frac{a}{d}Q_A \text{ and } s_1 = \frac{a^2}{d} \tag{9.17}$$

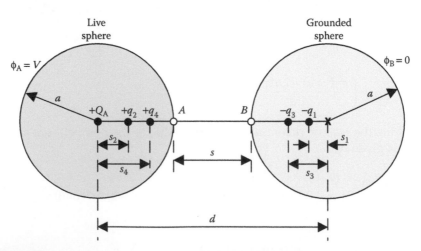

**FIGURE 9.5**
Method of successive imaging as applied to sphere-gap arrangement.

But the introduction of $-q_1$ will make the potential of live sphere different from $V$. Therefore, to keep the potential of live sphere equal to $V$, $+q_2$, which is the image of $-q_1$, is introduced within the live sphere such that the potential of live sphere due to $+q_2$ and $-q_1$ will be zero. As a result, the potential of live sphere due to $+Q_A$, $-q_1$ and $+q_2$ will be again $V$. The magnitude and location of $q_2$ are given by

$$q_2 = \frac{a}{d - s_1} q_1 = \frac{a^2}{d(d - s_1)} Q_A \text{ and } s_1 = \frac{a^2}{d - s_1} \tag{9.18}$$

But the introduction of $+q_2$ will make the potential of grounded sphere different from zero. Therefore, $-q_3$ is introduced within the grounded sphere as the image of $+q_2$ to make the potential of the grounded sphere equal to zero. Further, the introduction of $-q_3$ warrants introduction of $+q_4$ within the live sphere and so on. In this way, there will be an infinite series of charges within the two spheres: positive charges such as $+Q_A$, $+q_2$ and $+q_4$ within the live sphere and negative charges such as $-q_1$ and $-q_3$ within the grounded sphere. This method of taking successive image charges within the two spheres is known as the *method of successive imaging*. It may be seen from Equations 9.17 and 9.18 that each successive image charge is smaller in magnitude and gradually shifts towards the surface of the sphere within which it is located. In all practicality, it is adequate to take the first few images within the two spheres to achieve reasonably good accuracy in the computation of electric field. In the sphere-gap arrangement, maximum value of electric field intensity occurs at the so-called sparking tips of the spheres, namely, points $A$ and $B$, as shown in Figure 9.5. This maximum electric field intensity can be obtained as

$$E_{max} = \frac{V}{s} \left[ \frac{(s/a + 1) + \sqrt{(s/a + 1)^2 + 8}}{4} \right] \tag{9.19}$$

As discussed in Section 4.7, $E_{av} = V/s$

Therefore, field factor ($f$) for sphere-gap arrangement is as follows:

$$\frac{E_{max}}{E_{av}} = \left[ \frac{(s/a + 1) + \sqrt{(s/a + 1)^2 + 8}}{4} \right] \tag{9.20}$$

Variation of field factor ($f$) with gap distance ($s$) in the case of sphere-gap arrangement is presented in Table 9.1. It may be seen from Table 9.1 that the deviation from a uniform field ($f = 1$) for $s/a = 0.2$ is 6.8%, whereas that for $s/a = 1.0$ is 36.6%. Accuracy of voltage measurement by the sphere gap depends significantly on the degree of field non-uniformity between the two spheres. Hence, it is recommended in practice that the gap distance should not be made more than the radius of the spheres.

**TABLE 9.1**

Variation of Field Factor with Gap Distance for Sphere Gap

| s/a | 0.2 | 0.4 | 0.6 | 0.8 | 1.0 |
|-----|-----|-----|-----|-----|-----|
| f | 1.068 | 1.139 | 1.212 | 1.288 | 1.366 |

**PROBLEM 9.1**

Two spheres of 25 cm diameter have a gap distance of 2.5 cm between them. Determine the breakdown voltage of the sphere gap in air at standard temperature and pressure (STP).

*Solution:*
Given $s = 2.5$ cm and $a = (25/2) = 12.5$ cm.
  Therefore, $s/a = 2.5/12.5 = 0.2$. Correspondingly, field factor ($f$) = 1.068.
  $E_{max}$ corresponding to breakdown of air at STP is 30 kV$_p$/cm.

Therefore,

$$E_{av} = \frac{E_{max}}{f} = \frac{30}{1.068} = 28.09 \text{ kV}_p/\text{cm}$$

But,

$$E_{av} = \frac{V}{s}$$

Hence,

$$\frac{V}{2.5} = 28.09, \text{ or, } V = 70.22 \text{ kV}_p$$

### 9.3.2 Conducting Sphere in a Uniform Field

The problem of a conducting sphere placed in a uniform external field can be solved with the help of image charges, as discussed in Section 9.3. As shown in Figure 9.6, consider that the centre of the sphere is located at the origin $O$. The uniform external field is assumed to be created by two point charges $-Q$ and $+Q$ located at $(0,0,d)$ and $(0,0,-d)$, respectively. The field due to these two charges at the origin is given by

$$E_0 = \frac{2Q}{4\pi\varepsilon_0 d^2}\hat{u}_z \tag{9.21}$$

If $d$ is very large compared to the radius of the sphere $a$, then the field intensity will have approximately the value given by Equation 9.21 in the proximity of the sphere. The difference becomes negligibly small when $(d/a) \rightarrow \infty$. Then the configuration is equivalent to a conducting sphere in a uniform external field.

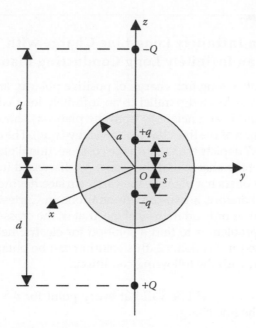

**FIGURE 9.6**
Conducting sphere in a uniform external field.

To solve this problem, two image charges $+q$ and $-q$ are placed at $(0,0,+s)$ and $(0,0,-s)$, respectively, where $s = (a^2/d)$. This will make the potential of the sphere to be zero. However, the sphere can also be considered as isolated, that is, electrically not connected to any other body, because the total image charge within the sphere is zero.

Then the potential at any point in the vicinity but outside the sphere can be written with the help of Equation 9.14 as follows:

$$\phi(r,\theta) = \frac{Q}{4\pi\varepsilon_0} \left[ \begin{array}{c} \dfrac{1}{(r^2 + d^2 + 2rd\cos\theta)^{1/2}} - \dfrac{a}{(r^2 d^2 + a^4 + 2rda^2\cos\theta)^{1/2}} \\[2ex] -\dfrac{1}{(r^2 + d^2 - 2rd\cos\theta)^{1/2}} + \dfrac{a}{(r^2 d^2 + a^4 - 2rda^2\cos\theta)^{1/2}} \end{array} \right] \tag{9.22}$$

The potential in the limiting condition of $d \to \infty$,

$$\phi(r,\theta) \approx -\frac{2Qr\cos\theta}{4\pi\varepsilon_0 d^2}\left(1 - \frac{a^3}{r^3}\right) \approx -E_0 r\cos\theta\left(1 - \frac{a^3}{r^3}\right) \tag{9.23}$$

In Equation 9.23, the first term $-E_0 r\cos\theta = -E_0 z$ is the potential due to the uniform external field and the second term is the potential due to the induced charge density on the surface of the sphere.

## 9.4 Image of an Infinitely Long Line Charge with Respect to an Infinitely Long Conducting Plane

Consider an infinitely long line charge of positive polarity and uniform line charge density $+\lambda_1$ is located parallel to an infinitely long conducting plane present at $y = 0$ and is at a height $y_1$ from the plane, as shown in Figure 9.7, such that the location of the line charge is given by $(x_1, y_1)$. The configuration as shown in Figure 9.7 depicts the view on the cross-sectional plane perpendicular to the length of the charge and the plane. As discussed in Section 9.2, the practical example of such a plane is the earth surface having zero potential. The problem is, therefore, a two-dimensional one in Cartesian coordinates, where the field varies only on the $x$–$y$ plane, that is, the cross-sectional plane.

Therefore, the problem is to find a solution for electric field in the region $y > 0$. As discussed in Section 9.2, this solution can be obtained by solving Laplace's equation with the following conditions:

1. The solution $\phi(x,y)$ will be valid at every point for $y > 0$ except the location of the line charge.

2. The potential at all the points on the conducting plane is zero, that is, $\phi(x,0) = 0$.

3. The potential approaches zero as the distance from the charge approaches infinity.

The electric potential at a distance $r$ from an infinitely long line charge is given by $Q/(2\pi\varepsilon_0)\ln(R/r)$, where $R \to \infty$. Considering the cylindrical symmetry of

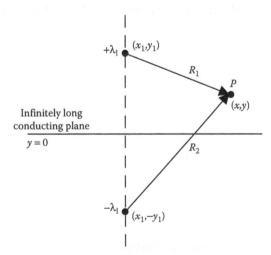

**FIGURE 9.7**
Infinitely long line charge near an infinitely long conducting plane.

the field due to an infinitely long line charge, it may be seen that the electric field due to an infinitely long line charge is dependent only on $r$-coordinate in cylindrical coordinate system and is independent of $\theta$ and $z$ coordinates. Hence, this electric potential satisfies Laplace's equation in cylindrical coordinates as follows:

$$\vec{\nabla}^2\phi = \frac{1}{r}\frac{\partial}{\partial r}\left(r\frac{\partial\phi}{\partial r}\right) = \frac{1}{r}\frac{\partial}{\partial r}\left(r\times-\frac{Q}{2\pi\varepsilon_0 r}\right) = \frac{1}{r}\frac{\partial}{\partial r}\left(-\frac{Q}{2\pi\varepsilon_0}\right) = 0 \qquad (9.24)$$

Similar to the discussion of Section 9.2, the infinitely long conducting plane is replaced by a fictitious line charge of the uniform line charge density $-\lambda_l$ located at $(x_1,-y_1)$, which is the image of the real line charge. Because the field due to an infinitely long line charge satisfies Laplace's equation, the field due to the real line charge and the image line charge will also satisfy Laplace's equation at all points for $y > 0$ except at the location of the real line charge.

Therefore, the electric potential at any point $P(x,y)$ will be given by

$$\phi_P = \frac{\lambda_l}{2\pi\varepsilon_0}\ln\frac{R_2}{R_1} \qquad (9.25)$$

where:

$$R_1 = \left[(x-x_1)^2 + (y-y_1)^2\right]^{1/2}$$

$$\text{and } R_2 = \left[(x-x_1)^2 + (y+y_1)^2\right]^{1/2}$$

From Equation 9.25, it is evident that the potential at a very large distance from the line charge will be zero and also that the potential at all the points on the conducting plane at $y = 0$ will be zero, because at every point on the plane $R_1 = R_2$.

In Cartesian coordinates for a two-dimensional system, the electric field intensity at any point $P(x,y)$ for $y > 0$ will be given by

$$\vec{E}_P = -\frac{\partial\phi_P}{\partial x}\hat{i} - \frac{\partial\phi_P}{\partial y}\hat{j} \qquad (9.26)$$

where:

$$E_{xP} = -\frac{\partial\phi_P}{\partial x} = \frac{Q}{2\pi\varepsilon_0}\left(\frac{x-x_1}{R_1^2} - \frac{x-x_1}{R_2^2}\right)$$

$$E_{yP} = -\frac{\partial\phi_P}{\partial y} = \frac{Q}{2\pi\varepsilon_0}\left(\frac{y-y_1}{R_1^2} - \frac{y+y_1}{R_2^2}\right)$$

For any point very near to the conducting plane, that is, $y \to 0$, $R_1 \approx R_2$. Hence, $E_{xP}$ will be zero. Thus

$$\lim_{y \to 0} \vec{E}_{yP} = \frac{\sigma_s}{\varepsilon_0} \hat{k} = \frac{-Qy_1}{\pi \varepsilon_0 \left[ (x - x_1)^2 + y_1^2 \right]^{3/2}} \hat{k}$$

$$\text{or, } \sigma_s = \frac{-Qy_1}{\pi \left[ (x - x_1)^2 + y_1^2 \right]^{3/2}} \tag{9.27}$$

Equation 9.27 gives the negative surface charge density induced on the conducting plane by the positive polarity real line charge.

From Equation 9.25, an expression for the equipotential surface can be obtained as follows:

$$\frac{R_2}{R_1} = \alpha$$

where:
  $\alpha$ is a constant

$$\text{or, } (x - x_1)^2 + (y + y_1)^2 = \alpha^2 \left[ (x - x_1)^2 + (y - y_1)^2 \right]$$

$$\text{or, } (x - x_1)^2 + \left( y - y_1 \frac{\alpha^2 + 1}{\alpha^2 - 1} \right)^2 = \left( \frac{2\alpha y_1}{\alpha^2 - 1} \right)^2 \tag{9.28}$$

Equation 9.28 shows that the equipotentials are circles of radius $\left| (2\alpha y_1)/(\alpha^2 - 1) \right|$ having centre at $x_1, (\alpha^2 + 1)/(\alpha^2 - 1)y_1$ on the cross-sectional plane. For the infinite conducting plane of zero potential $\alpha = 1$. For equipotentials above the infinite conducting plane $\alpha > 1$ and those below the infinite conducting plane $\alpha < 1$. In physical terms, the equipotentials are circular cylinders with axes parallel to the two line charges, as shown in Figure 9.8. As $\alpha$ increases the radius of the cylinder increases and the axis of the cylinder shifts further away from the line charge.

The equipotentials below the conducting planes do not have any physical meaning as the electric field is zero below the infinitely long conducting plane. But the equipotentials, as shown in Figure 9.8, give an important idea that two parallel cylinders having electric potentials $+V$ and $-V$ could be replaced by two infinitely long line charges. However, the line charges will not be located at the axes of the cylinders.

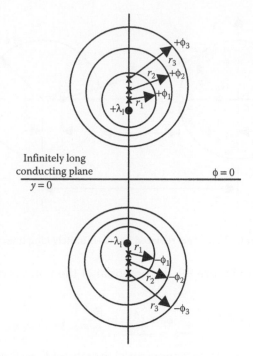

**FIGURE 9.8**

Equipotentials for an infinitely long charge and its image with respect to an infinitely long conducting plane.

## 9.5 Two Infinitely Long Parallel Cylinders

Electric field due to two parallel cylindrical transmission line conductors is the same as the field due to two infinitely long parallel cylinders. The cross-sectional view of the arrangement is shown in Figure 9.9. Electric field for this arrangement is two dimensional in Cartesian coordinates, because the field does not vary along the $z$-axis, which is along the length of the cylinders. Electric field varies only on the cross-sectional plane, which is taken as the $x$–$y$ plane. As discussed in Section 9.4, these two parallel cylinders having potential $+V$ and $-V$ could be replaced by two infinitely long line charges of the uniform line charge density $+\lambda_1$ and $-\lambda_1$ located within the respective cylinders, as shown in Figure 9.9. These two line charges together will create two cylindrical equipotential surfaces of radius $a$ having the specified potentials $+V$ and $-V$. The charges will be located at a distance $s$ from the axis of the respective cylinders. Therefore, the problem is to find the location of these charges.

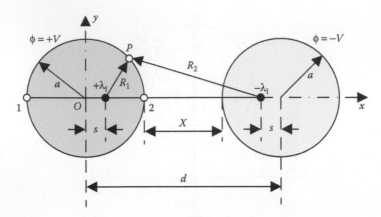

**FIGURE 9.9**
Two infinitely long parallel cylinders replaced by two infinitely long line charges.

With reference to Figure 9.9, the potential of the point $P$ on the surface of the cylinder is

$$\phi_P = \frac{\lambda_1}{2\pi\varepsilon_0} \ln \frac{R_2}{R_1}$$

For the cylinder surface to be equipotential the ratio of $R_2$ to $R_1$ must be constant. Considering the point 1 on the cylinder surface, as shown in Figure 9.9,

$$\phi_1 = \frac{\lambda_1}{2\pi\varepsilon_0} \ln \frac{d+a-s}{a+s} \tag{9.29}$$

and for the point 2 on the cylinder surface, as shown in Figure 9.9,

$$\phi_2 = \frac{\lambda_1}{2\pi\varepsilon_0} \ln \frac{d-a-s}{a-s} \tag{9.30}$$

But, $\phi_1 = \phi_2$. Hence, $(d+a-s)/(a+s) = (d-a-s)/(a-s) = (d-s)/a = a/s$ (by componendo–dividendo)

Therefore,

$$s^2 - sd + a^2 = 0$$

$$\text{or, } s = \frac{d}{2} \pm \frac{\sqrt{d^2 - 4a^2}}{2} \tag{9.31}$$

In the solution of $s$ as given by Equation 9.31, the additive expression has to be neglected, because in that case the image charge will be located outside the cylinder. Therefore,

$$s = \frac{d}{2} - \frac{\sqrt{d^2 - 4a^2}}{2} \tag{9.32}$$

For transmission lines, $d \gg a$ and hence $s \approx 0$, that is, the line charges are placed on the axes of the two cylinders.

Now, the potential at the point 2 on the cylinder surface, as shown in Figure 9.9, is $+V$. Hence,

$$\phi_2 = \frac{\lambda_1}{2\pi\varepsilon_0} \ln \frac{d-a-s}{a-s} = V$$

$$\text{or, } \lambda_1 = \frac{2\pi\varepsilon_0 V}{\ln\left[(d-a-s)/(a-s)\right]} \tag{9.33}$$

Equation 9.33 gives the magnitude of the uniform line charge density.

In the arrangement shown in Figure 9.9, maximum electric field intensity $(E_{max})$ occurs at the point 2, which is given by

$$E_{max} = \frac{\lambda_1}{2\pi\varepsilon_0}\left(\frac{1}{a-s} + \frac{1}{d-a-s}\right) = \frac{V}{\ln\left[(d-a-s)/(a-s)\right]}\left(\frac{1}{a-s} + \frac{1}{d-a-s}\right) \tag{9.34}$$

Right-hand side (RHS) of Equation 9.34 is in terms of the physical dimensions of the arrangement and the electric potential of the cylinders and hence can be computed in a straightforward manner.

Again, for the physical arrangement of Figure 9.9, $E_{av} = 2V/(d-2a)$.

Therefore, field factor

$$(f) = \frac{E_{max}}{E_{av}} = \frac{(d-2a)}{2\ln\left[(d-a-s)/(a-s)\right]}\left(\frac{1}{a-s} + \frac{1}{d-a-s}\right) \tag{9.35}$$

Putting the value of $s$ from Equation 9.32 in Equation 9.35 and simplifying it may be written that

$$f = \frac{\sqrt{(d/2a)^2 - 1}}{\ln\left[(d/2a) + \sqrt{(d/2a)^2 - 1}\right]} \tag{9.36}$$

For transmission lines, Equation 9.36 is often modified by putting $d = X + 2a$, which yields

$$f = \frac{\sqrt{(X/a)^2 + 4(X/a)}}{2\ln\left[(X/2a) + 1 + (1/2)\sqrt{(X/a)^2 + 4(X/a)}\right]} \tag{9.37}$$

Equation 9.37 represents the field factor as a function of the ratio of the gap distance between the two transmission line conductors $(X)$ and the radius of the conductors $(a)$.

For high-voltage transmission lines, $d \gg a$. As a result the field factor as given by Equation 9.36 reduces to

$$f = \frac{d}{2a \ln(d/a)} \tag{9.38}$$

$$\text{and } E_{av} = \frac{2V}{d-2a} \approx \frac{2V}{d}$$

Hence,

$$E_{max} = E_{av} \times f = \frac{2V}{d} \times \frac{d}{2a \ln(d/a)} = \frac{V}{a \ln(d/a)} \tag{9.39}$$

Capacitance per unit length between the two parallel cylinders can be obtained from Equations 9.32 and 9.33 as follows:

$$C = \frac{\lambda_1}{2V} = \frac{\pi \varepsilon_0}{\ln\left[(d-a-s)/(a-s)\right]} = \frac{\pi \varepsilon_0}{\ln\left[(d/2a) + \sqrt{(d/2a)^2 - 1}\right]} \tag{9.40}$$

Because $\ln(x + \sqrt{x^2 - 1}) = \cos h^{-1} x$, for $x > 1$ Equation 9.40 can be written as follows:

$$C = \frac{\pi \varepsilon_0}{\cos h^{-1}(d/2a)} \tag{9.41}$$

## PROBLEM 9.2

A long conductor of negligible radius is at a height 5 m from the earth surface and is parallel to it. It has a uniform line charge density of +1 nC/m. Find the electric potential and field intensity at a point 3 m below the line.

*Solution:*
The arrangement of the problem is shown in Figure 9.10. Because the conductor is considered to have negligible radius, the line charge is located on the axis of the conductor.

With reference to Equation 9.25, $\lambda_1 = 1$ nC/m, $R_1 = 3$ m and $R_2 = 7$ m.
Therefore,

$$\phi_P = \frac{10^{-9}}{2\pi \times 8.854 \times 10^{-12}} \ln \frac{7}{3} = 15.23 \text{ V}$$

Electric field intensity components at the point $P$ are obtained from Equation 9.26 as follows:

$$E_{xP} = 0$$

$$E_{yP} = \frac{10^{-9}}{2\pi \times 8.854 \times 10^{-12}} \left( \frac{2-5}{3^2} - \frac{2+5}{7^2} \right) = -8.56 \text{ V/m}.$$

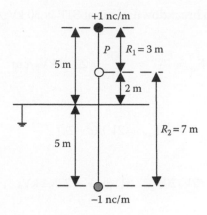

**FIGURE 9.10**
Pertaining to Problem 9.2.

## PROBLEM 9.3

Determine the breakdown voltage in air at standard temperature and pressure (STP) of a 20 cm diameter cylindrical electrode placed horizontally with its axis 20 cm above the earth surface.

*Solution:*

The arrangement of the problem is shown in Figure 9.11.

With reference to Equation 9.36 $d = 40$ cm and $a = 10$ cm. Hence, $(d/2a) = 2$. Therefore,

$$f = \frac{\sqrt{2^2 - 1}}{\ln\left(2 + \sqrt{2^2 - 1}\right)} = 1.315$$

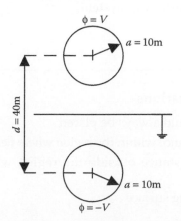

**FIGURE 9.11**
Pertaining to Problem 9.3.

$E_{max}$ corresponding to breakdown of air at STP is 30 kV$_p$/cm.
   Therefore,

$$E_{av} = \frac{E_{max}}{f} = \frac{30}{1.315} = 22.81 \text{ kV}_p/\text{cm}$$

But,

$$E_{av} = (2V)/(d - 2a)$$

Hence,

$$2V/20 = 22.81, \text{ or, } V = 228.1 \text{ kV}_p$$

## 9.6 Salient Features of Method of Images

1. Image charges are always located outside the region where the field is to be determined.

2. Depending on the type of problem, the magnitude of image charge may or may not be the same as that of the physical charge.

3. Depending on the type of problem, image charges may or may not be of polarity opposite to that of the physical charge.

4. In the image charge system, electric field is finite on both sides of the imaging surface. But in the physical system, electric field is non-zero and finite only on one side of the imaging surface. Because energy in electric field is proportional to volume integral of $E^2$, the electrostatic energy in the physical charge system is half of the electrostatic energy of the image charge system.

## Objective Type Questions

1. Image charge should always be placed
   a. At a finite distance within the region where field is to be computed
   b. At a finite distance outside the region where field is to be computed
   c. On the imaging surface
   d. At infinity

2. Image of a charge can be taken with respect to
   a. A finitely long conducting plane
   b. An infinitely long conducting plane
   c. A curved surface such as cylinder or sphere
   d. Both (b) and (c)

3. Electrostatic energy in the physical charge system only is
   a. Square of the electrostatic energy of the image charge system
   b. Double the electrostatic energy of the image charge system
   c. Equal to the electrostatic energy of the image charge system
   d. Half of the electrostatic energy of the image charge system

4. A point charge is placed in the wedge-shaped space between two grounded conducting planes making an angle 60° between them. Then the number of image charges will be
   a. 2
   b. 4
   c. 5
   d. 6

5. A point charge is placed in the space between two perpendicular conducting planes. Then
   a. All the three image charges will be of negative polarity
   b. Two image charges will be of negative polarity
   c. One image charge will be of negative polarity
   d. None of the image charges will be of negative polarity

6. Two long parallel cylinders of same radius with a specific separation distance can be replaced by two infinite line charges where
   a. Both the charges are located at the axes of the cylinders
   b. One charge is located at the axis of one cylinder, whereas the other is located away from the axis of the other cylinder
   c. Both the charges are located away from the axes of the cylinders
   d. Both the charges are located on the surface of the cylinders

7. Two identical spheres with a specific gap distance can be replaced by
   a. Two point charges located at the centres of the two spheres
   b. Two point charges located away from the centres of the two spheres
   c. Two point charges located on the surface of the two spheres
   d. Infinite number of point charges located within the two spheres

8. Method of successive images is used in the case of
   a. Two identical spheres
   b. A point charge near a grounded sphere
   c. Two long parallel cylinders
   d. A line charge near a long parallel grounded cylinder

9. Equipotentials due to two infinitely long parallel line charges will be
   a. Infinitely long cylinders of circular cross section
   b. Finite length cylinders of circular cross section
   c. Infinitely long cylinders of elliptical cross section
   d. Finite length cylinders of elliptical cross section

10. A point charge of magnitude $Q$ is placed at a distance $d$ from the centre of a grounded sphere of radius $a$ $(a < d)$. Then the magnitude of the image charge to replace the grounded sphere will be
   a. $(a/d)Q$
   b. $(a/d^2)Q$
   c. $(a^2/d)Q$
   d. $(a^2/d^2)Q$

**Answers:**    1) b; 2) d; 3) d; 4) c; 5) b; 6) c; 7) d; 8) a; 9) a; 10) a

# 10

## *Sphere or Cylinder in Uniform External Field*

**ABSTRACT**  If a conducting or dielectric object is present within the electric field caused by external sources, then the induced charges on such an object adds to the existing external field. If the size of the conducting or dielectric object is small compared to the external field region, then the electric field is modified only in the vicinity of such an object. In order to get the complete solution of the electric field in such configuration, the boundary conditions imposed by the conducting or dielectric object must also be satisfied. If the object is of well-defined geometric shape such as cylinder or sphere, and if the external field is considered to be uniform in space, then the complete solution for the electric field could be obtained analytically with the help of method of separation of variables. These analytical solutions are often used to validate numerical methodology and could also be used to estimate field modification due to small conducting or dielectric pieces present in large high-voltage insulation arrangement.

## 10.1 Introduction

Conducting and dielectric components are integral parts of any electrical equipment. If the size of the conducting or dielectric object is very small compared to dimensions of the field region where the object is located, then the object contributes to the field only in the domain near the object. In many cases, such objects are present as stray bodies in high-voltage insulation arrangement. As practical examples, one may cite a small piece of conductor or dielectric floating in liquid insulation of large volume in transformers, metallic dust particles floating in gaseous insulation within gas-insulated system (GIS) and so on. It is important to understand how the presence of a conducting or dielectric object modifies the external field in the vicinity of the object, because any enhancement of the electric field intensity due to the conducting or dielectric object may lead to unwanted discharge or in the worst case failure of the insulation system.

If it is assumed that the source charges (in practical arrangement, the electrodes or conductors with specific potentials) that produce the external field is located far away from the object under consideration, then they are unaffected by the presence of the object. Consequently, the field due to the source charges may be considered to be uniform at the location of the object. If the object is such that its shape is defined by well-known mathematical functions, for example, cylinders or spheres, then the complete solution for the electric field due to the source charges located at far away positions and the induced charges on the surface of the object could be obtained by solving Laplace's equation considering the field region to be free from any volume charge. However, in order to get the complete solution appropriate boundary conditions on the surface of the object, whether it is conducting or dielectric, need to be satisfied. One of the common methods of getting the analytical solution for a cylinder or sphere in a uniform external field is the method of separation of variables as described in this chapter.

## 10.2 Sphere in Uniform External Field

Consider a spherical object of radius $a$ within uniform external field, as shown in Figure 10.1. Because the boundary is a sphere of $r =$ constant, hence the system is best described in spherical coordinates, as shown in Figure 10.1. The uniform external field is given by $\vec{E}_0 = - E_0 \hat{u}_z$ and the potential at any point due to the external field is given by $E_0 r \cos\theta = E_0 z$ with respect to the centre of the sphere. In order to get the complete solution for the electric field in this system, Laplace's equation in spherical coordinates, as given in Equation 10.1, needs to be solved.

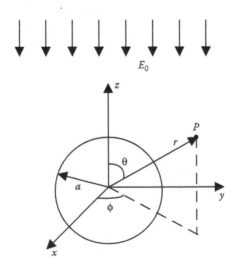

**FIGURE 10.1**
Sphere in uniform external field.

$$\frac{1}{r^2}\frac{\partial}{\partial r}\left(r^2\frac{\partial V}{\partial r}\right) + \frac{1}{r^2\sin\theta}\frac{\partial}{\partial\theta}\left(\sin\theta\frac{\partial V}{\partial\theta}\right) + \frac{1}{r^2\sin^2\theta}\frac{\partial^2 V}{\partial\phi^2} = 0 \qquad (10.1)$$

The field system has azimuthal symmetry with respect to the z-axis, that is, the field system does not change with the rotation around the z-axis. Therefore, z-axis is made the polar axis in the spherical coordinate system. Then the field is independent of coordinate φ and Laplace's equation reduces to

$$\frac{1}{r^2}\frac{\partial}{\partial r}\left(r^2\frac{\partial V}{\partial r}\right) + \frac{1}{r^2\sin\theta}\frac{\partial}{\partial\theta}\left(\sin\theta\frac{\partial V}{\partial\theta}\right) = 0 \qquad (10.2)$$

In order to separate the terms of the left-hand side (LHS) of Equation 10.2 into functions of only one variable, Equation 10.2 may be rewritten by multiplying $r^2$ as

$$\frac{\partial}{\partial r}\left(r^2\frac{\partial V}{\partial r}\right) + \frac{1}{\sin\theta}\frac{\partial}{\partial\theta}\left(\sin\theta\frac{\partial V}{\partial\theta}\right) = 0 \qquad (10.3)$$

Then LHS of Equation 10.3 is the sum of two terms, which are functions of only one variable each, that is, the first term is function of $r$ only, whereas the second term is function of θ only. The solution to Equation 10.3 can be obtained as the product of two functions of which one is dependent only on $r$ and the other is dependent only on θ.

Let the assumed solution be

$$V(r,\theta) = M(r)N(\theta) \qquad (10.4)$$

The assumed solution is convenient as the boundary lies at $r = $ constant.

Combining Equations 10.3 and 10.4

$$\frac{\partial}{\partial r}\left[r^2\frac{\partial M(r)N(\theta)}{\partial r}\right] + \frac{1}{\sin\theta}\frac{\partial}{\partial\theta}\left[\sin\theta\frac{\partial M(r)N(\theta)}{\partial\theta}\right] = 0$$

$$\text{or, } N(\theta)\frac{\partial}{\partial r}\left[r^2\frac{\partial M(r)}{\partial r}\right] + M(r)\frac{1}{\sin\theta}\frac{\partial}{\partial\theta}\left[\sin\theta\frac{\partial N(\theta)}{\partial\theta}\right] = 0$$

Dividing by $M(r)N(\theta)$,

$$\frac{1}{M(r)}\frac{d}{dr}\left[r^2\frac{dM(r)}{dr}\right] + \frac{1}{N(\theta)}\frac{1}{\sin\theta}\frac{d}{d\theta}\left[\sin\theta\frac{dN(\theta)}{d\theta}\right] = 0 \qquad (10.5)$$

The partial derivatives become total derivatives in Equation 10.5, as each term is dependent on only one coordinate.

The sum of two terms of the LHS of Equation 10.5 could be zero only when the two terms are separately equal to opposite and equal constant terms as given in Equation 10.6.

Equal and opposite separation constant solution:

$$\frac{1}{M(r)}\frac{d}{dr}\left[r^2\frac{dM(r)}{dr}\right] = +p \text{ and } \frac{1}{N(\theta)}\frac{1}{\sin\theta}\frac{d}{d\theta}\left[\sin\theta\frac{dN(\theta)}{d\theta}\right] = -p \quad (10.6)$$

where:

$p$ is a positive constant

Another solution is obtained when the separation constant is zero. Hence, the zero separation constant solution

$$\frac{1}{M(r)}\frac{d}{dr}\left[r^2\frac{dM(r)}{dr}\right] = 0 \text{ and } \frac{1}{N(\theta)}\frac{1}{\sin\theta}\frac{d}{d\theta}\left[\sin\theta\frac{dN(\theta)}{d\theta}\right] = 0 \quad (10.7)$$

Each of the above-mentioned two solutions is to be obtained separately.

*Determination of the zero separation constant solution:*

The first term of Equation 10.7 is

$$\frac{1}{M(r)}\frac{d}{dr}\left[r^2\frac{dM(r)}{dr}\right] = 0$$

where:

$M(r)$ is non-zero

Therefore,

$$\frac{d}{dr}\left[r^2\frac{dM(r)}{dr}\right] = 0, \text{ or, } r^2\frac{dM(r)}{dr} = C_0, \text{ or, } \frac{dM(r)}{dr} = \frac{C_0}{r^2}$$

Integrating and incorporating constants of integration

$$M(r) = \frac{C_{10}}{r} + C_{20} \quad (10.8)$$

Next the second term of Equation 10.7 is

$$\frac{1}{N(\theta)}\frac{1}{\sin\theta}\frac{d}{d\theta}\left[\sin\theta\frac{dN(\theta)}{d\theta}\right] = 0$$

where:

$N(\theta)$ is non-zero

Therefore,

$$\frac{1}{\sin\theta}\frac{d}{d\theta}\left[\sin\theta\frac{dN(\theta)}{d\theta}\right] = 0, \text{ or, } \frac{d}{d\theta}\left[\sin\theta\frac{dN(\theta)}{d\theta}\right] = 0$$

$$\text{or, } \sin\theta\frac{dN(\theta)}{d\theta} = A_0$$

Integrating and incorporating constant of integration

$$N(\theta) = A_{10} \ln\left(\tan\frac{\theta}{2}\right) + A_{20} \tag{10.9}$$

Equation 10.9 becomes undefined for $\theta = \pi$. But this is not feasible in the given system as potential must be a continuous function. Therefore, $A_{10}$ should be zero in Equation 10.9. Therefore,

$$N(\theta) = A_{20} \tag{10.10}$$

Then from Equations 10.4, 10.8 and 10.10, the zero separation constant solution can be obtained as

$$V(r,\theta) = \frac{C_1}{r} + C_2 \tag{10.11}$$

where:
$C_1 = A_{20}C_{10}$ and $C_2 = A_{20}C_{20}$

*Determination of the equal and opposite separation constant solution:*
  The first term of Equation 10.6 is

$$\frac{1}{M(r)}\frac{d}{dr}\left[r^2\frac{dM(r)}{dr}\right] = +p$$

$$\text{or, } \frac{d}{dr}\left[r^2\frac{dM(r)}{dr}\right] = +pM(r) \tag{10.12}$$

Putting $M(r) = Cr^n$ in Equation 10.12, we get

$$\frac{d}{dr}\left(r^2Cnr^{n-1}\right) = +pCr^n, \text{ or, } Cn(n+1)r^n = pCr^n, \text{ or, } n^2 + n - p = 0$$

Hence,

$$n = \frac{1}{2}\left(-1 \pm \sqrt{1+4p}\right) \tag{10.13}$$

The second term of Equation 10.6 is

$$\frac{1}{N(\theta)}\frac{1}{\sin\theta}\frac{d}{d\theta}\left[\sin\theta\frac{dN(\theta)}{d\theta}\right] = -p$$

$$\text{or, } \frac{1}{\sin\theta}\frac{d}{d\theta}\left[\sin\theta\frac{dN(\theta)}{d\theta}\right] = -pN(\theta) \tag{10.14}$$

Putting $N(\theta) = B\cos\theta$ in Equation 10.14,

$$\frac{d}{d\theta}\left(-B\sin^2\theta\right) = -pB\cos\theta\sin\theta, \text{ or, } p = 2.$$

Hence, from Equation 10.13 $n = +1, -2$.

$$\text{Therefore, } M(r) = C'r + \frac{C''}{r^{-2}} \text{ and } N(\theta) = B\cos\theta \qquad (10.15)$$

From Equations 10.4 and 10.15,

$$V(r,\theta) = \left(C_3 r + \frac{C_4}{r^2}\right)\cos\theta \qquad (10.16)$$

where:
 $C_3 = C'B$ and $C_4 = C''B$

The complete solution for potential function is uniquely given as a linear combination of the two solutions given by Equations 10.11 and 10.16.

$$V(r,\theta) = \frac{C_1}{r} + C_2 + \left(C_3 r + \frac{C_4}{r^2}\right)\cos\theta \qquad (10.17)$$

where:
 The constants are determined by satisfying the boundary conditions

It is evident from Equation 10.17 that the first term corresponds to a net charge on the sphere and the second term to a finite potential.

## 10.2.1 Conducting Sphere in Uniform Field

Consider that the sphere is a conducting one and is isolated and uncharged. Further, consider that the potential at the location of the centre of the sphere due to the external field is $V_0$.

Because the perturbing action of the sphere is negligible at a large distance from the sphere, the potential at a large distance from the sphere ($r \gg a$) is given by

$$V(r,\theta) = V_0 + E_0 r \cos\theta \qquad (10.18)$$

If the sphere is charged with a finite amount of charge $Q$, then

$$V(r,\theta) = \frac{Q}{4\pi\varepsilon_0 r} + V_0 + E_0 r \cos\theta \qquad (10.19)$$

In practical systems, floating metallic particles are usually not charged and hence Equation 10.18 is taken here for further discussion.

Comparing Equations 10.17 and 10.18 for $r \to \infty$, $C_2 = V_0$ and $C_3 = E_0$. $C_1$ will be zero for uncharged sphere.

Therefore, Equation 10.17 can be rewritten as

$$V(r,\theta) = V_0 + \left( E_0 r + \frac{C_4}{r^2} \right) \cos\theta \qquad (10.20)$$

On the conductor surface, that is, for $r = a$,

$$V(a,\theta) = V_0 + \left( E_0 a + \frac{C_4}{a^2} \right) \cos\theta \qquad (10.21)$$

But the conducting sphere surface is an equipotential and hence electric potential is independent of $\theta$ on the conductor surface.

Therefore, from Equation 10.21,

$$C_4 = -E_0 a^3$$

Hence, the complete solution for an electric potential in the domain $r > a$ is given by

$$V(r,\theta) = V_0 + E_0 \left( r - \frac{a^3}{r^2} \right) \cos\theta \qquad (10.22)$$

The $r$ and $\theta$ components of the electric field intensity could be obtained as follows:

$$E_r = -\frac{\partial V}{\partial r} = -E_0 \left( 1 + \frac{2a^3}{r^3} \right) \cos\theta \qquad (10.23)$$

$$E_\theta = -\frac{1}{r}\frac{\partial V}{\partial \theta} = E_0 \left( 1 - \frac{a^3}{r^3} \right) \sin\theta \qquad (10.24)$$

On the conducting sphere surface, tangential component of the electric field intensity must be zero as it is an equipotential surface. Equation 10.24 shows that for $r = a$, $E_\theta$ is zero, which in turn validates the solution obtained.

Again, on the conducting sphere surface, $E_r$ is the normal component of the electric field intensity, which is given by $E_r|_{r=a} = -3E_0 \cos\theta$. Thus, the maximum value of the electric field intensity on the surface of the conducting sphere is $3E_0$, that is, three times the strength of the uniform external field.

This is the reason why metallic dust particles should be avoided at all costs for GISs. Because the presence of metallic dust particles will increase the local electric field intensity three times, which will result into partial discharge within the GIS, that is very detrimental for the GIS operation.

Induced surface charge density on the surface of the conducting sphere may be obtained as follows:

$$\frac{\sigma_s}{\varepsilon_0} = E_r\bigg|_{r=a} = -3E_0\cos\theta, \text{ or, } \sigma_s = -3\varepsilon_0 E_0\cos\theta \qquad (10.25)$$

As stated earlier, the sphere may be charged with an additional charge $Q$, which is distributed uniformly on the sphere surface and its effect on the field could be found by superposition.

### 10.2.2 Dielectric Sphere in Uniform Field

In the case of dielectric sphere present in uniform external field, there will be two solutions to potential function, $V_i$ valid for the region within the sphere having dielectric of permittivity $\varepsilon_i$ and $V_e$ valid for the region outside the sphere having dielectric of permittivity $\varepsilon_e$. Therefore, from Equation 10.17

$$V_i(r,\theta) = \frac{C_{1i}}{r} + C_{2i} + \left(C_{3i}r + \frac{C_{4i}}{r^2}\right)\cos\theta \qquad (10.26)$$

and

$$V_e(r,\theta) = \frac{C_{1e}}{r} + C_{2e} + \left(C_{3e}r + \frac{C_{4e}}{r^2}\right)\cos\theta \qquad (10.27)$$

The potential at large distance $r$ $(r \gg a)$ from the sphere

$$V(r,\theta) = V_0 + E_0 r\cos\theta \qquad (10.28)$$

where:
   $V_0$ is the potential at the location of the centre of the sphere due to the
      external field

Comparing Equations 10.27 and 10.28 for $r \to \infty$, $C_{2e} = V_0$ and $C_{3e} = E_0$. $C_{1e}$ will be zero as a dielectric sphere is not considered to have any free charge.
   Hence, Equation 10.27 can be rewritten as

$$V_e(r,\theta) = V_0 + \left(E_0 r + \frac{C_{4e}}{r^2}\right)\cos\theta \qquad (10.29)$$

Inside the dielectric sphere electric potential must be finite at all the points. Hence, from Equation 10.26 $C_{1i} = C_{4i} = 0$, $C_{2i} = V_0$. Hence, Equation 10.26 can be rewritten as

$$V_i(r,\theta) = V_0 + C_{3i} r\cos\theta \qquad (10.30)$$

At $r = a$, both Equations 10.29 and 10.30 should yield the same electric potential. Therefore,

$$V_0 + \left( E_0 a + \frac{C_{4e}}{a^2} \right) \cos \theta = V_0 + C_{3i} a \cos \theta$$

$$\text{or, } E_0 a + \frac{C_{4e}}{a^2} = C_{3i} a \qquad (10.31)$$

On the dielectric–dielectric boundary, the normal component of the electric field intensity should be same on both sides of the boundary. For the spherical boundary, $r$-component of the electric field intensity is the normal component on the boundary. Hence,

$$-\varepsilon_i \left( \frac{\partial V_i}{\partial r} \right)_{r=a} = -\varepsilon_e \left( \frac{\partial V_e}{\partial r} \right)_{r=a}$$

$$\text{or, } \varepsilon_i C_{3i} = \varepsilon_e \left( E_0 - \frac{2C_{4e}}{a^3} \right) \qquad (10.32)$$

From Equations 10.31 and 10.32

$$C_{4e} = \frac{\varepsilon_e - \varepsilon_i}{2\varepsilon_e + \varepsilon_i} a^3 E_0$$

and

$$C_{3i} = \frac{3\varepsilon_e}{2\varepsilon_e + \varepsilon_i} E_0$$

Therefore, the complete solutions for potential functions inside and outside the dielectric sphere are given by

$$V_i(r, \theta) = V_0 + \frac{3\varepsilon_e}{2\varepsilon_e + \varepsilon_i} E_0 r \cos \theta \qquad (10.33)$$

and

$$V_e(r, \theta) = V_0 + \left( r + \frac{\varepsilon_e - \varepsilon_i}{2\varepsilon_e + \varepsilon_i} \frac{a^3}{r^2} \right) E_0 \cos \theta \qquad (10.34)$$

Noting that $r\cos \theta = z$, potential function inside the dielectric sphere can be written as

$$V_i(x, y, z) = V_0 + \frac{3\varepsilon_e}{2\varepsilon_e + \varepsilon_i} E_0 z \qquad (10.35)$$

Hence, the electric potential within the dielectric sphere varies in only z-direction, that is, the direction of the external field. Electric field intensity within the dielectric sphere will, therefore, have only the z-component, which is given by

$$E_{zi} = -\frac{\partial V_i}{\partial z} = -\frac{3\varepsilon_e}{2\varepsilon_e + \varepsilon_i} E_0 \qquad (10.36)$$

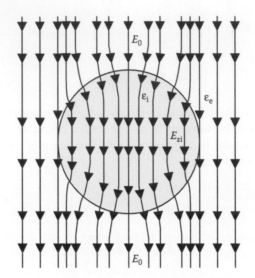

**FIGURE 10.2**
Electric field in and around dielectric sphere in uniform field.

Equation 10.36 shows that the magnitude of the electric field intensity within the dielectric sphere is constant. Typical field distribution in and around a dielectric sphere within uniform external field is shown in Figure 10.2.

Equation 10.36 also shows that if $\varepsilon_i < \varepsilon_e$, then $|E_{zi}| > E_0$. Consider the case of a spherical air bubble trapped within a moulded solid insulation of relative permittivity 4. If the magnitude of the electric field intensity in solid insulation at the location of the air bubble is $E_0$, then the magnitude of the electric field intensity within the air bubble will be $1.33E_0$. The operating electric field intensity within solid insulation is usually kept at a higher value as the solid insulation has a higher dielectric strength and hence such increase in field intensity within the air bubble often causes partial discharge within the air bubble as the dielectric strength of air is much lower than solid insulation.

## 10.3 Cylinder in Uniform External Field

Consider a long cylindrical object of radius $a$ within uniform external field, as shown in Figure 10.3. Because the boundary is a circle of $r =$ constant on the $x$–$y$ plane, the system is best described in cylindrical coordinates, as shown in Figure 10.3. The uniform external field is given by $\vec{E}_0 = -E_0\hat{i}$ and the potential at any point due to the external field is given by $E_0 r \cos\theta = E_0 x$ with respect to the axis of the cylinder. In order to get the complete solution for the electric field in this system, Laplace's equation in cylindrical coordinates as given in Equation 10.37 needs to be solved.

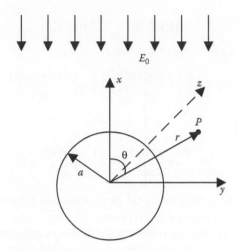

**FIGURE 10.3**
Cylinder in uniform external field.

$$\frac{1}{r}\frac{\partial}{\partial r}\left(r\frac{\partial V}{\partial r}\right)+\frac{1}{r^2}\frac{\partial^2 V}{\partial\theta^2}+\frac{\partial^2 V}{\partial z^2}=0 \qquad (10.37)$$

For this arrangement, the electric field distribution does not vary along the length of the cylinder, that is, along z-coordinate. Hence, Laplace's equation reduces to

$$\frac{1}{r}\frac{\partial}{\partial r}\left(r\frac{\partial V}{\partial r}\right)+\frac{1}{r^2}\frac{\partial^2 V}{\partial\theta^2}=0 \qquad (10.38)$$

Separating the terms of the LHS into functions of only one variable by multiplying $r^2$ with Equation 10.38, it may be written that

$$r\frac{\partial}{\partial r}\left(r\frac{\partial V}{\partial r}\right)+\frac{\partial^2 V}{\partial\theta^2}=0 \qquad (10.39)$$

The two terms on the LHS of Equation 10.39 are functions of only one variable each, that is, the first term is function of $r$ only, whereas the second term is function of $\theta$ only. The solution to Equation 10.39 can be obtained as the product of two functions of which one is dependent only on $r$ and the other is dependent only on $\theta$.

Let the assumed solution be

$$V(r,\theta) = M(r)N(\theta) \qquad (10.40)$$

The assumed solution is convenient as the boundary lies at $r$ = constant.
Combining Equations 10.39 and 10.40,

$$r\frac{\partial}{\partial r}\left[r\frac{\partial M(r)N(\theta)}{\partial r}\right]+\frac{\partial^2 M(r)N(\theta)}{\partial\theta^2}=0$$

$$\text{or, } N(\theta)\left\{r\frac{\partial}{\partial r}\left[r\frac{\partial M(r)}{\partial r}\right]\right\}+M(r)\frac{\partial^2 M(r)N(\theta)}{\partial\theta^2}=0$$

Dividing by $M(r)N(\theta)$,

$$\frac{r}{M(r)}\frac{d}{dr}\left[r\frac{dM(r)}{dr}\right]+\frac{1}{N(\theta)}\frac{d^2N(\theta)}{d\theta^2}=0 \tag{10.41}$$

The partial derivatives become total derivatives in Equation 10.41 as each term is dependent on only one coordinate.

As in the case of sphere in uniform field, zero separation constant solution and equal and opposite separation constant solution are to be obtained separately in this case, too.

*Determination of the zero separation constant solution:*

The first term of Equation 10.41 is

$$\frac{r}{M(r)}\frac{d}{dr}\left[r\frac{dM(r)}{dr}\right]=0$$

where:

  $M(r)$ is non-zero

Therefore,

$$\frac{d}{dr}\left[r\frac{dM(r)}{dr}\right]=0,\text{ or, } r\frac{dM(r)}{dr}=C_0,\text{ or, }\frac{dM(r)}{dr}=\frac{C_0}{r}$$

Integrating and incorporating constants of integration

$$M(r)=C_{10}\ln r+C_{20} \tag{10.42}$$

Next the second term of Equation 10.41 is

$$\frac{1}{N(\theta)}\frac{d^2N(\theta)}{d\theta^2}=0$$

where:

  $N(\theta)$ is non-zero

Therefore,

$$N(\theta)=A_{10}\theta+A_{20} \tag{10.43}$$

But, from Equations 10.42 and 10.43, it can be seen that there is discontinuity of potential at $r=0$ and $\theta=\infty$, which are not feasible in the given arrangement as potential must be a continuous function. Hence, $C_{10} = A_{10} = 0$ in Equations 10.42 and 10.43.

Therefore,

$$V(r,\theta) = C_{20}A_{20} = C_1 \qquad (10.44)$$

*Determination of the equal and opposite separation constant solution:*

The first term of Equation 10.41 is

$$\frac{r}{M(r)}\frac{d}{dr}\left[r\frac{dM(r)}{dr}\right] = +p$$

where:

$p$ is a positive constant

$$\text{or, } r^2\frac{d^2M(r)}{dr^2} + r\frac{dM(r)}{dr} = +pM(r) \qquad (10.45)$$

Substituting $M(r) = Cr^n$ in Equation 10.45, it may be obtained that

$$n(n-1) + n = p, \text{ or, } n = \pm\sqrt{p}$$

Hence,

$$M(r) = \frac{C'}{r^{\sqrt{p}}} + C''r^{\sqrt{p}} \qquad (10.46)$$

Again, the second term of Equation 10.41 is

$$\frac{1}{N(\theta)}\frac{d^2N(\theta)}{d\theta^2} = -p$$

$$\text{or, } \frac{d^2N(\theta)}{d\theta^2} = -pN(\theta) \qquad (10.47)$$

Substituting $N(\theta) = e^{a\theta}$ in Equation 10.47, it may be obtained that

$$a^2e^{a\theta} = -pe^{a\theta}, \text{ or, } a = \pm i\sqrt{p}$$

Hence,

$$N(\theta) = B\cos\left(\sqrt{p}\theta + \alpha\right) \qquad (10.48)$$

Equations 10.46 and 10.48 lead to

$$V(r,\theta) = M(r)N(\theta) = \left(\frac{C_2}{r^{\sqrt{p}}} + C_3r^{\sqrt{p}}\right)\cos\left(\sqrt{p}\theta + \alpha\right) \qquad (10.49)$$

where:

$C_2 = C'B$ and $C_3 = C''B$

From Equations 10.44 and 10.49, the complete solution for potential function at all values of $r$ and $\theta$ can be obtained as

$$V(r,\theta) = C_1 + \left(\frac{C_2}{r^{\sqrt{p}}} + C_3 r^{\sqrt{p}}\right)\cos\left(\sqrt{p}\,\theta + \alpha\right) \tag{10.50}$$

The potential at large distance $r$ ($r \gg a$) from the cylinder is given by

$$V(r,\theta) = V_0 + E_0 r\cos\theta \tag{10.51}$$

Matching Equations 10.50 and 10.51, $\sqrt{p} = 1$ and $\alpha = 0$.
  Hence, the complete solution as given by Equation 10.50 reduces to

$$V(r,\theta) = C_1 + \left(\frac{C_2}{r} + C_3 r\right)\cos\theta \tag{10.52}$$

### 10.3.1 Conducting Cylinder in Uniform Field

Comparing Equations 10.51 and 10.52, $C_1 = V_0$ and $C_3 = E_0$.
  Therfore,

$$V(r,\theta) = V_0 + \left(\frac{C_2}{r} + E_0 r\right)\cos\theta$$

On the conductor surface, that is, for $r = a$,

$$V(a,\theta) = V_0 + \left(\frac{C_2}{a} + E_0 a\right)\cos\theta \tag{10.53}$$

But the conducting cylinder surface is an equipotential and hence electric potential is independent of $\theta$ on the conductor surface.
  Therefore, from Equation 10.53,

$$C_2 = -E_0 a^2$$

Hence, the complete solution for electric potential in the domain $r > a$ is given by

$$V(r,\theta) = V_0 + E_0\left(r - \frac{a^2}{r}\right)\cos\theta \tag{10.54}$$

The $r$ and $\theta$ components of the electric field intensity could be obtained as follows:

$$E_r = -\frac{\partial V}{\partial r} = -E_0\left(1 + \frac{a^2}{r^2}\right)\cos\theta \tag{10.55}$$

$$E_\theta = -\frac{1}{r}\frac{\partial V}{\partial \theta} = E_0\left(1 - \frac{a^2}{r^2}\right)\sin\theta \tag{10.56}$$

Equation 10.56 shows that for $r = a$, $E_\theta$ is zero, that is, the tangential component of the electric field intensity is zero on the cylindrical conductor surface as it is an equipotential surface.

Again, on the conducting cylinder surface, $E_r$ is the normal component of the electric field intensity, which is given by $E_r\big|_{r=a} = -2E_0 \cos\theta$. Thus, the maximum value of the electric field intensity on the surface of the conducting cylinder is $2E_0$, that is, twice the magnitude of the uniform external field. Comparing this maximum electric field intensity with the value obtained for conducting sphere in uniform field, it may be seen that the enhancement of field intensity is more if the conducting object is spherical is shape.

Induced surface charge density on the surface of the conducting cylinder may be obtained as follows:

$$\frac{\sigma_s}{\varepsilon_0} = E_r\bigg|_{r=a} = -2E_0 \cos\theta, \text{ or, } \sigma_s = -2\varepsilon_0 E_0 \cos\theta \tag{10.57}$$

### 10.3.2 Dielectric Cylinder in Uniform Field

Potential function valid for the region within the cylinder having dielectric of permittivity $\varepsilon_i$ is

$$V_i(r,\theta) = C_{1i} + \left(\frac{C_{2i}}{r} + C_{3i}r\right)\cos\theta \tag{10.58}$$

and the potential function valid for the region outside the cylinder having dielectric of permittivity $\varepsilon_e$ is

$$V_e(r,\theta) = C_{1e} + \left(\frac{C_{2e}}{r} + C_{3e}r\right)\cos\theta \tag{10.59}$$

The potential at large distance $r$ ($r \gg a$) from the cylinder

$$V(r,\theta) = V_0 + E_0 r \cos\theta \tag{10.60}$$

where:

$V_0$ is the potential at the location of the axis of the cylinder due to the external field

Comparing Equations 10.59 and 10.60 for $r \to \infty$, $C_{1e} = V_0$ and $C_{3e} = E_0$.

Hence, Equation 10.59 can be rewritten as

$$V_e(r,\theta) = V_0 + \left(\frac{C_{2e}}{r} + E_0 r\right)\cos\theta \tag{10.61}$$

Inside the dielectric cylinder electric potential must be finite at all the points. Hence, from Equation 10.58, $C_{1i} = V_0$ and $C_{2i} = 0$. Hence, Equation 10.58 can be rewritten as

$$V_i(r,\theta) = V_0 + C_{3i} r \cos\theta \tag{10.62}$$

At any point on the dielectric cylinder surface, that is, for $r = a$, electric potential as may be obtained from Equations 10.61 and 10.62 must be unique. Hence,

$$\frac{C_{2e}}{a} + E_0 a = C_{3i} a \tag{10.63}$$

From the boundary condition of normal component of electric flux density at $r = a$,

$$-\varepsilon_i \left( \frac{\partial V_i}{\partial r} \right)\Big|_{r=a} = -\varepsilon_e \left( \frac{\partial V_e}{\partial r} \right)\Big|_{r=a}$$

$$\text{or, } \varepsilon_i C_{3i} = \varepsilon_e \left( -\frac{C_{2e}}{a^2} + E_0 \right) \tag{10.64}$$

From Equations 10.63 and 10.64

$$C_{2e} = \frac{\varepsilon_e - \varepsilon_i}{\varepsilon_e + \varepsilon_i} a^2 E_0$$

$$\text{and } C_{3i} = \frac{2\varepsilon_e}{\varepsilon_e + \varepsilon_i} E_0$$

Therefore, the complete solutions for potential functions inside and outside the dielectric cylinder are given by

$$V_i(r, \theta) = V_0 + \frac{2\varepsilon_e}{\varepsilon_e + \varepsilon_i} E_0 r \cos\theta \tag{10.65}$$

$$\text{and } V_e(r, \theta) = V_0 + \left( r + \frac{\varepsilon_e - \varepsilon_i}{\varepsilon_e + \varepsilon_i} \frac{a^2}{r} \right) E_0 \cos\theta \tag{10.66}$$

As $r\cos\theta = x$, potential function inside the dielectric cylinder can be written as

$$V_i(x, y) = V_0 + \frac{2\varepsilon_e}{\varepsilon_e + \varepsilon_i} E_0 x \tag{10.67}$$

Hence, the electric potential within the dielectric cylinder varies in only $x$-direction, that is, the direction of the external field. Electric field intensity within the dielectric cylinder will, therefore, have only the $x$-component, which is given by

$$E_{xi} = -\frac{\partial V_i}{\partial x} = -\frac{2\varepsilon_e}{\varepsilon_e + \varepsilon_i} E_0 \tag{10.68}$$

Similar to the case of dielectric sphere in the uniform field, Equation 10.68 shows that the magnitude of the electric field intensity within the dielectric cylinder is constant. Typical field distribution on the $x$–$y$ plane in and around a dielectric cylinder within uniform external field will be the same as that shown in Figure 10.2.

As in the case of dielectric sphere in uniform field, for dielectric cylinder in uniform field also $|E_{zi}| > E_0$ if $\varepsilon_i < \varepsilon_e$. If a cylindrical air bubble is trapped within a moulded solid insulation of relative permittivity 4, then the magnitude of the electric field intensity within the air bubble will be $1.6E_0$, where $E_0$ is the magnitude of the electric field intensity in solid insulation at the location of the air bubble. Comparing this result with the corresponding value in the case of dielectric sphere, it may be seen that field enhancement is more if the gas cavity in liquid or solid insulation is cylindrical in shape.

## Objective Type Questions

1. A conducting sphere is placed within an external field of the uniform field intensity $E_0$. Then the maximum value of field intensity on the surface of the conducting sphere will be

    a. 0

    b. $E_0$

    c. $2E_0$

    d. $3E_0$

2. A conducting cylinder is placed within an external field of the uniform field intensity $E_0$. Then the maximum value of field intensity on the surface of the conducting cylinder will be

    a. 0

    b. $E_0$

    c. $2E_0$

    d. $3E_0$

3. A dielectric sphere is placed within an external field of the uniform field intensity $E_0$. The external field intensity acts in the direction of the $x$-axis. Then the field intensity within the dielectric sphere will be in the direction of

    a. $x$-axis

    b. $y$-axis

    c. $z$-axis

    d. None of the above

4. A dielectric cylinder is placed within an external field of the uniform field intensity $E_0$. The external field intensity acts in the direction of y-axis. Then the field intensity within the dielectric cylinder will be in the direction of

   a.  x-axis

   b.  y-axis

   c.  z-axis

   d.  None of the above

5. A dielectric sphere is placed within an external field of the uniform field intensity $E_0$. Then the field intensity within the dielectric sphere will be

   a.  Zero

   b.  Equal to $E_0$

   c.  Constant but not equal to $E_0$

   d.  None of the above

6. A dielectric cylinder is placed within an external field of the uniform field intensity $E_0$. Then the field intensity within the dielectric cylinder will be

   a.  Zero

   b.  Equal to $E_0$

   c.  Constant but not equal to $E_0$

   d.  None of the above

7. A gas cavity is present in large volume of solid insulation. Then the ratio of field intensity within the gas cavity for spherical to that for cylindrical cavity will be

   a.  Zero

   b.  <1

   c.  1

   d.  >1

8. A conducting object is present in an external field of uniform field intensity. Then the ratio of maximum field intensity on the surface of the conducting object for spherical to that for cylindrical object will be

   a.  Zero

   b.  <1

   c.  1

   d.  >1

**Answers:**   1) d; 2) c; 3) a; 4) b; 5) c; 6) c; 7) b; 8) d

# 11

# *Conformal Mapping*

**ABSTRACT**   Conformal mapping is a useful technique in the area of complex analysis and has many applications in different practical configurations. There are several problems in engineering that can be expressed in terms of functions of a complex variable, which involve complicated geometries. In such cases, by choosing an appropriate mapping function, the problem having inconvenient geometry can be transformed into a problem having much more convenient geometry. Conformal mapping is a technique that allows one to take hard problems, map them onto a coordinate system, where they are convenient to solve, find the solution and then map the solution back to the original system. This chapter gives a brief discussion on conformal mapping and some of its applications in solving practical problems in electrostatics.

## 11.1 Introduction

Analytical solutions to many field problems, particularly Dirichlet problems, can be obtained using methods such as Fourier series and integral transforms. These methods are applicable only for simple regions and the solutions are either infinite series or improper integrals, which are difficult to evaluate. Closed-form solutions to many Dirichlet problems can be obtained using conformal mapping, which is a similarity transformation. If a function is harmonic, that is, it satisfies Laplace's equation $\vec{\nabla}^2 f = 0$, then the transformation of such a function via conformal mapping is also harmonic. Hence, equations in relation to any field that can be represented by a potential function can be solved with the help of conformal mapping. However, conformal mapping can only be employed in two-dimensional fields. If the solution for potential field is required in three-dimensional cases, then conformal mapping is applicable to only those configurations where the potential field is translationally invariant along any one of the three axes. The two-dimensional potential fields that can be solved by conformal mapping are static electric fields, static magnetic fields, static electric flow fields, stationary thermal flow fields, stationary hydrodynamic flow fields to name a few. According to Riemann mapping theorem any two regions with same connectivity may be conformally mapped to one another. But in practical applications, conformal mapping is used only in those cases where the maps take simpler, explicit forms, so that one may carry out actual calculations with those maps.

As the application of conformal mapping is limited to variables, which solve the Laplace's equation for two-dimensional fields, one such variable of practical interest is the electrostatic potential in a region of space that is free of charges. This chapter, therefore, focuses on application of conformal mapping to determine electrostatic potential field by solving two-dimensional Laplace's equation.

## 11.2 Basic Theory of Conformal Mapping

Conformal transformation is based on the properties of analytic functions. Let $z = x + iy$ be a complex variable such that the real and imaginary parts $x$ and $y$ are real-valued variables, and $f(z) = u(z) + iv(z) = u(x,y) + iv(x,y)$ be a complex-valued function such that the real and imaginary parts $u$ and $v$ are real- and single-valued functions of real-valued variables $x$ and $y$.

If the derivative of $f(z)$ exists at a point $z$, then the partial derivatives of $u$ and $v$ exist at that point and obey the Cauchy–Riemann equations as follows.

$$\frac{\partial u}{\partial x} = \frac{\partial v}{\partial y} \text{ and } \frac{\partial u}{\partial y} = -\frac{\partial v}{\partial x} \tag{11.1}$$

A function $f(z)$ is analytic at a point $z_0$ if its derivative $f'(z)$ exists not only at $z_0$, but at every point in the neighbourhood of $z_0$. It can also be shown that if $f(z)$ is analytic, the partial derivatives of $u$ and $v$ of all orders exist and are continuous functions of $x$ and $y$. Therefore,

$$\frac{\partial^2 u}{\partial x^2} = \frac{\partial}{\partial x}\left(\frac{\partial u}{\partial x}\right) = \frac{\partial}{\partial x}\left(\frac{\partial v}{\partial y}\right) = \frac{\partial^2 v}{\partial x \partial y} = \frac{\partial}{\partial y}\left(\frac{\partial v}{\partial x}\right) = \frac{\partial}{\partial y}\left(-\frac{\partial u}{\partial y}\right) = -\frac{\partial^2 u}{\partial^2 y}$$

$$\text{or, } \frac{\partial^2 u}{\partial x^2} + \frac{\partial^2 u}{\partial y^2} = 0 \tag{11.2}$$

In the same way, one may get,

$$\frac{\partial^2 v}{\partial x^2} + \frac{\partial^2 v}{\partial y^2} = 0 \tag{11.3}$$

Equations 11.2 and 11.3 show that both the functions $u(x,y)$ and $v(x,y)$ satisfy Laplace's equation.

Any function that has continuous second-order partial derivatives and satisfies Laplace's equation is called a *harmonic function*. Thus, both the real part, $u(x,y)$, and imaginary part, $v(x,y)$, of the complex function $f(z)$ are harmonic functions. If the function $f(z) = u(x,y) + iv(x,y)$ is analytic, then $u(x,y)$ and $v(x,y)$ are conjugate harmonic functions. If one of two harmonic functions is known, then the other can be found using Cauchy–Riemann equations.

Thus, both the conjugate harmonic functions $u(x,y)$ and $v(x,y)$ can be used to find the potential as they satisfy Laplace's equation.

## 11.2.1 Mapping of Shapes

From a different viewpoint, the complex function $f(z)$ can be considered as a tool for change of variables, that is, a transformation from the complex $z$-plane to the complex $w$-plane, as shown in Figure 11.1, where

$$z = x + iy \text{ and } w = u + iv$$

It can also be shown that if the function $f$ is analytic at a point $z = z_0$ on the $z$-plane, where the first-order derivative $f'(z_0)$ is non-zero, there exists a neighbourhood of the point $w_0$ in the $w$-plane, in which the function $w = f(z)$ has a unique inverse $z = F(w)$. The functions $f(z)$ and $F(w)$, therefore, define a change of variables from $(x,y)$ to $(u,v)$ and from $(u,v)$ to $(x,y)$, respectively.

On the $z$-plane, $dz = dx + idy$ and on the $w$-plane $dw = du + idv$.

Therefore,

$$|dz|^2 = dx^2 + dy^2 \tag{11.4}$$

$$\text{and } |dw|^2 = du^2 + dv^2 \tag{11.5}$$

Then, on the $z$-plane, square of the length element can be written as

$$dl^2 = dx^2 + dy^2 = |dz|^2 \tag{11.6}$$

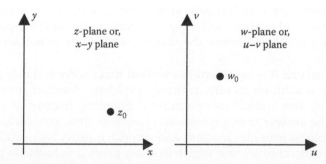

**FIGURE 11.1**
Mapping between $z$-plane and $w$-plane.

and, on the $w$-plane, square of the length element can be written as

$$dL^2 = du^2 + dv^2 = |dw|^2 \qquad (11.7)$$

Therefore, from Equations 11.6 and 11.7, it may be written that

$$\frac{dL}{dl} = \left|\frac{dw}{dz}\right| \qquad (11.8)$$

Thus, in the neighbourhood of each point in $z$-plane, if $w(z)$ is analytic and have a non-zero derivative, that is, finite slope at that point, then the ratio of length elements in two planes remains constant. The net result of this transformation is to change the dimensions in equal proportions and rotate each infinitesimal area in the neighbourhood of that point. In general, a linear transformation $w = f(z) = az + b$, where $a$ and $b$ are complex numbers, rotates by $\arg(a)$ in the anti-clockwise direction, dilates or compresses by $|a|$ and translates by $b$. Thus, the ratio of linear dimensions, which may also be represented as the angle, is preserved. As a result, conformal mapping is isogonic because it preserves angles. Hence, all curves in the $z$-plane that intersect each other at particular angles are mapped into curves in the $w$-plane that intersect each other at exactly the same angles. This property is most useful for electric field analysis as the equipotentials and the fieldlines, which are normal to each other in $z$-plane, are mapped to corresponding curves in $w$-plane, which are also mutually orthogonal.

Furthermore, $f'(z)F'(w) = |dw/dz||dz/dw| = 1$, which means that the inverse mapping is also conformal. Because of this uniqueness and conformal property of inverse mapping, solution obtained in the $w$-plane can be mapped back to $z$-plane.

When infinitesimally small region is considered, every shape in the $z$-plane is transformed into a similar shape in the $w$-plane, for example, a rectangle in the $z$-plane remains a rectangle in $w$-plane. However, shape will not be preserved in general, particularly in a large scale as the value of $|dw/dz|$ may vary considerably at different points in the $z$-plane. As a result rotation and scaling will vary from one point in the $z$-plane to its neighbouring point and hence the similarity of shape is not achieved for large regions.

At this juncture, it is pertinent to mention that conformal mapping does not provide a solution to any arbitrary problem. Another question that arises is why one should use conformal mapping instead of numerical methods. The answer to this question is two-fold: first, analytical solutions to field problems provides insight, and second, it provides useful approximations to difficult problems, which in many cases is valuable to practicing engineers.

## 11.2.2 Preservation of Angles in Conformal Mapping

As shown in Figure 11.2, two curves $A$ and $B$ intersect each other at an angle $\alpha$ at the point $z_i$ in the $z$-plane. With the help of the tangent vectors to the curves, the angle between the curves could be computed. Let, $t_{zA}$ and $t_{zB}$ be the tangent vectors to the curves $A$ and $B$, respectively. Then from the law of cosines it may be written that

$$\alpha = \cos^{-1}\left( \frac{\left|t_{zA}\right|^2 + \left|t_{zB}\right|^2 - \left|t_{zA} - t_{zB}\right|^2}{2\left|t_{zA}\right|\left|t_{zB}\right|} \right) \tag{11.9}$$

The corresponding transformed curves $A'$ and $B'$ intersect at an angle $\beta$ in the $w$-plane. Let $t'_{wA}$ and $t'_{wB}$ be the tangent vectors to the curves $A'$ and $B'$, respectively. Then $\beta$ can be obtained as

$$\beta = \cos^{-1}\left( \frac{\left|t'_{wA}\right|^2 + \left|t'_{wB}\right|^2 - \left|t'_{wA} - t'_{wB}\right|^2}{2\left|t'_{wA}\right|\left|t'_{wB}\right|} \right) \tag{11.10}$$

Let a curve is parameterized in $z$-plane by $z = z(p)$ and the complex analytic function $w = f[z(p)]$ defines the mapped curve in the $w$-plane. Then the application of chain rule to $w = f[z(p)]$ gives $t'_w = f'[z(p)]t_z(p)$. Because the curves intersect in $z$-plane at $z = z_i$, then $t'_{wA} = f'(z_i)t_{zA}$ and $t'_{wB} = f'(z_i)t_{zB}$. Because $f'(z_i) \neq 0$, Equation 11.10 can be rewritten as

$$\beta = \cos^{-1}\left[ \frac{\left|f'(z_i)t_{zA}\right|^2 + \left|f'(z_i)t_{zB}\right|^2 - \left|f'(z_i)t_{zA} - f'(z_i)t_{zB}\right|^2}{2\left|f'(z_i)t_{zA}\right|\left|f'(z_i)t_{zB}\right|} \right] \tag{11.11}$$

In Equation 11.11, the absolute value $\left|f'(z_i)\right|^2$ cancels from the numerator and denominator and Equation 11.11 gets reduced to

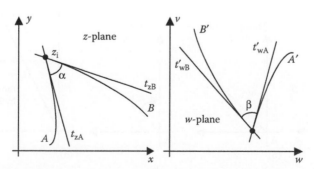

**FIGURE 11.2**
Preservation of angles in conformal mapping.

$$\beta = \cos^{-1}\left(\frac{\left|t_{zA}\right|^2 + \left|t_{zB}\right|^2 - \left|t_{zA} - t_{zB}\right|^2}{2\left|t_{zA}\right|\left|t_{zB}\right|}\right) \qquad (11.12)$$

From Equations 11.9 and 11.12, $\alpha = \beta$, which proves that angles are preserved in conformal mapping.

## PROBLEM 11.1

For the point $z = 1 + i$ in the $z$-plane, find the mapped point in the $w$-plane under the linear transformation $w = (1 + i)z + (2 + 2i)$.

*Solution:*
The given transformation function $w = f(z) = (1 + i)z + (2 + 2i) = \sqrt{2}e^{i(\pi/4)}z + (2 + 2i)$.

Hence, the transformation of the point $(1 + i)$ in the $z$-plane to the corresponding point in the $w$-plane can be obtained in three steps, as shown in Figure 11.3.

Step 1: The length $OP$ $\left(|z|\right)$ is multiplied by $|1+i| = \sqrt{2}$ to get the length $AB$, as shown in Figure 11.3b.

Step 2: The length $AB$ is rotated by an angle $(\pi/4)$ in the anti-clockwise direction to get the length $AC$, as shown in Figure 11.3c.

Step 3: The point $C$ is then translated by $(2 + 2i)$ to get the point $P'(2 + 4i)$ in the $w$-plane, which is the conformally mapped point corresponding to the point $P$ in the $z$-plane.

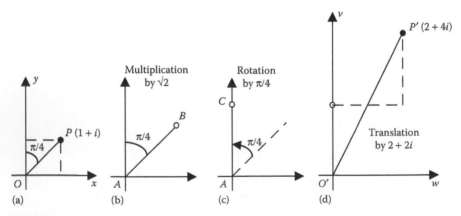

**FIGURE 11.3**
Pertaining to Problem 11.1.

## PROBLEM 11.2

Let $\Omega$ be the rectangular region in the $z$-plane bounded by $x = 1$, $y = 1$, $x = 3$ and $y = 2$. Find the mapped region $\Omega'$ in the $w$-plane under the linear transformation $w = (1 + i)z + (2 + 2i)$.

*Solution:*
Given, $w = f(z) = (1 + i)z + (2 + 2i) = (1 + i)(x + iy) + (2 + 2i) = (x - y + 2) + i(x + y + 2)$.

Hence, $u = x - y + 2$ and $v = x + y + 2$.

Therefore, for $x = 1$, $u = -y + 3$ and $v = y + 3$ or, $u + v = 6$, that is, the line $x = 1$ in the $z$-plane is mapped to the straight line $u + v = 6$ in the $w$-plane.

Similarly, for $y = 1$, $u = x + 1$ and $v = x + 3$ or, $u - v = -2$.

For $x = 3$, $u = -y + 5$ and $v = y + 5$ or, $u + v = 10$.

For $y = 2$, $u = x$ and $v = x + 4$ or, $u - v = -4$.

Therefore, the four straight lines in the $z$-plane defined by $x = 1$, $y = 1$, $x = 3$ and $y = 2$ are mapped to four straight lines defined by $u + v = 6$, $u - v = -2$, $u + v = 10$ and $u - v = -4$, respectively, in the $w$-plane. The mapping is shown in Figure 11.4. Under the linear transformation $w = az + b$, where $a = 1 + i$ and $b = 2 + 2i$, it may be seen that the rectangular region $\Omega$ in the $z$-plane is

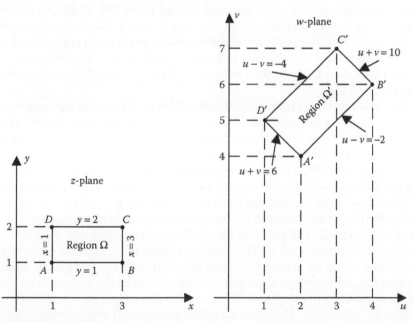

**FIGURE 11.4**
Pertaining to Problem 11.2.

translated by $b( = 2 + 2i)$, rotated by an angle $45°$ [ $= \arg(a) = \arg(1 + i)$] in the anti-clockwise direction and dilated by $\sqrt{2}( = |a| = |1 + i|)$ to another rectangular region $\Omega'$ in the $w$-plane.

## 11.3 Concept of Complex Potential

Let, $\phi(x,y)$ be a harmonic function in a domain $\Omega$. It is possible to define a harmonic conjugate function, $\psi(x,y)$, uniquely by Cauchy–Riemann equations in the same domain. Thus, an analytic function of $z = x + iy$ in the domain $\Omega$ can be written as

$$F(z) = \phi(x, y) + i\psi(x, y) \tag{11.13}$$

Consequently, $F(z)$ conformally maps the curves in the $z$-plane onto the corresponding curves in the $w$-plane and vice versa preserving the angles during mapping.

Because both the real and imaginary parts of $F(z)$, namely, $\phi(x,y)$ and $\psi(x,y)$, are harmonic functions, they satisfy Laplace's equation and hence either one of these two could be used to find potential. Thus, the complex analytic function $F(z)$ is known as *complex potential*. Laplace's equation is one of the most important partial differential equations in engineering and physics. The theory of solutions of Laplace's equation is known as *potential theory*. The concept of complex potential relates potential theory closely to complex analysis.

If $\phi(x,y)$ is considered to be real potential, then $\phi(x,y) = $ const represents equipotential lines in the $z$-plane. Because $\phi(x,y)$ and $\psi(x,y)$ are orthogonal, $\psi(x,y) = $ const represents electric fieldlines in the $z$-plane. For example, consider the complex potential function as $F(z) = Az + B = Ax + B + iAy$. Then the equipotential lines corresponding to $\phi(x,y) = Ax + B = $ const are straight lines parallel to $y$-axis and the electric fieldlines corresponding to $\psi(x,y) = Ay = $ const are straight lines parallel to $x$-axis.

Introduction of the concept of complex potential is advantageous in the following ways: (1) it is possible to handle equipotential and electric fieldlines simultaneously and (2) Dirichlet problems with difficult geometry of boundaries could be solved by conformal mapping by finding an analytic function $F(z)$, which maps a complicated domain $\Omega$ in the $z$-plane onto a simpler domain $\Omega'$ in the $w$-plane. The complex potential $F'(w)$ is solved in the $w$-plane by satisfying Laplace's equation along with the boundary conditions. Then the complex potential in the $z$-plane can be obtained by inverse transform from which the real potential is obtained as $\phi(x,y) = \text{Re}[F(z)]$.

This is a practicable way of solution as harmonic functions remain harmonic under conformal mapping.

---

## 11.4 Procedural Steps in Solving Problems Using Conformal Mapping

1. Find an analytic function $w = F(z)$ to map the original region $\Omega$ in the $z$-plane to the transformed region $\Omega'$ in the $w$-plane. The region $\Omega'$ should be a region for which explicit solutions to the problem at hand are known.

2. Transfer the boundary conditions from the boundaries of the region $\Omega$ in the $z$-plane to the boundaries of the transformed region $\Omega'$ in the $w$-plane.

3. Solve the problem and find the complex potential $F'(w)$ for the transformed region $\Omega'$ in the $w$-plane.

4. Map the solution $F'(w)$ for the region $\Omega'$ in the $w$-plane back to the complex potential $F(z)$ for the region $\Omega$ in the $z$-plane through inverse mapping.

The steps are schematically shown in Figure 11.5. The most important step is to find an appropriate mapping function $w = F(z)$, which fits the problem at hand. Once the right mapping function has been found, the problem is as good as solved.

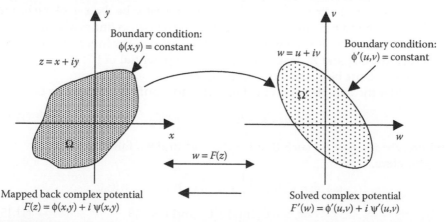

**FIGURE 11.5**
Schematic representation of solution of potential problem by conformal mapping.

## 11.5  Applications of Conformal Mapping in Electrostatic Potential Problems

Conformal mapping is a powerful method for solving boundary value problems in a two-dimensional potential theory through the transformation of a complicated region into a simpler region. Electric potential satisfies Laplace's equation in charge-free region. Therefore, electrostatic field that satisfies Laplace's equation in a two-dimensional region in $x$–$y$ plane, will also satisfy Laplace's equation in any plane to which the region may be transformed by an analytic complex potential function $F(z)$. For each value of complex $z = x + iy$, there is a corresponding value of complex $w = F(z)$. In other words, for every point in the $z$-plane, there is a corresponding point in the $w$-plane. As a result, the locus of any point in the $z$-plane will trace another path in $w$-plane. Let the locus in the $z$-plane maps onto a path $\phi'(u, v) = $ const in the $w$-plane, which corresponds to an equipotential and may also be the surface of a conductor. Then the problem can be solved in the $w$-plane incorporating the appropriate boundary condition, that is, the value of the conductor potential, and the results can be mapped back to $z$-plane to get the real potential and then the electric fieldlines can be obtained from the conjugate harmonic function. This section discusses some of the applications of conformal mapping in solving two-dimensional electrostatic potential problems.

### 11.5.1  Conformal Mapping of Co-Axial Cylinders

The cross-sectional view of a single-core cable is shown in Figure 11.6, where the co-axial cylindrical conductors are of infinite length in the direction normal to the plane of the paper. Hence, the field varies only in the cross-sectional plane and is translationally invariant in the direction of the length of the cable. Let the cross-sectional plane of the cable be the $x$–$y$ plane or the $z$-plane. Then the field in the region between the two cylindrical conductors can be found by conformal mapping. Let the radii of the inner and the outer conductors be $r_1$ and $r_2$, respectively, and the potential of the inner and the outer conductors be $V$ and zero, respectively.

Consider the complex analytical function for conformal mapping be

$$w = u + iv = C_1 \ln(z) + C_2 \tag{11.14}$$

where, $z = x + iy = re^{i\theta}$ such that $r = \sqrt{x^2 + y^2}$ and $\theta = \tan^{-1}(y/x)$

Therefore,

$$u + iv = C_1 \ln(re^{i\theta}) + C_2 = C_1 \ln r + C_2 + iC_1\theta$$

$$\text{or, } u = C_1 \ln(r) + C_2 \text{ and } v = C_1\theta \tag{11.15}$$

For the inner conductor, $x^2 + y^2 = r_1^2$ and hence it maps to a straight line $u_1 = $ constant parallel to $v$-axis in the $w$-plane. Similarly, the outer conductor

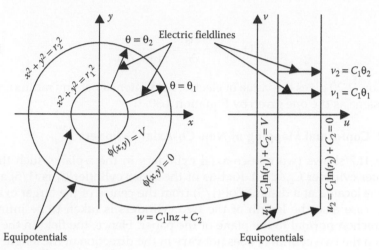

**FIGURE 11.6**
Conformal mapping of co-axial cylinders.

for which $x^2 + y^2 = r_2^2$ maps to another straight line $u_2 = $ constant parallel to $v$-axis in the $w$-plane, as shown in Figure 11.6. In other words, the field within the two cylindrical conductors in the $z$-plane is conformally mapped to field between two infinitely long parallel plates, that is, the field within a parallel plate capacitor, in the $w$-plane. From Figure 11.6 it may be seen that the orthogonality of the equipotentials in the form of circles and electric fieldlines in the form of radial lines in the $z$-plane are maintained in the $w$-plane, where the equipotentials are straight lines parallel to $v$-axis and the electric fieldlines are straight lines parallel to $u$-axis.

From the boundary conditions on the conductor surfaces

$$C_1 \ln r_1 + C_2 = V \tag{11.16}$$

$$\text{and } C_1 \ln r_2 + C_2 = 0 \tag{11.17}$$

From Equations 11.16 and 11.17,

$$C_1 = -\frac{V}{\ln(r_2/r_1)} \text{ and } C_2 = \frac{V \ln r_2}{\ln(r_2/r_1)} \tag{11.18}$$

The potential at any radius $r$ is given by $u = C_1 \ln r + C_2$. Correspondingly, in the $z$-plane

$$\phi(x,y) = -\frac{V \ln r}{\ln(r_2/r_1)} + \frac{V \ln r_2}{\ln(r_2/r_1)} = \frac{V \ln(r_2/r)}{\ln(r_2/r_1)} \tag{11.19}$$

Then,

$$E_r(x,y) = -\frac{\partial \phi}{\partial r} = \frac{V}{r \ln(r_2/r_1)} \tag{11.20}$$

Equation 11.20 gives the value of electric field intensity at any radius $r$, which is the same as the one given by Equation 4.30.

### 11.5.2 Conformal Mapping of Non-Co-Axial Cylinders

Figure 11.7 shows two non-co-axial cylinders in the z-plane, such that for the outer cylinder $C_2$, $|z| = 1$. Radius of the inner cylinder $C_1$ is (1/5) and its centre is located at a distance of (1/5) from the centre of the larger cylinder. In this case also the length of the two cylinders is taken to be infinite in the direction normal to the plane of the paper. Hence, the field in the space between the two cylinders does not vary in the direction of the length of the cylinders. Therefore, the cross-sectional plane is shown to be the z-plane in Figure 11.7. The inner cylinder is at a potential of $V$ while the outer cylinder is earthed. Direct solution of the field between the two cylinders is difficult in the z-plane. However, it is possible to conformally map the non-co-axial cylinders in the z-plane onto two co-axial cylinders in the w-plane keeping the boundary conditions, that is, boundary potentials, same.

In this transformation, the unit radius outer circle $C_2$ in the z-plane is mapped onto a unit radius circle $C_2'$ in the w-plane in such a way that the inner circle $C_1'$ becomes concentric with a radius $r_i$, as shown in Figure 11.7. The mapping function for this linear fractional transformation is

$$w = \frac{z-k}{kz-1} \tag{11.21}$$

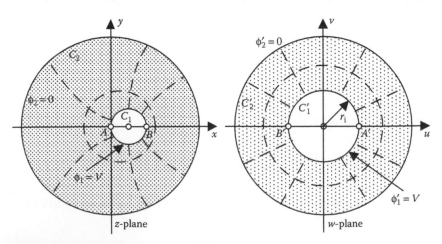

**FIGURE 11.7**
Conformal mapping of non-co-axial cylinders.

As shown in Figure 11.7, the two points on the inner circle $A(0,0)$ and $B(2/5,0)$ in the $z$-plane are mapped onto two points $A'(r_i,0)$ and $B'(-r_i,0)$ on the inner circle in the $w$-plane.

Hence, from Equation 11.21 for the points $A(0,0)$ and $A'(r_i,0)$ $r_i = (0 - k)/(0-1) = k$ and for the points $B\,(2/5,0)$ and $B'(-r_i,0)$ $-r_i = [(2/5) - k]/[(2k/5) - 1]$ $= (2 - 5r_i)/(2r_i - 5)$ or, $r_i^2 - 5r_i + 1 = 0$, or, $r_i = 4.79$ and $0.208$.

But, $r_i$ cannot be greater than 1 and hence, $r_i = 0.208$. Therefore, $k = r_i = 0.208$.

Thus, the mapping function for this problem is

$$w = \frac{z - 0.208}{0.208\,z - 1} \tag{11.22}$$

Writing the complex potential function in the $w$-plane as $F'(w) = a \ln w + b$, the real part of the complex potential can be written as

$$\phi'(u,v) = \mathrm{Re}[F'(w)] = a \ln|w| + b \tag{11.23}$$

Two conditions on boundary potentials in the $w$-plane are as follows: (1) $\phi' = 0$ for $|w| = 1$ and (2) $\phi' = V$ for $|w| = r_i$

Application of first boundary condition on Equation 11.23 yields

$$a\ln 1 + b = 0,\ \text{or},\ b = 0$$

Similarly, applying the second boundary condition on Equation 11.23 one would get

$$a\ln r_i + b = V,\ \text{or } \ln 0.208 = V,\ \text{or},\ a = -0.6368\ V$$

Thus, the desired solution for complex potential in the $z$-plane is

$$F(z) = -0.6368\ V\ln \frac{z - 0.208}{0.208z - 1} \tag{11.24}$$

The real potential within the two cylinders is then given by

$$\phi(x,y) = \mathrm{Re}[F(z)] = -0.6368\ V\ln\left|\frac{z - 0.208}{0.208z - 1}\right| \tag{11.25}$$

If the potentials are $+V$ and $-V$ instead of $V$ and 0, then from the first boundary condition

$$a\ln 1 + b = -V,\ \text{or } b = -V$$

and from the second boundary condition $a\ln r_i + b = V$, or, $a\ln 0.208 - V = V$ or, $a = -1.273\ V$.

Hence, the desired solution for complex potential in the z-plane is

$$F(z) = -1.273 \, V \ln \frac{z-0.208}{0.208z-1} - V = V\left(-1.273\ln\frac{z-0.208}{0.208z-1} - 1\right) \quad (11.26)$$

The real potential within the two cylinders is then given by

$$\phi(x,y) = \text{Re}[F(z)] = V\left(-1.273\ln\left|\frac{z-0.208}{0.208z-1}\right| - 1\right) \quad (11.27)$$

### 11.5.3 Conformal Mapping of Unequal Parallel Cylinders

Figure 11.8 shows two unequal parallel cylinders in the z-plane, such that for the larger cylinder $C_2$, $|z| = 1$. Radius of the smaller cylinder $C_1$ is $(1/2)$ and its centre is located at a distance of $(7/2)$ from the centre of the larger cylinder. In this case also the length of the two cylinders is taken to be infinite in the direction normal to the plane of the paper. Hence, the field in the space between the two parallel cylinders does not vary in the direction of the length of the cylinders. Therefore, the cross-sectional plane is shown to be the z-plane in Figure 11.8. The smaller cylinder is at a potential of $V$ whereas the larger cylinder is earthed. It is possible to map these two parallel cylinders onto two co-axial cylinders in the w-plane as follows.

In this transformation, too, the unit radius larger circle $C_2$ in the z-plane is mapped onto a unit radius circle $C_2'$ in the w-plane in such a way that the smaller circle $C_1'$ becomes concentric with a radius $r_i$, as shown in Figure 11.8. The mapping function for this linear fractional transformation is also

$$w = \frac{z-k}{kz-1} \quad (11.28)$$

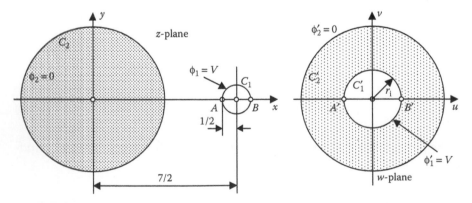

**FIGURE 11.8**
Conformal mapping of two unequal parallel cylinders.

However, as shown in Figure 11.8, the two points on the smaller circle $A(3,0)$ and $B(4,0)$ in the $z$-plane are mapped onto two points $A'(-r_i,0)$ and $B'(r_i,0)$ on the inner circle in the $w$-plane.

Hence, from Equation 11.26 for the points $A(3,0)$ and $A'(-r_i,0)$, $-r_i = (3-k)/(3k-1)$ and for the points $B(4,0)$ and $B'(r_i,0)$, $r_i = (4-k)/(4k-1) = (k-3)/(3k-1)$ or, $7k^2 - 26k + 7 = 0$, or, $k = 3.42$ and $0.292$.

For $k = 3.42$, $r_i = 0.046$ and for $k = 0.292$, $r_i = 21.84$.

But $r_i$ cannot be greater than 1 in the $w$-plane, so the solution is $k = 3.42$ and $r_i = 0.046$.

Writing the same complex potential function in the $w$-plane as $F'(w)=a\ln w + b$, as in Section 11.5.2, and applying the same boundary conditions for potential, as shown in Figure 11.8,

$$b = 0 \text{ and } a = \frac{V}{\ln r_i} = \frac{V}{\ln 0.046} = -0.3247\ V$$

Thus, the desired solution for complex potential in the $z$-plane is

$$F(z) = -0.3247\ V \ln \frac{z-3.42}{3.42z-1} \tag{11.29}$$

The real potential between the two unequal parallel cylinders is then given by

$$\phi(x,y) = \text{Re}[F(z)] = -0.3247\ V \ln \left| \frac{z-3.42}{3.42z-1} \right| \tag{11.30}$$

If the potentials are $+V$ and $-V$ instead of $V$ and $0$, then from the first boundary condition

$$a \ln 1 + b = -V, \text{ or } b = -V$$

and from the second boundary condition $a\ln r_i + b = V$, or, $a \ln 0.046 - V = V$ or, $a = -0.6494\ V$.

Hence, the desired solution for complex potential in the $z$-plane is

$$F(z) = -0.6494\ V\ln \frac{z-3.42}{3.42z-1} - V = V\left( -0.6494\ln \frac{z-3.42}{3.42z-1} - 1 \right) \tag{11.31}$$

The real potential between the two unequal parallel cylinders is then given by

$$\phi(x,y) = \text{Re}[F(z)] = V\left( -0.6494\ln \left| \frac{z-3.42}{3.42z-1} \right| - 1 \right) \tag{11.32}$$

### 11.5.3.1 Conformal Mapping of Equal Parallel Cylinders

With reference to Figure 11.8, if the radius of the cylinder 1, that is, $C_1$, is taken to be unity, then

$$k = 2.906 \text{ and } r_1 = 0.064$$

Hence, if the potentials of the two cylinders are $V$ and 0, respectively, then the desired solution for complex potential in the z-plane is

$$F(z) = -0.3638 \, V \ln \frac{z - 2.906}{2.906z - 1} \qquad (11.33)$$

The real potential between the two equal parallel cylinders is then given by

$$\phi(x, y) = \text{Re}[F(z)] = -0.3638 \, V \ln \left| \frac{z - 2.906}{2.906z - 1} \right| \qquad (11.34)$$

If the potentials of the two cylinders are $+V$ and $-V$, respectively, then the desired solution for complex potential in the z-plane is

$$F(z) = -0.7276 \, V \ln \frac{z - 2.906}{2.906z - 1} - V = V \left( -0.7276 \ln \frac{z - 2.906}{2.906z - 1} - 1 \right) \quad (11.35)$$

The real potential between the two equal parallel cylinders is then given by

$$\phi(x, y) = \text{Re}[F(z)] = V \left( -0.7276 \ln \left| \frac{z - 2.906}{2.906z - 1} \right| - 1 \right) \qquad (11.36)$$

---

## Objective Type Questions

1. Conformal mapping could be used in
   a. Two-dimensional system
   b. Axi-symmetric system
   c. Three-dimensional system
   d. All the above
2. In conformal mapping, the angle between intersecting curves
   a. Increases by 90°
   b. Decreases by 90°
   c. Increases by 180°
   d. Remains the same

3. Both real and imaginary parts of the complex potential $F(z)$ are harmonic functions. If $\text{Re}[F(z)]$ = constant represents equipotential lines, then $\text{Imag}[F(z)]$ = constant represents

   a. Electric flux lines

   b. Equipotential lines

   c. Zero potential line

   d. None of the above

4. In conformal mapping from z-plane to w-plane, which of the following operations are commonly performed?

   a. Scaling

   b. Rotation

   c. Translation

   d. All the above

5. By conformal mapping which of the following configurations in the z-plane could be mapped to co-axial cylinders in the w-plane?

   a. Non-co-axial cylinders

   b. Unequal parallel cylinders

   c. Equal parallel cylinders

   d. All the above

6. With the help of the complex analytic function $w = a \ln z + b$, which of the following configuration in the z-plane could be mapped to parallel plate capacitor in the w-plane?

   a. Co-axial cylinders

   b. Parallel cylinders

   c. Two equal spheres

   d. None of the above

7. In conformal mapping, a real-life configuration is mapped from

   a. z-plane to s-plane

   b. s-plane to z-plane

   c. z-plane to w-plane

   d. s-plane to w-plane

8. During conformal mapping from z-plane to w-plane

   a. Shape is preserved

   b. Angle of intersection is preserved

   c. Boundary conditions are preserved

   d. Both (b) and (c)

9. In conformal mapping of electrostatic field, complex potential is defined as $F(z) = \phi(x,y) + i\psi(x,y)$. If $\phi(x,y)$ is known, then $\psi(x,y)$ can be found from

   a. Gauss's law

   b. Cauchy–Riemann equations

   c. Divergence theorem

   d. Uniqueness theorem

10. In conformal mapping of electrostatic field, complex potential is defined as $F(z) = \phi(x,y) + i\psi(x,y)$. Then Laplace's equation is satisfied by

    a. $F(z)$

    b. $\phi(x,y)$

    c. $\psi(x,y)$

    d. Both (b) and (c)

11. In conformal mapping, equipotential and electric flux lines can be handled simultaneously by the introduction of the concept of

    a. Complex potential

    b. Complex electric field intensity

    c. Complex electric flux density

    d. None of the above

12. In spite of having numerical techniques, conformal mapping is used for electrostatic field analysis because it

    a. Can provide quicker solution to difficult problems

    b. Can provide more accurate solutions to difficult problems

    c. Can provide benchmark solutions for some problems which can be used for validation of numerical results

    d. All the above

**Answers:**   1) a; 2) d; 3) a; 4) d; 5) d; 6) a; 7) c; 8) d; 9) b; 10) d; 11) a; 12) c

# 12

## Graphical Field Plotting

**ABSTRACT** For a clear understanding of the concept of electric field, some method for describing it, both qualitatively and quantitatively, is needed. Graphical field plotting is one such method that is dynamic in nature and illustrates the vector nature of the electric field. Graphical field maps are commonly drawn for practical configurations, which may be considered as two-dimensional or which are axi-symmetric in nature. Typically, in graphical field plotting, electric fieldlines are drawn to provide information about the field. The strength of the field is indicated by the density of the field lines. A high density of electric fieldlines indicates a strong field and vice versa. Complementary information can also be conveyed by simultaneously drawing the equipotential lines. Field maps could also be used for obtaining approximate values of system parameters such as capacitance.

## 12.1 Introduction

Most of the practical problems have such complicated geometry that no exact method of finding the electric field is possible or feasible and approximate techniques are the only ones that can be used. Out of the several approximate techniques, numerical techniques are now extensively used to determine electric field distribution with high accuracy. Numerical techniques, which are widely used, will be discussed in detail in chapters 13 through 18. In this chapter, experimental and graphical field mapping methods are discussed. Experimental field mapping involve special equipments such as an electrolytic tank, a device for fluid flow, a conducting paper and an associated measuring system. The other mapping method is a graphical one and needs only paper and pencil. In both these methods, the exact value of the field quantities could not be determined, but accuracy level which is sufficient for practical engineering applications could be achieved. Graphical field plotting is economical compared to experimental method and is also capable of providing good accuracy when used with skill. Accuracy of the order of 5%–10% in capacitance determination could be achieved even by a non-expert simply by following the rules.

## 12.2 Experimental Field Mapping

Experimental method of field mapping is based on the analogy of stationary current field with static electric field, as presented in Table 12.1, rather than directly on measurement of electric field. If the medium between electrodes is isotropic, then volume conductivity and dielectric constant do not vary with position. Then current density ($J$) in the stationary current field and electric field intensity ($E$) and electric flux density ($D$) in the static electric field will be in the same direction. In other words, current density and electric fieldlines are the same. Thus, for a given electrode system, if a slightly conducting material, for example, a conducting paper or an electrolyte is placed instead of a dielectric material between the electrodes, then electric fieldlines and equipotential lines will remain the same.

It is well known that if one travels along a line through an electric field and measures electric scalar potential $V$ as one goes, then the negative of the rate of change of $V$ is equal to the component of electric field intensity $E$ in the direction of travel. In other words,

$$\vec{E} = -\frac{\partial V}{\partial l}\hat{u}_l \tag{12.1}$$

If $-(\partial V/\partial l)$ is maximum, then it gives the value of $E$ itself. If electric potential does not change with position, then the path of travel is at right angles to the electric field and is along an equipotential. Thus, electric field could be mapped by a voltmeter that will measure potential difference and two metal rods acting as probes. The probes are connected to the terminals of the voltmeter and are placed in various positions in an electric field to monitor the potential differences between the positions of the two probes.

For the determination of equipotential lines one probe is kept still, while the other probe is moved. In whichever position of the moving probe the voltmeter registers a zero reading; the potential of the moving probe is same as that of the standstill probe. By marking each such position, equipotentials could be traced.

**TABLE 12.1**

Analogy between Static Electric Field and Stationary Current Field

| Static Electric Field | Stationary Current Field |
|---|---|
| Electric flux | Electric current |
| $\vec{D} = \varepsilon\vec{E}$ | $\vec{J} = \kappa\vec{E}$ |
| Dielectric constant | Volume conductivity |
| $Q = \oint_s \vec{D}.d\vec{s}$ | $I = \oint_s \vec{J}.d\vec{s}$ |

For tracing electric fieldlines the two probes are kept at a constant separation distance and one probe is rotated around the other. The position of the rotating probe where the voltmeter registers a maximum reading, the electric field is changing at its maximum rate. Hence, the electric field at that location of the rotating probe is parallel to the line joining the two probes. By repeating this measurement process at several positions, the electric field could be mapped.

Because a real-life voltmeter draws a current, however small it may be, the measurement of the potential differences using voltmeter could not be done with vacuum or air as the medium. In practice, measurement is carried out for the electric field that is set up in a medium, which is slightly conducting.

Commonly slightly conducting paper, for example, a paper impregnated with carbon, is used. Because the paper is slightly conducting, the electric field due to the charged electrodes is almost the same as the one that would be produced in air or vacuum with similar geometry. At the same time, the paper is sufficiently conducting to supply the small current needed by the voltmeter.

Alternately, electrolytic tank setup is used, which consists of a specially fabricated insulating tray. A large sheet of laminated graph paper is pasted on the base plate of the tray. The tray is then half-filled with an electrolyte and the height of the electrolyte is kept same throughout the tray. Metallic electrodes are placed in the electrolytic tank, which are shaped to conform to the boundaries of the problem, and appropriate potential difference between the electrodes is maintained.

## 12.3 Field Mapping Using Curvilinear Squares

Field mapping by curvilinear squares is a graphical method based on the orthogonal property of a pair of conjugate harmonic functions and also on the geometric considerations. This method is suitable for mapping only those fields in which there is no variation of field in the direction normal to the plane of the sketch, that is, the field is two dimensional in nature. Many practical electric field problems may be considered as two dimensional, for example, the co-axial cylindrical system or a pair of long parallel wires. In these cases, the field remains same in all cross-sectional planes. It is a fact that no real system is infinitely long, but the idealization is a useful one for electric field analysis and visualization.

In this method, the field region of interest is discretized into a network of curvilinear squares formed by flux or fieldlines and equipotentials. Curvilinear square is a planar geometric figure that is different from a true square, as its sides are slightly curved and slightly unequal, but which approaches a true square as its dimensions become small. A typical curvilinear square is shown

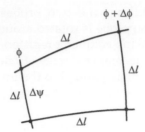

**FIGURE 12.1**
A typical curvilinear square.

in Figure 12.1. The field map thus obtained is unique for a given problem and helps in understanding the behaviour of electric field through visualization. The method of curvilinear square is capable of handling problems with complicated boundaries. A curvilinear field map is also independent of field property coefficients and could be directly applied from one physical field to another if an analogy exists between the concerned fields.

Theoretically, curvilinear field mapping is based on Cauchy–Riemann relations, which ensures that Laplace's equation is satisfied by a conjugate pair of harmonic functions in any orthogonal coordinate system. Hence, this method utilizes the fieldline coordinate representation of the electric field such that the electric field is always tangent to the fieldlines and depends only on the distribution of fieldlines and equipotentials.

### 12.3.1 Foundations of Field Mapping

Construction of field map using curvilinear squares is based on some significant features of the electric field as described below:

1. A conductor boundary is one of the equipotentials.
2. Equipotential and electric field intensity (or electric flux density) are normal to each other. As a conductor boundary is an equipotential, the electric field intensity and electric flux density vectors are always perpendicular to the conductor boundaries.
3. Electric flux lines (often termed as _streamlines_) originate and terminate on charges. Hence, in the case of a homogeneous and charge-free dielectric medium, electric flux lines originate and terminate on conductor boundaries.

Figure 12.2 shows two co-axial cylindrical conductor boundaries having a specified potential difference ($V$) and extending 1 m into the plane of the paper. A fieldline is considered to leave the boundary with more positive electric potential making an angle of 90° with the boundary at the point X. If the line is extended following the rule that it is always perpendicular to the

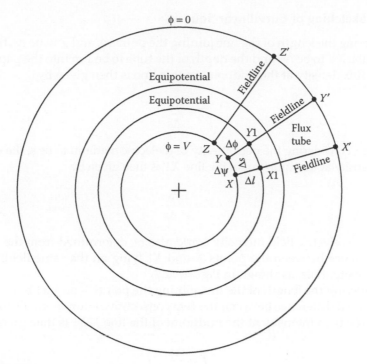

**FIGURE 12.2**
Field map between two co-axial cylinders.

equipotentials and if the dielectric medium is considered to be homogeneous and charge free, then the fieldline will terminate normally on the boundary of the less positive conductor at the point $X'$, as shown in Figure 12.2. In a similar manner, another fieldline could be drawn in such a way that it starts from the point $Y$ on the more positive conductor boundary and terminates on the point $Y'$ on the less positive conductor boundary. As the fieldlines are drawn perpendicular to the equipotentials everywhere, electric field intensity and electric flux density will be tangent to a fieldline everywhere on it. Consequently, no electric flux can cross any fieldline thus drawn. Therefore, if there is a charge of $\Delta Q$ on the surface of the conductor between the points $X$ and $Y$, then a flux of $\Delta \psi = \Delta Q$ will originate in this region and must terminate on the surface of the other conductor boundary between the points $X'$ and $Y'$. Such a pair of fieldlines is known as a *flux tube* as it seems to carry flux from one conductor to the other without losing any flux in between the two conductors. For the simplification of the interpretation of the field map, another flux tube $YZ$ may be drawn in such a way that the same amount of flux is carried in the flux tubes $XY$ and $YZ$. The method of determination of dimensions of the curvilinear square for drawing such flux tubes is described in the next section.

### 12.3.2 Sketching of Curvilinear Squares

Considering the length of the line joining the points $X$ and $Y$ to be $\Delta s$, the flux in the tube $XY$ to be $\Delta \psi$ and the depth of the tube to be 1 m into the paper, the electric flux density at the midpoint of this line is then given by

$$D = \frac{\Delta \psi}{\Delta s} \tag{12.2}$$

Therefore, considering the permittivity of dielectric medium to be $\varepsilon$, the electric field intensity at the midpoint of the line $XY$ is then given by

$$E = \frac{1}{\varepsilon} \frac{\Delta \psi}{\Delta s} \tag{12.3}$$

Alternately, electric field intensity could also be determined from the potential difference between the points $X$ and $X1$ lying on the same fieldline on two equipotentials, as shown in Figure 12.2.

Considering the length of the line joining the points $X$ and $X1$ to be $\Delta l$ and the potential difference between the two consecutive equipotentials to be $\Delta \phi$, the electric field intensity at the midpoint of the line $X$-$X1$ is then given by

$$E = \frac{\Delta \phi}{\Delta l} \tag{12.4}$$

Considering $\Delta s$ and $\Delta l$ to be small, the two values of electric field intensity, as given by Equations 12.3 and 12.4, may be taken to be equal. Hence,

$$\frac{1}{\varepsilon} \frac{\Delta \psi}{\Delta s} = \frac{\Delta \phi}{\Delta l}$$

$$\text{or,} \quad \frac{\Delta l}{\Delta s} = \varepsilon \frac{\Delta \phi}{\Delta \psi} \tag{12.5}$$

For sketching the field map, consider the following: (1) homogeneous dielectric having a constant permittivity $\varepsilon$; (2) constant amount of electric flux per tube, that is, $\Delta \psi$ is constant and (3) constant potential difference between two consecutive equipotentials, that is, $\Delta \phi$ is constant. Then from Equation 12.5, $\Delta l / \Delta s = $ constant. In other words, the ratio of the distance between fieldlines as measured along an equipotential and the distance between equipotentials as measured along a fieldline must be maintained constant and not the individual lengths. The simplest ratio of lengths that can be maintained is unity, so that $\Delta l = \Delta s$. Then the field region is divided into curvilinear squares by the fieldlines and equipotentials.

The field map thus obtained is composed of curvilinear squares of the same kind such that each square has the same potential difference across it and also has the same amount of flux through it. For a given $\Delta \phi$ and $\Delta \psi$, the sides of a curvilinear square are thus inversely proportional to electric field intensity. For a non-uniform field, electric field intensity varies with location and hence $\Delta l$ and $\Delta s$ vary with the strength of electric field. In the region of higher field strength, $\Delta l$ and $\Delta s$ are to be kept small, that is, the squares are to be made smaller in size where the magnitude of the field intensity is high. On the other hand, the squares are made larger in size in the field region where the field intensity is low.

It may be recalled that the product of electric charge and electric potential difference is the energy of electric field. Moreover, electric charge and electric flux has a one to one correspondence. Thus, for a field map, if $\Delta\phi$ and $\Delta\psi$ are kept constant, then their product remains constant and hence, the energy of the electric field remains constant. Therefore, curvilinear squares having the same ratio, as given by Equation 12.5, have the same energy stored in electric field regardless of the size of the square. A curvilinear square can thus be scaled up or down keeping the energy stored in the curvilinear square unaltered as long as the ratio given by Equation 12.5 remains unaltered.

### 12.3.3 Construction of Curvilinear Square Field Map

The fieldlines and equipotentials are typically drawn on the original sketch, which shows the conductor boundaries. Arbitrarily, one fieldline is begun from a point on the surface of the more positive conductor with a suitable value of $\Delta l$ and an equipotential is drawn perpendicular to the fieldline with a value of $\Delta s = \Delta l$. Then another fieldline is added to complete the curvilinear square. The field map is then gradually extended throughout the field region of interest. As the field map is extended, the condition of orthogonality of fieldline and equipotential should be kept paramount, even if this results in some squares with ratios other than unity. Construction of a satisfactory field map using curvilinear squares is a trial-and-error process that involves continuous adjustment and refinement. Typically, field maps are started as a coarse map having large curvilinear squares. Then the field map is fine-tuned through successive subdivisions to form a dense field map having higher accuracy. In the process of subdivision, the lengths between consecutive fieldlines as well as equipotentials are kept equal. Before starting the construction of a field map, it is a judicious practice to examine the geometry of the system and take advantage of any symmetry that may exist in the system under consideration. This is because of the fact the lines of symmetry serve as boundaries with no flux crossing and thereby separate regions of similar field maps.

### 12.3.4 Capacitance Calculation from Field Map

Once the field map is drawn, it is possible to determine the capacitance per unit length between the two conductors using the field map. It is well known that capacitance between two conductors having a potential difference of

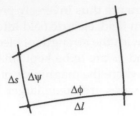

**FIGURE 12.3**
An isolated curvilinear rectangle.

*V* is given by $C = Q/V$, where *Q* is the charge on the conductor. Applying Gauss's law on a Gaussian surface enclosing the conductor having more positive potential, $Q = \psi$, where $\psi$ is the flux coming out of the conductor. Thus, $C = \psi/V$.

To calculate the capacitance with the help of curvilinear rectangle, consider first an isolated curvilinear rectangle, as shown in Figure 12.3. Let the flux through it be $\Delta\psi$ and the potential difference across it be $\Delta\phi$. Considering the curvilinear rectangle to be small, the flux density may be assumed uniform within the curvilinear rectangle, so that

$$\Delta\psi = \varepsilon E \Delta s \times 1 \tag{12.6}$$

where the depth is taken to be 1 m into the plane of the field map.

Electric field intensity (*E*) and the potential difference ($\Delta\phi$) are related as

$$\Delta\phi = E \times \Delta l \tag{12.7}$$

Combining Equations 12.6 and 12.7

$$\Delta\psi = \varepsilon \Delta s \times \frac{\Delta\phi}{\Delta l}$$

Therefore, the capacitance of the small curvilinear rectangle, which may be taken as a small field cell, is given by

$$\Delta C = \frac{\Delta\psi}{\Delta\phi} = \varepsilon \frac{\Delta s}{\Delta l} \tag{12.8}$$

The total amount of flux ($\psi$) emanating from one conductor and terminating on the other conductor may be obtained by adding all the small amounts of flux ($\Delta\psi$) through each flux tube, so that

$$\psi = \sum_{N_\psi} \Delta\psi = N_\psi \Delta\psi \tag{12.9}$$

where:

$\Delta\psi$ is assumed to be same for each flux tube

$N_\psi$ is the number of flux tubes in parallel, that is, the number of curvilinear rectangles in parallel

The total potential difference between the two conductors $(V)$ may be obtained by adding all the small amounts of potential differences $(\Delta\phi)$ between consecutive equipotentials starting from one conductor and finishing at the other conductor, that is,

$$V = \sum_{N_\phi} \Delta\phi = N_\phi \Delta\phi \tag{12.10}$$

where:

$\Delta\phi$ is assumed to be same between any two consecutive equipotentials

$N_\phi$ is the number of equipotentials (including the two conductors) minus one, that is, the number of curvilinear rectangles in series between the two conductors

Thus, capacitance per unit length of the two conductors is given by

$$C = \frac{\psi}{V} = \frac{N_\psi}{N_\phi} \frac{\Delta\psi}{\Delta\phi} = \frac{N_\psi}{N_\phi} \frac{\varepsilon \Delta s}{\Delta l} = \varepsilon \frac{N_\psi}{N_\phi} \tag{12.11}$$

where $\Delta s = \Delta l$, considering the ratio of the lengths to be unity, that is, considering curvilinear squares.

Hence, the determination of capacitance from the field map involves counting of curvilinear squares in two directions, one in series between the two conductors and the other in parallel around either conductor.

## 12.4 Field Mapping in Multi-Dielectric Media

From Equation 12.5, it may be seen that for the same value of electric flux per tube and same potential difference between two consecutive equipotentials,

$$\frac{\Delta l}{\Delta s} \propto \varepsilon \tag{12.12}$$

Thus, in the case of a two-dielectric configuration, as shown in Figure 12.4, the ratio of the sides of curvilinear element is to be made proportional to

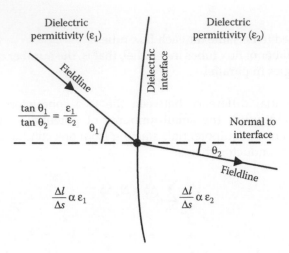

**FIGURE 12.4**
Mapping in a two-dimensional configuration with multi-dielectric media.

the relative permittivity of the dielectric medium in which the field map is drawn. In other words, curvilinear rectangles are to be used.

Moreover, the deviation of the fieldlines takes place at the boundary between the two dielectric media, as shown in Figure 12.4, which is given for charge-free dielectric media by

$$\frac{\tan\theta_1}{\tan\theta_2} = \frac{\varepsilon_1}{\varepsilon_2} \tag{12.13}$$

For two-dimensional configurations comprising multi-dielectric media, the field map is first drawn in the field region where there is only one dielectric media. Then the directions of the fieldlines are changed at the boundary between the two dielectric media according to Equation 12.13. Subsequently, the ratio of the sides of the curvilinear rectangles is changed as per Equation 12.12 and the field map is extended into the field region comprising a different dielectric medium. In this way, the field map could be obtained in configurations comprising several dielectric media.

## 12.5 Field Mapping in Axi-Symmetric Configuration

Consider a curvilinear rectangle in an axi-symmetric configuration, as shown in Figure 12.5. Let the radial distance of the centroid of the curvilinear rectangle from the axis of rotational symmetry be $r$.

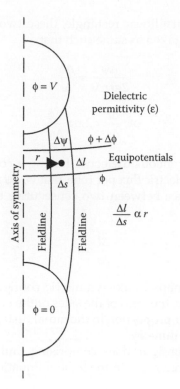

**FIGURE 12.5**
Field mapping in an axi-symmetric configuration.

Considering the flux through the rectangle to be $\Delta\psi$ and assuming the square to be small, the electric flux density can be taken to be uniform within the rectangle and is given by

$$D = \frac{\Delta\psi}{2\pi r \Delta s}$$

$$\text{or, } E = \frac{\Delta\psi}{2\pi \varepsilon r \Delta s} \tag{12.14}$$

Alternately, the electric field intensity as obtained from the potential difference between two consecutive equipotentials, which are the two sides of the rectangle perpendicular to the fieldlines, is given by

$$E = \frac{\Delta\phi}{\Delta l}$$

Considering a small curvilinear rectangle, these two values of the electric field intensity could be taken as same such that

$$\frac{\Delta\psi}{2\pi\varepsilon r\Delta s} = \frac{\Delta\phi}{\Delta l}$$

$$\text{or, } \frac{\Delta l}{\Delta s} = 2\pi\varepsilon r\frac{\Delta\phi}{\Delta\psi} \tag{12.15}$$

Considering (1) homogeneous dielectric having a constant permittivity $\varepsilon$; (2) constant amount of electric flux per tube, that is, $\Delta\psi$ is constant and (3) constant potential difference between two consecutive equipotentials, that is, $\Delta\phi$ is constant,

$$\frac{\Delta l}{\Delta s} \propto r \tag{12.16}$$

Hence, to draw field maps in axi-symmetric configurations comprising a homogeneous dielectric, the ratio of the sides of the curvilinear rectangles is to be increased in direct proportion to the radial distance of the square from the axis of rotational symmetry.

For axi-symmetric configurations comprising multiple dielectric media, Equation 12.16 is to be rewritten in the light of Equation 12.15 as

$$\frac{\Delta l}{\Delta s} \propto \varepsilon r \tag{12.17}$$

Therefore, for multi-dielectric media in axi-symmetric configurations, the ratio of the sides of the curvilinear rectangles is to be increased not only in direct proportion to the radial distance of the square from the axis of rotational symmetry but also in direct proportion to the relative permittivity of the dielectric medium in which the field map is drawn. The directions of the fieldlines at the boundary between the two dielectric media are to be changed according to Equation 12.13.

---

## Objective Type Questions

1. Experimental method of field mapping is carried out using
   a. Highly conducting paper
   b. Slightly conducting paper
   c. Non-conducting paper
   d. None of the above

2. Graphical method of field mapping using curvilinear squares or rectangles is suitable for
   a. Two-dimensional and three-dimensional systems
   b. Two-dimensional and axi-symmetric systems
   c. Only two-dimensional system
   d. Only axi-symmetric system

3. Which one of the following is most commonly used for graphical field mapping in a two-dimensional system with homogeneous dielectric medium?
   a. Rectilinear square
   b. Rectilinear rectangle
   c. Curvilinear square
   d. Curvilinear rectangle

4. For field mapping in a two-dimensional system with homogeneous dielectric medium, the ratio of the distance between fieldlines as measured along an equipotential and the distance between equipotentials as measured along a fieldline is to be
   a. Kept constant
   b. Made proportional to dielectric constant of the medium
   c. Made inversely proportional to the dielectric constant of the medium
   d. Chosen arbitrarily

5. For field mapping in a two-dimensional system with multiple dielectric media, the ratio of the distance between fieldlines as measured along an equipotential and the distance between equipotentials as measured along a fieldline is to be
   a. Kept constant
   b. Made proportional to dielectric constant of the medium
   c. Made inversely proportional to the dielectric constant of the medium
   d. Chosen arbitrarily

6. For field mapping in an axi-symmetric system with homogeneous dielectric medium, the ratio of the distance between fieldlines as measured along an equipotential and the distance between equipotentials as measured along a fieldline is to be
   a. Kept constant
   b. Made proportional to dielectric constant of the medium
   c. Made inversely proportional to the dielectric constant of the medium
   d. Made proportional to the radial distance from the axis of symmetry

7. For field mapping in an axi-symmetric system with multiple dielectric media, the ratio of the distance between fieldlines as measured
   along an equipotential and the distance between equipotentials as
   measured along a fieldline is to be

   a. Kept constant

   b. Made proportional to dielectric constant of the medium

   c. Made proportional to the radial distance from the axis of
      symmetry

   d. Made proportional to the product of dielectric constant of the
      medium and the radial distance from the axis of symmetry

8. Accuracy that could be achieved in capacitance determination by
   graphical field plotting is typically of the order of

   a. 0.5%–1%

   b. 1%–2%

   c. 5%–10%

   d. 10%–20%

9. Which one of the following is most commonly used for graphical
   field mapping in a two-dimensional system with multiple dielectric
   media?

   a. Rectilinear square

   b. Rectilinear rectangle

   c. Curvilinear square

   d. Curvilinear rectangle

10. Determination of capacitance from the field map involves counting
    of curvilinear squares or rectangles

    a. In series between the two conductors only

    b. In parallel around either conductor only

    c. In series between the two conductors and also in parallel around
       either conductor

    d. None of the above

**Answers:**    1) b; 2) b; 3) c; 4) a; 5) b; 6) d; 7) d; 8) c; 9) d; 10) c

# Bibliography

1. E. Weber, 'Mapping of fields', *Electrical Engineering*, Dec, pp. 1563–1570, 1934.
2. H. Poritsky, 'Graphical field plotting methods in engineering', *AIEE Transactions*, Vol. 57, pp. 727–732, 1938.
3. R.J.W. Koopman, 'Mapping of electric fields into curvilinear squares', *Transactions of Kansas Academy of Science*, Vol. 41, pp. 233–236, 1938.
4. E.O. Willoughby, 'Some applications of field plotting', *Journal of the Institution of Electrical Engineers—Part III: Radio and Communication Engineering*, Vol. 93, No. 24, pp. 275–293, 1946.
5. K.F. Sander and J.G. Yates, 'The accurate mapping of electric fields in an electrolytic tank', *Proceedings of the IEE—Part II: Power Engineering*, Vol. 100, No. 74, pp. 167–175, 1953.
6. H. Diggle and E.R. Hartill, 'Some applications of the electrolytic tank to engineering design problems', *Proceedings of the IEE—Part II: Power Engineering*, Vol. 101, No. 82, pp. 349–364, 1954.
7. L. Tasny-Tschiassny, 'The approximate solution of electric-field problems with the aid of curvilinear nets', *Proceedings of the IEE—Part C: Monographs*, Vol. 104, No. 5, pp. 116–129, 1957.
8. J.D. Horgan and J.A. Pesavento, 'The accurate determination of capacitance', *Electrical Engineering*, Vol. 77, No. 6, p. 513, 1958.
9. S.Y. King, 'The electric field near bundle conductors', *Proceedings of the IEE—Part C: Monographs*, Vol. 106, No. 10, pp. 200–206, 1959.
10. M.M. Sakr and B. Salvage, 'Electric stresses at conducting surfaces located in the field between plane parallel electrodes: Experiments with an electrolytic tank', *Proceedings of the Institution of Electrical Engineers*, Vol. 111, No. 6, pp. 1179–1181, 1964.
11. E.S. Ip, 'Electrostatic fields of transformer-bushing insulators', *Proceedings of the Institution of Electrical Engineers*, Vol. 114, No. 11, pp. 1729–1733, 1967.
12. R.B. Goldner, 'Rules for field plotting in a class of two-dimensional inhomogeneous conductors', *Proceedings of IEEE*, Vol. 56, No. 8, pp. 1367–1368, 1968.
13. S. Ramo, J.R. Whinnery and T. van Duzer, *Fields and Waves in Communication Electronics*, 3rd ed., John Wiley & Sons, New York, 1994.
14. M.N.O. Sadiku, *Elements of Electromagnetics*, 5th ed., Oxford University Press, New York, Oxford 2009.
15. S. Tou, *Visualization of Fields and Applications in Engineering*, John Wiley & Sons, Chichester, UK, 2011.

# 13

---

## Numerical Computation of Electric Field

**ABSTRACT**   In a high-voltage equipment, as a thumb rule, the cost of insulation increases with the cube of voltage rating. However, practical experience shows that insulation failure is the most frequent cause of major breakdown of electrical equipment. Hence, it is of paramount importance that the insulation in electrical equipment withstands the electric field stresses with adequate safety margin. For this purpose, it is necessary to determine the electric field distribution precisely. An inherent advantage of analytic solutions is the exactness. However, analytic solutions for the electric field are available only for problems having simple configurations. When the complexities of theoretical formulae make analytic solution extremely difficult, if not impossible, then one has to take resort to non-analytic methods. Graphical, experimental and analog methods are applicable to solve fewer problems, which have less complexity. With the advent of fast digital computers, numerical methods came into prominence. Numerical methods make it possible to solve practical problems, which are very complex in nature, when appropriate procedural steps are followed. Numerical methods are commonly based on the solutions of partial differential equations or integral equations with some analytic simplification for easy implementation.

---

## 13.1 Introduction

The design of the insulation of high-voltage apparatus between phases and earth and also between the phases is based on the knowledge of electric field distribution and the dielectric properties of the combination of insulating materials used in the system. The principal aim is that the insulation should withstand the electric stresses with adequate reliability and at the same time the insulation should not be over dimensioned.

It is well known that the withstand voltage of the external insulation of apparatus designed with non-self restoring insulation is determined by the maximum value of electric field intensity within the insulation system. Further, corona discharges are eliminated by proper design of high-voltage shielding electrodes. Thus, a comprehensive study of the electric field distribution in and around high-voltage equipment is of great practical importance.

High-voltage equipment, in practice, is, in most of the cases, subjected to AC field of frequency 50 or 60 Hz. These fields may be approximated as quasi-static as the wavelength is much longer compared to the dimension of the components involved. Because of this, the electrostatic field calculation is possible by the different methods in use.

*Mathematically, an electric field calculation problem may be formulated as follows.*

The purpose is to determine, at each point within the field region of interest (ROI), the value of potential $\phi(x,y,z)$ and that of the electric field intensity $\vec{E}(x,y,z)$ are to be determined, which are related as

$$\vec{E}(x,y,z) = -\vec{\nabla}\phi \tag{13.1}$$

In order to do that, either Laplace's equation for systems without any source of charge in the field region,

$$\vec{\nabla}^2\phi = 0 \tag{13.2}$$

or, Poisson's equation for systems with sources of charge in the field region,

$$\vec{\nabla}^2\phi = -\frac{\rho_v}{\varepsilon} \tag{13.3}$$

are required to be solved.

The solutions of these equations are called *boundary value problems*, whereby the boundary conditions are specified by means of the given potential of electrode (Dirichlet's problem) or by the given value of electric field intensity (Neumann's problem).

## 13.2 Methods of Determination of Electric Field Distribution

The methods that are employed for determination of electric field are detailed in Figure 13.1.

The analytical methods can only be applied to the cases, where the electrode or dielectric boundaries are of simple geometrical forms such as cylinders, spheres and so on. In other words, in this method, the boundaries are required to be defined exclusively by known mathematical functions. The results obtained are very accurate. But, as it is obvious, this method cannot be applied to complex problems. However, the results obtained by analytical methods for standard configurations are used still today to validate the results obtained by some other approximate methods such as numerical methods.

Earlier experimental as well as graphical methods were used to get a fair idea about the nature of field distribution in some practical cases. However, these methods are greatly limited in their areas of usage and the errors

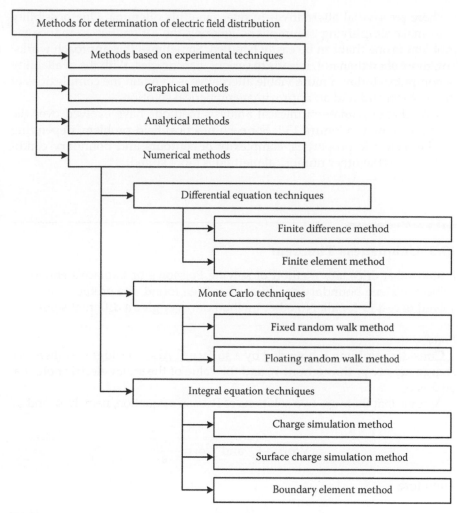

**FIGURE 13.1**
Different methods for the determination of the electric field distribution.

involved are usually very high for any complex problem to be taken directly for design purposes.

Nowadays, in more and more engineering problems, it is found that it is necessary to obtain approximate numerical solutions rather than exact closed-form solutions. The governing equations and boundary conditions for these problems could be written without too much effort, but it may be seen immediately that no simple analytical solution can be found. The difficulty in these engineering problems lies in the fact that either the geometry or some other feature of the problem is irregular. Analytical solutions to this type of problems seldom exist; yet these are the kinds of problems that engineers need to solve.

There are several alternatives to overcome this dilemma. One possibility is to make simplifying assumptions ignoring the difficulties to reduce the problem to one that can be easily handled. Sometimes this approach works; but, more often than not, it leads to serious inaccuracies. With the availability of computers today, a more viable alternative is to retain the complexities of the problem and find an approximate numerical solution.

Several approximate numerical analysis methods have evolved over the years, as shown in Figure 13.1. For each practical field problem, depending on the dielectric properties, complexity of contours and boundary conditions, one or the other numerical method is more suited.

## 13.3 Uniqueness Theorem

It states that once any method of solving Poisson's or Laplace's equations subject to given boundary conditions has been found, the problem has been solved once and for all. No other method can ever give a different solution.

*Proof:*

Consider a volume $V$ bounded by a surface $S$. Also consider that there is a charge density $\rho_v$ throughout $V$, and the value of the scalar electric potential on $S$ is $\phi_s$.

Assume that there are two solutions of Poisson's equation, namely, $\phi_1$ and $\phi_2$. Then

$$\vec{\nabla}^2\phi_1 = -\frac{\rho_v}{\varepsilon} \text{ and } \vec{\nabla}^2\phi_2 = -\frac{\rho_v}{\varepsilon}$$

Therefore,

$$\vec{\nabla}^2(\phi_1 - \phi_2) = 0 \tag{13.4}$$

Now, each solution must also satisfy the boundary conditions. It is to be noted here that one particular point cannot have two different electric potentials, as the work done to move a unit positive charge from infinity to that point is unique. Let the value of $\phi_1$ on the boundary is $\phi_{1s}$ and the value of $\phi_2$ on the boundary is $\phi_{2s}$ and they must be identical to $\phi_s$. Therefore,

$$\phi_{1s} = \phi_{2s} = \phi_s$$

$$\text{or, } \phi_{1s} - \phi_{2s} = 0$$

For any scalar $\phi$ and any vector $\vec{D}$, the following vector identity can be written.

$$\vec{\nabla}(\phi\vec{D}) \equiv \phi(\vec{\nabla}\cdot\vec{D}) + \vec{\nabla}\phi\cdot\vec{D} \tag{13.5}$$

Consider the scalar as $(\phi_1 - \phi_2)$ and the vector as $\vec{\nabla}(\phi_1 - \phi_2)$. Then from Identity 13.5,

$$\vec{\nabla} \cdot \left[ (\phi_1 - \phi_2)\vec{\nabla}(\phi_1 - \phi_2) \right] \equiv (\phi_1 - \phi_2)\left[ \vec{\nabla} \cdot \vec{\nabla}(\phi_1 - \phi_2) \right]$$
$$+ \vec{\nabla}(\phi_1 - \phi_2) \cdot \vec{\nabla}(\phi_1 - \phi_2) \quad (13.6)$$

Now, integrating throughout $V$ enclosed by $S$,

$$\int_V \vec{\nabla} \cdot \left[ (\phi_1 - \phi_2)\vec{\nabla}(\phi_1 - \phi_2) \right] dv \equiv \int_V (\phi_1 - \phi_2)\left[ \vec{\nabla} \cdot \vec{\nabla}(\phi_1 - \phi_2) \right] dv$$
$$+ \int_V \vec{\nabla}(\phi_1 - \phi_2) \cdot \vec{\nabla}(\phi_1 - \phi_2) dv$$
$$\equiv \int_V (\phi_1 - \phi_2)\left[ \vec{\nabla}^2(\phi_1 - \phi_2) \right] dv \quad (13.7)$$
$$+ \int_V \left[ \vec{\nabla}(\phi_1 - \phi_2) \right]^2 dv$$

Applying divergence theorem to the left-hand side of Identity 13.7,

$$\int_V \vec{\nabla} \cdot \left[ (\phi_1 - \phi_2)\vec{\nabla}(\phi_1 - \phi_2) \right] dv = \int_S (\phi_{1S} - \phi_{2S})\vec{\nabla}(\phi_1 - \phi_2) ds = 0 \quad (13.8)$$

as $\phi_{1s} = \phi_{2s}$ on the specified surface $S$.

On the right-hand side of Identity 13.7, $\vec{\nabla}^2(\phi_1 - \phi_2) = 0$ from Equation 13.4. Hence, Identity 13.7 reduces to

$$\int_V \left[ \vec{\nabla}(\phi_1 - \phi_2) \right]^2 dv = 0 \quad (13.9)$$

Because $\left[ \vec{\nabla}(\phi_1 - \phi_2) \right]^2$ cannot be negative, the integrand must be zero everywhere, so that the integral may be zero.

Hence,

$$\left[ \vec{\nabla}(\phi_1 - \phi_2) \right]^2 = 0 \text{ or, } \vec{\nabla}(\phi_1 - \phi_2) = 0 \quad (13.10)$$

Again, if the gradient of $(\phi_1 - \phi_2)$ is zero everywhere, then

$$\phi_1 - \phi_2 = \text{constant} \quad (13.11)$$

This constant may be evaluated by considering a point on the boundary surface $S$, so that

$$\phi_1 - \phi_2 = \phi_{1s} - \phi_{2s} = 0$$

$$\text{or, } \phi_1 = \phi_2$$

which means that the two solutions are identical.

However, in practice, if the same problem is solved by using different numerical techniques, the results are not exactly the same. This is due to the fact that the errors in a particular numerical method are often problem dependent and hence the results are not exactly same in all the methods. Therefore, this is not a violation of the Uniqueness theorem.

## 13.4 Procedural Steps in Numerical Electric Field Computation

The following are the procedural steps that need to be followed for most of the numerical electric field computation methods.

At first, the ROI needs to be identified. ROI is the region where the solution for electric field is to be obtained. For example, normally the field solution is not needed within the electrode volume or below the earth surface. Hence, for an isolated electrode and the earth surface, the ROI will be a region between the electrode surface and the earth surface, as shown in Figure 13.2. Before the ROI is identified, the geometries of the components that comprise the field system need to be defined. This step is nowadays done with the help of computer-aided design (CAD) software.

The subsequent procedural step is to discretize the entire ROI or the boundaries to create the nodes where the solution of field will be obtained. Ideally, one should find the field solution at each and every point within the ROI. But it will result in immense computational burden and hence the field solution is obtained at discrete nodes. This step is called *discretization* and is often done with the help of mesh generators, which are software modules

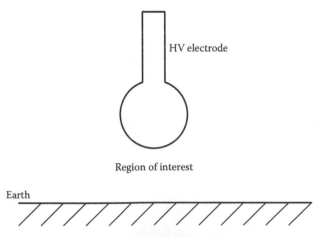

**FIGURE 13.2**
Depiction of the region of interest for electric field computation.

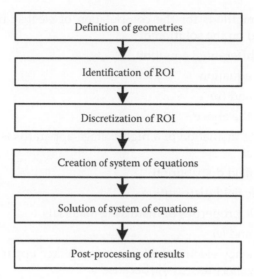

**FIGURE 13.3**
Procedural steps in numerical electric field computation.

that create the mesh within the entire ROI or on the boundaries. In order that the electric field solution can be obtained at any specific location within the ROI, a pre-defined variation of electric field between successive nodes is assumed. In fact, this assumption is a root cause of inaccuracy of the numerical method.

The next step is to create the system of equations based on the numerical method that is being employed. Subsequently, the system of equations is solved using a suitable solver. The solver needs to be chosen depending on the nature of the coefficient matrix that is being created by the specific numerical method. This solution gives the results for the unknown field quantities at the predefined nodes. Finally, the results at any desired location is computed using the assumed variation of electric field between the nodes, which is termed as *post-processing of results*. The procedural steps are depicted in Figure 13.3.

## Objective Type Questions

1. Methods applicable to solve electric field problems having less complexity are
   a. Graphical method
   b. Experimental method
   c. Analytical method
   d. All the above

2. Numerical methods for the computation of electric field are commonly based on the solution of
   a. Partial differential equation
   b. Integral equation
   c. Both (a) and (b)
   d. None of the above
3. Design of the insulation of a high-voltage apparatus is based on the knowledge of
   a. Electric field distribution
   b. Magnetic field distribution
   c. Dielectric properties of insulating materials
   d. Both (a) and (c)
4. Power frequency electric field in a high-voltage apparatus could be approximated as quasi-static because
   a. The wavelength is much longer compared to the dimension of the components involved
   b. The wavelength is slightly longer compared to the dimension of the components involved
   c. The wavelength is slightly shorter compared to the dimension of the components involved
   d. The wavelength is much shorter compared to the dimension of the components involved
5. Boundary value problems where the boundary conditions are specified by means of the given value of electric potential are called
   a. Gauss's problem
   b. Dirichlet's problem
   c. Poisson's problem
   d. Laplace's problem
6. Boundary value problems where the boundary conditions are specified by means of the given value of electric field intensity are called
   a. Gauss's problem
   b. Neumann's problem
   c. Poisson's problem
   d. Laplace's problem

7. Results of numerical methods are commonly validated using the results obtained from

   a. Experimental method
   b. Analytical method
   c. Graphical method
   d. All the above

8. Accurate solutions of electric field problems having irregular geometries are obtained by

   a. Making simplifying assumptions ignoring the difficulties to reduce the problem
   b. Finding an approximate numerical solution by retaining the complexities of the problem
   c. Both (a) and (b)
   d. None of the above

9. According to uniqueness theorem, any electric field problem has been solved once and for all if particular equations subject to the given boundary conditions are solved, which are

   a. Gauss's equation
   b. Laplace's equation
   c. Poisson's equation
   d. Both (b) and (c)

10. In numerical computation of electric field, the discretization of field region is carried out to

    a. Obtain field solution at each and every point within the field region
    b. Obtain field solution at discrete nodes within the field region
    c. Reduce the computational burden
    d. Both (b) and (c)

**Answers:**   1) d; 2) c; 3) d; 4) a; 5) b; 6) b; 7) b; 8) b; 9) d; 10) d

# 14

# Numerical Computation of High-Voltage Field by Finite Difference Method

**ABSTRACT** Numerical methods for solving Laplace's equation by expressing it in finite-difference form have been attempted by researchers way back in the mid-1950s. However, it is only in the mid-1960s, with the advent of relatively high-speed computers, that this numerical method of solution became a feasible proposition. Finite difference method (FDM), as it is commonly known, could be used to study electrostatic field distribution in two-dimensional (2D), axi-symmetric as well as in three-dimensional (3D) cases. This chapter presents detailed discussions on the FDM formulations as applied to all the above-mentioned cases. The system of FDM equations, the accuracy criteria as well as the technique to handle unbounded field region problems have been discussed thoroughly. Suitable application examples of 2D and axi-symmetric case studies are also given that show typical FDM grids for easy comprehension of FDM methodology.

## 14.1 Introduction

The principle of finite difference method (FDM) is to discretize the entire region under study and solve for unknown potentials a set of coupled simultaneous linear equations, which approximate Laplace's or Poisson's equations. In fact, this is the objective of most of the numerical-field computation techniques that are being used at present.

In FDM, for a two-dimensional (2D) system, the entire region of interest (ROI) is discretized using either rectangles or squares. In a three-dimensional (3D) system, the discretization is done using either rectangular parallelepipeds or cubes. Most commonly, electric potential is assumed to vary linearly between two successive nodes. However, this is not mandatory. Any other type variation, for example, quadratic or polynomial, may also be assumed. But, a complex nature of potential variation increases the computational burden greatly and may not always give improved accuracy. If the electric potential is assumed to vary linearly, as it is commonly considered, then the nodes need to be closely spaced where the field varies significantly in space. This is generally the case near the electrodes or dielectric boundaries, particularly in the cases of contours having sharp corners. On the other hand, in the region away from the electrodes or dielectric boundaries, where the field does not change rapidly in space, the nodes may be spaced relatively widely apart.

For multi-dielectric problems, care should be taken during discretization to make sure that only one dielectric is present between two consecutive nodes. This is achieved by arranging one layer of nodes along the dielectric–dielectric interface. This aspect will be taken up in more details in a Sections 14.4 and 14.5 in this chapter.

## 14.2 FDM Equations in 3D System for Single-Dielectric Medium

As stated earlier, in a 3D system, discretization is done using either rectangular parallelepipeds or cubes. In such cases, one particular node is connected to six neighbouring nodes, as shown in Figure 14.1. As it is assumed that electric potential varies linearly between two successive nodes, it is obvious that the potential of that particular node will be related to potentials of the six connected nodes. FDM equation for any unknown node potential is developed in terms of potentials of the connected nodes by satisfying Laplace's equation. FDM equation thus developed is a linear equation, that is an approximation of Laplace's equation, which is a second-order partial differential equation.

Because the nodal distances in a practical system are unequal, the following approach is normally taken for the development of the FDM equations. After discretization, the largest nodal distance ($h$) is identified within the ROI. Then

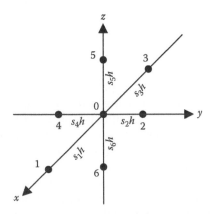

**FIGURE 14.1**
Unequal nodal distances for FDM equation development in three-dimensional (3D) system.

all the other nodal distances are represented as a fraction of that largest nodal distance as $s_x h$, where $s_x < 1$. This is done because the factor $s_x$ is a dimensionless quantity and the FDM equation is developed in terms of potentials of the six connected nodes and the dimensionless factors $s_x$. Therefore, the developed FDM equation becomes a linear equation involving electric potential only.

As shown in Figure 14.1, the unknown potential of node 0 will be formulated in terms of potentials of the six connected nodes 1 through 6. Electric potential being a continuous function within the ROI and the nodal distances being not large, the Taylor series can be applied for the determination of the potential of any one connected node from the potential of node 0. Taylor series in 3D system is expressed as follows:

$$
f(x+a,y+b,z+c) = f(x,y,z) + a\frac{\partial}{\partial x}f(x,y,z) + b\frac{\partial}{\partial y}f(x,y,z)
$$

$$
+ c\frac{\partial}{\partial z}f(x,y,z) + \frac{a^2}{2!}\frac{\partial^2}{\partial x^2}f(x,y,z) + \frac{b^2}{2!}\frac{\partial^2}{\partial y^2}f(x,y,z) \quad (4.1)
$$

$$
+ \frac{c^2}{2!}\frac{\partial^2}{\partial z^2}f(x,y,z) + \cdots
$$

Applying the Taylor series expansion between nodes 1 and 0 considering the potential of node 0 ($V_0$) as $f(x,y,z)$ and the node potential of node 1 ($V_1$) as $f(x + a,y + b,z + c)$, so that $a = s_1 h$, $b = c = 0$ and neglecting higher order terms,

$$
V_1 = V_0 + s_1 h\left.\frac{\partial V}{\partial x}\right|_0 + \frac{(s_1 h)^2}{2}\left.\frac{\partial^2 V}{\partial x^2}\right|_0 \quad (14.2)
$$

Similarly, applying the Taylor series expansion between nodes 0 and 3, such that $a = -s_3 h$, $b = c = 0$.

$$
V_3 = V_0 - s_3 h\left.\frac{\partial V}{\partial x}\right|_0 + \frac{(s_3 h)^2}{2}\left.\frac{\partial^2 V}{\partial x^2}\right|_0 \quad (14.3)
$$

Eliminating $\partial V/\partial x$ from Equations 14.2 and 14.3,

$$
\left.\frac{\partial^2 V}{\partial x^2}\right|_0 = \frac{\left[(V_1/s_1) + (V_3/s_3)\right] - V_0\left[(1/s_1) + (1/s_3)\right]}{h^2(s_1 + s_3)/2} \quad (14.4)
$$

Similarly, between nodes 2 and 4 in the $y$-direction,

$$
\left.\frac{\partial^2 V}{\partial y^2}\right|_0 = \frac{\left[(V_2/s_2) + (V_4/s_4)\right] - V_0\left[(1/s_2) + (1/s_4)\right]}{h^2(s_2 + s_4)/2} \quad (14.5)
$$

and between nodes 5 and 6 in the z-direction,

$$\frac{\partial^2 V}{\partial z^2}\bigg|_0 = \frac{\left[(V_5/s_5)+(V_6/s_6)\right]-V_0\left[(1/s_5)+(1/s_6)\right]}{h^2(s_5+s_6)/2} \tag{14.6}$$

Now, Laplace's equation in the Cartesian coordinates at node 0,

$$\frac{\partial^2 V}{\partial x^2}\bigg|_0 + \frac{\partial^2 V}{\partial y^2}\bigg|_0 + \frac{\partial^2 V}{\partial z^2}\bigg|_0 = 0 \tag{14.7}$$

Therefore, from Equations 14.4 through 14.7, eliminating $h$, the FDM equation for the unknown node potential $V_0$ is obtained as

$$V_0 = \frac{\left\{\begin{array}{l} \left[1/(s_1+s_3)\right]\left[(V_1/s_1)+(V_3/s_3)\right]+\left[1/(s_2+s_4)\right]\left[(V_2/s_2)+(V_4/s_4)\right] \\ +\left[1/(s_5+s_6)\right]\left[(V_5/s_5)+(V_6/s_6)\right] \end{array}\right\}}{(1/s_1s_3)+(1/s_2s_4)+(1/s_5s_6)} \tag{14.8}$$

For equal nodal distances in a 3D system, $s_1 = s_2 = s_3 = s_4 = s_5 = s_6 = 1$. Therefore, Equation 14.8 reduces to

$$V_0 = \frac{1}{6}(V_1+V_2+V_3+V_4+V_5+V_6) \tag{14.9}$$

For a 2D system, as shown in Figure 14.2, with unequal nodal distances, the FDM Equation 14.8 reduces to

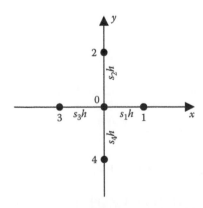

**FIGURE 14.2**
Unequal nodal distances for FDM equation development in 2D system.

$$V_0 = \frac{\left[1/(s_1 + s_3)\right]\left[(V_1/s_1) + (V_3/s_3)\right] + \left[1/(s_2 + s_4)\right]\left[(V_2/s_2) + (V_4/s_4)\right]}{(1/s_1 s_3) + (1/s_2 s_4)} \quad (14.10)$$

For equal nodal distances in a 2D system, the FDM Equation 14.10 reduces to

$$V_0 = \frac{1}{4}(V_1 + V_2 + V_3 + V_4) \quad (14.11)$$

**PROBLEM 14.1**

Consider the 3D arrangement with single dielectric having six given planes *a*, *b*, *c*, *d*, *e* and *f*, as shown in Figure 14.3. Write the FDM equations for the unknown node potentials $V_{01}$ and $V_{02}$. Given $h_2 = 0.6\ h_1$.

*Solution:*

In this case, for both the nodes $0_1$ and $0_2$, $s_1 = s_2 = s_3 = s_4 = 1$ and $s_5 = s_6 = (h_2/h_1) = 0.6$.

$$V_{01} = \frac{\left\{\begin{array}{l}\left[1/(1+1)\right]\left[(V_6/1) + (V_{22}/1)\right] + \left[1/(1+1)\right]\left[(V_{02}/1) + (V_{34}/1)\right] \\ + \left[1/(0.6+0.6)\right]\left[(V_{30}/0.6) + (V_{14}/0.6)\right]\end{array}\right\}}{(1/1) + (1/1) + \left[1/(0.6 \times 0.6)\right]} \quad (14.12)$$

and

$$V_{02} = \frac{\left\{\begin{array}{l}\left[1/(1+1)\right]\left[(V_7/1) + (V_{23}/1)\right] + \left[1/(1+1)\right]\left[(V_{33}/1) + (V_{01}/1)\right] \\ + \left[1/(0.6+0.6)\right]\left[(V_{31}/0.6) + (V_{15}/0.6)\right]\end{array}\right\}}{(1/1) + (1/1) + \left[1/(0.6 \times 0.6)\right]} \quad (14.13)$$

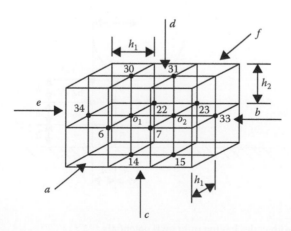

**FIGURE 14.3**
3D arrangement with two nodes having unknown potentials.

## PROBLEM 14.2

For the 2D system with single dielectric, as shown in Figure 14.4, write the FDM equations for the unknown node potentials. Boundary node potentials are given in the figure.

*Solution:*

From Figure 14.4, it may be seen that for nodes 1, 2, 4 and 5, the nodal distances are equal, that is, $h = r/2$. For node 3, $s_1 = s_3 = s_4 = 1$ and $s_2$ can be computed trigonometrically as 0.268, as $s_2h = (r - r\cos\theta)$ and $\sin\theta = (h/r)$. Moreover, for nodes 1 and 2, symmetry with respect to the central plane is to be considered, such that for node 1 another node on the left-hand side has to be considered whose potential will be equal to $V_4$ and similarly for node 2.

$$V_1 = \frac{1}{4}(V_4 + 100 + V_4 + V_2)$$

$$V_2 = \frac{1}{4}(V_5 + V_1 + V_5 + 0)$$

$$V_3 = \frac{\left\{ \begin{array}{c} \left[1/(1+1)\right]\left[(70/1) + (100/1)\right] + \left[1/(0.268+1)\right] \\ \left[(100/0.268) + (V_4/1)\right] \end{array} \right\}}{(1/1) + \left[1/(0.268 \times 1)\right]} \qquad (14.14)$$

$$V_4 = \frac{1}{4}(52 + V_3 + V_1 + V_5)$$

$$V_5 = \frac{1}{4}(36 + V_4 + V_2 + 0)$$

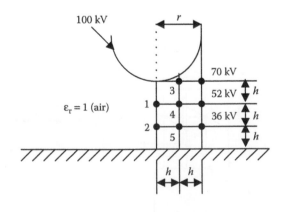

**FIGURE 14.4**
2D arrangement with nodes having unknown potentials.

## 14.3 FDM Equations in Axi-Symmetric System for Single-Dielectric Medium

When the field is expressed in cylindrical coordinates $(r,\theta,z)$ and the field distribution is independent of '$\theta$', then the field distribution is said to be axi-symmetric or rotationally symmetric, for example, insulators, bushings and so on. A typical diagram of an axi-symmetric object is shown in Figure 14.5.

To determine the electric field distribution in an axi-symmetric system, Laplace's equation in cylindrical coordinates, as given below, needs to be solved.

$$\frac{\partial^2 V}{\partial r^2} + \frac{1}{r}\frac{\partial V}{\partial r} + \frac{1}{r^2}\frac{\partial^2 V}{\partial \theta^2} + \frac{\partial^2 V}{\partial z^2} = 0 \qquad (14.15)$$

In an axi-symmetric system, $V$ is independent of $\theta$, so that Equation 14.15 reduces to

$$\frac{\partial^2 V}{\partial r^2} + \frac{1}{r}\frac{\partial V}{\partial r} + \frac{\partial^2 V}{\partial z^2} = 0 \qquad (14.16)$$

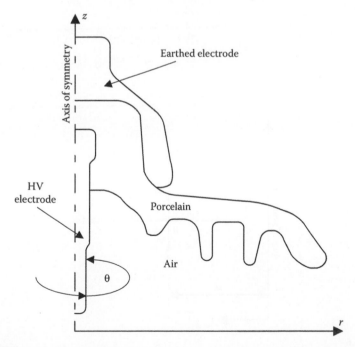

**FIGURE 14.5**
Typical axi-symmetric insulator geometry.

### 14.3.1 FDM Equation for a Node Lying Away from the Axis of Symmetry

In FDM, for an axi-symmetric system, the ROI is discretized using either rectangles or squares. In such cases, one particular node is connected to four neighbouring nodes, as shown in Figure 14.6. As it is assumed that electric potential varies linearly between two successive nodes, it is obvious that the potential of that particular node will be related to potentials of the four connected nodes. FDM equation for any unknown node potential is developed in terms of potentials of the connected nodes by satisfying Laplace's equation in cylindrical coordinates. Figure 14.6 shows node 0 with unknown potential lying at a certain distance away from the axis of symmetry. In such a case, the radial distance of node 0 from the axis of symmetry is also taken as multiple ($sh$) of the largest nodal distance $h$.

From the Taylor series expansion between nodes 1 and 0 in the $r$-direction, neglecting higher order terms,

$$V_1 = V_0 + s_1 h \left.\frac{\partial V}{\partial r}\right|_0 + \frac{(s_1 h)^2}{2} \left.\frac{\partial^2 V}{\partial r^2}\right|_0 \tag{14.17}$$

Similarly, from the Taylor series expansion between nodes 0 and 3 in the $r$-direction,

$$V_3 = V_0 - s_3 h \left.\frac{\partial V}{\partial r}\right|_0 + \frac{(s_3 h)^2}{2} \left.\frac{\partial^2 V}{\partial r^2}\right|_0 \tag{14.18}$$

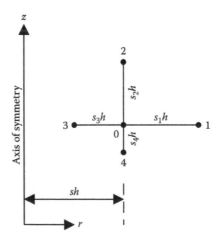

**FIGURE 14.6**
Unequal nodal distances for FDM equation development in axi-symmetric system for a node lying away from the axis of symmetry.

Eliminating $\partial V/\partial r$ from Equations 14.17 and 14.18,

$$\frac{\partial^2 V}{\partial r^2}\bigg|_0 = \frac{\left[(V_1/s_1)+(V_3/s_3)\right]-V_0\left[(1/s_1)+(1/s_3)\right]}{h^2(s_1+s_3)/2} \tag{14.19}$$

Eliminating $\partial^2 V/\partial r^2$ from Equations 14.17 and 14.18,

$$\frac{\partial V}{\partial r}\bigg|_0 = \frac{(V_1/s_1^2)-(V_3/s_3^2)-V_0\left[(1/s_1^2)-(1/s_3^2)\right]}{h\left[(1/s_1)+(1/s_3)\right]} \tag{14.20}$$

Considering the fact that the radial distance of node 0 from the axis of symmetry is $r = sh$,

$$\frac{1}{r}\frac{\partial V}{\partial r}\bigg|_0 = \frac{1}{sh}\frac{\partial V}{\partial r}\bigg|_0 = \frac{(V_1/s_1^2)-(V_3/s_3^2)-V_0\left[(1/s_1^2)-(1/s_3^2)\right]}{sh^2\left[(1/s_1)+(1/s_3)\right]} \tag{14.21}$$

Similarly, between nodes 2 and 4 in the z-direction,

$$\frac{\partial^2 V}{\partial z^2}\bigg|_0 = \frac{\left[(V_2/s_2)+(V_4/s_4)\right]-V_0\left[(1/s_2)-(1/s_4)\right]}{h^2(s_2+s_4)/2} \tag{14.22}$$

Putting the relevant expressions from Equations 14.19, 14.21 and 14.22 in Laplace's Equation (14.16), the FDM equation for the unknown node potential $V_0$ can be obtained as

$$V_0 = \frac{\left(\begin{array}{c}\left[1/(s_1+s_3)\right]\left\{\left[(2s+s_3)/s_1\right]V_1+\left[(2s-s_1)/s_3\right]V_3\right\} \\ +\left[2s/(s_2+s_4)\right]\left[(V_2/s_2)+(V_4/s_4)\right]\end{array}\right)}{\left[(2s+s_3-s_1)/s_1 s_3\right]+(2s/s_2 s_4)} \tag{14.23}$$

For an axi-symmetric arrangement with equal nodal distances, that is, $s_1 = s_2 = s_3 = s_4 = 1$, Equation 14.23 reduces to

$$V_0 = \frac{1}{4}(V_1+V_2+V_3+V_4)+\frac{V_1-V_3}{8s} \tag{14.24}$$

It may be seen from the above equation that for a node lying on the axis of symmetry, that is, for $s = 0$, Equation 14.24 is not valid. Hence, the FDM equation for a node lying on the axis of symmetry needs to be developed separately.

### 14.3.2 FDM Equation for a Node Lying on the Axis of Symmetry

Figure 14.7 shows an axi-symmetric nodal arrangement with unequal nodal distances where node 0 having unknown node potential is lying on the axis of symmetry.

For an axi-symmetric system along the axis of symmetry, the electric flux lines are tangent to the axis. In other words, as $r \to 0$, $\partial V/\partial r \to 0$.

Therefore, applying L'Hospital's rule,

$$\underset{r \to 0}{\text{Lt}} \frac{1}{r} \frac{\partial V}{\partial r} = \frac{\partial^2 V}{\partial r^2}$$

Hence, for a node lying on the axis of symmetry, Laplace's Equation 14.16 is modified to

$$2 \frac{\partial^2 V}{\partial r^2} + \frac{\partial^2 V}{\partial z^2} = 0 \tag{14.25}$$

From the Taylor series expansion between the nodes 1 and 0 in the $r$-direction, neglecting higher order terms,

$$V_1 = V_0 + s_1 h \left. \frac{\partial V}{\partial r} \right|_0 + \frac{(s_1 h)^2}{2} \left. \frac{\partial^2 V}{\partial r^2} \right|_0 \tag{14.26}$$

But, as $r \to 0$, $\partial V/\partial r \to 0$. Therefore, from Equation 14.26

$$\left. \frac{\partial^2 V}{\partial r^2} \right|_0 = \frac{2(V_1 - V_0)}{(s_1 h)^2} \tag{14.27}$$

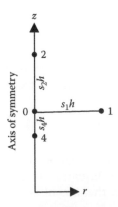

**FIGURE 14.7**
Axi-symmetric system with unequal nodal distances with a node lying on the axis of symmetry.

Following Equation 14.22 applying Taylor series between the nodes 2, 0 and 4 in the z-direction,

$$\left.\frac{\partial^2 V}{\partial z^2}\right|_0 = \frac{\left[(V_2/s_2)+(V_4/s_4)\right]-V_0\left[(1/s_2)+(1/s_4)\right]}{h^2(s_2+s_4)/2} \tag{14.28}$$

Then satisfying Laplace's Equation 14.25 with the help of Equations 14.27 and 14.28, the FDM equation for a node lying on the axis of symmetry can be obtained as follows:

$$V_0 = \frac{(2V_1/s_1^2)+\left[1/(s_2+s_4)\right]\left[(V_2/s_2)+(V_4/s_4)\right]}{(2/s_1^2)+(1/s_2 s_4)} \tag{14.29}$$

For equal nodal distances, that is, for $s_1 = s_2 = s_4 = 1$, Equation 14.29 reduces to

$$V_0 = \frac{1}{6}(4V_1 + V_2 + V_4) \tag{14.30}$$

**PROBLEM 14.3**
For the axi-symmetric arrangement with equal nodal distances, as shown in Figure 14.8, write the FDM equations for the unknown node potentials.

*Solution:*
It may be noted from Figure 14.8 that nodes 1 and 2 lie on the axis of symmetry, whereas nodes 3, 4, 5 and 6 lie away from the axis of symmetry.

$$V_1 = \frac{1}{6}(4V_3 + 100 + V_2)$$

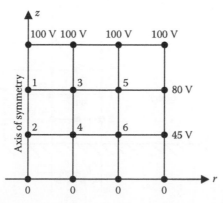

**FIGURE 14.8**
Nodal arrangement pertaining to Problem 14.3.

$$V_2 = \frac{1}{6}(4V_4 + V_1 + 0)$$

$$V_3 = \frac{1}{4}(V_5 + 100 + V_1 + V_4) + \frac{V_5 - V_1}{8 \times 1} \text{ as } s = 1$$

$$V_4 = \frac{1}{4}(V_6 + V_3 + V_2 + 0) + \frac{V_6 - V_2}{8 \times 1} \text{ as } s = 1$$

$$V_5 = \frac{1}{4}(80 + 100 + V_3 + V_6) + \frac{80 - V_3}{8 \times 2} \text{ as } s = 2$$

$$V_6 = \frac{1}{4}(45 + V_5 + V_4 + 0) + \frac{45 - V_4}{8 \times 2} \text{ as } s = 2$$

## PROBLEM 14.4

For the axi-symmetric system with single dielectric, as shown in Figure 14.9, write the FDM equations for the unknown node potentials. Boundary node potentials are given in the figure.

*Solution:*

It may be noted from Figure 14.9 that for nodes 1, 2, 4 and 5 the nodal distances are equal, out of which nodes 1 and 2 lie on the axis of symmetry. Nodes 3, 4 and 5 lie away from the axis of symmetry, out of which the nodal distances for node 3 are unequal such that $s_1 = s_3 = s_4 = 1$ and $s_2 = 0.268$.

$$V_1 = \frac{1}{6}(4V_4 + 100 + V_2)$$

$$V_2 = \frac{1}{6}(4V_5 + V_1 + 0)$$

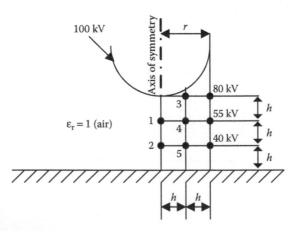

**FIGURE 14.9**
Nodal arrangement pertaining to Problem 14.4.

$$V_3 = \cfrac{\begin{pmatrix} \left[1/(1+1)\right]\left\{\left[(2\times1+1)/1\right]80+\left[(2\times1-1)/1\right]100\right\} \\ +\left[(2\times1)/(0.268+1)\right]\left[(100/0.268)+(V_4/1)\right] \end{pmatrix}}{\left[(2\times1+1-1)/(1\times1)\right]+\left[(2\times1)/(0.268\times1)\right]}$$

$$V_4 = \frac{1}{4}(55+V_3+V_1+V_5)+\frac{55-V_1}{8\times1} \text{ as } s=1$$

$$V_5 = \frac{1}{4}(40+V_4+V_2+0)+\frac{40-V_2}{8\times1} \text{ as } s=1$$

## 14.4 FDM Equations in 3D System for Multi-Dielectric Media

Because the commonly used assumption in FDM is linear variation of electric potential between two successive nodes, it is imperative that there should not be two different dielectric media between two successive nodes. In other words, during discretization, it should be ensured that one set of nodes will always be on the dielectric interface. Figure 14.10 shows one such nodal distribution with unequal nodal distances. The $y$–$z$ plane is considered to be the dielectric interface and one set of nodes is on the dielectric interface. For

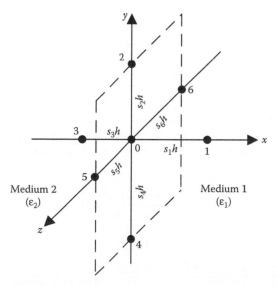

**FIGURE 14.10**
Nodal arrangement for FDM equation development in 3D system with multi-dielectric media.

all the nodes that lie within either medium 1 or medium 2 there will be only that dielectric between any two successive nodes and hence the FDM equations for a single-dielectric medium could be used for such nodes. But for the nodes lying on the dielectric interface it is not the case. As shown in Figure 14.10, between nodes 0 and 1 there is medium 1, and between nodes 0 and 3 there is medium 2. Therefore, FDM equation needs to be developed for the nodes that lie on the dielectric interface applying suitable boundary conditions.

For Laplacian field, that is, considering that the dielectric boundary does not have any free charge present on it, the necessary boundary condition is that the normal component of flux density remains constant on both the sides of the dielectric interface. For the nodal arrangement shown in Figure 14.10, the $x$-component of electric flux density is the normal component on the dielectric interface as the $y$–$z$ plane is the dielectric interface. Hence,

$$\vec{D}_x = \vec{D}'_x$$

$$\text{or,} \quad \varepsilon_1 \vec{E}_x = \varepsilon_2 \vec{E}'_x$$

$$\text{or,} \quad -\varepsilon_1 \frac{\partial V}{\partial x} = -\varepsilon_2 \frac{\partial V'}{\partial x} \tag{14.31}$$

$$\text{or,} \quad \frac{\partial V'}{\partial x} = \frac{\varepsilon_1}{\varepsilon_2} \frac{\partial V}{\partial x} = K \frac{\partial V}{\partial x}$$

where:

$$K = \frac{\varepsilon_1}{\varepsilon_2}$$

Here, it is to be noted that two potential functions need to be considered on the two sides of the dielectric interface as shown in Equation 14.31. The potential function $V$ is valid for medium 1 and $V'$ is valid for medium 2.

Accordingly, two Laplace's equations, one with $V$ and the other with $V'$, need to be satisfied in this case, as given in Equations 14.32a and 14.32b. Equation 14.32a is valid for medium 1, and Equation 14.32b is valid for medium 2.

$$\frac{\partial^2 V}{\partial x^2} + \frac{\partial^2 V}{\partial y^2} + \frac{\partial^2 V}{\partial z^2} = 0 \tag{14.32a}$$

$$\frac{\partial^2 V'}{\partial x^2} + \frac{\partial^2 V'}{\partial y^2} + \frac{\partial^2 V'}{\partial z^2} = 0 \tag{14.32b}$$

Now, from the Taylor series expansion between the pair of nodes, the following expressions are obtained:

$$V_1 = V_0 + s_1 h \left.\frac{\partial V}{\partial x}\right|_0 + \frac{s_1^2 h^2}{2}\left.\frac{\partial^2 V}{\partial x^2}\right|_0 \quad \text{between nodes 1 and 0} \qquad (14.33\text{a})$$

$$V_3 = V_0 - s_3 h \left.\frac{\partial V'}{\partial x}\right|_0 + \frac{s_3^2 h^2}{2}\left.\frac{\partial^2 V'}{\partial x^2}\right|_0 \quad \text{between nodes 0 and 3} \qquad (14.33\text{b})$$

$$V_2 = V_0 + s_2 h \left.\frac{\partial V}{\partial y}\right|_0 + \frac{s_2^2 h^2}{2}\left.\frac{\partial^2 V}{\partial y^2}\right|_0 \quad \text{between nodes 2 and 0} \qquad (14.33\text{c})$$

$$V_4 = V_0 - s_4 h \left.\frac{\partial V}{\partial y}\right|_0 + \frac{s_4^2 h^2}{2}\left.\frac{\partial^2 V}{\partial y^2}\right|_0 \quad \text{between nodes 0 and 4} \qquad (14.33\text{d})$$

$$V_5 = V_0 + s_5 h \left.\frac{\partial V}{\partial z}\right|_0 + \frac{s_5^2 h^2}{2}\left.\frac{\partial^2 V}{\partial z^2}\right|_0 \quad \text{between nodes 5 and 0} \qquad (14.33\text{e})$$

$$V_6 = V_0 - s_6 h \left.\frac{\partial V}{\partial z}\right|_0 + \frac{s_6^2 h^2}{2}\left.\frac{\partial^2 V}{\partial z^2}\right|_0 \quad \text{between nodes 0 and 6} \qquad (14.33\text{f})$$

It is to be mentioned here that nodes 0, 2, 4, 5 and 6 lie on the dielectric interface. According to the boundary conditions on dielectric–dielectric interface, electric potential and tangential component of electric field remain constant on the dielectric interface.

From Equation 14.33a

$$\left.\frac{\partial^2 V}{\partial x^2}\right|_0 = \frac{V_1 - V_0 - s_1 h (\partial V / \partial x)|_0}{s_1^2 h^2 / 2} \qquad (14.34)$$

and from Equation 14.33b

$$\left.\frac{\partial^2 V'}{\partial x^2}\right|_0 = \frac{V_3 - V_0 + s_3 h (\partial V' / \partial x)|_0}{s_3^2 h^2 / 2} = \frac{V_3 - V_0 + s_3 h K (\partial V / \partial x)|_0}{s_3^2 h^2 / 2} \qquad (14.35)$$

From Equations 14.33c and 14.33d along the $y$-direction on the $y$–$z$ plane,

$$\left.\frac{\partial^2 V}{\partial y^2}\right|_0 = \frac{\left[(V_2/s_2) + (V_4/s_4)\right] - V_0\left[(1/s_2) + (1/s_4)\right]}{h^2(s_2 + s_4)/2} = \left.\frac{\partial^2 V'}{\partial y^2}\right|_0 \qquad (14.36)$$

and from Equations 14.33e and 14.33f along the $z$-direction on the $y$–$z$ plane,

$$\left.\frac{\partial^2 V}{\partial z^2}\right|_0 = \frac{\left[(V_5/s_5) + (V_6/s_6)\right] - V_0\left[(1/s_5) + (1/s_6)\right]}{h^2(s_5 + s_6)/2} = \left.\frac{\partial^2 V'}{\partial z^2}\right|_0 \qquad (14.37)$$

From Equation 14.32a, that is, Laplace's equation that is valid for medium 1,

$$
\frac{V_1 - V_0 - s_1 h (\partial V/\partial x)\big|_0}{s_1^2 h^2/2} + \frac{\left[(V_2/s_2) + (V_4/s_4)\right] - V_0\left[(1/s_2) + (1/s_4)\right]}{h^2(s_2 + s_4)/2}
$$
$$
+ \frac{\left[(V_5/s_5) + (V_6/s_6)\right] - V_0\left[(1/s_5) + (1/s_6)\right]}{h^2(s_5 + s_6)/2} = 0 \tag{14.38}
$$

and from Equation 14.32b, that is, Laplace's equation that is valid for medium 2,

$$
\frac{V_3 - V_0 + s_3 h K (\partial V/\partial x)\big|_0}{s_3^2 h^2/2} + \frac{\left[(V_2/s_2) + (V_4/s_4)\right] - V_0\left[(1/s_2) + (1/s_4)\right]}{h^2(s_2 + s_4)/2}
$$
$$
+ \frac{\left[(V_5/s_5) + (V_6/s_6)\right] - V_0\left[(1/s_5) + (1/s_6)\right]}{h^2(s_5 + s_6)/2} = 0 \tag{14.39}
$$

Eliminating $\partial V/\partial x\big|_0$ from Equations 14.38 and 14.39,

$$
V_0 = \frac{\left( \begin{array}{c} (V_1/s_1) + (V_3/s_3 K) + \left[(s_1 K + s_3)/K\right] \\ \left\{\left[1/(s_2 + s_4)\right]\left[(V_2/s_2) + (V_4/s_4)\right] + \left[1/(s_5 + s_6)\right]\left[(V_5/s_5) + (V_6/s_6)\right]\right\} \end{array} \right)}{(1/s_1) + (1/s_3 K) + \left[(s_1 K + s_3)/K\right]\left[(1/s_2 s_4) + (1/s_5 s_6)\right]} \tag{14.40}
$$

Equation 14.40 is the FDM equation for the unknown potential of a node lying on the dielectric interface in the 3D system with unequal nodal distances.

For equal nodal distances in the 3D multi-dielectric system, that is, when $s_1 = s_2 = s_3 = s_4 = s_5 = s_6 = 1$, Equation 14.40 reduces to

$$
V_0 = \frac{V_1\left[2K/(K+1)\right] + V_2 + \left[2V_3/(K+1)\right] + V_4 + V_5 + V_6}{6} \tag{14.41}
$$

For a 2D multi-dielectric system with unequal nodal distances, the FDM equation for the unknown potential of a node lying on the dielectric interface will be as follows, where the dielectric interface is considered to be along the $y$-axis, as shown in Figure 14.11.

$$
V_0 = \frac{(V_1/s_1) + (V_3/s_3 K) + \left[(s_1 K + s_3)/K(s_2 + s_4)\right]\left[(V_2/s_2) + (V_4/s_4)\right]}{(1/s_1) + (1/s_3 K) + \left[(s_1 K + s_3)/K s_2 s_4\right]} \tag{14.42}
$$

For equal nodal distances in the 2D multi-dielectric system, that is, when $s_1 = s_2 = s_3 = s_4 = 1$, Equation 14.42 reduces to

$$
V_0 = \frac{V_1\left[2K/(K+1)\right] + V_2 + \left[2V_3/(K+1)\right] + V_4}{4} \tag{14.43}
$$

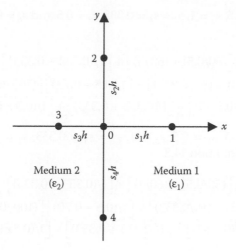

**FIGURE 14.11**
Nodal arrangement for FDM equation development in 2D system with multi-dielectric media.

## PROBLEM 14.5

For the 3D arrangement with two different dielectric media, as shown in Figure 14.12, write the FDM equations for nodes 1, 2 and 3. The known node potentials are as follows: $V_{13} = V_{23} = V_{33} = 100$ V, $V_{14} = V_{24} = V_{34} = 0$ V, $V_{11} = V_{21} = V_{31} = 50$ V, $V_{12} = V_{22} = V_{32} = 60$ V and $V_{15} = V_{35} = 55$ V. Given that $\varepsilon_1 = 4$ and $\varepsilon_2 = 1$.

*Solution:*
In this problem, the largest nodal distance is $h$.

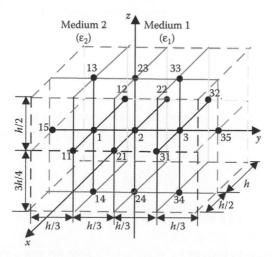

**FIGURE 14.12**
**(See colour insert.)** Nodal arrangement pertaining to Problem 14.5.

For node 1: $s_1 = 0.5$, $s_3 = 1$, $s_2 = s_4 = 0.333$, $s_5 = 0.5$ and $s_6 = 0.75$. Therefore, as per Equation 14.8

$$V_1 = \frac{\left\{\begin{array}{l} \left[1/(0.5+1)\right]\left[(50/0.5)+(60/1)\right]+\left[1/(0.333+0.333)\right] \\ \left[(V_2/0.333)+(55/0.333)\right]+\left[1/(0.5+0.75)\right]\left[(100/0.5)+(0/0.75)\right] \end{array}\right\}}{\left[1/(0.5+1)\right]+\left[1/(0.333+0.333)\right]+\left[1/(0.5\times0.75)\right]}$$

Similarly for node 3: $s_1 = 0.5$, $s_3 = 1$, $s_2 = s_4 = 0.333$, $s_5 = 0.5$ and $s_6 = 0.75$. Therefore, as per Equation 14.8

$$V_3 = \frac{\left\{\begin{array}{l} \left[1/(0.5+1)\right]\left[(50/0.5)+(60/1)\right]+\left[1/(0.333+0.333)\right] \\ \left[(55/0.333)+(V_2/0.333)\right]+\left[1/(0.5+0.75)\right]\left[(100/0.5)+(0/0.75)\right] \end{array}\right\}}{\left[1/(0.5+1)\right]+\left[1/(0.333+0.333)\right]+\left[1/(0.5\times0.75)\right]}$$

For node 2, as per Equation 14.40, $V_1 = V_3$, $V_3 = V_1$, $V_2 = V_{23} = 100$ V, $V_4 = V_{24} = 0$ V, $V_5 = V_{21} = 50$ V and $V_6 = V_{22} = 60$ V and $s_1 = s_3 = 0.333$, $s_2 = 0.5$, $s_4 = 0.75$, $s_5 = 0.5$, $s_6 = 1$ and $K = (\varepsilon_1/\varepsilon_2) = 4$.

$$V_2 = \frac{\left(\begin{array}{c} (V_3/0.333)+\left[V_1/(0.333\times4)\right]+\left[(0.333\times4+0.333)/4\right] \\ \left\{\begin{array}{l}\left[1/(0.5+0.75)\right]\left[(100/0.5)+(0/0.75)\right] \\ +\left[1/(0.5+1)\right]\left[(50/0.5)+(60/1)\right]\end{array}\right\} \end{array}\right)}{\left(\begin{array}{c}(1/0.333)+\left[1/(0.333\times4)\right]+\left[(0.333\times4+0.333)/4\right] \\ \left\{\left[1/(0.5+0.75)\right]+\left[1/(0.5\times1)\right]\right\}\end{array}\right)}$$

## PROBLEM 14.6

For the 2D multi-dielectric configuration with series dielectric arrangement, as shown in Figure 14.13, write the FDM equations for the nodes having unknown potentials.

*Solution:*
For nodes 1, 2 and 3, symmetry of the configuration has to be considered with respect to the central plane. The nodal distances are equal for these three nodes. For node 1:

$$V_1 = \frac{1}{4}(V_5 + 100 + V_5 + V_2)$$

For node 2: As per Equation 14.43, $V_1 = V_1$, $V_2 = V_6$, $V_3 = V_3$, $V_4 = V_6$ and $K = (1/3)$.

$$V_2 = \frac{\left[(2\times0.333)/(0.333+1)\right]V_1 + V_6 + \left[2V_3/(0.333+1)\right]+V_6}{4}$$

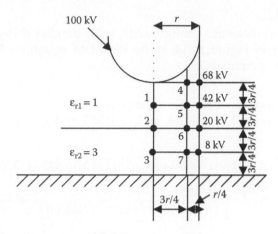

**FIGURE 14.13**
Nodal arrangement pertaining to Problem 14.6.

For node 3:

$$V_3 = \frac{1}{4}(V_7 + V_2 + V_7 + 0)$$

For node 4: The nodal distances are unequal such that $s_1 = (1/3)$, $s_2 = 0.882$ and $s_3 = s_4 = 1$ because the largest nodal distance is $(3r/4)$. Then as per Equation 14.10

$$V_4 = \frac{\left\{ \begin{array}{l} \left[1/(0.333 + 1)\right]\left[(68/0.333) + (100/1)\right] + \left[1/(0.882 + 1)\right] \\ \left[(100/0.882) + (V_2/1)\right] \end{array} \right\}}{\left[1/(0.333 \times 1)\right] + \left[1/(0.882 \times 1)\right]}$$

For node 5: The nodal distances are unequal such that $s_1 = (1/3)$, $s_2 = 1$ and $s_3 = s_4 = 1$. Then as per Equation 14.10

$$V_5 = \frac{\left[1/(0.333 + 1)\right]\left[(42/0.333) + (V_1/1)\right] + \left[1/(1 + 1)\right]\left[(V_4/1) + (V_6/1)\right]}{\left[1/(0.333 \times 1)\right] + \left[1/(1 \times 1)\right]}$$

For node 6: As per Equation 14.42, $V_1 = V_5$, $V_2 = V_2$, $V_3 = V_7$, $V_4 = 20$ and $K = (1/3)$. Nodal distance factors are $s_1 = s_2 = s_3 = 1$ and $s_4 = (1/3)$.

$$V_6 = \frac{\left\{ \begin{array}{l} (V_5/1) + \left[V_7/(1 \times 0.333)\right] + \left[(1 \times 0.333 + 1)/0.333(1 + 0.333)\right] \\ \left[(V_2/1) + (V_4/0.333)\right] \end{array} \right\}}{(1 + 1) + \left[1/(1 \times 0.333)\right] + \left[(1 \times 0.333 + 1)/(0.333 \times 1 \times 0.333)\right]}$$

For node 7: The nodal distances are unequal such that $s_1 = (1/3)$ and $s_2 = s_3 = s_4 = 1$. Then as per Equation 14.10

$$V_7 = \frac{\left[1/(0.333 + 1)\right]\left[(8/0.333) + (V_3/1)\right] + \left[1/(1 + 1)\right]\left[(V_6/1) + (0/1)\right]}{\left[1/(0.333 \times 1)\right] + \left[1/(1 \times 1)\right]}$$

## PROBLEM 14.7

For the 2D multi-dielectric configuration with parallel dielectric arrangement, as shown in Figure 14.14, write the FDM equations for the nodes having unknown potentials.

*Solution:*

The largest nodal distance for this arrangement is $h$.

Therefore, for node 1: $s_1 = 1$, $s_2 = s_4 = 0.333$ and $s_3 = 0.5$. Then as per Equation 14.10

$$V_1 = \frac{\left\{ \begin{array}{l} \left[1/(1+0.5)\right]\left[(66/1)+(V_3/0.5)\right]+\left[1/(0.333+0.333)\right] \\ \left[(100/0.333)+(V_2/0.333)\right] \end{array} \right\}}{\left[1/(1\times0.5)\right]+\left[1/(0.333\times0.333)\right]}$$

For node 2: $s_1 = 1$, $s_2 = s_4 = 0.333$ and $s_3 = 0.5$. Then as per Equation 14.10

$$V_2 = \frac{\left[1/(1+0.5)\right]\left[(33/1)+(V_4/0.5)\right]+\left[1/(0.333+0.333)\right]\left[(V_1/0.333)+(0/0.333)\right]}{\left[1/(1\times0.5)\right]+\left[1/(0.333\times0.333)\right]}$$

For node 3: As per Equation 14.42, $V_1 = V_1$, $V_2 = 100$, $V_3 = V_5$, $V_4 = V_4$ and $K = (1/4)$. Nodal distance factors are $s_1 = s_3 = 0.5$ and $s_2 = s_4 = 0.333$.

$$V_3 = \frac{\left\{ \begin{array}{l} (V_1/0.5)+\left[V_5/(0.5\times0.25)\right]+\left[(0.5\times0.25+0.5)/0.25(0.333+0.333)\right] \\ \left[(100/0.333)+(V_4/0.333)\right] \end{array} \right\}}{(1/0.5)+\left[1/(0.5\times0.25)\right]+\left[(0.5\times0.25+0.5)/(0.25\times0.333\times0.333)\right]}$$

For node 4: As per Equation 14.42, $V_1 = V_2$, $V_2 = V_3$, $V_3 = V_6$, $V_4 = 0$ and $K = (1/4)$. Nodal distance factors are $s_1 = s_3 = 0.5$ and $s_2 = s_4 = 0.333$.

$$V_4 = \frac{\left\{ \begin{array}{l} (V_2/0.5)+\left[V_6/(0.5\times0.25)\right]+\left[(0.5\times0.25+0.5)/0.25(0.333+0.333)\right] \\ \left[(V_3/0.333)+(0/0.333)\right] \end{array} \right\}}{(1/0.5)+\left[1/(0.5\times0.25)\right]+\left[(0.5\times0.25+0.5)/(0.25\times0.333\times0.333)\right]}$$

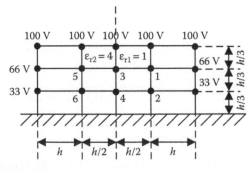

**FIGURE 14.14**
Nodal arrangement pertaining to Problem 14.7.

For node 5: $s_1 = 0.5$, $s_2 = s_4 = 0.333$ and $s_3 = 1$. Then as per Equation 14.10

$$V_5 = \frac{\left\{\begin{array}{l} \left[1/(0.5+1)\right]\left[(V_3/0.5)+(66/1)\right]+\left[1/(0.333+0.333)\right] \\ \left[(100/0.333)+(V_6/0.333)\right] \end{array}\right\}}{\left[1/(0.5+1)\right]+\left[1/(0.333\times0.333)\right]}$$

For node 6: $s_1 = 0.5$, $s_2 = s_4 = 0.333$ and $s_3 = 1$. Then as per Equation 14.10

$$V_6 = \frac{\left[1/(0.5+1)\right]\left[(V_4/0.5)+(33/1)\right]+\left[1/(0.333\times0.333)\right]\left[(V_5/0.333)+(0/0.333)\right]}{\left[1/(0.5+1)\right]+\left[1/(0.333\times0.333)\right]}$$

---

## 14.5 FDM Equations in Axi-Symmetric System for Multi-Dielectric Media

### 14.5.1 For Series Dielectric Media

In this case, the dielectric interface is considered to be normal to the axis of symmetry. FDM equations need to be developed for a node lying on the dielectric interface. This node could be away from the axis of symmetry and could also be on the axis of symmetry.

#### 14.5.1.1 For the Node on the Dielectric Interface Lying Away from the Axis of Symmetry

For the nodal arrangement shown in Figure 14.15, the z-component of electric flux density is the normal component on the dielectric interface that is parallel to the r-axis.

$$\vec{D}_z = \vec{D}_z'$$

$$\text{or, } \varepsilon_1\vec{E}_z = \varepsilon_2\vec{E}_z'$$

$$\text{or, } -\varepsilon_1\frac{\partial V}{\partial z} = -\varepsilon_2\frac{\partial V'}{\partial z} \tag{14.44}$$

$$\text{or, } \frac{\partial V'}{\partial z} = \frac{\varepsilon_1}{\varepsilon_2}\frac{\partial V}{\partial z} = K\frac{\partial V}{\partial z}$$

where:

$$K = \frac{\varepsilon_1}{\varepsilon_2}$$

Here, it is to be noted again that the potential function $V$ is valid for medium 1 and $V'$ is valid for medium 2.

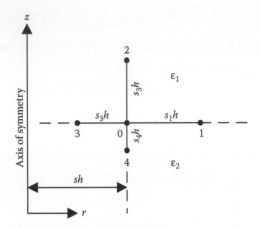

**FIGURE 14.15**
Nodal arrangement for FDM equation development in axi-symmetric system with multi-dielectric media in series dielectric arrangement when the node is lying away from the axis of symmetry.

Accordingly, two Laplace's equations, one with $V$ and the other with $V'$, need to be satisfied in this case, as given in Equations 14.45a and 14.45b. Equation 14.45a is valid for medium 1, and Equation 14.45b is valid for medium 2.

$$\frac{\partial^2 V}{\partial r^2} + \frac{1}{r}\frac{\partial V}{\partial r} + \frac{\partial^2 V}{\partial z^2} = 0 \qquad (14.45a)$$

$$\frac{\partial^2 V'}{\partial r^2} + \frac{1}{r}\frac{\partial V'}{\partial r} + \frac{\partial^2 V'}{\partial z^2} = 0 \qquad (14.45b)$$

Now, from the Taylor series expansion between the pair of nodes, the following expressions are obtained

$$V_1 = V_0 + s_1 h \left.\frac{\partial V}{\partial r}\right|_0 + \frac{s_1^2 h^2}{2}\left.\frac{\partial^2 V}{\partial r^2}\right|_0 \quad \text{between nodes 1 and 0} \qquad (14.46a)$$

$$V_3 = V_0 - s_3 h \left.\frac{\partial V}{\partial r}\right|_0 + \frac{s_3^2 h^2}{2}\left.\frac{\partial^2 V}{\partial r^2}\right|_0 \quad \text{between nodes 0 and 3} \qquad (14.46b)$$

$$V_2 = V_0 + s_2 h \left.\frac{\partial V}{\partial z}\right|_0 + \frac{s_2^2 h^2}{2}\left.\frac{\partial^2 V}{\partial z^2}\right|_0 \quad \text{between nodes 2 and 0} \qquad (14.46c)$$

$$V_4 = V_0 - s_4 h \left.\frac{\partial V'}{\partial z}\right|_0 + \frac{s_4^2 h^2}{2}\left.\frac{\partial^2 V'}{\partial z^2}\right|_0 \quad \text{between nodes 0 and 4} \qquad (14.46d)$$

It is to be noted here that nodes 1, 0 and 3 lie on the dielectric interface. According to the boundary conditions on dielectric–dielectric interface, electric potential and tangential component of electric field remain constant on the dielectric interface.

From Equations 14.46a and 14.46b along the *r*-direction on the dielectric interface,

$$\left.\frac{\partial^2 V}{\partial r^2}\right|_0 = \frac{\left[(V_1/s_1) + (V_3/s_3)\right] - V_0\left[(1/s_1) + (1/s_3)\right]}{h^2(s_1 + s_3)/2} = \left.\frac{\partial^2 V'}{\partial r^2}\right|_0 \tag{14.47}$$

From Equations 14.46a and 14.46b considering the radial distance of the node 0 from the axis of symmetry to be *sh*, that is, $r = sh$,

$$\left.\frac{1}{r}\frac{\partial V}{\partial r}\right|_0 = \left.\frac{1}{sh}\frac{\partial V}{\partial r}\right|_0 = \frac{(V_1/s_1^2) - (V_3/s_3^2) - V_0\left[(1/s_1^2) - (1/s_3^2)\right]}{sh^2\left[(1/s_1) + (1/s_3)\right]} = \left.\frac{1}{r}\frac{\partial V'}{\partial r}\right|_0 \tag{14.48}$$

From Equation 14.46c

$$\left.\frac{\partial^2 V}{\partial z^2}\right|_0 = \frac{V_2 - V_0 - s_2 h(\partial V/\partial z)|_0}{s_2^2 h^2/2} \tag{14.49}$$

and from Equation 14.46d

$$\left.\frac{\partial^2 V'}{\partial z^2}\right|_0 = \frac{V_4 - V_0 + s_4 h(\partial V'/\partial z)|_0}{s_4^2 h^2/2} = \frac{V_4 - V_0 + s_4 hK(\partial V/\partial z)|_0}{s_4^2 h^2/2} \tag{14.50}$$

From Equations 14.45a, 14.47, 14.48 and 14.49,

$$\frac{\left[(V_1/s_1) + (V_3/s_3)\right] - V_0\left[(1/s_1) + (1/s_3)\right]}{h^2(s_1 + s_3)/2}$$

$$+ \frac{(V_1/s_1^2) - (V_3/s_3^2) - V_0\left[(1/s_1^2) - (1/s_3^2)\right]}{sh^2\left[(1/s_1) + (1/s_3)\right]} + \frac{V_2 - V_0 - s_2 h(\partial V/\partial z)|_0}{s_2^2 h^2/2} = 0 \tag{14.51}$$

and from Equations 14.45b, 14.47, 14.48 and 14.50,

$$\frac{\left[(V_1/s_1) + (V_3/s_3)\right] - V_0\left[(1/s_1) + (1/s_3)\right]}{h^2(s_1 + s_3)/2}$$

$$+ \frac{(V_1/s_1^2) - (V_3/s_3^2) - V_0\left[(1/s_1^2) + (1/s_3^2)\right]}{sh^2\left[(1/s_1) + (1/s_3)\right]} + \frac{V_4 - V_0 + s_4 hK(\partial V/\partial z)|_0}{s_4^2 h^2/2} = 0 \tag{14.52}$$

Eliminating $\partial V/\partial z|_0$ from Equations 14.51 and 14.52,

$$V_0 = \frac{\left[\begin{array}{l}\left(\{(Ks_2+s_4)[2+(s_3/s)]\}/s_1(s_1+s_3)\right)V_1 + (2K/s_2)V_2 \\ + \left(\{(Ks_2+s_4)[2-(s_1/s)]\}/s_3(s_1+s_3)\right)V_3 + (2/s_4)V_4\end{array}\right]}{\left[2(Ks_2+s_4)/s_1s_3\right] + \{[(s_3-s_1)(Ks_2+s_4)]/ss_1s_3\} + (2K/s_2) + (2/s_4)} \quad (14.53)$$

Equation 14.53 is the FDM equation for the unknown potential of a node lying on the dielectric interface, where the node is away from the axis of symmetry, in an axi-symmetric system with unequal nodal distances having series dielectric arrangement (Figure 14.15).

For equal nodal distances, that is, for $s_1 = s_2 = s_3 = s_4 = 1$, Equation 14.53 reduces to

$$V_0 = \frac{1}{4}\left(V_1 + \frac{2K}{K+1}V_2 + V_3 + \frac{2}{K+1}V_4\right) + \frac{V_1 - V_3}{8s} \quad (14.54)$$

and for single-dielectric system, that is, for $K = 1$, with equal nodal distances Equation 14.54 reduces to

$$V_0 = \frac{1}{4}(V_1 + V_2 + V_3 + V_4) + \frac{V_1 - V_3}{8s} \quad (14.55)$$

### 14.5.1.2 *For the Node on the Dielectric Interface Lying on the Axis of Symmetry*

Laplace's equation in medium 1:

$$2\frac{\partial^2 V}{\partial r^2} + \frac{\partial^2 V}{\partial z^2} = 0 \quad (14.56)$$

Laplace's equation in medium 2:

$$2\frac{\partial^2 V'}{\partial r^2} + \frac{\partial^2 V'}{\partial z^2} = 0 \quad (14.57)$$

As per Equation 14.44

$$\frac{\partial V'}{\partial z} = K\frac{\partial V}{\partial z}$$

where:

$$K = \frac{\varepsilon_1}{\varepsilon_2}$$

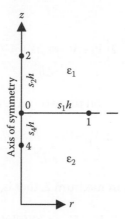

**FIGURE 14.16**
Nodal arrangement for FDM equation development in axi-symmetric system with multi-dielectric media in series dielectric arrangement when the node is lying on the axis of symmetry.

Now, from the Taylor series expansion between the pair of nodes as shown in Figure 14.16, the following expressions are obtained:

$$V_1 = V_0 + s_1 h \left.\frac{\partial V}{\partial r}\right|_0 + \frac{s_1^2 h^2}{2} \left.\frac{\partial^2 V}{\partial r^2}\right|_0 \quad \text{between nodes 1 and 0} \qquad (14.58a)$$

$$V_2 = V_0 + s_2 h \left.\frac{\partial V}{\partial z}\right|_0 + \frac{s_2^2 h^2}{2} \left.\frac{\partial^2 V}{\partial z^2}\right|_0 \quad \text{between nodes 2 and 0} \qquad (14.58b)$$

$$V_4 = V_0 - s_4 h \left.\frac{\partial V'}{\partial z}\right|_0 + \frac{s_4^2 h^2}{2} \left.\frac{\partial^2 V'}{\partial z^2}\right|_0 \quad \text{between nodes 0 and 4} \qquad (14.58c)$$

In Equation 14.58a,

$$\left.\frac{\partial V}{\partial r}\right|_0 \to 0, \text{ as } r \to 0$$

Hence,

$$2\left.\frac{\partial^2 V}{\partial r^2}\right|_0 = \frac{4(V_1 - V_0)}{s_1^2 h^2} \qquad (14.59)$$

From Equation 14.58b,

$$\left.\frac{\partial^2 V}{\partial z^2}\right|_0 = \frac{2\left[V_2 - V_0 - s_2 h (\partial V/\partial z)|_0\right]}{s_2^2 h^2}$$

From Equation 14.58c,

$$\left.\frac{\partial^2 V'}{\partial z^2}\right|_0 = \frac{2\left[V_4 - V_0 + s_4 hK(\partial V/\partial z)|_0\right]}{s_4^2 h^2}$$

Therefore, from Laplace's equation in medium 1, that is, Equation 14.56,

$$\frac{4(V_1 - V_0)}{s_1^2 h^2} + \frac{2\left[V_2 - V_0 - s_2 h(\partial V/\partial z)|_0\right]}{s_2^2 h^2} = 0 \tag{14.60}$$

and from Laplace's equation in medium 2, that is, Equation 14.57,

$$\frac{4(V_1 - V_0)}{s_1^2 h^2} + \frac{2\left[V_4 - V_0 + s_4 hK(\partial V/\partial z)|_0\right]}{s_4^2 h^2} = 0 \tag{14.61}$$

Eliminating $\partial V/\partial z|_0$ from Equations 14.60 and 14.61,

$$V_0 = \frac{\left[2(Ks_2 + s_4)/s_1^2\right]V_1 + (K/s_2)V_2 + (1/s_4)V_4}{\left[2(Ks_2 + s_4)/s_1^2\right] + (K/s_2) + (1/s_4)} \tag{14.62}$$

Equation 14.62 is the FDM equation for the unknown potential of a node lying on the dielectric interface, where the node is on the axis of symmetry, in an axi-symmetric system with unequal nodal distances having series dielectric arrangement (Figure 14.16).

For equal nodal distances, that is, for $s_1 = s_2 = s_4 = 1$, Equation 14.62 reduces to

$$V_0 = \frac{2(K+1)V_1 + KV_2 + V_4}{2(K+1) + K + 1} \tag{14.63}$$

and for a single-dielectric system, that is, for $K = 1$, with equal nodal distances Equation 14.63 reduces to

$$V_0 = \frac{1}{6}(4V_1 + V_2 + V_4)$$

### 14.5.2 For Parallel Dielectric Media

In this case, the dielectric interface is considered to be parallel to the axis of symmetry. FDM equations need to be developed for a node lying on the dielectric interface. This node is away from the axis of symmetry, as in this case the dielectric interface cannot be on the axis of symmetry.

For the nodal arrangement shown in Figure 14.17, the $r$-component of electric flux density is the normal component on the dielectric interface that is parallel to the $z$-axis.

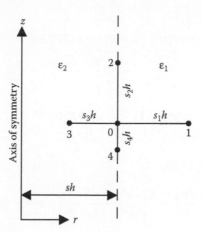

**FIGURE 14.17**
Nodal arrangement for FDM equation development in axi-symmetric system with multi-dielectric media in parallel dielectric arrangement when the node is lying away from the axis of symmetry.

$$\vec{D}_r = \vec{D}_r'$$

$$\text{or,} \quad \frac{\partial V'}{\partial r} = K \frac{\partial V}{\partial r} \tag{14.64}$$

where:

$$K = \frac{\varepsilon_1}{\varepsilon_2}$$

Now, from the Taylor series expansion between the pair of nodes, the expressions that may be obtained are as follows:

$$V_1 = V_0 + s_1 h \left.\frac{\partial V}{\partial r}\right|_0 + \left.\frac{s_1^2 h^2}{2} \frac{\partial^2 V}{\partial r^2}\right|_0 \quad \text{between nodes 1 and 0} \tag{14.65a}$$

$$V_3 = V_0 - s_3 h \left.\frac{\partial V'}{\partial r}\right|_0 + \left.\frac{s_3^2 h^2}{2} \frac{\partial^2 V'}{\partial r^2}\right|_0 \quad \text{between nodes 0 and 3} \tag{14.65b}$$

$$V_2 = V_0 + s_2 h \left.\frac{\partial V}{\partial z}\right|_0 + \left.\frac{s_2^2 h^2}{2} \frac{\partial^2 V}{\partial z^2}\right|_0 \quad \text{between nodes 2 and 0} \tag{14.65c}$$

$$V_4 = V_0 - s_4 h \left.\frac{\partial V}{\partial z}\right|_0 + \left.\frac{s_4^2 h^2}{2} \frac{\partial^2 V}{\partial z^2}\right|_0 \quad \text{between nodes 0 and 4} \tag{14.65d}$$

From Equation 14.65a

$$\left.\frac{\partial^2 V}{\partial r^2}\right|_0 = \frac{V_1 - V_0 - s_1 h (\partial V/\partial r)|_0}{s_1^2 h^2/2} \tag{14.66}$$

and from Equation 14.65b

$$\left.\frac{\partial^2 V'}{\partial r^2}\right|_0 = \frac{V_3 - V_0 + s_3 h\, K (\partial V/\partial r)|_0}{s_3^2 h^2/2} \tag{14.67}$$

Again, from Equation 14.65a

$$\left.\frac{1}{r}\frac{\partial V}{\partial r}\right|_0 = \frac{V_1 - V_0}{ss_1 h^2} - \frac{s_1 h}{2sh}\left.\frac{\partial^2 V}{\partial r^2}\right|_0 \quad \text{as } r = sh \tag{14.68}$$

and from Equation 14.65b

$$\left.\frac{1}{r}\frac{\partial V'}{\partial r}\right|_0 = \frac{V_0 - V_3}{ss_3 h^2} + \frac{s_3 h}{2sh}\left.\frac{\partial^2 V'}{\partial r^2}\right|_0 \quad \text{as } r = sh \tag{14.69}$$

From Equations 14.65c and 14.65d along the z-direction on the dielectric interface,

$$\left.\frac{\partial^2 V}{\partial z^2}\right|_0 = \frac{\left[(V_2/s_2) + (V_4/s_4)\right] - V_0\left[(1/s_2) + (1/s_4)\right]}{h^2(s_2 + s_4)/2} = \left.\frac{\partial^2 V'}{\partial z^2}\right|_0 \tag{14.70}$$

Satisfying Laplace's equation in medium 1:

$$\frac{\partial^2 V}{\partial r^2} + \frac{1}{r}\frac{\partial V}{\partial r} + \frac{\partial^2 V}{\partial z^2} = 0$$

$$\frac{V_1 - V_0 - s_1 h(\partial V/\partial r)|_0}{s_1^2 h^2/2} + \frac{V_1 - V_0}{ss_1 h^2} - \frac{s_1 h}{2sh}\frac{V_1 - V_0 - s_1 h(\partial V/\partial r)|_0}{s_1^2 h^2/2}$$

$$+ \frac{\left[(V_2/s_2) + (V_4/s_4)\right] - V_0\left[(1/s_2) + (1/s_4)\right]}{h^2(s_2 + s_4)/2} = 0 \tag{14.71}$$

Similarly, satisfying Laplace's equation in medium 2:

$$\frac{\partial^2 V'}{\partial r^2} + \frac{1}{r}\frac{\partial V'}{\partial r} + \frac{\partial^2 V'}{\partial z^2} = 0$$

$$\frac{V_3 - V_0 + s_3 h\, K(\partial V/\partial r)|_0}{s_3^2 h^2/2} + \frac{V_0 - V_3}{ss_3 h^2} + \frac{s_3 h}{2sh}\frac{V_3 - V_0 + s_3 h\, K(\partial V/\partial r)|_0}{s_3^2 h^2/2}$$

$$+ \frac{\left[(V_2/s_2) + (V_4/s_4)\right] - V_0\left[(1/s_2) + (1/s_4)\right]}{h^2(s_2 + s_4)/2} = 0 \tag{14.72}$$

Eliminating $\partial V/\partial r\big|_0$ from Equations 14.71 and 14.72,

$$V_0 = \frac{\left(\begin{array}{c}\left[K/s_1(2s-s_1)\right]V_1 + \left[1/(2s+s_3)\right]V_3 + \left[1/(s_2+s_4)\right] \\ \left\{\left[s_1K/(2s-s_1)\right] + \left[s_3/(2s+s_3)\right]\right\}\left[(V_2/s_2)+(V_4/s_4)\right]\end{array}\right)}{\left[K/s_1(2s-s_1)\right]+\left[1/(2s+s_3)\right]+(1/s_2s_4)\left\{\left[s_1K/(2s-s_1)\right]+\left[s_3/(2s+s_3)\right]\right\}} \quad (14.73)$$

Equation 14.73 is the FDM equation for unknown potential of a node lying on the dielectric interface, where the node is away from the axis of symmetry, in an axi-symmetric system with unequal nodal distances having a parallel dielectric arrangement.

For equal nodal distances, that is, when $s_1 = s_2 = s_3 = s_4 = 1$, Equation 14.73 reduces to

$$V_0 = \frac{\left[\begin{array}{c}\left[K/(2s-1)\right]V_1 + \left[1/(2s+1)\right]V_3 \\ + (1/2)\left(\left\{\left[K/(2s-1)\right]+\left[1/(2s+1)\right]\right\}\left[V_2+V_4\right]\right)\end{array}\right]}{2\left\{\left[K/(2s-1)\right]+\left[1/(2s+1)\right]\right\}} \quad (14.74)$$

and for a single-dielectric system, that is, $K = 1$, with equal nodal distances Equation 14.74 reduces to

$$V_0 = \frac{1}{4}(V_1 + V_2 + V_3 + V_4) + \frac{V_1 - V_3}{8s}$$

## PROBLEM 14.8

For the axi-symmetric multi-dielectric configuration with series dielectric arrangement, as shown in Figure 14.18, write the FDM equations for the nodes having unknown potentials.

*Solution:*

In this configuration, the largest nodal distance is $(2r/3)$. Therefore, the respective nodal distance factors are calculated based on this largest nodal distance.

For node 1: As per Equation 14.29, $V_1 = V_5$, $V_2 = 100$, $V_4 = V_2$, $s_1 = 0.5$ and $s_2 = s_4 = 1$.

$$V_1 = \frac{\left[(2\times V_5)/0.5^2\right]+\left[1/(1+1)\right]\left[(100/1)+(V_2/1)\right]}{(2/0.5^2)+\left[1/(1\times 1)\right]}$$

For node 2: As per Equation 14.62, $V_1 = V_6$, $V_2 = V_1$, $V_4 = V_3$, $s_1 = 0.5$, $s_2 = s_4 = 1$ and $K = (1/4)$.

$$V_2 = \frac{\left[2(0.25\times 1+1)/0.5^2\right]V_6 + (0.25/1)V_1 + (1/1)V_3}{\left[2(0.25\times 1+1)/0.5^2\right]+(0.25/1)+(1/1)}$$

For node 3: As per Equation 14.29, $V_1 = V_7$, $V_2 = V_2$, $V_4 = 0$, $s_1 = 0.5$ and $s_2 = s_4 = 1$.

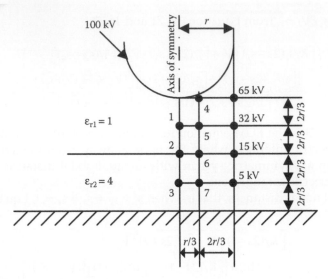

**FIGURE 14.18**
Nodal arrangement pertaining to Problem 14.8.

$$V_3 = \frac{\left[(2 \times V_7)/0.5^2\right] + \left[1/(1+1)\right]\left[(V_2/1) + (0/1)\right]}{(2/0.5^2) + \left[1/(1 \times 1)\right]}$$

For node 4: The nodal distances are unequal such that $s_1 = 1$, $s_2 = 0.086$, $s_3 = 0.5$, $s_4 = 1$ and $s = 0.5$; $V_1 = 65$, $V_2 = 100$, $V_3 = 100$ and $V_4 = V_5$. Then as per Equation 14.23

$$V_4 = \frac{\left(\begin{array}{l}\left[1/(1+0.5)\right]\left\{\left[(2\times0.5+0.5)/1\right]65 + \left[(2\times0.5-1)/0.5\right]100\right\} \\ + \left[(2\times0.5)/(0.086+1)\right]\left[(100/0.086) + (V_5/1)\right]\end{array}\right)}{\left[(2\times0.5+0.5-1)/(1\times0.5)\right] + \left[(2\times0.5)/(0.086\times1)\right]}$$

For node 5: The nodal distances are unequal such that $s_1 = s_2 = s_4 = 1$, $s_3 = 0.5$ and $s = 0.5$; $V_1 = 32$, $V_2 = V_4$, $V_3 = V_1$ and $V_4 = V_6$. Then as per Equation 14.23

$$V_5 = \frac{\left(\begin{array}{l}\left[1/(1+0.5)\right]\left\{\left[(2\times0.5+0.5)/1\right]32 + \left[(2\times0.5-1)/0.5\right]V_1\right\} \\ + \left[(2\times0.5)/(1+1)\right]\left[(V_4/1) + (V_6/1)\right]\end{array}\right)}{\left[(2\times0.5+0.5-1)/(1\times0.5)\right] + \left[(2\times0.5)/(1\times1)\right]}$$

For node 6: As per Equation 14.53, $V_1 = 15$, $V_2 = V_5$, $V_3 = V_2$, $V_4 = V_7$, $s_1 = s_2 = s_4 = 1$, $s_3 = 0.5$, $s = 0.5$ and $K = (1/4)$.

$$V_6 = \cfrac{\left(\begin{array}{l}\{(0.25\times1+1)[2+(0.5/0.5)]/[1\times(1+0.5)]\}\times15+[(2\times0.25)/1]\,V_5 \\[2mm] +\{(0.25\times1+1)[2-(1/0.5)]/[0.5\times(1+0.5)]\}\,V_2+(2/1)V_7\end{array}\right)}{\left\{\begin{array}{l}[2(0.25\times1+1)/(1\times0.5)]+[(0.5-1)(0.25\times1+1)/(0.5\times1\times0.5)] \\[2mm] +[(2\times0.25)/1]+(2/1)\end{array}\right\}}$$

For node 7: The nodal distances are unequal such that $s_1 = s_2 = s_4 = 1$, $s_3 = 0.5$ and $s = 0.5$; $V_1 = 5$, $V_2 = V_6$, $V_3 = V_3$ and $V_4 = 0$. Then as per Equation 14.23

$$V_7 = \cfrac{\left(\begin{array}{l}[1/(1+0.5)]\{[(2\times0.5+0.5)/1]5+[(2\times0.5-1)/0.5]\,V_3\} \\[2mm] +[(2\times0.5)/(1+1)][(V_6/1)+(0/1)]\end{array}\right)}{[(2\times0.5+0.5-1)/(1\times0.5)]+[(2\times0.5)/(1\times1)]}$$

## PROBLEM 14.9

For the axi-symmetric multi-dielectric configuration with parallel dielectric arrangement, as shown in Figure 14.19, write the FDM equations for the nodes having unknown potentials.

*Solution:*

In this configuration the largest nodal distance is $h$. Therefore, the respective nodal distance factors are calculated based on this largest nodal distance.

For node 1: The nodal distances are unequal such that $s_1 = s_2 = s_4 = 1$, $s_3 = 0.5$ and $s = 1.833$; $V_1 = 60$, $V_2 = 100$, $V_3 = V_3$ and $V_4 = V_2$. Then as per Equation 14.23

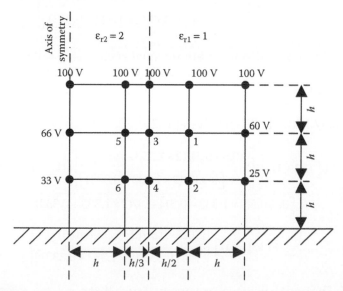

**FIGURE 14.19**
Nodal arrangement pertaining to Problem 14.9.

$$V_1 = \frac{\left(\begin{array}{c}\left[1/(1+0.5)\right]\left\{\left[(2\times1.833+0.5)/1\right]60+\left[(2\times1.833-1)/0.5\right]V_3\right\} \\ +\left[(2\times1.833)/(1+1)\right]\left[(100/1)+(V_2/1)\right]\end{array}\right)}{\left[(2\times1.833+0.5-1)/(1\times0.5)\right]+\left[(2\times1.833)/(1\times1)\right]}$$

For node 2: The nodal distances are unequal such that $s_1 = s_2 = s_4 = 1$, $s_3 = 0.5$ and $s = 1.833$; $V_1 = 25$, $V_2 = V_1$, $V_3 = V_4$ and $V_4 = 0$. Then as per Equation 14.23

$$V_2 = \frac{\left(\begin{array}{c}\left[1/(1+0.5)\right]\left\{\left[(2\times1.833+0.5)/1\right]25+\left[(2\times1.833-1)/0.5\right]V_4\right\} \\ +\left[(2\times1.833)/(1+1)\right]\left[(V_1/1)+(0/1)\right]\end{array}\right)}{\left[(2\times1.833+0.5-1)/(1\times0.5)\right]+\left[(2\times1.833)/(1\times1)\right]}$$

For node 3: The nodal distances are unequal such that $s_1 = 0.5$, $s_2 = s_4 = 1$, $s_3 = 0.333$ and $s = 1.333$; $V_1 = V_1$, $V_2 = 100$, $V_3 = V_5$, $V_4 = V_4$ and $K = (1/2)$. Then as per Equation 14.73

$$V_3 = \frac{\left[\begin{array}{c}\left[0.5/0.5(2\times1.333-0.5)\right]V_1+\left[1/(2\times1.333+0.333)\right]V_5 \\ +\left[1/(1+1)\right]\left\{\begin{array}{c}\left[(0.5\times0.5)/(2\times1.333-0.5)\right] \\ +\left[0.333/(2\times1.333+0.333)\right]\end{array}\right\}\left[(100/1)+(V_4/1)\right]\end{array}\right]}{\left(\begin{array}{c}\left[0.5/0.5(2\times1.333-0.5)\right]+\left[1/(2\times1.333+0.333)\right] \\ +\left[1/(1\times1)\right]\left\{\begin{array}{c}\left[(0.5\times0.5)/(2\times1.333-0.5)\right] \\ +\left[0.333/(2\times1.333+0.333)\right]\end{array}\right\}\end{array}\right)}$$

For node 4: The nodal distances are unequal such that $s_1 = 0.5$, $s_2 = s_4 = 1$, $s_3 = 0.333$ and $s = 1.333$; $V_1 = V_2$, $V_2 = V_3$, $V_3 = V_6$, $V_4 = 0$ and $K = (1/2)$. Then as per Equation 14.73

$$V_4 = \frac{\left[\begin{array}{c}\left[0.5/0.5(2\times1.333-0.5)\right]V_2+\left[1/(2\times1.333+0.333)\right]V_6 \\ +\left[1/(1+1)\right]\left(\left\{\begin{array}{c}\left[(0.5\times0.5)/(2\times1.333-0.5)\right] \\ +\left[0.333/(2\times1.333+0.333)\right]\end{array}\right\}\left[(V_3/1)+(0/1)\right]\right)\end{array}\right]}{\left(\begin{array}{c}\left[0.5/0.5(2\times1.333-0.5)\right]+\left[1/(2\times1.333+0.333)\right] \\ +\left[1/(1\times1)\right]\left\{\begin{array}{c}\left[(0.5\times0.5)/(2\times1.333-0.5)\right] \\ +\left[0.333/(2\times1.333+0.333)\right]\end{array}\right\}\end{array}\right)}$$

For node 5: The nodal distances are unequal such that $s_2 = s_3 = s_4 = 1$, $s_1 = 0.333$ and $s = 1$; $V_1 = V_3$, $V_2 = 100$, $V_3 = 66$ and $V_4 = V_6$. Then as per Equation 14.23

$$V_5 = \frac{\left( \begin{array}{l} \left[1/(0.333+1)\right] \left\{ \left[(2\times1+1)/0.333\right] V_3 + \left[(2\times1-0.333)/1\right]\times66 \right\} \\ + \left[(2\times1)/(1+1)\right] \left[(100/1)+(V_6/1)\right] \end{array} \right)}{\left[(2\times1+1-0.333)/(0.333\times1)\right] + \left[(2\times1)/(1\times1)\right]}$$

For node 6: The nodal distances are unequal such that $s_2 = s_3 = s_4 = 1$, $s_1 = 0.333$ and $s = 1$; $V_1 = V_4$, $V_2 = V_5$, $V_3 = 33$ and $V_4 = 0$. Then as per Equation 14.23

$$V_6 = \frac{\left( \begin{array}{l} \left[1/(0.333+1)\right] \left\{ \left[(2\times1+1)/0.333\right] V_4 + \left[(2\times1-0.333)/1\right]\times33 \right\} \\ + \left[(2\times1)/(1+1)\right] \left[(V_5/1)+(0/1)\right] \end{array} \right)}{\left[(2\times1+1-0.333)/(0.333\times1)\right] + \left[(2\times1)/(1\times1)\right]}$$

## 14.6 Simulation Details

### 14.6.1 Discretization

In FDM, for the 2D system, the entire ROI, that is, the region where the field distribution is required to be calculated, is discretized using either rectangles or squares. In the 3D system, discretization is done using either rectangular parallelepipeds or cubes. Because the potential is commonly assumed to vary linearly between two successive nodes, the nodes need to be closely spaced where the field varies significantly in space. This is generally the case near the electrodes or dielectric boundaries, particularly in the cases of contours having sharp corners. On the other hand, in the region away from the electrodes or dielectric boundaries, where the field does not change rapidly in space, the nodes may be spaced relatively widely apart.

For multi-dielectric problems, care should be taken during discretization to make sure that only one dielectric is present between two consecutive nodes. This is achieved by arranging one layer of nodes along the dielectric–dielectric interface.

The finite difference model of a problem gives a point-wise approximation to the governing equations, for example, Laplace's equation. This model is formed by writing difference equations for an array of grid points called *nodes*, which is improved as more nodes are used in the simulation. With the help of FDM, one can treat some fairly difficult problems; but for problems having irregular geometries or an unusual specification of boundary conditions, the FDM becomes hard to use.

As an example of how FDM might be used to represent a complex geometrical shape, consider the high-voltage insulator cross section shown in Figure 14.20. A finite difference mesh would reasonably cover the insulator

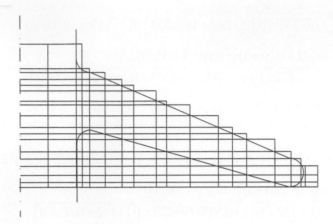

**FIGURE 14.20**
Discretization of insulator geometry by FDM mesh.

volume, but the boundaries must be approximated by a series of horizontal and vertical lines or *stair steps*. This results in poor approximation of the curved insulator boundary.

## 14.6.2 Simulation of an Unbounded Field Region

FDM is well suited for simulating bounded field regions, that is, field regions having well-defined boundaries. For unbounded field regions, a major difficulty in the implementation of FDM is the placement of nodes in the space, which is far away from the components that affect the field distribution. This difficulty is surmounted by placing a fictitious boundary, as shown in Figure 14.21, at a location relatively distant from the components influencing the field. This fictitious boundary is placed with the assumption that the pair of nodes on both sides of this boundary has same potential, for example, the potential of the nodes 1, 3, 5 and 7 are assumed to be same as that of the nodes 2, 4, 6 and 8, respectively. If this fictitious boundary is placed in a region where the field does not vary rapidly in space, then the imposition of this fictitious boundary does not incorporate any significant error in the field computation.

## 14.6.3 Accuracy Criteria

The accuracy of simulation is dependent on the nature of discretization of the field region and hence it is important to determine the simulation accuracy using certain well-accepted criteria as detailed below.

The *potential error* on the electrode boundaries can be determined at a number of checkpoints on the electrode surface between two consecutive nodes. Such check points are often called *control points*. This potential error is defined

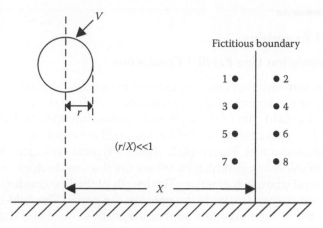

**FIGURE 14.21**
Simulation of unbounded field region.

as the difference between the known potential of the electrode and the computed potential at the control point. From such calculations one can determine the average or the maximum or the mean squared value of the potential error.

The error in the electric field intensity is usually higher than the potential error. Hence, compared to the potential error the *deviation angle* on the electrode surface is a more sensitive indicator of the simulation accuracy. The deviation angle is defined as the angular deviation of the electric stress vector at the control point on the electrode surface from the direction of the normal to its surface.

In multi-dielectric systems, the discrepancy in the tangential electric stress at the control points on the dielectric interface can be computed. Another criterion for checking the simulation accuracy is to compute the discrepancy in the normal flux density at the control point on the dielectric interface. For a good simulation, such discrepancies should be small.

### 14.6.4 System of FDM Equation

In FDM, the potential of any node is related to either four connected nodes in the 2D system or six connected nodes in the 3D system. Hence, if a field region is discretized using $N$ ($N \gg 1$) number of nodes, then the system of FDM equation will be an $N \times N$ matrix. But in each row of this matrix, there will be non-zero value in only five or seven elements depending on the dimension of the simulated system and all the other elements out of $N$ elements of each row of the matrix will be zero. Hence, it is obvious that the system of FDM equations generates a highly sparse matrix.

Hence, it is advisable to solve the system of FDM equations by an iterative technique such as Gauss–Seidel method rather than using a direct method such as Gaussian elimination. In the iterative methods, suitable technique is to be employed to achieve accelerated convergence.

## 14.7 FDM Examples

### 14.7.1 Transmission Line Parallel Conductors

Length of the transmission line conductors can be considered to be very large compared to the diameter of the conductors as well as the conductor spacings. Hence, the field due to three-phase transmission line conductors could be computed as a 2D case, as the field is assumed to be not varying along the length of the conductors. Figure 14.22 shows a typical arrangement of three-phase transmission line conductors, where the line conductors are arranged in an equilateral triangle formation. The length of the line conductor is taken in the direction perpendicular to the plane of the paper. The field is assumed to be computed at the mid-span, so that the earth surface and the transmission towers are considered to be far away from the conductors. This is an example of unbounded field region problem and hence fictitious boundaries, as discussed in Section 14.6.2, need to be introduced, as shown in Figure 14.22. FDM grid, as shown in Figure 14.22, depicts smaller nodal distances where the field is expected to be higher, whereas the nodal distances are higher elsewhere. This example involves only one dielectric as the conductors are surrounded by

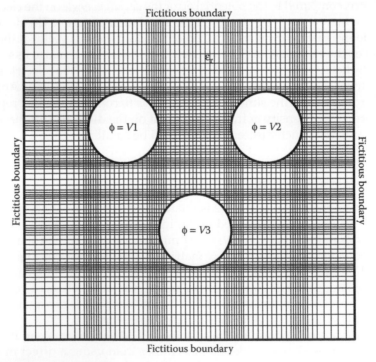

**FIGURE 14.22**
FDM grid for three-phase transmission line conductors.

air. It should be stressed here that all the FDM grids shown in section 14.7 are suggestive only. Users may use other grids using their discretion.

### 14.7.2 Post-Type Insulator

Figure 14.23 shows the FDM grid for a simple post-type insulator stressed between two electrodes. It is an axi-symmetric system with two dielectric media. It is also an example of unbounded field region problem and hence a fictitious boundary on the right has been introduced, as shown in Figure 14.23. There should be two more fictitious boundaries that need to be introduced, one at the top and one at the bottom. But it is left to the imagination of the reader. Coarse and finer meshes are taken depending on the nature of field concentration. Typically finer mesh size is to be taken near the boundaries and as one goes away from the boundary relatively higher mesh size could be taken without compromising accuracy.

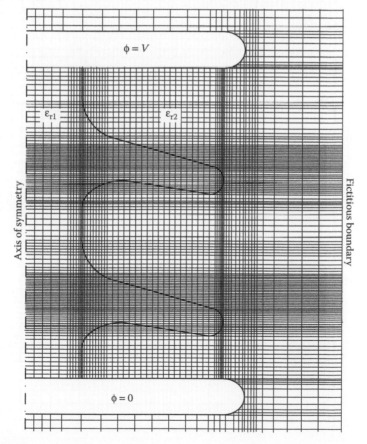

**FIGURE 14.23**
FDM grid for a simple post-type insulator.

### 14.7.3 Disc-Type Insulator

Disc-type insulators are widely used for supporting high-voltage transmission line conductors from the tower cross-arm. A realistic disc-type insulator and the associated FDM grid are shown in Figure 14.24. It is an axi-symmetric case study with two dielectric media, typically porcelain ($\varepsilon_{r1}$) as solid insulating material surrounded by air ($\varepsilon_{r2}$). As it is an unbounded field region problem, three fictitious boundaries have been introduced for field computation, as shown in Figure 14.24. As stated in the earlier examples, the nodal distances are to be suitably chosen in relation to field distribution, so that high accuracy is achieved. Contrary to post-type insulators, the high-voltage end of the disc-type insulator is at the bottom (the pin, as shown in Figure 14.24) and the earthed end is at the top (the cap, as shown in Figure 14.24).

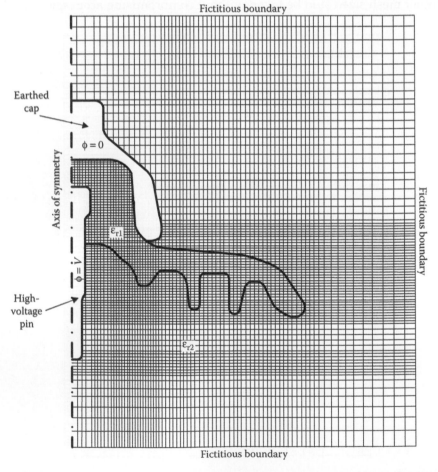

**FIGURE 14.24**
FDM grid for a disc-type insulator.

## Objective Type Questions

1. Finite difference method is based on
   a. Differential equation technique
   b. Integral equation technique
   c. Monte Carlo technique
   d. Experimental technique

2. FDM equations may be derived from
   a. Logarithmic series expansion
   b. Binomial series expansion
   c. Taylor series expansion
   d. Fourier series expansion

3. FDM is best suited to solve numerically
   a. Laplace's equation
   b. Poisson's equation
   c. Gauss's law in differential form
   d. Both (a) and (b)

4. For equally spaced nodes in a 2D system, FDM equation is given by
   a. $V_0 = (V_1 + V_2 + V_3 + V_4 + V_0)/5$
   b. $V_0 = (V_1 + V_2 + V_3 + V_4 + V_0)/4$
   c. $V_0 = (2V_1 + V_2 + V_3 + V_4)/4$
   d. $V_0 = (V_1 + V_2 + V_3 + V_4)/4$

5. Between any two connected nodes in FDM, electric field intensity is commonly assumed to be
   a. Linearly varying
   b. Constant
   c. Inversely varying
   d. Exponentially varying

6. FDM equations are most commonly solved by
   a. Iterative technique
   b. Gaussian elimination
   c. LU decomposition
   d. Both (b) and (c)

7. For equally spaced nodes in a 3D system, FDM equation is given by
   a. $V_0 = (V_1 + V_2 + V_3 + V_4)/4$
   b. $V_0 = (V_1 + V_2 + V_3 + V_4 + V_5 + V_6)/6$

   c.  $V_0 = (V_1 + V_2 + V_3 + V_4 + V_5 + V_6)/4$

   d.  $V_0 = (V_1 + V_2 + V_3 + V_4)/6$

8. Matrix representation of the system of equations to be solved in FDM gives

   a.  Diagonal matrix

   b.  Unit matrix

   c.  Sparse matrix

   d.  Nearly full matrix

9. The FDM equation $V_0 = (V_1 + V_2 + V_3 + V_4)/4 + (V_1 - V_3)/8s$ is valid for

   a.  2D system with equal nodal distances

   b.  2D system with unequal nodal distances

   c.  Axi-symmetric system with equal nodal distances

   d.  Axi-symmetric system with unequal nodal distances

10. Simulation of unbounded configurations extended up to infinity gives rise to difficulties in

   a.  Finite difference method

   b.  Finite element method

   c.  Charge simulation method

   d.  Both (a) and (b)

11. The nodal FDM equation $V_0 = (4V_1 + V_2 + V_4)/6$ for equally spaced nodes is valid for a node

   a.  In a 2D system

   b.  In a 3D system

   c.  On the axis in an axi-symmetric system

   d.  On dielectric boundary in a 2D system

12. For a given configuration in FDM, if the number of nodes is increased, then accuracy

   a.  Remains unchanged

   b.  Decreases

   c.  Increases

   d.  Oscillates

13. In simulation of multi-dielectric problem by FDM, two successive nodes are placed in such a way that

   a.  Both lie in the same dielectric

   b.  Each one lies in different dielectric

    c.  One on the boundary and other one in a dielectric

    d.  Both (a) and (c)

14. On the conductor boundary, simulation accuracy is checked by calculating

    a.  Potential error

    b.  Continuity of normal component of electric flux density

    c.  Deviation angle

    d.  Both (a) and (c)

15. For a given 2D configuration in FDM, if the inter-node spacing of equally spaced nodes is increased, then the results

    a.  Get better

    b.  Get worse

    c.  Remain unchanged

    d.  Oscillate

16. On the dielectric–dielectric interface, simulation accuracy is checked by calculating

    a.  Potential error

    b.  Continuity of normal component of electric flux density

    c.  Deviation angle

    d.  Both (a) and (c)

17. To achieve higher simulation accuracy as well as faster computation, discretization of the field region should be done in such a way that

    a.  The nodes are closely spaced everywhere

    b.  The nodes are sparsely spaced everywhere

    c.  The nodes are closely spaced where the field varies sharply and are sparsely spaced elsewhere

    d.  The nodes are sparsely spaced where the field varies sharply and are closely spaced elsewhere

18. In general, the nodes need to be spaced unequally

    a.  Near the electrode boundaries

    b.  Near the dielectric–dielectric boundaries

    c.  Far away from the boundaries

    d.  Both (a) and (b)

**Answers:**  1) a;  2) c;  3) a;  4) d;  5) b;  6) a;  7) b;  8) c;  9) c;  10) d;  11) c;  12) c;
              13) d;  14) d;  15) b;  16) b;  17) c;  18) d

## Bibliography

1. G. Shortley, R. Weller, P. Darby and E.H. Gamble, 'Numerical solution of axisymmetrical problems with applications to electrostatics and torsion', *Journal of Applied Physics*, Vol. 18, pp. 116–129, 1947.
2. D.N. De G. Allen and S.C.R. Dennis, 'The application of relaxation methods to the solution of differential equations in three dimensions', *Quarterly Journal of Mechanics Applied Mathematics*, Vol. 4, pp. 199–208, 1951.
3. H.E. Kulsrud, 'A practical technique for the determination of the optimum relaxation factor of successive over-relaxation method', *Communication Association for Computing Machinery*, Vol. 4, pp. 184–187, 1961.
4. B.A. Carré, 'The determination of the optimum accelerating factor for successive over-relaxation', *Computer Journal*, Vol. 4, pp. 73–78, 1961.
5. K.J. Binns and P.J. Lawrenson, *Analysis and Computation of Electric and Magnetic Field Problems*, Pergamon Press, Oxford, 1963.
6. G.E. Forsythe and W.R. Wasow, *Finite Difference Methods for Partial Differential Equations*, Wiley, New York, 1965.
7. D.F. Binns, 'Calculation of field factor for a vertical sphere gap taking account of surrounding earthed surfaces', *Proceedings of IEEE*, Vol. 112, pp. 1575–1582, 1965.
8. M.V. Schneider, 'Computation of impedance and attenuation of TEM-lines by finite difference methods', *IEEE Transactions on Microwave Theory and Techniques*, Vol. 13, pp. 793–800, 1965.
9. M.J. Billings, B.P. Nellist and P. Swarbrick, 'Investigation into new designs for h.v. insulators using synthetic materials', *Proceedings of IEEE*, Vol. 113, No. 10, pp. 1643–1648, 1966.
10. R.H. Galloway, H.McL. Ryan and M.F Scott, 'Calculation of electric fields by digital computer', *Proceedings of IEEE*, Vol. 114, No.6, pp. 824–829, 1967.
11. J.T. Storey and M.J. Billings, 'General digital-computer program for the determination of 3-dimensional electrostatic axially symmetric fields', *Proceedings of IEEE*, Vol. 114, No. 10, pp. 1551–1555, 1967.
12. J.W. Duncan, 'The accuracy of finite-difference solutions of Laplace's equation', *IEEE Transactions on Microwave Theory and Techniques*, Vol. 15, No. 10, pp. 575–582, 1967.
13. J.T. Storey and M.J. Billings, 'Determination of the 3-dimensional electrostatic field in a curved bushing', *Proceedings of IEEE*, Vol. 116, No. 4, pp. 639–643, 1969.
14. J.T. Storey, 'The determination of axially symmetric fields by digital computation', *IEEE Transactions on Electrical Insulation*, Vol. 4, No. 2, pp. 23–30, 1969.
15. M.F. Scott, J.M. Mattingley, M. John and H.M. Ryan, 'Computation of electric fields: Recent developments and practical applications', *IEEE Transactions on Electrical Insulation*, Vol. 9, No. 1, pp. 18–25, 1974.
16. B. Aldefeld, 'Computer-aided design of electromagnetic actuators using finite difference techniques', *IEEE Transactions on Magnetics*, Vol. 12, No. 6, p. 1047, 1976.
17. B. Tomescu and F.M. Tomescu, 'Electrostatic field computation in non-rectangular configurations', *Revue Roumaine des Science Techniques Series Electrotechnique et Energetique*, Vol. 22, No. 4, pp. 491–498, 1977.

18. G. Molinari, M.R. Podesta, G. Sciutto and A. Viviani, 'Finite difference method with irregular grid and transformed discretization metric', *IEEE PES Winter Meeting*, New York, January 29–February 3, 1978.
19. E.S. Kolechitskii, A.A. Filippov and O.V. Firsova, 'Methods for calculating electric fields of high-voltage equipment', *Soviet Electrical Engineering*, Vol. 51, No. 4, pp. 17–22, 1980.
20. E.F. Fuch and K. Senske, 'Comparison of iterative solutions of the finite difference method with measurements as applied to Poisson's and the diffusion equations', *IEEE PES Winter Meeting*, Atlanta, GA, February 1–6, 1981.
21. F. Donazzi and G. Luoni, 'Comparison of different methods in computing electrostatic fields', *IEEE Transactions on Magnetics*, Vol. 19, No. 6, pp. 2596–2599, 1983.
22. M. Hizal, 'Computation of the electric field at a solid/gas interface in the presence of surface and volume charges', *IEEE Proceedings A*, Vol. 133, No. 9, pp. 577–586, 1986.
23. A. Christ and H. Hartnagel, 'Three-dimensional finite-difference method for the analysis of microwave device embedding', *IEEE Transactions on MTT*, Vol. 35, pp. 688–696, 1987.
24. A.A. Read and D.T. Stephenson, 'Computers and computer graphics in teaching electromagnetic fields', *Proceedings of Frontiers in Education Conference*, IEEE Publication, Terre Haute, IN, pp. 523–530, October 24–27, 1987.
25. M.J. Sablik, D. Golimowski, J.R. Sharber and J.D. Winningham, 'Computer simulation of a 360° field-of-view top-hat electrostatic analyser', *Review of Scientific Instruments*, Vol. 59, No. 1, pp. 146–155, 1988.
26. M. Abdel-Salam and M.T. El-Mohandes, 'Combined method based on finite differences and charge simulation for calculating electric fields', *IEEE Transactions on Industry Applications*, Vol. 25, No. 6, pp. 1060–1066, 1989.
27. D. Hollmann, S. Haffa and W. Wiesbeck, 'Improved analysis of dielectric resonator coupling using a three-dimensional finite difference method', *Proceedings of 21st European Microwave Conference*, Stuttgart, Germany, Vol. 1, pp. 565–570, September 9–12, Microwave Exhibitions and Publishers, Richmond, UK, 1991.
28. N. Femia and V. Tucci, 'Computer models of electrical discharges', *Proceedings of International Conference on Computation in Electromagnetics*, London, IEE Conference Publication No. 350, pp. 119–122, November 25–27, 1991.
29. R.K. Gordon, M.D. Tew and A.Z. Elsherbeni, 'An efficient finite difference method for finding the electric potential in regions with small perturbations', *Proceedings of IEEE Antennas and Propagation Society International Symposium*, Chicago, IL, June 18–25, IEEE Publication, New Jersey, 1992.
30. A. Bouziane, K. Hidaka, J.E. Jones, A.R. Rowlands, M.C. Tapiamacioglu and R.T. Waters, 'Paraxial corona discharge. II. Simulation and analysis', *IEE Proceedings – Science, Measurement and Technology*, Vol. 141, No. 3, pp. 205–214, 1994.
31. O. Sucre and D. Suster, 'Solution of unbounded bi-dimensional boundary-value problems with the finite-difference method', *Proceedings of the International Congress on Numerical Methods in Engineering and Applied Sciences*, CIMENICS, Mérida, Venezuela, pp. 231–238, March, Venezuelan Society of Numerical Methods in Engineering, Caracas, Venezuela, 1996.
32. E. Lami, F. Mattachini, R. Sala and H. Vigl, 'A mathematical model of electrostatic field in wires-plate electrostatic precipitators', *Journal of Electrostatics*, Vol. 39, No. 1, pp. 1–21, 1997.

33. L. Li and E. Wang, 'An improved finite difference method for calculating electric field in high voltage vacuum interrupter', *Proceedings of 20th International Symposium on Discharges and Electrical Insulation in Vacuum*, Tours, France, pp. 519–522, July 1–5, IEEE Publication, 2002.
34. M.A. Elhirbawy, L.S. Jennings, S.M. Al Dhalaan and W.W.L. Keerthipala, 'Practical results and finite difference method to analyze the electric and magnetic field coupling between power transmission line and pipeline', *Proceedings of the 2003 International Symposium on Circuits and Systems*, Bangkok, Thailand, Vol. 3, pp. 431–438, May 25–28, IEEE Publication,2003.

# 15

## Numerical Computation of High-Voltage Field by Finite Element Method

**ABSTRACT**   Finite element method (FEM) is a general procedure that may be adapted to approximate the solution to various partial differential equations. Two such equations in electric field analysis are Laplace's and Poisson's equations. Over the course of last five decades, FEM has been proved to be very efficient in electric field analysis in two-dimensional (2D), axi-symmetric and three-dimensional (3D) systems having multiple dielectric media. Basic FEM formulations in all these systems have been discussed in this chapter. The variational approach of minimizing the functional, namely, potential energy of electric field, has been taken as the basis of FEM formulation. Details of finite elements that are mostly used in 2D and 3D systems have been presented. The chapter also highlights the procedural steps in FEM along with the nuances of its implementations such as system of FEM equations and its solution, refinement and acceptability of FEM mesh and sources of error in FEM. Use of FEM in design cycle has been also discussed from a practical viewpoint.

## 15.1 Introduction

Finite element method (FEM) is a numerical analysis technique to obtain solutions to the differential equations that describe, or approximately describe, a wide variety of physical problems ranging from solid, fluid and soil mechanics, to electromagnetism or dynamics. The underlying premise of the FEM is that a complicated region of interest can be subdivided into a series of smaller sub-regions in which the differential equations are approximately solved. By assembling the set of equations for each sub-region, the behaviour over the entire region of interest is determined.

It is difficult to state the exact origin of the FEM, because the basic concepts have evolved over a period of 100 or more years. The term *finite element* was first coined by Clough in 1960. In the early 1960s, FEM was used for approximate solution of problems in stress analysis, fluid flow, heat transfer and some other areas. In the late 1960s and early 1970s, the application of FEM was extended to a much wider variety of engineering problems. Significant advances in mathematical treatments, including the development of new elements, and convergence studies were made in 1970s. Most of the commercial FEM software packages originated in the 1970s and 1980s. FEM is one of the most important developments in computational methods to occur in the twentieth century. The method has evolved from one with applications in structural engineering at the beginning to a widely utilized and richly varied computational approach for many scientific and technological areas at present.

## 15.2 Basics of FEM

Using the FEM, the region of interest is discretized into smaller sub-regions called *elements*, as shown in Figure 15.1, and the solution is determined in terms of discrete values of some primary field variables, for example, electric potential, at the nodes. The governing equation, for example, Laplace's or Poisson's equations, is now applied to the domain of a single element. At the element level, the solution to the governing equation is replaced by a continuous function approximating the distribution of the field variable $\phi$ over the element domain, expressed in terms of the unknown nodal values $\phi_1$, $\phi_2$ and $\phi_3$ of the solution $\phi$. A system of equations in terms of $\phi_1$, $\phi_2$ and $\phi_3$ can then be formulated for the element. Once the element equations have been determined, the elements are assembled to form the entire region of interest. Assembly is accomplished using the basic rule that the value of the field variable at a node must be the same for each element that shares that node. The solution $\phi$ to the problem becomes a piecewise approximation, expressed in

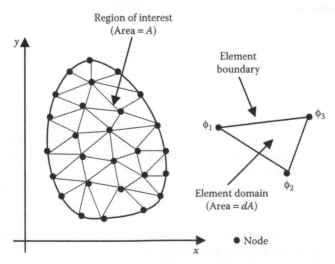

**FIGURE 15.1**
Depiction of region of interest, element and nodes for FEM formulation.

terms of the nodal values of ϕ. The assembly procedure results in a system of linear algebraic equations.

Several approaches can be used to transform the physical formulation of the problem to its finite element discrete analogue. If the physical formulation of the problem is known as a differential equation, for example, Laplace's or Poisson's equations, then the most popular method of its finite element formulation is the Galerkin method. If the physical problem can be formulated as minimization of a functional, then the variational formulation of the finite element equations is usually used. For problems in high-voltage fields, the functional turns out to be the energy stored in the electric field.

A third and even more versatile approach to derive element properties is known as the *weighted residuals approach*. The weighted residuals approach begins with the governing equations of the problem and proceeds without relying on a variational statement. This approach is advantageous because it makes it possible to extend the FEM to problems where no functional is available.

## 15.3 Procedural Steps in FEM

In general terms, the main steps of the finite element solution procedure are as follows.

1. At the beginning the region of interest is discretized into finite elements.

2. Suitable functions are considered to interpolate the field variables over the element.

3. The matrix equation for the finite element is formed relating the nodal values of the unknown field variables to other physical parameters.

4. Global equation system is formed for the entire region of interest by assembling all the element equations. Element connectivities are used for the assembly process. Boundary conditions, which are not accounted in element equations, are imposed before the solution of equations.

5. The finite element global equation system is solved to get the nodal values of the sought field variables.

6. In many cases, additional parameters need to be calculated after the solution of global equation system. For example, in high-voltage field problems, electric field intensity, electric flux density and charges are of interest in addition to electric potential, which are obtained after solution of the global equation system.

## 15.4 Variational Approach towards FEM Formulation

For high-voltage field problems, the principle of minimum potential energy is used in this approach. The principle of minimum potential energy can be stated as: Out of all possible potential functions, $\phi(x,y,z)$, the one which minimizes the total potential energy is the potential solution that will satisfy equilibrium, and will be the actual potential due to the applied field forces.

Thus, a potential function that will minimize the functional, that is, potential energy, is desired. Minimization of functionals falls within the field of variational calculus. In most cases, an exact function is impossible to determine, necessitating the use of approximate numerical methods. Minimization of potential energy in a finite element formulation is carried out using the energy approach. FEM develops the equations from simple element shapes, in which the unknowns of the solution are the potentials at the nodes. The calculus of variations enables the energy equation to be reduced to a set of simultaneous equations with the nodal potentials as the unknown quantities.

### 15.4.1 FEM Formulation in a 2D System with Single-Dielectric Medium

The potential energy in a two-dimensional (2D) electric field is given by

$$U_{\text{total}} = \iint_A \frac{1}{2} \varepsilon_0 \varepsilon_r \left| \vec{E} \right|^2 \cdot l \cdot dA \tag{15.1}$$

$$\text{or, } U_{\text{total}} = \iint_A \frac{1}{2} \varepsilon_0 \varepsilon_r \left| -\vec{\nabla}\phi \right|^2 \cdot l \cdot dA \tag{15.2}$$

where:
$E$ = electric field intensity
$\phi$ = electric potential
$l$ = length normal to the area $A$ (usually considered as unity for 2D field)
$\varepsilon_0$ = permittivity of free space
$\varepsilon_r$ = relative permittivity of dielectric

The integration of Equation 15.1 must be carried out over the area $A$, which is identical to the field region under consideration, as shown in Figure 15.1. Because this area must be finite, FEM cannot be applied to the problems with *open fields* without modifications.

To apply FEM, the region of interest is to be discretized by so-called finite elements, as shown in Figure 15.1. If a region of interest is divided into elements such that the continuity of electric potential between elements is enforced, then the total potential energy is equal to the sum of the individual energies of each element. For $N$ number of elements, the total potential energy can then be stated as follows:

$$U_{total} = \sum_{e=1}^{N} U(e) \tag{15.3}$$

To minimize the total potential energy, $U$, of the entire region of interest, $U(e)$ must be minimized for each element. Seeking a set of nodal potentials for each element will minimize $U(e)$. Observe that the functional $U(e)$ is a function only of the nodal potentials. Using calculus of variations, an extremization of $U(e)$ occurs when the vector of the first partial derivatives with respect to $\phi$ is zero.

The simplest 2D element is the linear triangular element, as shown in Figure 15.2. For this element, there are three nodes at the vertices of the triangle, which are numbered around the element in the anti-clockwise direction. Electric potential $\phi$ is assumed to be varying linearly within the element such that

$$\phi = \alpha_1 + \alpha_2 x + \alpha_3 y \tag{15.4}$$

Hence,

$$E_x = -\frac{\partial \phi}{\partial x} = -\alpha_2 \text{ and } E_y = -\frac{\partial \phi}{\partial y} = -\alpha_3 \tag{15.5}$$

Thus, for this element, the electric field intensity components are constant throughout the element. As a result, this type of element is also known *constant stress element (CST)*.

Now, considering a triangular element, as shown in Figure 15.2

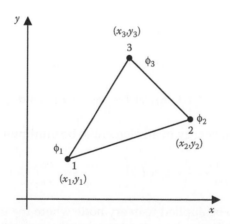

**FIGURE 15.2**
Linear triangular element.

$$\phi_1 = \alpha_1 + \alpha_2 x_1 + \alpha_3 y_1$$

$$\phi_2 = \alpha_1 + \alpha_2 x_2 + \alpha_3 y_2 \qquad (15.6)$$

$$\phi_3 = \alpha_1 + \alpha_2 x_3 + \alpha_3 y_3$$

Therefore, from Equation 15.6

$$\alpha_2 = \frac{\begin{vmatrix} 1 & \phi_1 & y_1 \\ 1 & \phi_2 & y_2 \\ 1 & \phi_3 & y_3 \end{vmatrix}}{\begin{vmatrix} 1 & x_1 & y_1 \\ 1 & x_2 & y_2 \\ 1 & x_3 & y_3 \end{vmatrix}}$$

$$\text{or, } \alpha_2 = \frac{1}{D}[\phi_1(y_2 - y_3) + \phi_2(y_3 - y_1) + \phi_3(y_1 - y_2)] \qquad (15.7)$$

Where:

$$D = (x_2 y_3 - x_3 y_2) + (x_3 y_1 - x_1 y_3) + (x_1 y_2 - x_2 y_1)$$
$$= 2 \text{ times the area of the triangle} \qquad (15.8)$$

Similarly,

$$\alpha_3 = \frac{1}{D}[\phi_1(x_3 - x_2) + \phi_2(x_1 - x_3) + \phi_3(x_2 - x_1)] \qquad (15.9)$$

The magnitude of the electric field intensity within an element $T$,

$$\left|\vec{E}_T\right| = \sqrt{E_x^2 + E_y^2} = \sqrt{\alpha_2^2 + \alpha_3^2} \qquad (15.10)$$

Hence, the electric potential energy in an element $T$

$$U_T = \frac{1}{2}\varepsilon_0\varepsilon_r\left|\vec{E}_T\right|^2 A \cdot l = \frac{1}{2}\varepsilon_0\varepsilon_r A \cdot l \cdot (\alpha_2^2 + \alpha_3^2) \qquad (15.11)$$

For electric potential energy in an element to be minimum,

$$\frac{\partial U_T}{\partial \phi_1} = \frac{1}{2}\varepsilon_0\varepsilon_r A \cdot l \cdot 2\left(\alpha_2 \frac{\partial \alpha_2}{\partial \phi_1} + \alpha_3 \frac{\partial \alpha_3}{\partial \phi_1}\right) = 0 \qquad (15.12)$$

Equation 15.12 is to be applied to every node where the unknown potential is to be determined. It may be noted here that the node under consideration may belong to more than one element. Then Equation 15.12 is to be applied for all such elements considering the node under consideration as node 1 and the other two nodes of the element being node 2 and node 3 taken in the anti-clockwise direction.

Now,

$$\frac{\partial \alpha_2}{\partial \phi_1} = \frac{y_2 - y_3}{D}, \frac{\partial \alpha_3}{\partial \phi_1} = \frac{x_3 - x_2}{D} \text{ and } A = D/2$$

Therefore, from Equation 15.12

$$\frac{1}{2}\varepsilon_0\varepsilon_r\frac{D}{2}\cdot l\cdot 2\left(\alpha_2\frac{y_2-y_3}{D}+\alpha_3\frac{x_3-x_2}{D}\right)=0 \tag{15.13}$$

$$\text{or, } \frac{1}{2}\varepsilon_0\varepsilon_r\cdot l\cdot\left[\alpha_2(y_2-y_3)+\alpha_3(x_3-x_2)\right]=0$$

Hence, from Equations 15.7, 15.9 and 15.13

$$\frac{\varepsilon_0\varepsilon_r l}{2D}\{[\phi_1(y_2-y_3)+\phi_2(y_3-y_1)+\phi_3(y_1-y_2)][y_2-y_3] \tag{15.14a}$$
$$+[\phi_1(x_3-x_2)+\phi_2(x_1-x_3)+\phi_3(x_2-x_1)][x_3-x_2]\}=0$$

$$\text{or, } \frac{\varepsilon_r}{D}\{\phi_1\left[(x_3-x_2)^2+(y_2-y_3)^2\right]+\phi_2\left[(x_3-x_2)(x_1-x_3)+(y_2-y_3)(y_3-y_1)\right]$$
$$+\phi_3\left[(x_3-x_2)(x_2-x_1)+(y_2-y_3)(y_1-y_2)\right]\}=0 \tag{15.14b}$$

$$\text{or, } K_{1T}\phi_1+K_{2T}\phi_2+K_{3T}\phi_3=0 \tag{15.15}$$

where:

$$K_{1T}=\frac{\varepsilon_{rT}}{D_T}\left[(x_{3T}-x_{2T})^2+(y_{2T}-y_{3T})^2\right]$$

$$K_{2T}=\frac{\varepsilon_{rT}}{D_T}[(x_{3T}-x_{2T})(x_{1T}-x_{3T})+(y_{2T}-y_{3T})(y_{3T}-y_{1T})] \tag{15.16}$$

$$K_{3T}=\frac{\varepsilon_{rT}}{D_T}[(x_{3T}-x_{2T})(x_{2T}-x_{1T})+(y_{2T}-y_{3T})(y_{1T}-y_{2T})]$$

where:
Subscript $T$ denotes the element number
$D_T$ is twice the area of the element, as given by Equation 15.8
$\varepsilon_{rT}$ is the permittivity of the dielectric within the element

Discretization using triangular elements is usually done is such a way that one particular node is connected to either six other nodes in hexagonal connectivity, as shown in Figure 15.3a, or to eight other nodes in octagonal

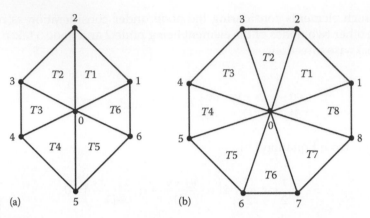

**FIGURE 15.3**
Nodal connectivity – (a) 6 elements (hexagonal) and (b) 8 elements (octagonal).

connectivity, as shown in Figure 15.3b. For hexagonal connectivity, an equation may be formed involving potentials of all the six nodes surrounding the node '0' applying Equation 15.15. In such case, for every element, node 0 of Figure 15.3 is considered to be node 1 of Equation 15.15 and the other two nodes are considered to be node 2 and node 3 in the anti-clockwise direction. Application of Equation 15.15 thus results in six simultaneous linear equations, the summation of which may be represented as follows.

$$F_1\phi_1 + F_2\phi_2 + F_3\phi_3 + F_4\phi_4 + F_5\phi_5 + F_6\phi_6 + F_0\phi_0 = 0 \qquad (15.17)$$

where:

$$F_1 = K_{2(T1)} + K_{3(T6)}$$

$$F_2 = K_{2(T2)} + K_{3(T1)}$$

$$F_3 = K_{2(T3)} + K_{3(T2)}$$

$$F_4 = K_{2(T4)} + K_{3(T3)}$$

$$F_5 = K_{2(T5)} + K_{3(T4)}$$

$$F_6 = K_{2(T6)} + K_{3(T5)}$$

$$(15.18)$$

$$\text{and } F_0 = \sum_{T=1}^{6} K_{1T}$$

Application of Equation 15.17 to all the nodes having unknown potential will generate the FEM system of simultaneous linear equations, which needs to be solved for determining the node potentials. Equations 15.17 and 15.18 could be suitably modified for octagonal nodal connectivity. Here, it may also be noted that FEM formulation as described above automatically takes into account the unequal elemental sizes as the coefficients as in Equation 15.18 are all computed in terms of nodal coordinates that may have any numerical value.

## 15.4.2 FEM Formulation in 2D System with Multi-Dielectric Media

For computing electric field in multi-dielectric media, triangular elements are so positioned that any given triangular element comprises only one dielectric medium. In other words, a set of nodal points are to be placed on the interface between two dielectrics, as shown in Figure 15.4. Hence, the coefficients $K_{1T}$, $K_{2T}$ and $K_{3T}$ for any node are to be calculated depending on its nodal position (i.e., 1, 2 or 3) in an element considering the proper value of $\varepsilon_r$.

While applying Equation 15.17 for the nodal connectivity shown in Figure 15.4, the following modifications need to be made for $F_2$, $F_5$ and $F_0$ keeping the others unchanged.

$$F_2 = K_{2(T2)} + K_{3(T1)}, \; F_5 = K_{2(T5)} + K_{3(T4)} \text{ and } F_0 = \sum_{T=1}^{6} K_{1T}$$

where:

$$K_{2(T2)} = \frac{\varepsilon_{r2}}{D_{T2}}[(x_3 - x_2)(x_0 - x_3) + (y_2 - y_3)(y_3 - y_0)]$$

$$K_{3(T1)} = \frac{\varepsilon_{r1}}{D_{T1}}[(x_2 - x_1)(x_1 - x_0) + (y_1 - y_2)(y_0 - y_1)]$$

$$K_{2(T5)} = \frac{\varepsilon_{r1}}{D_{T5}}[(x_6 - x_5)(x_0 - x_6) + (y_5 - y_6)(y_6 - y_0)]$$

$$K_{3(T4)} = \frac{\varepsilon_{r2}}{D_{T4}}[(x_5 - x_4)(x_4 - x_0) + (y_4 - y_5)(y_0 - y_4)]$$

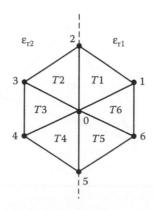

**FIGURE 15.4**
Elemental discretization for multi-dielectric media.

For the computation of $F_0$, $K_{1T}$ is to be calculated considering $\varepsilon_{r1}$ for the elements 1, 5 and 6 and considering $\varepsilon_{r2}$ for the elements 2, 3 and 4, using Equation 15.16. For example, for elements $T3$ and $T6$, respectively, the expressions for $K_{1T}$ will be as follows:

$$K_{1(T3)} = \frac{\varepsilon_{r2}}{D_{T3}}\left[(x_4 - x_3)^2 + (y_3 - y_4)^2\right]$$

$$\text{and } K_{1(T6)} = \frac{\varepsilon_{r1}}{D_{T6}}\left[(x_1 - x_6)^2 + (y_6 - y_1)^2\right]$$

Here, it may be noted that no separate formulation is required for multi-dielectric media in FEM in contrast to FDM.

### 15.4.3 FEM Formulation in Axi-Symmetric System

As already discussed, the electric potential energy in a triangular element is

$$U_e = \frac{1}{2}\varepsilon_0\varepsilon_r\left|\vec{E}\right|^2 \cdot A \cdot l \tag{15.19}$$

where:
  $(A \cdot l)$ is the volume of the element

For an axi-symmetric system, this volume is created due to the rotation of a triangular element around the axis of symmetry (Figure 15.5). The area of the triangle being $A$, $l$ should then be the mean length of rotation, that is, $2\pi$ times the radial distance of the centroid of the triangle.

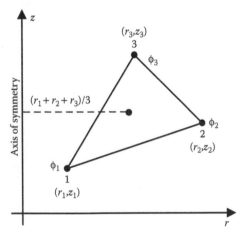

**FIGURE 15.5**
Triangular element for an axi-symmetric formulation.

Therefore,

$$l = 2\pi \frac{(r_1 + r_2 + r_3)}{3} \tag{15.20}$$

Putting this expression for $l$ in Equation 15.14b

$$\frac{\varepsilon_0 \varepsilon_r}{2D} \cdot \frac{2\pi(r_1 + r_2 + r_3)}{3} \left\{ \phi_1 \left[ (r_3 - r_2)^2 + (z_2 - z_3)^2 \right] \right.$$

$$+ \phi_2 [(r_3 - r_2)(r_1 - r_3) + (z_2 - z_3)(z_3 - z_1)] \tag{15.21}$$

$$\left. + \phi_3 [(r_3 - r_2)(r_2 - r_1) + (z_2 - z_3)(z_1 - z_2)] \right\} = 0$$

Therefore, from the above equation in an axi-symmetric system,

$$K_{1T}\phi_1 + K_{2T}\phi_2 + K_{3T}\phi_3 = 0$$

where:

$$K_{1T} = \frac{R\varepsilon_{rT}}{D_T} \left[ (r_{3T} - r_{2T})^2 + (z_{2T} - z_{3T})^2 \right]$$

$$K_{2T} = \frac{R\varepsilon_{rT}}{D_T} \left[ (r_{3T} - r_{2T})(r_{1T} - r_{3T}) + (z_{2T} - z_{3T})(z_{3T} - z_{1T}) \right] \tag{15.22}$$

$$K_{3T} = \frac{R\varepsilon_{rT}}{D_T} \left[ (r_{3T} - r_{2T})(r_{2T} - r_{1T}) + (z_{2T} - z_{3T})(z_{1T} - z_{2T}) \right]$$

$$\text{and } R = (r_1 + r_2 + r_3)$$

For the axi-symmetric system with multi-dielectric media, the modifications to be brought in are the same as those described for a 2D formulation discussed in Section 15.4.2.

### 15.4.4 Shape Function, Global and Natural Coordinates

Solution of FEM equations gives the field variables at the nodes, but the field variables at different points inside the element are also needed. For this purpose, shape functions are used to interpolate the nodal values of field variables to compute the corresponding values at arbitrary points inside the elements.

Consider that one vertex of the triangle is moved holding the other two fixed. Then both the deformed and non-deformed triangles can be drawn on top of one another, as shown in Figure 15.6a.

Assuming the deformation to be linear, the areas of the deformed triangles can be calculated from the base and height of the triangle. As the base remains constant, the area is a linear function of the height, that is, the displacement of the node within the triangle. As all the three nodes could be displaced, so three linear functions need to be written to describe the displacement of an

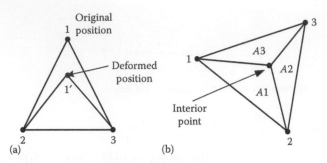

**FIGURE 15.6**
(a) Triangle in non-deformed and deformed states and (b) interpolation at an interior point.

interior point due to the displacement of each of the nodes. The displacement of the interior point will be computed by summing the displacement due to each of the three nodes of the triangle.

As shown in Figure 15.6b, the interior point divides the triangle into three sub-triangles. All three nodes may move and hence the motion of the interior point is some combination of their displacements. Let $A1$, $A2$ and $A3$ be the area of each of sub-triangles and $A$ be the total area of the triangular element. Then $A = A1 + A2 + A3$.

Therefore, the shape functions are defined as follows:

$$L_1 = \frac{A1}{A}, L_2 = \frac{A2}{A} \text{ and } L_3 = \frac{A3}{A} \tag{15.23}$$

In terms of the nodal coordinates, the shape functions are given by

$$L_1 = \frac{1}{2A}\left[(x_1y_2 - x_2y_1)+(y_1 - y_2)x+(x_2-x_1)y\right]$$

$$L_2 = \frac{1}{2A}\left[(x_2y_3 - x_3y_2)+(y_2 - y_3)x+(x_3-x_2)y\right]$$

$$L_3 = \frac{1}{2A}\left[(x_3y_1 - x_1y_3)+(y_3 - y_1)x+(x_1-x_3)y\right] \tag{15.24}$$

$$A = \frac{1}{2}[(x_2y_3 - x_3y_2)+(x_3y_1 - x_1y_3)+(x_1y_2 - x_2y_1)] = \frac{1}{2}\det\begin{vmatrix} 1 & x_1 & y_1 \\ 1 & x_2 & y_2 \\ 1 & x_3 & y_3 \end{vmatrix}$$

The same shape functions can be used to compute the coordinates of a point interior to the triangle:

$$x = L_1x_1 + L_2x_2 + L_3x_3$$
$$y = L_1y_1 + L_2y_2 + L_3y_3 \tag{15.25}$$

From Equation 15.25, it is obvious that the values of the shape functions for node 1 are (1,0,0), for node 2 are (0,1,0) and for node 3 are (0,0,1). The shape functions are not independent of one another because $L_1 + L_2 + L_3 = 1$. Hence, if two shape functions are known, then it is possible to compute the third. Introducing the natural coordinates within the triangle $(\xi, \eta)$ such that

$$L_1 = \xi, L_2 = \eta$$

Then                                                                                    (15.26)

$$L_3 = 1 - \xi - \eta$$

The representation of natural coordinates is shown in Figure 15.7.

A local coordinate system that relies on the element geometry for its definition and whose coordinates range between zero and unity within the element is known as the *natural coordinate system*. Natural coordinate systems have the property that one particular coordinate has unit value at one node of the element and zero value at the other nodes. Use of natural coordinates is advantageous in deriving interpolation functions and also in the development of curved-sided elements.

Substituting $\xi$ and $\eta$ in Equation 15.25, the relationship between the global coordinates $(x,y)$ and the natural coordinates $(\xi,\eta)$ can be obtained as given below:

$$x = (x_1 - x_3)\xi + (x_2 - x_3)\eta + x_3$$
$$y = (y_1 - y_3)\xi + (y_2 - y_3)\eta + y_3$$                                              (15.27a)

The above equation can be used to compute the natural coordinates. If any point within the triangle is given, then the coordinates of the vertices $(x_1,y_1)$, $(x_2,y_2)$ and $(x_3,y_3)$ are known along with the coordinate of the given point $(x,y)$. Therefore, Equation 15.27 can be solved for $\xi$ and $\eta$. Subsequently, if the field variable, say

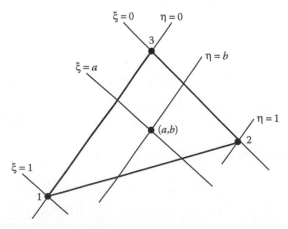

**FIGURE 15.7**
Representation of natural coordinates.

electric potential, is known at the nodes, then the natural coordinates can be used to compute the field variable for the point at $(x,y)$. For example,

$$\phi(x,y) = (\phi_1 - \phi_3)\xi + (\phi_2 - \phi_3)\eta + \phi_3 \qquad (15.27\text{b})$$

### 15.4.5 Derivation of Field Variables Using Natural Coordinates

The potential within the element can therefore be viewed as a function of $(x,y)$ or $(\xi,\eta)$. Therefore, using the chain rule of derivatives,

$$\begin{Bmatrix} \dfrac{\partial\phi}{\partial\xi} \\[2mm] \dfrac{\partial\phi}{\partial\eta} \end{Bmatrix} = \begin{bmatrix} \dfrac{\partial x}{\partial\xi} & \dfrac{\partial y}{\partial\xi} \\[2mm] \dfrac{\partial x}{\partial\eta} & \dfrac{\partial y}{\partial\eta} \end{bmatrix} \begin{Bmatrix} \dfrac{\partial\phi}{\partial x} \\[2mm] \dfrac{\partial\phi}{\partial y} \end{Bmatrix} = J \begin{Bmatrix} \dfrac{\partial\phi}{\partial x} \\[2mm] \dfrac{\partial\phi}{\partial y} \end{Bmatrix} \qquad (15.28)$$

where:
  $J$ is the Jacobian matrix of the transformation

From Equation 15.27a, the Jacobian can be obtained as

$$J = \begin{bmatrix} (x_1 - x_3) & (y_1 - y_3) \\ (x_2 - x_3) & (y_2 - y_3) \end{bmatrix} \quad \text{and} \quad J^{-1} = \frac{1}{2A}\begin{bmatrix} (y_2 - y_3) & -(y_1 - y_3) \\ -(x_2 - x_3) & (x_1 - x_3) \end{bmatrix} \qquad (15.29)$$

where:

$$\det J = (x_1 - x_3)(y_2 - y_3) - (x_2 - x_3)(y_1 - y_3) \qquad (15.30)$$
$$= 2(\text{area of the triangular element})$$

From Equations 15.28 and 15.29,

$$\begin{Bmatrix} \dfrac{\partial\phi}{\partial x} \\[2mm] \dfrac{\partial\phi}{\partial y} \end{Bmatrix} = J^{-1}\begin{Bmatrix} \dfrac{\partial\phi}{\partial\xi} \\[2mm] \dfrac{\partial\phi}{\partial\eta} \end{Bmatrix} = \frac{1}{2A}\begin{bmatrix} (y_2 - y_3) & -(y_1 - y_3) \\ -(x_2 - x_3) & (x_1 - x_3) \end{bmatrix}\begin{Bmatrix} \dfrac{\partial\phi}{\partial\xi} \\[2mm] \dfrac{\partial\phi}{\partial\eta} \end{Bmatrix} \qquad (15.31)$$

From Equations 15.27b and 15.31,

$$\begin{Bmatrix} -E_x \\ -E_y \end{Bmatrix} = \frac{1}{D}\begin{bmatrix} (y_2 - y_3) & -(y_1 - y_3) \\ -(x_2 - x_3) & (x_1 - x_3) \end{bmatrix}\begin{bmatrix} (\phi_1 - \phi_3) \\ (\phi_2 - \phi_3) \end{bmatrix} \qquad (15.32)$$

where:
  $D = 2A$

From Equation 15.32

$$E_x = \frac{1}{D}[(y_1 - y_3)(\phi_2 - \phi_3) - (y_2 - y_3)(\phi_1 - \phi_3)]$$

$$\text{(15.33a)}$$

$$= \frac{1}{D}[\phi_1(y_3 - y_2) + \phi_2(y_1 - y_3) + \phi_3(y_2 - y_1)]$$

$$\text{and } E_y = \frac{1}{D}[(x_2 - x_3)(\phi_1 - \phi_3) - (x_1 - x_3)(\phi_2 - \phi_3)]$$

$$\text{(15.33b)}$$

$$= \frac{1}{D}[\phi_1(x_2 - y_3) + \phi_2(x_3 - x_1) + \phi_3(x_1 - x_2)]$$

Equations 15.33a and 15.33b thus give $\vec{E}_x = -\alpha_2$ and $\vec{E}_y = -\alpha_3$, which are same as those obtained in Equations 15.5, 15.7 and 15.9.

## PROBLEM 15.1

The nodal coordinates of a linear triangular element are (1,1), (3,2) and (2,3) mm, respectively. The potentials at the three nodes are as follows: node 1, 100 V; node 2, 150 V and node 3, 200 V. Calculate the potential and the electric field intensity components at the point $P(2,2)$ mm within the triangular element.

*Solution:*

$$A = \frac{1}{2}[(x_2 y_3 - x_3 y_2) + (x_3 y_1 - x_1 y_3) + (x_1 y_2 - x_2 y_1)]$$

$$= \frac{1}{2}[(3 \times 3 - 2 \times 2) + (2 \times 1 - 1 \times 3) + (1 \times 2 - 3 \times 1)] = \frac{3}{2}$$

$$L_1 = \frac{1}{2A}\left[(x_1 y_2 - x_2 y_1) + (y_1 - y_2)x + (x_2 - x_1)y\right]$$

$$= \frac{2}{2 \times 3}[(1 \times 2 - 3 \times 1) + (1 - 2)2 + (3 - 1)2] = \frac{1}{3}$$

$$L_2 = \frac{1}{2A}[(x_2 y_3 - x_3 y_2) + (y_2 - y_3)x + (x_3 - x_2)y]$$

$$= \frac{2}{2 \times 3}[(3 \times 3 - 2 \times 2) + (2 - 3)2 + (2 - 3)2] = \frac{1}{3}$$

$$L_3 = \frac{1}{2A}\left[(x_3 y_1 - x_1 y_3) + (y_3 - y_1)x + (x_1 - x_3)y\right]$$

$$= \frac{2}{2 \times 3}[(2 \times 1 - 1 \times 3) + (3 - 1)2 + (1 - 2)2] = \frac{1}{3}$$

Therefore,

$$\phi(2,2) = \frac{1}{3} \times 100 + \frac{1}{3} \times 150 + \frac{1}{3} \times 200 = 150 \text{ V}$$

$$\alpha_2 = \frac{1}{D}[\phi_1(y_2 - y_3) + \phi_2(y_3 - y_1) + \phi_3(y_1 - y_2)]$$

$$= \frac{1}{2A}[100(2 - 3) + 150(3 - 2) + 200(1 - 2)]$$

$$= -50 \text{ V/mm}$$

$$\alpha_3 = \frac{1}{D}[\phi_1(x_3 - x_2) + \phi_2(x_1 - x_3) + \phi_3(x_2 - x_1)]$$

$$= \frac{1}{2A}[100(2 - 3) + 150(1 - 2) + 200(3 - 1)]$$

$$= 50 \text{ V/mm}$$

Hence,

$$\vec{E}_x(2,2) = -\alpha_2 = 50 \text{ V/mm and } \vec{E}_y(2,2) = -\alpha_3 = -50 \text{ V/mm}$$

It is to be noted here that the nodes 1(1,1), 2(3,2) and 3(2,3) are taken in the anti-clockwise direction. If these three nodes are taken in the clockwise direction, for example, nodes 1(1,1), 2(2,3) and 3(3,2), then the area of the triangle becomes negative. This is why the nodes of a triangular element have to be taken always in the anti-clockwise direction.

### 15.4.6 Other Types of Elements for 2D and Axi-Symmetric Systems

#### 15.4.6.1 Quadratic Triangular Element

There are six nodes on this element, as shown in Figure 15.8. Three nodes are at the three vertices and three nodes are at the middle of the three sides of the triangle.

Electric potential is assumed to be a quadratic function of global coordinates $(x,y)$ within the element such that

$$\phi = \alpha_1 + \alpha_2 x + \alpha_3 y + \alpha_4 x^2 + \alpha_5 xy + \alpha_6 y^2 \tag{15.34}$$

Thus, the electric field intensity components are as follows:

$$E_x = -\frac{\partial \phi}{\partial x} = -(\alpha_2 + 2\alpha_4 x + \alpha_5 y) \text{ and } E_y = -\frac{\partial \phi}{\partial y} = -(\alpha_3 + 2\alpha_6 y + \alpha_5 x) \tag{15.35}$$

Thus, the electric field intensity varies linearly within the triangular element. Hence, this type of element is known *linear stress triangle*, which gives better results than the CST.

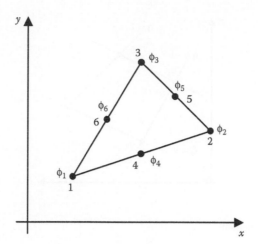

**FIGURE 15.8**
Quadratic triangular element.

In terms of the natural coordinates, the six shape functions for this element are as follows:

$$L_1 = \xi(2\xi - 1); \; L_2 = \eta(2\eta - 1); \; L_3 = \zeta(2\zeta - 1);$$
$$L_4 = 4\xi\eta; \; L_5 = 4\eta\zeta; \; L_6 = 4\zeta\xi \tag{15.36}$$

Where:

$$\zeta = 1 - \xi - \eta$$

Electric potential function within the element is written as

$$\phi = L_1\phi_1 + L_2\phi_2 + L_3\phi_3 + L_4\phi_4 + L_5\phi_5 + L_6\phi_6 \tag{15.37}$$

### 15.4.6.2 Linear Quadrilateral Element

In this element, there are four nodes at the four corners of a quadrilateral, as shown in Figure 15.9. In the natural coordinate system $(\xi, \eta)$, the four shape functions are given as follows.

$$L_1 = \frac{1}{4}(1 - \xi)(1 - \eta); \; L_2 = \frac{1}{4}(1 + \xi)(1 - \eta)$$
$$L_3 = \frac{1}{4}(1 + \xi)(1 + \eta); \; L_4 = \frac{1}{4}(1 - \xi)(1 + \eta) \tag{15.38}$$

At any point inside the element, $L_1 + L_2 + L_3 + L_4 = 0$.

In a generalized way, the shape function can be written as

$$L_i = \frac{1}{4}(1 + \xi\xi_i)(1 + \eta\eta_i), \; i = 1, \ldots, 4$$

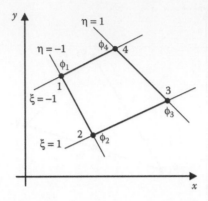

**FIGURE 15.9**
Linear quadrilateral element.

The electric potential within the element varies as

$$\phi = L_1\phi_1 + L_2\phi_2 + L_3\phi_3 + L_4\phi_4 \tag{15.39}$$

### 15.4.6.3 Quadratic Quadrilateral Element

There are eight nodes in this element, four nodes at the four corners of the quadrilateral and four nodes at the middle of the four sides, as shown in Figure 15.10.

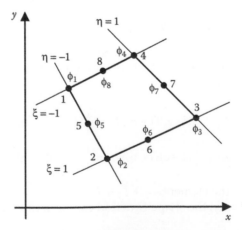

**FIGURE 15.10**
Quadratic quadrilateral element.

The eight shape functions can be represented in natural coordinates as follows:

$$L_1 = \frac{1}{4}(1-\xi)(\eta-1)(\xi+\eta+1); \; L_2 = \frac{1}{4}(1+\xi)(\eta-1)(\eta-\xi+1)$$

$$L_3 = \frac{1}{4}(1+\xi)(1+\eta)(\xi+\eta-1); \; L_4 = \frac{1}{4}(\xi-1)(\eta+1)(\xi-\eta+1)$$

(15.40)

$$L_5 = \frac{1}{2}(1-\xi^2)(1-\eta); \; L_6 = \frac{1}{2}(1+\xi)(1-\eta^2)$$

$$L_7 = \frac{1}{2}(1-\xi^2)(1+\eta); \; L_8 = \frac{1}{2}(1-\xi)(1-\eta^2)$$

At any point inside the element, $\sum_{i=1}^{8} L_i = 0$.

The electric potential function within the element is given by $\phi = \sum_{i=1}^{8} L_i \phi_i$.

For an axi-symmetric system, all the elements resemble a ring having its axis same as the axis of symmetry of the system, whose cross section will be as per the figures shown in the respective subsections hereabove.

## PROBLEM 15.2

For the 2D configuration having a single-dielectric medium, as shown in Figure 15.11, write the FEM equations for the nodes having unknown node potentials. Consider the finite elements to be linear triangular elements.

*Solution:*
Out of the 12 nodes shown in Figure 15.11, the potential of only two nodes are unknown, namely, nodes 6 and 7. Both these nodes belong to six elements in the mesh, as shown in Figure 15.11. Assuming the coordinate of node 9 to

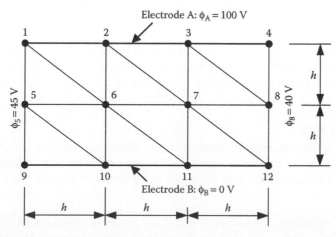

**FIGURE 15.11**
Pertaining to Problem 15.2.

be (0,0) and considering $h = 1$, the coordinates of the other nodes are determined accordingly.

For node 6:

$$F_1\phi_1 + F_2\phi_5 + F_3\phi_{10} + F_4\phi_{11} + F_5\phi_7 + F_6\phi_2 + F_0\phi_6 = 0$$

Where:

$\phi_1 = 100$ V, $\phi_5 = 45$ V, $\phi_{10} = 0$ V, $\phi_{11} = 0$ V, $\phi_7 = ?$, $\phi_2 = 100$ V, $\phi_6 = ?$

$$F_1 = K_{2(T-6-1-5)} + K_{3(T-6-2-1)}; \ F_2 = K_{2(T-6-5-10)} + K_{3(T-6-1-5)}$$

$$F_3 = K_{2(T-6-10-11)} + K_{3(T-6-5-10)}; F_4 = K_{2(T-6-11-7)} + K_{3(T-6-10-11)}$$

$$F_5 = K_{2(T-6-7-2)} + K_{3(T-6-11-7)}; \ F_6 = K_{2(T-6-2-1)} + K_{3(T-6-7-2)}$$

$$\text{and} \ \ F_0 = K_{1(T-6-1-5)} + K_{1(T-6-5-10)} + K_{1(T-6-10-11)} + K_{1(T-6-11-7)}$$

$$+ K_{1(T-6-7-2)} + K_{1(T-6-2-1)}$$

For node 7:

$$F_1\phi_2 + F_2\phi_6 + F_3\phi_{11} + F_4\phi_{12} + F_5\phi_8 + F_6\phi_3 + F_0\phi_7 = 0$$

Where:

$\phi_2 = 100$ V, $\phi_6 = ?$, $\phi_{11} = 0$ V, $\phi_{12} = 0$ V, $\phi_8 = 40$ V, $\phi_3 = 100$ V, $\phi_7 = ?$

$$F_1 = K_{2(T-7-2-6)} + K_{3(T-7-3-2)}; \ F_2 = K_{2(T-7-6-11)} + K_{3(T-7-2-6)}$$

$$F_3 = K_{2(T-7-11-12)} + K_{3(T-7-6-11)}; \ F_4 = K_{2(T-7-12-8)} + K_{3(T-7-11-12)}$$

$$F_5 = K_{2(T-7-8-3)} + K_{3(T-7-12-8)}; \ F_6 = K_{2(T-7-3-2)} + K_{3(T-7-8-3)}$$

$$\text{and} \ \ F_0 = K_{1(T-7-2-6)} + K_{1(T-7-6-11)} + K_{1(T-7-11-12)} + K_{1(T-7-12-8)}$$

$$+ K_{1(T-7-8-3)} + K_{1(T-7-3-2)}$$

From the above two equations $\phi_6$ and $\phi_7$ can be solved. The coefficients are to be computed using the expressions given in Equation 15.16. In this problem, the value of $\varepsilon_r$ is not specified and hence it may be taken as 1 for all the elements as there is only one dielectric medium to be considered.

**PROBLEM 15.3**

For the 2D configuration having two dielectric media, as shown in Figure 15.12, write the FEM equations for the nodes having unknown node potentials. Consider the finite elements to be linear triangular elements.

*Solution:*

The same equations for nodes 6 and 7, as given in Problem 15.2, are to be solved here for determining the values of $\phi_6$ and $\phi_7$. However, the values of $\varepsilon_r$ are to be chosen in the following manner for different elements:

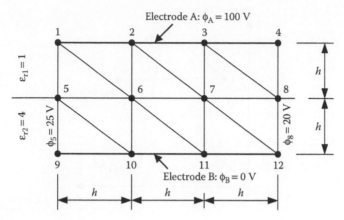

**FIGURE 15.12**
Pertaining to Problem 15.3.

$\varepsilon_{r1} = 1$ for the elements $(T\text{-}6\text{-}1\text{-}5)$, $(T\text{-}6\text{-}2\text{-}1)$, $(T\text{-}7\text{-}2\text{-}6)$, $(T\text{-}7\text{-}3\text{-}2)$ and $(T\text{-}7\text{-}8\text{-}3)$ and

$\varepsilon_{r2} = 4$ for the elements $(T\text{-}6\text{-}5\text{-}10)$, $(T\text{-}6\text{-}10\text{-}11)$, $(T\text{-}7\text{-}6\text{-}11)$, $(T\text{-}7\text{-}11\text{-}12)$ and $(T\text{-}7\text{-}12\text{-}8)$

### 15.4.7 FEM Formulation in 3D System

The potential energy in a 3D electric field is given by

$$U_{\text{total}} = \iiint_V \frac{1}{2} \varepsilon_0 \varepsilon_r \left| \vec{E} \right|^2 dV \tag{15.41}$$

$$\text{or, } U_{\text{total}} = \iiint_V \frac{1}{2} \varepsilon_0 \varepsilon_r \left| -\nabla \phi \right|^2 dV \tag{15.42}$$

To apply FEM, the region of interest is to be discretized by solid finite elements. For $N$ number of solid elements, the total potential energy can then be stated as follows:

$$U_{\text{total}} = \sum_{e=1}^{N} U(e)$$

To minimize the total potential energy, $U$, of the entire region of interest, $U(e)$ must be minimized for each solid element.

The simplest three-dimensional (3D) solid element is the linear tetrahedron element, as shown in Figure 15.13. For this element, there are four nodes at the four corners of the tetrahedron, which are numbered in such a way that the first three nodes are arranged in the anti-clockwise direction when viewed from node 4, for example, 1, 2 and 3 nodes of Figure 15.13 are arranged in the anti-clockwise direction when viewed from node 4.

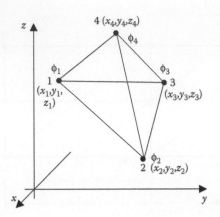

**FIGURE 15.13**
Linear tetrahedral element.

Electric potential $\phi$ is assumed to be varying linearly within the element such that

$$\phi = \alpha_1 + \alpha_2 x + \alpha_3 y + \alpha_4 z \tag{15.43}$$

Hence,

$$E_x = -\frac{\partial \phi}{\partial x} = -\alpha_2, E_y = -\frac{\partial \phi}{\partial y} = -\alpha_3 \text{ and } E_z = -\frac{\partial \phi}{\partial z} = -\alpha_4 \tag{15.44}$$

Electric field intensity components are constant within a linear tetrahedral element. Hence, it is called a CST in high-voltage (HV) field computation.

Again, the potentials at the four corners of the element are given by

$$\phi_1 = \alpha_1 + \alpha_2 x_1 + \alpha_3 y_1 + \alpha_4 z_1$$
$$\phi_2 = \alpha_1 + \alpha_2 x_2 + \alpha_3 y_2 + \alpha_4 z_2$$
$$\phi_3 = \alpha_1 + \alpha_2 x_3 + \alpha_3 y_3 + \alpha_4 z_3 \tag{15.45}$$
$$\phi_4 = \alpha_1 + \alpha_2 x_4 + \alpha_3 y_4 + \alpha_4 z_4$$

Hence,

$$\alpha_2 = \frac{\begin{vmatrix} 1 & \phi_1 & y_1 & z_1 \\ 1 & \phi_2 & y_2 & z_2 \\ 1 & \phi_3 & y_3 & z_3 \\ 1 & \phi_4 & y_4 & z_4 \end{vmatrix}}{\begin{vmatrix} 1 & x_1 & y_1 & z_1 \\ 1 & x_2 & y_2 & z_2 \\ 1 & x_3 & y_3 & z_3 \\ 1 & x_4 & y_4 & z_4 \end{vmatrix}} = \frac{\Delta_2}{\Delta} \tag{15.46}$$

where:

$\Delta$ = six times the volume of the element and

$$\Delta_2 = -\phi_1 \begin{vmatrix} 1 & y_2 & z_2 \\ 1 & y_3 & z_3 \\ 1 & y_4 & z_4 \end{vmatrix} + \phi_2 \begin{vmatrix} 1 & y_1 & z_1 \\ 1 & y_3 & z_3 \\ 1 & y_4 & z_4 \end{vmatrix} - \phi_3 \begin{vmatrix} 1 & y_1 & z_1 \\ 1 & y_2 & z_2 \\ 1 & y_4 & z_4 \end{vmatrix} + \phi_4 \begin{vmatrix} 1 & y_1 & z_1 \\ 1 & y_2 & z_2 \\ 1 & y_3 & z_3 \end{vmatrix} \tag{15.47}$$

$$= -\phi_1 \Delta_{1yz} + \phi_2 \Delta_{2yz} - \phi_3 \Delta_{3yz} + \phi_4 \Delta_{4yz}$$

where:

$\Delta_{1yz}, \Delta_{2yz}, \Delta_{3yz}$ and $\Delta_{4yz}$ are two times the area of triangles opposite to the nodes 1, 2, 3 and 4, respectively, when these triangles are projected to the $y$–$z$ plane

Therefore,

$$\alpha_2 = \frac{-\phi_1 \Delta_{1yz} + \phi_2 \Delta_{2yz} - \phi_3 \Delta_{3yz} + \phi_4 \Delta_{4yz}}{\Delta} \tag{15.48}$$

Similarly,

$$\alpha_3 = \frac{\phi_1 \Delta_{1zx} - \phi_2 \Delta_{2zx} + \phi_3 \Delta_{3zx} - \phi_4 \Delta_{4zx}}{\Delta} \text{ and}$$

$$\alpha_4 = \frac{-\phi_1 \Delta_{1xy} + \phi_2 \Delta_{2xy} - \phi_3 \Delta_{3xy} + \phi_4 \Delta_{4xy}}{\Delta} \tag{15.49}$$

where:

$\Delta_{1zx}, \Delta_{2zx}, \Delta_{3zx}$ and $\Delta_{4zx}$ are two times the area of triangles opposite to the nodes 1, 2, 3 and 4, respectively, when these triangles are projected to the $z$–$x$ plane

$\Delta_{1xy}, \Delta_{2xy}, \Delta_{3xy}$ and $\Delta_{4xy}$ are two times the area of triangles opposite to the nodes 1, 2, 3 and 4, respectively, when these triangles are projected to the $x$–$y$ plane

Now, electric potential energy in a tetrahedral element is given by

$$U_e = \frac{1}{2} \varepsilon_0 \varepsilon_r \left| \vec{E} \right|^2 V = \frac{1}{2} \varepsilon_0 \varepsilon_r V \left( E_x^2 + E_y^2 + E_z^2 \right)$$

$$= \frac{1}{2} \varepsilon_0 \varepsilon_r V \left( \alpha_2^2 + \alpha_3^2 + \alpha_4^2 \right) \tag{15.50}$$

For electric potential energy in an element to be minimum,

$$\frac{\partial U_e}{\partial \phi_1} = \frac{1}{2} \varepsilon_0 \varepsilon_r V \times 2 \left( \alpha_2 \frac{\partial \alpha_2}{\partial \phi_1} + \alpha_3 \frac{\partial \alpha_3}{\partial \phi_1} + \alpha_4 \frac{\partial \alpha_4}{\partial \phi_1} \right) = 0 \tag{15.51}$$

$$\text{or, } \varepsilon_0 \varepsilon_r V \left( \alpha_2 \frac{\partial \alpha_2}{\partial \phi_1} + \alpha_3 \frac{\partial \alpha_3}{\partial \phi_1} + \alpha_4 \frac{\partial \alpha_4}{\partial \phi_1} \right) = 0 \tag{15.52}$$

In Equation 15.52

$$\frac{\partial \alpha_2}{\partial \phi_1} = -\frac{\Delta_{1yz}}{\Delta}, \frac{\partial \alpha_3}{\partial \phi_1} = \frac{\Delta_{1zx}}{\Delta} \text{ and } \frac{\partial \alpha_4}{\partial \phi_1} = -\frac{\Delta_{1xy}}{\Delta} \tag{15.53}$$

Therefore, from Equations 15.48, 15.49, 15.52 and 15.53

$$\varepsilon_0 \varepsilon_r \frac{\Delta}{6} \left[ -\frac{\Delta_{1yz}}{\Delta} \left( \frac{-\Delta_{1yz}\phi_1 + \Delta_{2yz}\phi_2 - \Delta_{3yz}\phi_3 + \Delta_{4yz}\phi_4}{\Delta} \right) \right.$$
$$+ \frac{\Delta_{1zx}}{\Delta} \left( \frac{\Delta_{1zx}\phi_1 - \Delta_{2zx}\phi_2 + \Delta_{3zx}\phi_3 - \Delta_{4zx}\phi_4}{\Delta} \right) \tag{15.54}$$
$$\left. - \frac{\Delta_{1xy}}{\Delta} \left( \frac{-\Delta_{1xy}\phi_1 + \Delta_{2xy}\phi_2 - \Delta_{3xy}\phi_3 + \Delta_{4xy}\phi_4}{\Delta} \right) \right] = 0$$

or,

$$\frac{\varepsilon_r}{6\,\Delta} \left\{ \phi_1 \left[ (\Delta_{1yz})^2 + (\Delta_{1zx})^2 + (\Delta_{1xy})^2 \right] \right.$$
$$- \phi_2 [\Delta_{1yz}\Delta_{2yz} + \Delta_{1zx}\Delta_{2zx} + \Delta_{1xy}\Delta_{2xy}]$$
$$+ \phi_3 [\Delta_{1yz}\Delta_{3yz} + \Delta_{1zx}\Delta_{3zx} + \Delta_{1xy}\Delta_{3xy}] \tag{15.55}$$
$$\left. - \phi_4 [\Delta_{1yz}\Delta_{4yz} + \Delta_{1zx}\Delta_{4zx} + \Delta_{1xy}\Delta_{4xy}] \right\} = 0$$

Equation 15.55 can be represented as

$$k_{1e}\phi_1 + k_{2e}\phi_2 + k_{3e}\phi_3 + k_{4e}\phi_4 = 0 \tag{15.56}$$

where:
subscript $e$ denotes the element number and

$$k_{1e} = \frac{\varepsilon_{re}}{6\,\Delta_e} \left\{ [\Delta_{(1yz)e}]^2 + [\Delta_{(1zx)e}]^2 + [\Delta_{(1xy)e}]^2 \right\}$$

$$k_{2e} = -\frac{\varepsilon_{re}}{6\,\Delta_e} [\Delta_{(1yz)e}\Delta_{(2yz)e} + \Delta_{(1zx)e}\Delta_{(2zx)e} + \Delta_{(1xy)e}\Delta_{(2xy)e}]$$

$$\tag{15.57}$$

$$k_{3e} = \frac{\varepsilon_{re}}{6\,\Delta_e} [\Delta_{(1yz)e}\Delta_{(3yz)e} + \Delta_{(1zx)e}\Delta_{(3zx)e} + \Delta_{(1xy)e}\Delta_{(3xy)e}]$$

$$k_{4e} = -\frac{\varepsilon_{re}}{6\,\Delta_e} [\Delta_{(1yz)e}\Delta_{(4yz)e} + \Delta_{(1zx)e}\Delta_{(4zx)e} + \Delta_{(1xy)e}\Delta_{(4xy)e}]$$

### 15.4.7.1 Natural Coordinates of Linear Tetrahedral Element

The natural coordinates and the shape functions of a linear tetrahedral element are described in terms of the volumes, as shown in Figure 15.14. Any point $P$ within the tetrahedral element (1-2-3-4) subdivides it into four sub-tetrahedra as shown. Then the shape functions are given by Equation 15.58,

$$L_1 = \frac{V(P-1-2-3)}{V(1-2-3-4)}, L_2 = \frac{V(P-1-2-4)}{V(1-2-3-4)}$$

$$L_3 = \frac{V(P-1-3-4)}{V(1-2-3-4)} \text{ and } L_4 = \frac{V(P-2-3-4)}{V(1-2-3-4)}$$

(15.58)

where:

$V$ denotes volume such that $L_1 + L_2 + L_3 + L_4 = 1$
Denoting, $L_1 = \xi$, $L_2 = \eta$, $L_3 = \zeta$, $L_4 = 1 - \xi - \eta - \zeta$

Then the natural coordinates $(\xi, \eta, \zeta)$ of the four corner nodes are as follows: node 1 (0,0,0), node 2 (0,0,1), node 3 (0,1,0) and node 4 (1,0,0), as shown in Figure 3.14.

Coordinate transformation within the element is done using the following expressions:

$$x = L_1 x_1 + L_2 x_2 + L_3 x_3 + L_4 x_4$$
$$y = L_1 y_1 + L_2 y_2 + L_3 y_3 + L_4 y_4$$
$$z = L_1 z_1 + L_2 z_2 + L_3 z_3 + L_4 z_4$$
$$L_1 + L_2 + L_3 + L_4 = 1$$

(15.59)

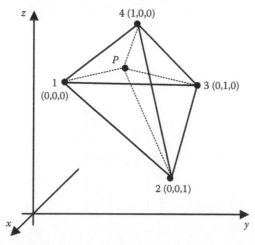

**FIGURE 15.14**
Natural coordinates of linear tetrahedral element.

The same shape functions are also used to describe potential function within the tetrahedral element, such that

$$\phi = L_1\phi_1 + L_2\phi_2 + L_3\phi_3 + L_4\phi_4 \tag{15.60}$$

Equation 15.59 can be solved to give

$$L_1 = \frac{1}{\Delta}\left\{ +\begin{vmatrix} x_2 & x_3 & x_4 \\ y_2 & y_3 & y_4 \\ z_2 & z_3 & z_4 \end{vmatrix} - x\begin{vmatrix} 1 & 1 & 1 \\ y_2 & y_3 & y_4 \\ z_2 & z_3 & z_4 \end{vmatrix} + y\begin{vmatrix} 1 & 1 & 1 \\ x_2 & x_3 & x_4 \\ z_2 & z_3 & z_4 \end{vmatrix} - z\begin{vmatrix} 1 & 1 & 1 \\ x_2 & x_3 & x_4 \\ y_2 & y_3 & y_4 \end{vmatrix}\right\}$$

$$L_2 = \frac{1}{\Delta}\left\{ -\begin{vmatrix} x_1 & x_3 & x_4 \\ y_1 & y_3 & y_4 \\ z_1 & z_3 & z_4 \end{vmatrix} + x\begin{vmatrix} 1 & 1 & 1 \\ y_1 & y_3 & y_4 \\ z_1 & z_3 & z_4 \end{vmatrix} - y\begin{vmatrix} 1 & 1 & 1 \\ x_1 & x_3 & x_4 \\ z_1 & z_3 & z_4 \end{vmatrix} + z\begin{vmatrix} 1 & 1 & 1 \\ x_1 & x_3 & x_4 \\ y_1 & y_3 & y_4 \end{vmatrix}\right\} \tag{15.61}$$

$$L_3 = \frac{1}{\Delta}\left\{ +\begin{vmatrix} x_1 & x_2 & x_4 \\ y_1 & y_2 & y_4 \\ z_1 & z_2 & z_4 \end{vmatrix} - x\begin{vmatrix} 1 & 1 & 1 \\ y_1 & y_2 & y_4 \\ z_1 & z_2 & z_4 \end{vmatrix} + y\begin{vmatrix} 1 & 1 & 1 \\ x_1 & x_2 & x_4 \\ z_1 & z_2 & z_4 \end{vmatrix} - z\begin{vmatrix} 1 & 1 & 1 \\ x_1 & x_2 & x_4 \\ y_1 & y_2 & y_4 \end{vmatrix}\right\}$$

$$L_4 = \frac{1}{\Delta}\left\{ -\begin{vmatrix} x_1 & x_2 & x_3 \\ y_1 & y_2 & y_3 \\ z_1 & z_2 & z_3 \end{vmatrix} + x\begin{vmatrix} 1 & 1 & 1 \\ y_1 & y_2 & y_3 \\ z_1 & z_2 & z_3 \end{vmatrix} - y\begin{vmatrix} 1 & 1 & 1 \\ x_1 & x_2 & x_3 \\ z_1 & z_2 & z_3 \end{vmatrix} + z\begin{vmatrix} 1 & 1 & 1 \\ x_1 & x_2 & x_3 \\ y_1 & y_2 & y_3 \end{vmatrix}\right\}$$

where:

$\Delta$ is six times the volume of the tetrahedron defined by the nodes 1-2-3-4 and is given in Equation 15.46

Jacobian matrix:

$$\begin{Bmatrix} \dfrac{\partial \phi}{\partial \xi} \\ \dfrac{\partial \phi}{\partial \eta} \\ \dfrac{\partial \phi}{\partial \zeta} \end{Bmatrix} = \begin{bmatrix} \dfrac{\partial x}{\partial \xi} & \dfrac{\partial y}{\partial \xi} & \dfrac{\partial z}{\partial \xi} \\ \dfrac{\partial x}{\partial \eta} & \dfrac{\partial y}{\partial \eta} & \dfrac{\partial z}{\partial \eta} \\ \dfrac{\partial x}{\partial \zeta} & \dfrac{\partial y}{\partial \zeta} & \dfrac{\partial z}{\partial \zeta} \end{bmatrix} \begin{Bmatrix} \dfrac{\partial \phi}{\partial x} \\ \dfrac{\partial \phi}{\partial y} \\ \dfrac{\partial \phi}{\partial z} \end{Bmatrix} = J\begin{Bmatrix} \dfrac{\partial \phi}{\partial x} \\ \dfrac{\partial \phi}{\partial y} \\ \dfrac{\partial \phi}{\partial z} \end{Bmatrix} \tag{15.62}$$

Hence,

$$\begin{Bmatrix} \dfrac{\partial \phi}{\partial x} \\ \dfrac{\partial \phi}{\partial y} \\ \dfrac{\partial \phi}{\partial z} \end{Bmatrix} = J^{-1}\begin{Bmatrix} \dfrac{\partial \phi}{\partial \xi} \\ \dfrac{\partial \phi}{\partial \eta} \\ \dfrac{\partial \phi}{\partial \zeta} \end{Bmatrix}$$

**PROBLEM 15.4**

For the linear tetrahedral element shown in Figure 15.15, find the potential of the point *P*.

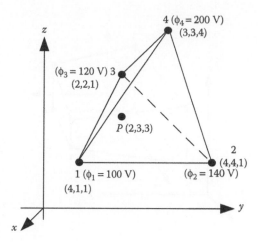

**FIGURE 15.15**
Pertaining to Problem 15.4.

*Solution:*
As per Equation 15.61 the four shape functions are

$$L_1 = -\frac{1}{3}, L_2 = 0, L_3 = \frac{2}{3} \text{ and } L_4 = \frac{2}{3}$$

Consequently, the potential of the point

$$P = -\frac{1}{3} \times 100 + 0 \times 140 + \frac{2}{3} \times 120 + \frac{2}{3} \times 200 = 180 \text{ V}$$

It is to be noted here that the value of $\Delta$ in this case is 18, that is, the volume of the tetrahedron is three units. If the position of nodes 2 and 3 are interchanged, so that nodes 1, 2 and 3 are arranged in the clockwise direction when viewed from node 4, then the volume becomes −3, that is, it becomes negative.

### 15.4.7.2 Linear Hexahedral Element

The generalization of a 2D quadrilateral in a 3D system is a hexahedron. In finite element literature, it is also known as *brick*. A hexahedron is topologically equivalent to a cube. It has eight corners, twelve edges and six faces. Elements having this geometry are extensively used in modelling 3D solids in FEM.

The eight corners of a hexahedron element are locally numbered 1, 2,..., 8, as shown in Figure 15.16. In order to guarantee a positive volume, or, more precisely, a positive Jacobian determinant at every point, the rules for node numbering of the hexahedral element are as follows:

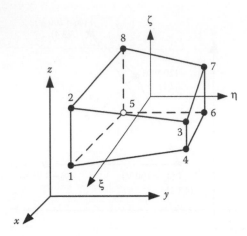

**FIGURE 15.16**
Linear hexahedral element.

1. Consider one corner, say node 1, and one face pertaining to that corner, say 1-2-3-4. Note that for a given corner, there are three possible faces meeting at that corner.
2. Number the other three corners as 2,3,4 traversing the face 1-2-3-4 in the counterclockwise direction while one looks at that face from the opposite one, that is, face 5-6-7-8.
3. Number the corners of the face directly opposite to 1-2-3-4 as 5,6,7,8, respectively, traversing the face 5-6-7-8 in counterclockwise direction while one looks at that face from the opposite one, that is, face 1-2-3-4.

The natural coordinates for this geometry are called $\xi$, $\eta$ and $\zeta$, and are also called *isoparametric hexahedral coordinates*. The natural coordinates vary from −1 on one face to +1 on the opposite face, taking the value zero on the *median* face. As in the case of other type of elements, this particular choice of limits is made to facilitate the use of the standard Gauss integration formulae. The natural coordinates $(\xi, \eta, \zeta)$ of the eight nodes are as follows: node 1(1,−1,−1), node 2(1,−1,1), node 3(1,1,1), node 4(1,1,−1), node 5(−1,−1,−1), node 6(−1,1,−1), node 7(−1,1,1) and node 8(−1,−1,1).

The shape functions are given as follows:

$$L_i = \frac{1}{8}(1+\xi\xi_i)(1+\eta\eta_i)(1+\zeta\zeta_i), \, i = 1,\ldots,8 \tag{15.63}$$

where:
$(\xi_i, \eta_i, \zeta_i)$ are the natural coordinates of the $i$th node, as given above

The following relationship holds good for the shape functions: $\sum_{i=1}^{8} L_i(\xi, \eta, \zeta) = 1$. Coordinate mapping is done using the following equations:

$$x = \sum_{i=1}^{8} L_i x_i, \ y = \sum_{i=1}^{8} L_i y_i \text{ and } z = \sum_{i=1}^{8} L_i z_i \qquad (15.64)$$

Electric potential at any point within the element is given by $\phi = \sum_{i=1}^{8} L_i \phi_i$.

### 15.4.7.3 Isoparametric Element

All the types of elements discussed in this chapter are known as isoparametric element. The term *isoparametric* arises from the fact that the parametric description used to describe the variation of the unknown field parameter within an element is exactly the same as that used to map the geometry of the element from the global coordinates to the natural coordinates. The main advantage of isoparametric formulation is that the element equations need only be evaluated in natural coordinate system. Thus, for each element in the mesh the integrals can be evaluated by a standard procedure.

However, it is not necessary to use shape functions of the same order for describing the geometry and the field variable in an element. If the geometry is described by a lower order model than the field variable, then the element is called *sub-parametric* element. On the other hand, if the geometry is described by a higher order shape function, then the element is called a *super-parametric* element.

### 15.4.8 Mapping of Finite Elements

Mapping between natural and global coordinates is an important issue in FEM, as all calculations are performed in natural coordinates. Such mapping results in geometric transformation of the elements.

Figure 15.17 shows the mapping of a triangle of arbitrary side length into an isosceles triangle.

Figure 15.18 shows how a quadrilateral of arbitrary side length can be mapped to unit square using linear quadrilateral element.

Figure 15.19 shows how a curved quadrilateral can be mapped to a unit square using quadratic quadrilateral element.

Figure 15.20 shows the mapping of a hexahedron of arbitrary side length into a unit cube using a quadratic hexahedral element.

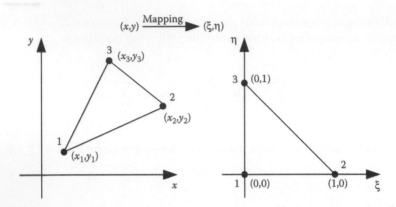

**FIGURE 15.17**
Mapping of linear triangular element.

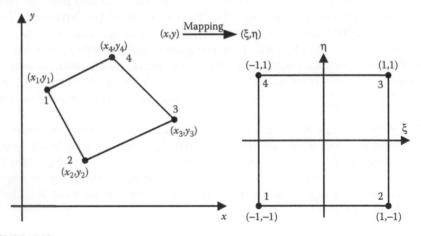

**FIGURE 15.18**
Mapping of linear quadrilateral element.

## 15.5 Features of Discretization in FEM

In FEM the continuous domain is replaced by a series of simple, intercon-nected elements whose field variable characteristics are comparatively easy to compute. In true sense, these elements are connected to each other along their boundaries but the assumption that the elements are connected only at their nodes is made in order to perform a theoretical approximation. A wide variety of element types in two and three dimensions are now available. It is a duty of the person doing the analysis to determine not only the appropriate type of elements for the problem at hand, but also the density required to sufficiently approximate the solution. It is essential to apply engineering judgement.

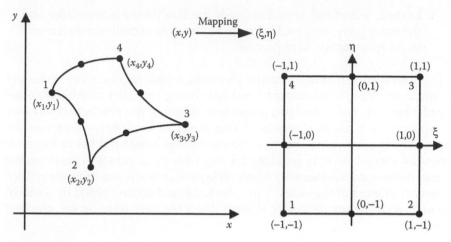

**FIGURE 15.19**
Mapping of curved quadrilateral element.

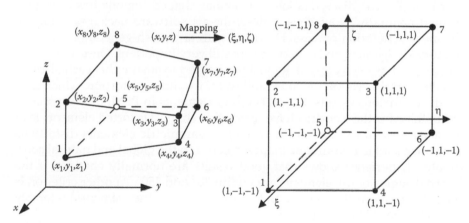

**FIGURE 15.20**
Mapping of a hexahedral element.

## 15.5.1 Refinement of FEM Mesh

In FEM mesh, every element has a size ($h$) and an order ($p$). Either reducing the element size ($h$) or increasing the element order ($p$) reduces the error in FEM. Consequently, there are three basic approaches towards mesh refinement in FEM: the $h$, $p$ and the $h$–$p$ methods.

1. In the $h$ method, the element order ($p$) is kept constant and the mesh is refined by making the size ($h$) smaller.

2. On the other hand, in the $p$ method, the element size ($h$) is kept constant and the element order $p$ is increased for mesh refinement.

3. In the *h–p* method, simultaneously the size (*h*) is made smaller and the order (*p*) is increased to create higher order small-sized elements in the mesh refinement process.

It is often claimed that higher order elements, which require more nodes per element, results in less computational time using a smaller number of large-sized elements. But in real life, geometries defining the practical objects are complex, which anyway require a fine mesh to accurately discretize the geometry. In such cases, the mesh size is usually small and hence the error does not exceed what is required for engineering accuracy. Therefore, the use of higher order *h* elements offers no benefit over the use of lower order *h* elements in most of the cases. Thus, the *h* method accompanied by a robust quadrilateral or hex generator is most often the best solution for practical design jobs.

### 15.5.2 Acceptability of Element after Discretization

Traditionally, the discretization of irregular-shaped regions has been performed manually. Nowadays, state-of-the-art software packages automate the mesh generation process. However, with any mesh-generation package, the user's judgement and experience are still very important. Once a finite element mesh has been created, it must be checked to ensure that each element satisfies certain criteria for acceptability, for example, distortion, which may produce spurious results. For all types of elements in FEM, the best results are obtained if the elements have reasonable shape. Distorted elements lead to major inaccuracies, as in the case of isoparametric elements distortions very often lead to non-unique mapping between the global and natural coordinates. Experience shows that good results are normally obtained if the internal angles of the elements are within 30° and 150°. Another criterion is the ratio between the longest and shortest sides of the element. Preferably this ratio should be smaller than 5:1.

Figure 15.21 shows a few elements having very bad shape that need to be avoided in FEM mesh.

### 15.6 Solution of System of Equations in FEM

Applications of the FEM to practical systems lead to large systems of simultaneous linear algebraic equations, which are symmetric, positive definite and sparse. Many solution methods make use of these properties to provide fast and efficient computation algorithms, which are now implemented in nearly all finite element packages. Only half of the matrix including diagonal entries needs to be stored because of the symmetry. Positive definite matrices are

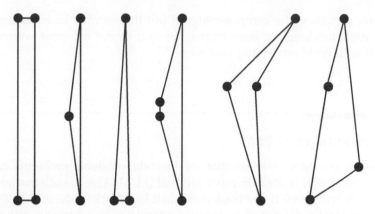

**FIGURE 15.21**
Elements having distorted shape.

characterized by large positive entries on the main diagonal. As a result, solution can be carried out without pivoting. Storage and computations could be economized using sparsity. Solution methods for simultaneous linear equation systems can be broadly divided into two groups: direct methods and iterative methods. Direct solution methods are usually used for problems of moderate size. For large problems, iterative methods are preferable as they require less computing time. The choice of solution method is very much dependent on the size of the problem as well as the type of analysis.

### 15.6.1 Sources of Error in FEM

There are three main sources of error in a typical FEM solution, namely, discretization error, formulation error and numerical error.

Discretization error results from transforming the continuous physical region of interest into a finite element model, and can be related to modelling the boundary shape, the boundary conditions and so on. In many problems, poor geometry representation causes serious discretization error. Discretization error can be effectively reduced by the refinement of FEM mesh.

Formulation error results from the use of elements that do not precisely describe the behaviour of the physical problem. For example, a particular finite element might be formulated on the assumption that electric potential varies in a linear manner over the domain. Such an element will produce no formulation error when it is used to model a linearly varying electric potential, but would create a significant formulation error if it is used to represent a quadratic or cubic varying electric potential. The magnitude of this error depends on the size of the elements relative to the nature of variation of field variables. Formulation error in most physical problems reduces as the element size decreases.

Numerical error occurs as a result of numerical calculation procedures, and includes truncation errors and round off errors. This is a function of the computer accuracy, the computer algorithm, the number of equations and the element subdivision. Both truncation and round off error sources are reduced with good modelling practices.

## 15.7 Advantages of FEM

Early work on numerical solution of boundary-valued problems can be traced to the use of finite difference method (FDM). Use of such method was reported by Southwell in his book published long back in the mid-1940s. The FDM is generally restricted to simple geometries in which an orthogonal grid is possible to construct. For irregular geometries, a global transformation of the governing equations (e.g. Poisson's equation in HV fields) must be made to create an orthogonal computational domain. Moreover, the implementation of boundary conditions in FDM is often cumbersome.

The beginning of the FEM actually stems from the difficulties associated with using FDM for solving difficult, geometrically irregular problems. Unlike FDM, which envisions the solution region as an array of grid points, the FEM envisions the solution region as made up of many small, interconnected sub-regions or elements. A finite element model of a problem gives a piecewise approximation to the governing equations. The basic premise of the FEM is that a solution region can be analytically modelled or approximated by replacing it with an assemblage of discrete elements. Because these elements can be put together in a variety of ways, they can be used to represent exceedingly complex shapes.

For the high-voltage insulator problem, the finite element model gives a good approximation of the region of interest using the simplest 2D element, that is, the linear triangular element, as shown in Figure 15.22. In FEM, a better approximation of the boundary shape is obtained because the curved boundary is represented by straight lines of any inclination. However, it is not intended here to suggest that finite element models are decidedly better for all problems. The only purpose of the example is to demonstrate that the FEM is particularly well suited for problems with complex geometries.

### 15.7.1 Using FEM in the Design Cycle

Using FEM analysis in the design cycle of a product is advantageous. FEM can be used to determine the real-life behaviour of a new design concept under various practical conditions, and therefore to make possible refinement prior to the creation of drawings in computer-aided design (CAD), when changes are inexpensive. Once a detailed CAD model has been developed, FEM can be used to analyze the design in detail, which saves time and

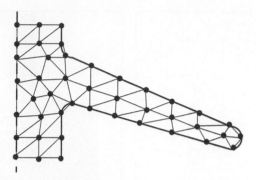

**FIGURE 15.22**
Modelling of HV insulator using triangular element.

money by reducing the number of prototypes required. Further an existing product, which is experiencing a field problem or is being improved, can be analyzed to speed up the change in engineering design and reduce its cost. In addition, FEM analysis can now be performed on increasingly affordable personal computers. However, FEM analysis can reduce product testing, but cannot totally replace it. It is important to note here that an inexperienced user of FEM can deliver incorrect answers, on which significant and expensive decisions will be based. FEM is a demanding tool, in that the analyst must be proficient not only in subject being solved, but also in mathematics, computer applications and especially the FEM itself.

## 15.8  FEM Examples

### 15.8.1  Circuit Breaker Contacts

Figure 15.23 shows the typical contact arrangement of a high-voltage circuit breaker. In order to study the arcing in the contact arrangement, it is necessary to know the electric field distribution for such arrangement. It is an axi-symmetric system. Figure 15.23a shows a triangular mesh comprising relatively larger triangular elements that could be used for FEM analysis. If the mesh size does not provide acceptable accuracy, then the discretization could be refined, as shown in Figure 15.23b. Typically, the triangles are smaller in the region where the field is expected to be high and the variation of field is non-uniform. On the other hand, in those regions where the field intensity magnitude as well as non-uniformity is lower, triangular elements of larger size are used. As a thumb rule, the element size is smaller near the boundaries and the element size becomes progressively larger as the distance from the boundary increases.

**FIGURE 15.23**
Modelling of circuit breaker contact using (a) coarse triangular mesh and (b) fine triangular mesh.

### 15.8.2 Cylindrical Insulator

A vertical cylindrical insulator with end electrodes is shown in Figure 15.24. These end electrodes also act as corona shields for high-voltage applications. This system is an axi-symmetric unbounded field region problem. In such cases, a fictitious boundary has to be considered for FEM analysis. In this example, a spherical boundary is assumed as shown. The basic rule of assuming the fictitious boundary is that the boundary should be considered at a location where the field variation is very small in space.

### 15.8.3 Porcelain Bushing of Transformer

Bushings are used in those cases where a live conductor has to pass through an earthed body or through a body of a different potential. In the case of transformers, the windings are normally within the tank which is earthed. Hence, the live conductor has to pass through the earthed tank for making connection to the winding. Hence, bushings are necessary for transformers. Figure 15.25 shows a typical porcelain bushing used in transformers. It is an axi-symmetric arrangement with unbounded region outside the porcelain outer cover. The central conductor has a solid insulation layer over it. The space between this solid insulation and the outer porcelain cover is filled with transformer oil. Typically, failure of bushings occurs in two different ways: (1) by means of puncture of insulation between the live conductor and the

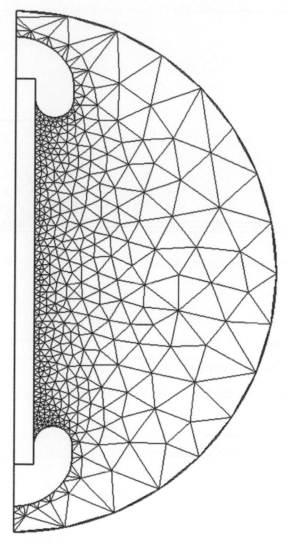

**FIGURE 15.24**
Modelling of vertical cylindrical insulator incorporating fictitious boundary.

earthed tank where the radial distance between them is minimum and (2) by
means of flashover that occurs along the surface of the porcelain outer cover.
But in both the cases such a failure results in a terminal short circuit of the
transformer, which is a major fault in any power system and is catastrophic
from the viewpoint of the transformer health. Therefore, it is very important to know the electric field distribution in and around bushings, so that
unwanted field concentration does not take place that may lead to failure.
Typical FEM discretization of a transformer porcelain bushing is shown in
Figure 15.25, which need to be refined depending on the accuracy required.

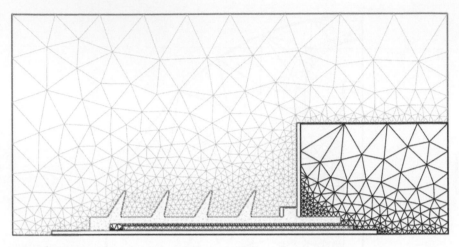

**FIGURE 15.25**
Modelling of porcelain bushing of transformer.

## Objective Type Questions

1. Finite element method is based on
   a.  Taylor series expansion
   b.  Fourier transform
   c.  Integral equation solution
   d.  Differential equation solution
2. In principle, finite element method is closer to
   a.  Charge simulation method
   b.  Finite difference method
   c.  Monte Carlo method
   d.  Fourier transform
3. In 2D FEM formulation, which one of the following elements is most commonly used?
   a.  Straight-line element
   b.  Ring element
   c.  Triangular element
   d.  Tetrahedral element

4. In finite element method, which one of the following is true?
   a. Only conductor boundaries are discretized
   b. Only dielectric boundaries are discretized
   c. Both (a) and (b)
   d. Entire field region is discretized

5. In FEM, if number of elements is increased, then error
   a. Decreases
   b. Increases
   c. Remains unchanged
   d. Oscillates

6. Poisson's equation can be solved conveniently by
   a. Finite element method
   b. Finite difference method
   c. Charge simulation method
   d. Surface charge simulation method

7. In FEM formulation, within each element which one of the following parameters is considered to vary linearly, in general?
   a. Electric potential
   b. Electric field intensity
   c. Electric flux density
   d. Both (a) and (b)

8. In FEM, electric field intensity is determined by
   a. Numerical integration
   b. Numerical differentiation
   c. Series expansion
   d. Analytical expression

9. Which one of the following is the basis of variational approach in FEM formulation?
   a. Energy within the field region is maximum
   b. Field intensity within the field region is maximum
   c. Energy within the field region is minimum
   d. Field intensity within the field region is minimum

10. In FEM, higher number of elements are needed to simulate the region where field intensity
    a. Varies sharply
    b. Remains constant

    c. Varies slowly

    d. Is zero

11. FEM formulations give rise to a coefficient matrix that by property is a

    a. Sparse matrix

    b. Nearly full matrix

    c. Diagonal matrix

    d. Unit matrix

12. Which one of the following is a linear element?

    a. 3-node triangular element

    b. 6-node triangular element

    c. 8-node quadrilateral element

    d. 20-node hexahedral element

13. Which one of the following is a quadratic element?

    a. 3-node triangular element

    b. 6-node triangular element

    c. 4-node quadrilateral element

    d. 8-node hexahedral element

14. The system of simultaneous linear algebraic equations as obtained by the application of FEM in large practical problem is

    a. Symmetric

    b. Sparse

    c. Positive definite

    d. All of the above

15. Errors that arise in a typical FEM solution are

    a. Discretization error

    b. Formulation error

    c. Numerical error

    d. All of the above

16. To obtain good results in FEM analysis, the ratio between the longest and shortest sides of the element should preferably be

    a. Less than 20:1

    b. Less than 15:1

    c. Less than 10:1

    d. Less than 5:1

17. For refinement of FEM mesh, most commonly used technique is

   a. To make the size of the element smaller keeping the element order constant

   b. To increase the element order keeping the size of the element same

   c. To decrease the size of the element along with the increase in element order

   d. To re-arrange the nodes without altering the size and order of the element

18. As the size of the element decreases

   a. Discretization error increases

   b. Discretization error decreases

   c. Formulation error decreases

   d. Both (b) and (c)

19. In FEM analysis, a curved quadrilateral can be mapped to a unit square element by the use of

   a. Quadratic triangular element

   b. Linear quadrilateral element

   c. Quadratic quadrilateral element

   d. Quadratic hexahedral element

20. In FEM analysis, a hexahedron of arbitrary side length can be mapped to a unit cube element by the use of

   a. Linear tetrahedral element

   b. Quadratic tetrahedral element

   c. Linear hexahedral element

   d. Quadratic hexahedral element

21. For a finite element, if the shape function defining the geometry is of the same order as that used for defining the field variable, then the element is known as

   a. Super-parametric

   b. Iso-parametric

   c. Sub-parametric

   d. None of the above

22. In terms of the shape function used for defining the geometry as well as field variable in a finite element, the element could be

   a. Super-parametric

   b. Iso-parametric

   c. Sub-parametric

   d. All of the above

23. For a linear hexahedral element, if three shape functions are denoted as $L_1 = \xi$, $L_2 = \eta$ and $L_3 = \zeta$, then the fourth shape function $L_4$ is given by

   a. $\xi - \eta - \zeta$

   b. $\xi + \eta + \zeta$

   c. $1 - \xi - \eta - \zeta$

   d. $1 + \xi + \eta + \zeta$

24. For a linear triangular element, if the three shape functions are denoted as $L_1$, $L_2$ and $L_3$, then

   a. $L_1 + L_2 + L_3 = 1$

   b. $L_1 + L_2 + L_3 = 0$

   c. $L_1 + L_2 + L_3 = -1$

   d. $L_1 = L_2 + L_3$

25. For a linear triangular element, if two shape functions are denoted as $L_1 = \xi$ and $L_2 = \eta$, then the third shape function $L_3$ is given by

   a. $\xi - \eta$

   b. $\xi + \eta$

   c. $1 - \xi - \eta$

   d. $1 + \xi + \eta$

26. In FEM analysis, the approach using the minimization of potential energy is known as

   a. Galerkin's formulation

   b. Variational formulation

   c. Weighted residual formulation

   d. None of the above

27. In high-voltage field analysis, the solution of global system of FEM equations usually gives the nodal values of

   a. Electric potential

   b. Electric field intensity

   c. Electric flux density

   d. Electric charge

**Answers:**   1) d; 2) b; 3) c; 4) d; 5) a; 6) a; 7) a; 8) b; 9) c; 10) a; 11) a; 12) a; 13) b; 14) d; 15) d; 16) d; 17) a; 18) d; 19) c; 20) d; 21) b; 22) d; 23) c; 24) a; 25) c; 26) b; 27) a

# Bibliography

1. N. Flatabo and H. Riege, 'Automatic calculation of electric fields', *1st ISH Symposium*, Munich, Germany, pp. 17–22, 1972.
2. B.H. McDonald and A. Wexler, 'Finite element solution of unbounded field problems', *IEEE Transactions on Microwave Theory and Techniques*, Vol. 20, No. 12, pp. 841–847, 1972.
3. O.W. Andersen, 'Laplacian electrostatic field calculations by finite elements with automatic grid generation', *IEEE Transactions on Power Apparatus and Systems*, Vol. 92, No. 5, pp. 1485–1492, 1973.
4. A. Di Napoli, C. Mazzetti and U. Ratti, *A Computerized Program for the Numerical Solution of Harmonic Fields with Boundary Condition Non-Properly Defined*, Compumag, Oxford, 1976.
5. O.W. Andersen, 'Finite element solution of complex potential electric fields', *IEEE Transactions on Power Apparatus and Systems*, Vol. 96, No. 4, pp. 1156–1161, 1977.
6. A. Di Napoli and C. Mazzetti, 'Electrostatic and electromagnetic field computation for the H.V. resistive divider design', *IEEE Transactions on Power Apparatus and Systems*, Vol. 98, No. 1, pp. 197–206, 1979.
7. O.W. Andersen, 'Two stage solution of three dimensional electrostatic fields by finite differences and finite elements', *IEEE Transactions on Power Apparatus and Systems*, Vol. 100, No. 8, pp. 3714–3721, 1981.
8. M.V.K. Chari, A. Konrad, J. D'Angelo and M.A. Palmo, 'Finite element computation of three-dimensional electrostatic and magnetostatic field problems', *IEEE Transactions on Magnetics*, Vol. 19, No. 6, pp. 2321–2232, 1983.
9. J.F. Hoburg and J.L. Davis, 'A student-oriented finite element program for electrostatic potential problems', *IEEE Transactions on Education*, Vol. 26, No. 4, pp. 138–142, 1983.
10. N. Burais, L. Krahenbuhl and A. Nicolas, 'Potential distribution simulation in high voltage measurement system', *IEEE Transactions on Magnetics*, Vol. 21, No. 6, pp. 2392–2395, 1985.
11. G.A. Kallio and D.E. Stock, 'Computation of electrical conditions inside wire-duct electrostatic precipitators using a combined finite-element, finite-difference technique', *Journal of Applied Physics*, Vol. 59, No. 6, pp. 1799–1806, 1986.
12. J. Penman and M.D. Grieve, 'Self-adaptive finite-element techniques for the computation of inhomogeneous Poissonian fields', *IEEE Transactions on Industry Applications*, Vol. 24, No. 6, pp. 1042–1049, 1988.
13. H. Yamashita, K. Shinozaki and E. Nakamae, 'A boundary-finite element method to compute directly electric field intensity with high accuracy', *IEEE Transactions on Power Delivery*, Vol. 3, No. 4, pp. 1754–1760, 1988.
14. O.W. Andersen, 'PC-based field calculations for electric power applications', *IEEE Computer Applications in Power*, Vol. 2, No. 4, pp. 22–25, 1989.
15. J.R. Brauer, H. Kalfaian and H. Moreines, 'Dynamic electric fields computed by finite elements', *IEEE Transactions on Industry Applications*, Vol. 25, No. 6, pp. 1088–1092, 1989.
16. R. Lakshmipathi, Y.N. Rao and G.S. Rao, 'Computation of electrostatic field distribution in a cable by least-squares smoothing finite element method', *Communications in Applied Numerical Methods*, Vol. 5, No. 1, pp. 15–22, 1989.

17. S. Cristina, G. Dinelli and M. Feliziani, 'Numerical computation of corona space charge and V-I characteristic in DC electrostatic precipitators', *IEEE Transactions on Industry Applications*, Vol. 27, No. 1, pp. 147–153, 1991.
18. C. Xiang and C. Li, 'A finite element algorithm of plotting electric force line in two-dimensional electrostatic field computation', *IEEE Transactions on Magnetics*, Vol. 28, No. 2, pp. 1789–1792, 1992.
19. H. Yamashita, E. Nakamae, T. Okano, M.S.A.A. Hammam, C. Burns and G. Adams, 'A color graphics display of the field intensity around the insulator on 13.2 kV distribution lines', *IEEE Transactions on Electrical Insulation*, Vol. 8, No. 4, pp. 1696–1702, 1993.
20. A. Wu and M.D. Driga, 'Mathematical formulation of 2D and 3D finite and boundary element model for transient electric fields in high performance capacitors', *IEEE Transactions on Magnetics*, Vol. 29, No. 1, pp. 1088–1092, 1993.
21. M. Abdel-Salam and Z. Al-Hamouz, 'Novel finite-element analysis of space-charge modified fields', *IEEE Proceedings Science Measurement and Technology*, Vol. 141, No. 5, pp. 369–378, 1994.
22. C. Xiang, J. Liu, Y. Xie, R. He, G. Zhang and C. Yang, 'Design of insulated structure for load-ratio voltage power transformer by finite element method', *IEEE Transactions on Magnetics*, Vol. 30, No. 5, pp. 2944–2947, 1994.
23. Q. Chen, A. Konrad and S. Baronijan, 'Asymptotic boundary conditions for axisymmetric finite element electrostatic analysis', *IEEE Transactions on Magnetics*, Vol. 30, No. 6, pp. 4335–4337, 1994.
24. S. Cristina and M. Feliziani, 'Calculation of ionized fields in DC electrostatic precipitators in the presence of dust and electric wind', *IEEE Transactions on Industry Applications*, Vol. 31, No. 6, pp. 1446–1451, 1995.
25. D. Xingqi and A. Tongyi, 'A new FEM approach for open boundary Laplace's problem', *IEEE Transactions on Microwave Theory and Techniques*, Vol. 44, No. 1, pp. 157–160, 1996.
26. A. Konrad and M. Graovac, 'The finite element modelling of conductors and floating potentials', *IEEE Transactions on Magnetics*, Vol. 32, No. 5, pp. 4329–4331, 1996.
27. S.E. Asenjo, O.N. Morales and E.A. Valdenegro, 'Solution of low frequency complex fields in polluted insulators by means of the finite element method', *IEEE Transactions on Dielectrics and Electrical Insulation*, Vol. 4, No. 1, pp. 10–16, 1997.
28. Z.M. Al-Hamouz, M. Abdel-Salam and A.M. Al-Shehri, 'Inception voltage of corona in bipolar ionized fields-effect on corona power loss', *IEEE Transactions on Industry Applications*, Vol. 34, No. 1, pp. 57–65, 1998.
29. Z.M. Al-Hamouz, M. Abdel-Salam and A. Mufti, 'Improved calculation of finite-element analysis of bipolar corona including ion diffusion', *IEEE Transactions on Industry Applications*, Vol. 34, No. 1, pp. 301–309, 1998.
30. J.Q. Feng and D.A. Hays, 'A finite-element analysis of the electrostatic force on a uniformly charged dielectric sphere resting on a dielectric-coated electrode in a detaching electric field', *IEEE Transactions on Industry Applications*, Vol. 34, No. 1, pp. 84–91, 1998.
31. D.C. Faircloth and N.L. Allen, 'Calculations based on measurements of charge deposited by a streamer on a PTFE surface', *IEEE Transactions on Dielectrics and Electrical Insulation*, Vol. 10, No. 2, pp. 291–294, 2003.
32. O.C. Zienkiewicz and R.L. Taylor, *Finite Element Method*, Elsevier Butterworth-Heinemann, 2005.

33. S.J. Han, J. Zou, S.Q. Gu, J.L. He and J.S. Yuan, 'Calculation of the potential distribution of high voltage metal oxide arrester by using an improved semi-analytic finite element method', *IEEE Transactions on Magnetics*, Vol. 41, No. 5, pp. 1392–1395, 2005.

34. B.S. Ram, 'Three-dimensional electrostatic field analysis of perpendicular cylinders separated by a gap', *IEEE Transactions on Power Delivery*, Vol. 11, No. 1, pp. 521–522, 2006.

35. M. Boutaayamou, R.V. Sabariego and P. Dular, 'A perturbation method for the 3D finite element modelling of electrostatically driven MEMS', *Sensors*, Vol. 8, No. 2, pp. 994–1003, 2008.

36. L.N. Rossi, V.C. Silva, L.B. Martinho and J.R. Cardoso, 'A geometrical approach of 3-D FEA for educational purposes applied to electrostatic fields', *IEEE Transactions on Magnetics*, Vol. 44, No. 6, pp. 1674–1677, 2008.

37. H. Daochun, R. Jiangjun, C. Wei, L. Tianwei, W. Yuanhang and L. Jia, 'Flashover prevention on high-altitude HVAC transmission line insulator strings', *IEEE Transactions on Dielectrics and Electrical Insulation*, Vol. 16, No. 1, pp. 88–98, 2009.

38. M. Gaber, J. Pihler, M. Stegne and M. Trlep, 'Flashover condition for a special three-electrode spark gap design', *IEEE Transactions on Power Delivery*, Vol. 25, No. 1, pp. 500–507, 2010.

39. P. Maity, N. Gupta, V. Parameswaran and S. Basu, 'On the size and dielectric properties of the interphase in epoxy-alumina nanocomposite', *IEEE Transactions on Dielectrics and Electrical Insulation*, Vol. 17, No. 6, pp. 1665–1675, 2010.

40. N. Farnoosh, K. Adamiak and G.S.P. Castle, '3-D numerical simulation of particle concentration effect on a single-wire ESP performance for collecting polydispersed particles', *IEEE Transactions on Dielectrics and Electrical Insulation*, Vol. 18, No. 1, pp. 211–220, 2011.

41. X.B. Bian, D.Y. Yu, M. Xiaobo, M. Macalpine, W. Liming, G. Zhicheng, Y. Wenjun and Z. Shuzhen, 'Corona-generated space charge effects on electric field distribution for an indoor corona cage and a monopolar test line', *IEEE Transactions on Dielectrics and Electrical Insulation*, Vol. 18, No. 5, pp. 1767–1778, 2011.

42. W.N. Fu, S.L. Ho, S. Niu and J.G. Zhu, 'Comparison study of finite element methods to deal with floating conductors in electric field', *IEEE Transactions on Magnetics*, Vol. 48, No. 2, pp. 351–354, 2012.

# 16

# Numerical Computation of High-Voltage Field by Charge Simulation Method

**ABSTRACT** An alternative approach towards solving Laplace's or Poisson's equations by differential equation techniques is to take integrals of these equations using discrete charges or by subdividing the interfaces into subsections of charges. Conventional charge simulation method (CSM) is based on the usage of discrete charges. This chapter discusses in detail the simulation of single-dielectric as well as multi-dielectric arrangements using CSM for both symmetric and asymmetric systems. Different types of charge configurations, accuracy criteria and the factors affecting simulation accuracy have been presented at length. Formulations incorporating complex charges for computing field in systems having potentials with time-phase differences or for computing capacitive-resistive fields including volume, as well as surface, conductions has been thoroughly discussed. As the insulation in high-voltage equipment is often stressed by transient excitations, such cases have also been critically examined. Developments that have taken place over the years to improve the performance of CSM have also been included in this chapter. Two-dimensional, axi-symmetric as well as three-dimensional case studies of practical significance have been explained for the proper understanding of the simulation technique.

## 16.1 Introduction

The principle of finite difference method (FDM) and finite element method (FEM) is to provide the entire region of interest (ROI) into a large number of sub-regions, and solve for unknown potentials a set of coupled simultaneous linear equations, which approximate Laplace's or Poisson's equations. Compared to these two methods, only boundary surfaces, that is, electrode surfaces and dielectric interfaces, are subdivided and charges are taken as unknowns in charge simulation method (CSM). First, it follows that the amount of human time and effort needed for subdivision is greatly reduced in CSM. Second, the electric field strength can be given explicitly in CSM without any numerical differentiation of the potential, which results in significant reduction in error. The second characteristic is very important because the field strength is usually more important for the design of an insulating system than electric potential.

The earlier attempts for numerical field solutions employing CSM were reported by Loeb et al. in 1950 and then by Abou-Seada and Nasser [1]. Subsequently, in a comprehensive paper Singer, Steinbigler and Weiss presented the details of CSM [2]. Since then, many refinements to the original method have been proposed and CSM has evolved into a very powerful and efficient tool for computing electric fields in HV equipment. CSM is very simple and is applicable to systems having more than one dielectric medium. This method is also suitable for three-dimensional (3D) fields with or without symmetry.

## 16.2 CSM Formulation for Single-Dielectric Medium

The basic principle of conventional CSM is very simple. For the calculation of electric fields, the distributed charges on the surface of the electrode are replaced by $N$ number of fictitious charges placed inside the electrode, as shown in Figure 16.1. The fictitious charges are placed inside the electrode to avoid singularity problem. In general, the fictitious charges are to be always placed outside the ROI, as the field is ideally required to be determined at all the points within the ROI. If the fictitious charges are placed within the ROI, then at the location of the fictitious charges singularity arises because at these points, the distance between the charge and the point at which the field solution is required becomes zero.

The types and positions of these fictitious charges are predetermined, that is, user defined, but their magnitudes are unknown. In order to determine

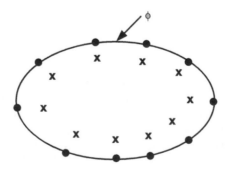

**✗** Fictitious charges: $j = 1, \ldots, N$

**●** Contour points: $i = 1, \ldots, N$

**FIGURE 16.1**
Fictitious charges and contour points for CSM formulation in single-dielectric medium.

their magnitude, some collocation points, which are called *contour points*, are selected on the surface of electrode. In the conventional CSM, the number of contour points is chosen to be equal to the number of fictitious charges. Then it is required that at any one of these contour points the potential resulting from superposition of effects of all the fictitious charges is equal to the known electrode potential. Let $Q_j$ be the $j$th fictitious charge and $\phi$ be the known potential of the electrode. Then according to the superposition principle,

$$\sum_{j=1}^{N} P_{ij}Q_j = \phi \qquad (16.1)$$

where:

$P_{ij}$ is the potential coefficient, that is, the potential at the point $i$ due to a unit charge at the location $j$, which can be evaluated analytically for different types of fictitious charges by satisfying Laplace's equation

When Equation 16.1 is applied to $N$ number of contour points, it leads to the following system of $N$ linear equations for $N$ unknown fictitious charges:

$$P_{11}Q_1 + P_{12}Q_2 + \cdots + P_{1j}Q_j + \cdots + P_{1N}Q_N = \phi$$

$$P_{21}Q_1 + P_{22}Q_2 + \cdots + P_{2j}Q_j + \cdots + P_{2N}Q_N = \phi$$

$$\vdots$$

$$P_{i1}Q_1 + P_{i2}Q_2 + \cdots + P_{ij}Q_j + \cdots + P_{iN}Q_N = \phi \qquad (16.2)$$

$$\vdots$$

$$P_{N1}Q_1 + P_{N2}Q_2 + \cdots + P_{Nj}Q_j + \cdots + P_{NN}Q_N = \phi$$

In matrix form, Equation 16.2 can be written as

$$
\begin{bmatrix}
P_{11} & P_{12} \ldots P_{1j} \ldots P_{1N} \\
P_{21} & P_{22} \ldots P_{2j} \ldots P_{2N} \\
\vdots \\
P_{i1} & P_{i2} \ldots P_{ij} \ldots P_{iN} \\
\vdots \\
P_{N1} & P_{N2} \ldots P_{Nj} \ldots P_{NN}
\end{bmatrix}_{N \times N}
\begin{bmatrix}
Q_1 \\
Q_2 \\
\vdots \\
Q_i \\
\vdots \\
Q_N
\end{bmatrix}_{N \times 1}
= [\phi]_{N \times 1} \qquad (16.3)
$$

where:

[$P$] = potential coefficient matrix

[$\phi$] = column vector of known potential of contour points

Equation 16.3 is solved for the unknown fictitious charges. As soon as the required fictitious charge system is determined, the potential and the field intensity at any point within the ROI can be calculated. Although the potential is found by Equation 16.1, the electric field intensities are calculated by superposition of all the stress vector components. For example, in the Cartesian co-ordinate system, the three superimposed field components at any point $i$ are given as follows.

$$E_{x,i} = -\sum_{j=1}^{N} \frac{\partial P_{ij}}{\partial x} Q_j = -\sum_{j=1}^{N} F_{x,ij} Q_j \qquad (16.4)$$

$$E_{y,i} = -\sum_{j=1}^{N} \frac{\partial P_{ij}}{\partial y} Q_j = -\sum_{j=1}^{N} F_{y,ij} Q_j \qquad (16.5)$$

$$\text{and } E_{z,i} = -\sum_{j=1}^{N} \frac{\partial P_{ij}}{\partial z} Q_j = -\sum_{j=1}^{N} F_{z,ij} Q_j \qquad (16.6)$$

where:
$F_{x,ij}$, $F_{y,ij}$ and $F_{z,ij}$ are the electric field intensity coefficients in the $x$, $y$ and $z$ directions, respectively, that is, the components in the $x$, $y$ and $z$ directions, respectively, of electric field intensity at the point $i$ for a unit charge at the location $j$

In many cases, the effect of the ground plane is to be considered for electric field calculation. This plane can be taken into account by the introduction of image charge.

### 16.2.1 Formulation for Floating Potential Electrodes

Floating potential conductors are often present in high-voltage system, the most common example being condenser bushings. If floating electrodes are present, whose potentials are constant but unknown, then the boundary condition that is imposed for field computation is given below.

$$\phi_{i+1} - \phi_i = 0, \text{ for } i = 1, \dots, N-1 \qquad (16.7)$$

Moreover, a supplementary condition is included such that the sum of fictitious charges for each floating electrode is zero.

Then the system of equation that is obtained will be as follows:

$$
\begin{bmatrix}
1 & 1 & \cdots & 1 & \cdots & 1 \\
(P_{21}-P_{11}) & (P_{22}-P_{12}) & \cdots (P_{2j}-P_{1j}) & \cdots (P_{2N}-P_{1N}) \\
(P_{31}-P_{21}) & (P_{32}-P_{22}) & \cdots (P_{3j}-P_{2j}) & \cdots (P_{3N}-P_{2N}) \\
\vdots \\
(P_{(i+1)1}-P_{i1}) & (P_{(i+1)2}-P_{i2}) & \cdots (P_{(i+1)j}-P_{ij}) & \cdots (P_{(i+1)N}-P_{iN}) \\
\vdots \\
(P_{N1}-P_{(N-1)1}) & (P_{N2}-P_{(N-1)2}) & \cdots (P_{Nj}-P_{(N-1)j}) & \cdots (P_{NN}-P_{(N-1)N})
\end{bmatrix}_{N \times N}
\begin{bmatrix}
Q_1 \\
Q_2 \\
\vdots \\
Q_i \\
\vdots \\
Q_N
\end{bmatrix}_{N \times 1}
= [0]_{N \times 1} \quad (16.8)
$$

If the floating electrode has a net charge, then the supplementary condition is included such that the sum of its fictitious charges is equal to the known net charge value ($Q_E$). In Equation 16.8 the first row is then modified as follows

$$
Q_1 + Q_2 + \cdots + Q_j + \cdots + Q_N = Q_E \quad (16.9)
$$

## 16.3 CSM Formulation for Multi-Dielectric Media

The field computation for multi-dielectric system is somewhat complicated due to the fact that the dipoles are realigned in dielectric media under the influence of the applied voltage. Such realignment of dipoles produces a net surface charge on the dielectric interface. Thus, in addition to the electrodes, each dielectric interface needs to be simulated by fictitious charges. Here, it is important to note that the dielectric boundary does not correspond to an equipotential surface. Moreover, it must be possible to calculate the electric field on both sides of the dielectric boundary.

It has been mentioned earlier that the fictitious charges should be outside the ROI. In the case of electrodes this has been achieved by placing the charges within the electrodes. But, for the dielectric–dielectric interface, both the sides are within the ROI. Hence, any fictitious charge placed on either side of the interface would cause singularity problem. This issue is solved by placing two charges for every contour point on the dielectric–dielectric interface. For solving the field within the dielectric A, the set of charges placed within dielectric B are considered and vice versa.

In the simple example shown in Figure 16.2, there are $N_1$ number of charges and contour points to simulate the electrode, of which $N_A$ are on the side of dielectric A and $(N_1 - N_A)$ are on the side of dielectric B. These $N_1$ charges are valid for field calculation in both the dielectrics. At the dielectric interface, there are $N_2$ contour points sequentially numbered from $(N_1 + 1, \ldots, N_1 + N_2)$,

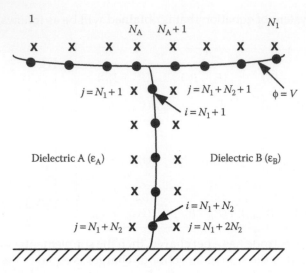

**FIGURE 16.2**
Arrangement of fictitious charges for multi-dielectric media.

with $N_2$ charges $(N_1 + 1,...,N_1 + N_2)$ in dielectric A valid for dielectric B and $N_2$ charges $(N_1 + N_2 + 1,...,N_1 + 2N_2)$ in dielectric B valid for dielectric A. Altogether there are $(N_1 + N_2)$ number of contour points and $(N_1 + 2N_2)$ number of fictitious charges.

In order to determine the fictitious charges, a system of equations is formulated by imposing the following boundary conditions.

1. At each contour point on the electrode surface, the potential must be equal to the known electrode potential. This condition is also known as *Dirichlet's condition* on the electrode surface.
2. At each contour point on the dielectric interface, the potential and the normal component of flux density must be same when computed from either side of the boundary.

Thus, the application of the first boundary condition to contour points 1 to $N_1$ yields the following equations.

$$\sum_{j=1}^{N_1} P_{ij}Q_j + \sum_{j=N_1+N_2+1}^{N_1+2N_2} P_{ij}Q_j = V \qquad ...i = 1, N_A \tag{16.10}$$

$$\text{and} \sum_{j=1}^{N_1} P_{ij}Q_j + \sum_{j=N_1+1}^{N_1+N_2} P_{ij}Q_j = V \qquad ...i = N_A +1, N_1 \tag{16.11}$$

Again, the application of the second boundary condition for potential and normal flux density to contour points $N_1 + 1$ to $N_1 + N_2$ on the dielectric interface results in the following equations.

From potential continuity condition:

$$\sum_{j=N_1+1}^{N_1+N_2} P_{ij}Q_j - \sum_{j=N_1+N_2+1}^{N_1+N_2} P_{ij}Q_j = 0 \qquad ...i = N_1+1, N_1+N_2 \qquad (16.12)$$

From continuity condition of normal flux density $D_n$:

$$D_{nA}(i) - D_{nB}(i) = 0 \qquad ...i = N_1+1, N_1+N_2 \qquad (16.13)$$

Equation 16.13 can be expanded as follows.

$$(\varepsilon_A - \varepsilon_B)\sum_{j=1}^{N_1} F_{n,ij}Q_j - \varepsilon_B \sum_{j=N_1+1}^{N_1+N_2} F_{n,ij}Q_j + \varepsilon_A \sum_{j=N_1+N_2+1}^{N_1+2N_2} F_{n,ij}Q_j = 0 ... i = N_1+1, N_1+N_2 \qquad (16.14)$$

where:
   $F_{n,ij}$ is the field coefficient in the direction normal to the dielectric boundary
      at the respective contour point
   $\varepsilon_A$ and $\varepsilon_B$ are the permittivities of dielectrics A and B, respectively

Equations 16.10 through 16.14 are solved to determine the unknown ficti-tious charges. These equations can be presented in matrix form, as shown in Matrix 16.M.1

*Matrix 16.M.1 System of equations for CSM in multi-dielectric media.*

## 16.4 Types of Fictitious Charges

The successful application of the CSM requires a proper choice of the types of fictitious charges. Point and line charges of infinite and semi-infinite lengths were used in the initial works on this method. Singer et al. [2] introduced ring charges and finite length line charges. Subsequently, a large variety of differ-ent charge configurations have been proposed. These other types of charge configurations include elliptic cylindrical charge, axi-spheroidal charge, plane-sheet charge, disk charge, ring-segment charge, volume charges, shell and annular plate charges as well as variable density line charge.

In general, the choice of type of fictitious charge to be used depends on the complexity of the physical system and the available computational facilities. The potential and field coefficients for point and line charges are given by simple expressions and require very small computation time. For complex charge configuration, such coefficients may have to be computed numerically. On the other hand, a smaller number of charges may be used if complex charge configurations are employed, which reduces the overall memory requirement and computation time. In practice, most of the HV systems can be successfully simulated by using point, line and ring charges or a suitable combination of these charges.

### 16.4.1 Point Charge

Point charge is the simplest of all types of fictitious charges. It can be used in two-dimensional (2D) as well as 3D configurations. Figure 16.3 shows the point charge $Q_j$ along with its image with respect to the $x$–$y$ plane in 3D system.

Then, the potential at the point $i$ due to the point charge $Q_j$ and its image is given by

$$\phi_{ij} = \frac{Q_j}{4\pi\varepsilon}\left(\frac{1}{r_1} - \frac{1}{r_2}\right) \tag{16.15}$$

where:

$$r_1 = \left[(x_i - x_j)^2 + (y_i - y_j)^2 + (z_i - z_j)^2\right]^{1/2}$$

$$r_2 = \left[(x_i - x_j)^2 + (y_i - y_j)^2 + (z_i + z_j)^2\right]^{1/2}$$

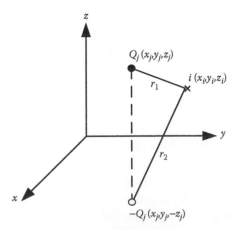

**FIGURE 16.3**
Point charge configuration along with its image.

Putting $Q_j = 1$ in Equation 16.15, the expression for potential coefficient is given by

$$P_{ij} = \frac{1}{4\pi\varepsilon}\left(\frac{1}{r_1} - \frac{1}{r_2}\right) \tag{16.16}$$

Expressions for the electric field intensity coefficients are as follows:

$$F_{x,ij} = -\frac{\partial P_{ij}}{\partial x} = \frac{1}{4\pi\varepsilon}\left(\frac{x_i - x_j}{r_1^{3/2}} - \frac{x_i - x_j}{r_2^{3/2}}\right) \tag{16.17}$$

$$F_{y,ij} = -\frac{\partial P_{ij}}{\partial y} = \frac{1}{4\pi\varepsilon}\left(\frac{y_i - y_j}{r_1^{3/2}} - \frac{y_i - y_j}{r_2^{3/2}}\right) \tag{16.18}$$

$$F_{z,ij} = -\frac{\partial P_{ij}}{\partial z} = \frac{1}{4\pi\varepsilon}\left(\frac{z_i - z_j}{r_1^{3/2}} - \frac{z_i + z_j}{r_2^{3/2}}\right) \tag{16.19}$$

### 16.4.2 Infinite Length Line Charge

Infinite length line charges are used in 2D configurations, particularly for simulating long conductors in the case of transmission lines, cables and so on. Figure 16.4 shows the infinite length line charge $Q_j$ along with its image with respect to the $x$–$z$ plane. In this configuration, electric field is considered to be independent of $z$-axis, that is, the length of the long conductors, while the field varies in the $x$–$y$ plane, which is normal to the length of the long conductors.

In a 2D system, all computations are performed for unit length of the system under consideration. Hence, $Q_j$ is the charge per unit length for the infinite length line charge.

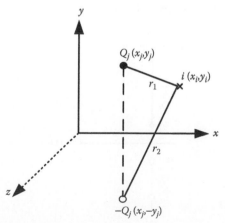

**FIGURE 16.4**
Infinite length line charge configuration along with its image.

The expression for potential coefficient is then given by

$$P_{ij} = \frac{1}{2\pi\varepsilon} \ln \frac{r_2}{r_1} \tag{16.20}$$

where:

$$r_1 = \left[ (x_i - x_j)^2 + (y_i - y_j)^2 \right]^{1/2}$$

$$r_2 = \left[ (x_i - x_j)^2 + (y_i + y_j)^2 \right]^{1/2}$$

Expressions for the electric field intensity coefficients are as follows:

$$F_{x,ij} = \frac{1}{2\pi\varepsilon} \left[ \frac{(x_i - x_j)}{r_1^2} - \frac{(x_i - x_j)}{r_2^2} \right] \tag{16.21}$$

$$F_{y,ij} = \frac{1}{2\pi\varepsilon} \left[ \frac{(y_i - y_j)}{r_1^2} - \frac{(y_i + y_j)}{r_2^2} \right] \tag{16.22}$$

### 16.4.3 Finite Length Line Charge

Finite length line charges of uniform charge density are used in axi-symmetric configurations, particularly for simulating cylindrical geometries in the case of bushings, circuit breakers and so on. Figure 16.5 shows the finite length line charge along with its image. Finite length line charges of uniform charge density are commonly placed on the z-axis, that is, the axis of symmetry. Let the magnitude of the finite length line charge be $Q_j$ and the length of the charge be $(z_{j2} - z_{j1})$, as shown in Figure 16.5. Then considering uniform charge density, charge per unit length is $[Q_j/(z_{j2} - z_{j1})]$. The expressions for potential and electric field intensity coefficients were first developed by Steinbigler et al. [2] and are given below.

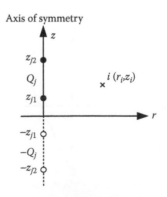

**FIGURE 16.5**
Finite length line charge along with its image.

The expression for potential coefficient is given by

$$P_{ij} = \frac{1}{4\pi\varepsilon(z_{j2} - z_{j1})}\left[\ln\frac{(z_{j2} - z_i + \gamma_1)(z_{j1} + z_i + \gamma_2)}{(z_{j1} - z_i + \delta_1)(z_{j2} + z_i + \delta_2)}\right]$$ (16.23)

where:

$$\gamma_1 = \sqrt{r_i^2 + (z_{j2} - z_i)^2}$$

$$\delta_1 = \sqrt{r_i^2 + (z_{j1} - z_i)^2}$$

$$\gamma_2 = \sqrt{r_i^2 + (z_{j1} + z_i)^2}$$

$$\delta_2 = \sqrt{r_i^2 + (z_{j2} + z_i)^2}$$

The expressions for the electric field intensity coefficients are as follows:

$$F_{r,ij} = \frac{1}{4\pi\varepsilon(z_{j2} - z_{j1})}\left(\frac{z_{j2} - z_i}{r_i\gamma_1} - \frac{z_{j1} - z_i}{r_i\delta_1} + \frac{z_{j1} + z_i}{r_i\gamma_2} - \frac{z_{j2} + z_i}{r_i\delta_2}\right)$$ (16.24)

$$F_{z,ij} = \frac{1}{4\pi\varepsilon(z_{j2} - z_{j1})}\left(\frac{1}{\gamma_1} - \frac{1}{\delta_1} - \frac{1}{\gamma_2} + \frac{1}{\delta_2}\right)$$ (16.25)

## 16.4.4 Ring Charge

Ring charges of uniform charge density are used in axi-symmetric configurations, particularly for simulating spherical- and cylindrical-shaped geometries. Figure 16.6 shows the ring charge along with its image. Ring charges of uniform charge density are commonly placed with their axes on the z-axis, that is, the axis of symmetry. Let the magnitude of the ring charge be $Q_j$ and the radius of the ring charge be $r_j$, as shown in Figure 16.6. Then considering

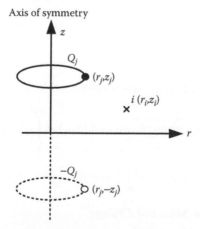

**FIGURE 16.6**
Ring charge of uniform charge density along with its image.

uniform charge density, charge per unit length is $[Q_j/(2\pi r_j)]$. The expressions for potential and electric field intensity coefficients were first developed by Singer et al. [2] and are given below.

The expression for potential coefficient is given by

$$P_{ij} = \frac{1}{2\pi^2\varepsilon}\left[\frac{K(k_1)}{\alpha_1} - \frac{K(k_2)}{\alpha_2}\right] \tag{16.26}$$

where:

$$\alpha_1 = \sqrt{(r_j + r_i)^2 + (z_j - z_i)^2}$$

$$\alpha_2 = \sqrt{(r_j + r_i)^2 + (z_j + z_i)^2}$$

$K(k_1)$ and $K(k_2)$ are elliptic integrals of first kind such that

$$K(k_1) = \int_0^{\pi/2} \frac{d\theta}{\sqrt{1 - k_1^2 \sin^2 \theta}} \text{ and } K(k_2) = \int_0^{\pi/2} \frac{d\theta}{\sqrt{1 - k_2^2 \sin^2 \theta}}$$

$$k_1 = \frac{2\sqrt{r_i r_j}}{\alpha_1} \text{ and } k_2 = \frac{2\sqrt{r_i r_j}}{\alpha_2}$$

The expressions for the electric field intensity coefficients are as follows:

$$F_{r,ij} = \frac{1}{4\pi^2\varepsilon r_i}\left(\begin{array}{c}\left\{\dfrac{\left[(r_j^2 - r_i^2) + (z_j + z_i)^2\right]E(k_2) - \beta_2^2 K(k_2)}{\alpha_2 \beta_2^2}\right\} \\ -\left\{\dfrac{\left[(r_j^2 - r_i^2) + (z_j - z_i)^2\right]E(k_1) - \beta_1^2 K(k_1)}{\alpha_1 \beta_1^2}\right\}\end{array}\right) \tag{16.27}$$

$$F_{z,ij} = \frac{1}{2\pi^2\varepsilon}\left[\frac{(z_j - z_i)E(k_1)}{\alpha_1 \beta_1^2} - \frac{(z_j + z_i)E(k_2)}{\alpha_2 \beta_2^2}\right] \tag{16.28}$$

where:

$E(k_1)$ and $E(k_2)$ are elliptic integrals of second kind such that

$$E(k_1) = \int_0^{\pi/2} \sqrt{1 - k_1^2 \sin^2 \theta}\ d\theta \text{ and } E(k_2) = \int_0^{\pi/2} \sqrt{1 - k_2^2 \sin^2 \theta}\ d\theta$$

$$\beta_1 = \left[(r_j - r_i)^2 + (z_j - z_i)^2\right]^{1/2} \text{ and } \beta_2 = \left[(r_j - r_i)^2 + (z_j + z_i)^2\right]^{1/2}$$

### 16.4.5 Arbitrary Line Segment Charge

Arbitrary line segment charge having uniform charge density is used to simulate 3D geometries. Figure 16.7 shows an arbitrary line segment charge

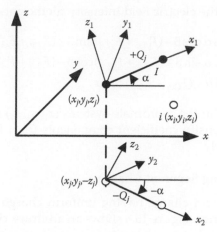

**FIGURE 16.7**
Arbitrary line segment charge along with its image.

along with its image. Let the magnitude of the line segment charge be $Q_j$ and the length of the line segment charge be $l$, as shown in Figure 16.7. The expressions for potential and electric field intensity coefficients were first developed by Utmischi [3] and are given below.

The expression for potential coefficient is given by

$$P_{ij} = \frac{1}{4\pi\varepsilon l} \ln \left\{ \frac{\left[ l - x_1 + \sqrt{z_1^2 + y_1^2 + (l - x_1^2)} \right]}{\left( -x_1 + \sqrt{z_1^2 + y_1^2 + x_1^2} \right)} \frac{\left( -x_2 + \sqrt{z_2^2 + y_2^2 + x_2^2} \right)}{\left[ l - x_2 + \sqrt{z_2^2 + y_2^2 + (l - x_2^2)} \right]} \right\}$$ (16.29)

In the development of this expression, the following coordinate transformations are carried out: (1) displacement of the origin to the points $(x_j, y_j, z_j)$ and $(x_j, y_j, -z_j)$; (2) rotation of $Q_j$ by $\alpha$ and $-Q_j$ by $-\alpha$ in the $x$–$z$ plane and (3) rotation of $Q_j$ and $-Q_j$ by an angle $\beta$ in the $x$–$y$ plane. Thus,

$$x_1 = \left[ (x_i - x_j)\cos\beta + (y_i - y_j)\sin\beta \right]\cos\alpha + (z_i - z_j)\sin\alpha$$

$$y_1 = -(x_i - x_j)\sin\beta + (y_i - y_j)\cos\beta$$ (16.30)

$$z_1 = -\left[ (x_i - x_j)\cos\beta + (y_i - y_j)\sin\beta \right]\sin\alpha + (z_i - z_j)\cos\alpha$$

and

$$x_2 = \left[ (x_i - x_j)\cos\beta + (y_i - y_j)\sin\beta \right]\cos\alpha - (z_i + z_j)\sin\alpha$$

$$y_2 = -(x_i - x_j)\sin\beta + (y_i - y_j)\cos\beta$$ (16.31)

$$z_2 = \left[ (x_i - x_j)\cos\beta + (y_i - y_j)\sin\beta \right]\sin\alpha + (z_i + z_j)\cos\alpha$$

The expressions for the electric field intensity coefficients are as follows:

$$F_{x,ij} = (F_{x1,ij} + F_{x2,ij})\cos\alpha\cos\beta - (F_{y1,ij} + F_{y2,ij})\sin\beta - (F_{z1,ij} + F_{z2,ij})\sin\alpha\cos\beta$$
$$F_{y,ij} = (F_{x1,ij} + F_{x2,ij})\cos\alpha\sin\beta + (F_{y1,ij} + F_{y2,ij})\cos\beta - (F_{z1,ij} + F_{z2,ij})\sin\alpha\sin\beta \quad (16.32)$$
$$F_{z,ij} = (F_{x1,ij} + F_{x2,ij})\sin\alpha + (F_{z1,ij} + F_{z2,ij})\cos\alpha$$

Differentiation of $P_{ij}$ in the coordinate systems $(x_1, y_1, z_1)$ and $(x_2, y_2, z_2)$ gives the electric field intensity coefficient components $(F_{x1,ij}, F_{y1,ij}, F_{z1,ij})$ and $(F_{x2,ij}, F_{y2,ij}, F_{z2,ij})$, respectively.

### 16.4.6 Arbitrary Ring Segment Charge

Arbitrary ring segment charge having uniform charge density is used to simulate 3D geometries. Figure 16.8 shows an arbitrary ring segment charge along with its image. Let the magnitude of the line segment charge be $Q_j$ and the radius of the ring segment charge be $r_0$, as shown in Figure 16.8.

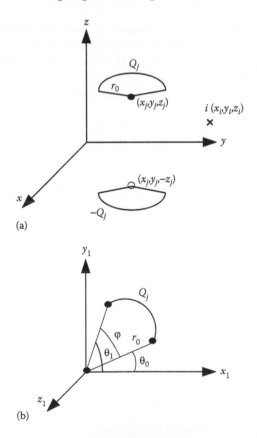

**FIGURE 16.8**
Arbitrary ring segment charge configuration for a 3D field computation: (a) ring segment charge along with its image and (b) ring segment charge after coordinate transformation.

The expressions for potential and electric field intensity coefficients were first developed by Utmischi [3] and are given below.

In the development of the expressions for potential and electric field coefficients, the following coordinate transformations are carried out: (1) displacement of the origin to the points $(x_j, y_j, z_j)$ and $(x_j, y_j, -z_j)$; (2) rotation of $Q_j$ by $\alpha$ and $-Q_j$ by $-\alpha$ in the $x$–$z$ plane and (3) rotation of $Q_j$ and $-Q_j$ by an angle $\beta$ in the $x$–$y$ plane. Thus, the coordinate systems $(x_1, y_1, z_1)$ and $(x_2, y_2, z_2)$ are assigned to the ring segment charge $Q_j$ and its image $-Q_j$, which are to be derived as per Equations 16.30 and 16.31, respectively.

The expression for the potential coefficient is given as

$$P_{ij} = \frac{1}{4\pi\varepsilon(\theta_1 - \theta_0)} \int_{\theta_0}^{\theta_1} \left( \frac{1}{d_1} - \frac{1}{d_2} \right) d\varphi \tag{16.33}$$

where:

$$d_1 = (x_1^2 + y_1^2 + z_1^2 + r_0^2 - 2x_1 r_0 \cos\varphi - 2y_1 r_0 \sin\varphi)^{1/2}$$

$$d_2 = (x_2^2 + y_2^2 + z_2^2 + r_0^2 - 2x_2 r_0 \cos\varphi - 2y_2 r_0 \sin\varphi)^{1/2}$$

The electric field intensity components are determined as per Equation 16.32 where:

$$F_{x1,ij} = \frac{1}{4\pi\varepsilon(\theta_1 - \theta_0)} \int_{\theta_0}^{\theta_1} \frac{(x_1 - r_0 \cos\varphi)}{d_1^3} d\varphi$$

$$F_{y1,ij} = \frac{1}{4\pi\varepsilon(\theta_1 - \theta_0)} \int_{\theta_0}^{\theta_1} \frac{(y_1 - r_0 \cos\varphi)}{d_1^3} d\varphi$$

$$F_{z1,ij} = \frac{1}{4\pi\varepsilon(\theta_1 - \theta_0)} \int_{\theta_0}^{\theta_1} \frac{z}{d_1^3} d\varphi$$

$(F_{x2,ij}, F_{y2,ij}, F_{z2,ij})$ are determined in the same way for the image charge. Integrations are generally carried out numerically.

## 16.5 CSM with Complex Fictitious Charges

In order to calculate the field for a sinusoidal applied voltage, the calculations can be performed as a DC field in so far as the applied voltage does not change so fast that electromagnetic treatment is required. Then the instantaneous field strength is merely dependent on the applied voltage at that time instant. Thus, the conventional CSM with real fictitious charges can be used to compute AC fields for three-phase systems. It has been shown that the field distribution for sinusoidal applied voltage can be calculated in an efficient way by the use of complex fictitious charges. This is permitted because the fictitious charges also change sinusoidally with an angular frequency same as that of the applied voltage. Hence, by the use of complex fictitious charges, Equation 16.1 is modified as follows.

$$\sum_{j=1}^{N} P_{ij}\bar{Q}_j = \bar{\phi} \tag{16.34}$$

where a bar on a variable represents a complex quantity. Application of Equation 16.34 to $N$ number of contour points consists of a set of simultaneous linear equations for complex unknown charges $\bar{Q}_j$ with real coefficients as given below in matrix form.

$$[P]_{N\times N}[\bar{Q}]_N = [\bar{\phi}]_N \tag{16.35}$$

Equation 16.35 is solved to find complex solutions for the fictitious charges.

To explain the above technique in a detailed manner, consider the case of Figure 16.9, which shows four conductors of which three are energized from a three-phase AC source, while the fourth one is grounded. Let $V_{ph}$ be the phase voltage of the three-phase source. Again, let there be $N$ number of complex fictitious charges and contour points, respectively, for each conductor. The charges and the contour points are numbered as follows, $1,\ldots,N$ for conductor $A$; $N+1,\ldots,2N$ for conductor $B$; $2N+1,\ldots,3N$ for conductor $C$ and $3N+1,\ldots,4N$ for conductor $G$. Then the application of Equation 16.34 to all these contour points gives the following equations.

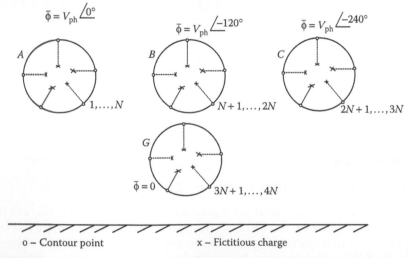

o – Contour point          x – Fictitious charge

**FIGURE 16.9**
Application of CSM with complex fictitious charge for AC field calculation.

For conductor *A*:

$$\sum_{j=1}^{4N} P_{ij}\overline{Q}_j = V_{ph}\angle 0°, \quad \text{for } i = 1,\ldots,N \tag{16.36}$$

For conductor *B*:

$$\sum_{j=1}^{4N} P_{ij}\overline{Q}_j = V_{ph}\angle -120°, \quad \text{for } i = N+1,\ldots,2N \tag{16.37}$$

For conductor *C*:

$$\sum_{j=1}^{4N} P_{ij}\overline{Q}_j = V_{ph}\angle -240°, \quad \text{for } i = 2N+1,\ldots,3N \tag{16.38}$$

For conductor *G*:

$$\sum_{j=1}^{4N} P_{ij}\overline{Q}_j = 0, \quad \text{for } i = 3N+1,\ldots,4N \tag{16.39}$$

Equations 16.36 through 16.39 can be expressed in matrix form, as shown in Matrix 16.M.2:

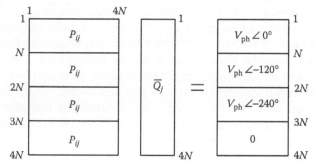

*Matrix 16.M.2 System of equations for CSM application in three-phase conductor arrangement using complex fictitious charges.*

Matrix 16.M.2 is solved for the unknown complex fictitious charges $\overline{Q}_j$.

## 16.6 Capacitive-Resistive Field Computation by CSM

Normally high-voltage equipment is insulated with materials of such a high resistivity that it can be treated as infinite for field calculation. In such cases, the field distribution is purely capacitive. But for lower values of volume or surface resistivity, the field distribution is capacitive-resistive or even resistive depending on the value of resistivity. In the case of capacitive field distribution, the instantaneous field is independent of waveform of applied voltage. But a very distinctive feature of capacitive-resistive fields is their time dependency and dependency on the waveform of applied voltage. Hence, capacitive-resistive field calculation including volume or surface resistivity is very important in studying DC and low frequency fields, impulse fields, contaminated insulators, voltage dividers, cables and so on.

Bachmann [4] first proposed a technique based on CSM for capacitive-resistive field calculation. In his method, as the first step, the capacitive field distribution is calculated by CSM assuming resistivity to be infinite. Then the electrode–electrode and dielectric interface–electrode capacitances are calculated from this capacitive field distribution. After this, as the second step, an equivalent $R$-$C$ network is constructed, which comprises these capacitances and surface resistances. Finally, the voltage distribution for the capacitive-resistive field is calculated from the $R$-$C$ network. This method has two major drawbacks. First, the capacitances between the dielectric interface and electrode are dependent on the field distribution and hence, are not identical in capacitive and capacitive-resistive fields. Second, the calculation of field intensities from the $R$-$C$ network is very laborious and results in significant errors.

Takuma et al. [5] first proposed a method for direct simulation of the instantaneous capacitive-resistive field distribution with fictitious charges. Their method based on CSM and employing complex fictitious charges is generally extended, so that any capacitive-resistive field including volume resistance or surface resistance can be calculated, when the field distribution is Laplacian in the region except on the boundaries. Use of complex fictitious charges as well as appropriate boundary conditions permits the simulation of non-linear and transient problems also. Singer [6] has used complex charges and Fourier integrals to calculate the impulse stresses of conductive dielectrics. Use of discrete as well as area complex charges have been reported for capacitive-resistive field calculation.

### 16.6.1 Capacitive-Resistive Field Computation Including Volume Resistance

For capacitive-resistive field calculation including volume resistance, the principle of the method is that the field effect of the true charges produced by volume resistance is incorporated by means of complex fictitious charges in the CSM.

If the volume charge density is $\sigma_v$, then

$$\vec{\nabla} \cdot (\varepsilon \bar{E}) = \sigma_v \tag{16.40}$$

where:
$\bar{E}$ is the electric field intensity

Again, if the current density through the volume of the dielectric is $\bar{J}$, then

$$\vec{\nabla}\cdot\bar{J} = \vec{\nabla}\cdot\left(\frac{\bar{E}}{\rho_v}\right) = -\frac{\partial\sigma_v}{\partial t} \tag{16.41}$$

where:

$\rho_v$ is the volume resistivity and is constant, that is, independent of $\bar{E}$

Now, if $\varepsilon$ is independent of time $t$ and $E$, then Equation 16.40 can be modified as follows:

$$\vec{\nabla}\cdot\left(\varepsilon\frac{\partial\bar{E}}{\partial t}\right) = \frac{\partial\sigma_v}{\partial t} \tag{16.42}$$

Equations 16.41 and 16.42 lead to

$$\vec{\nabla}\cdot\left(\frac{\bar{E}}{\rho_v} + \varepsilon\frac{\partial\bar{E}}{\partial t}\right) = 0 \tag{16.43}$$

For, AC fields of angular frequency $\omega$, $\bar{E} = E_m\sin\omega t$. Hence,

$$\frac{\partial\bar{E}}{\partial t} = j\omega\bar{E}$$

Thus, Equation 16.43 can be rewritten as

$$\vec{\nabla}\cdot\left(\frac{\bar{E}}{\rho_v} + j\omega\varepsilon\bar{E}\right) = 0$$

$$\text{or, } \vec{\nabla}\cdot\left[\left(\frac{1}{\rho_v} + j\omega\varepsilon\right)\bar{E}\right] = 0 \tag{16.44}$$

Equation 16.44 shows that the fields including volume resistivity $\rho_v$ can be computed by replacing the permittivity $\varepsilon$ in purely capacitive field with the complex permittivity $\bar{\varepsilon}$ such that,

$$\vec{\nabla}\cdot(\bar{\varepsilon}\cdot\bar{E}) = 0 \tag{16.45}$$

where:

$$\left(\bar{\varepsilon} = \varepsilon + \frac{1}{j\omega\rho_v}\right)$$

Again, if $\bar{\varepsilon}$ is constant in the region of field calculation, then Equation 16.45 becomes Laplace's equation as given below.

$$\vec{\nabla}\cdot\bar{E} = 0$$

Equation 16.45 permits the use of CSM for capacitive-resistive field calculation including volume resistance. However, from the above discussion, it
becomes clear that in fields containing volume resistance, CSM cannot be
applied to problems where $\varepsilon$ or $\rho_v$ is dependent on the electric field. This is
because in such cases the field distribution cannot be expressed by superposing solutions of Laplace's equation.

The above method can be explained explicitly as described below. Consider
a two-dielectric arrangement, as shown in Figure 16.10. In Figure 16.10, the
two-dielectric media are assumed to have volume resistivities of $\rho_{vA}$ and $\rho_{vB}$,
respectively. The charges and the contour points are numbered in the same
way as that given in Section 16.3. However, in Section 16.3, only real fictitious
charges were taken for capacitive field calculation. But, for capacitive-resistive
field calculation including volume resistance, complex fictitious charges are
employed in place of real fictitious charges. The system of equations to be
solved for unknown charges is derived by imposing the boundary conditions
on the electrode surfaces and on the dielectric interfaces. The resulting equations with complex treatment are as follows.

1. Dirichlet's condition on the electrode surface:

$$\bar{\phi}(i) = \bar{V} \tag{16.46}$$

Equation 16.46 can be expanded for all the contour points on the
electrode surface in the following way.

$$\sum_{j=1}^{N_1} P_{ij}\bar{Q}_j + \sum_{j=N_1+N_2+1}^{N_1+2N_2} P_{ij}\bar{Q}_j = \bar{V} \quad \ldots i = 1,\ldots,N_A \tag{16.47}$$

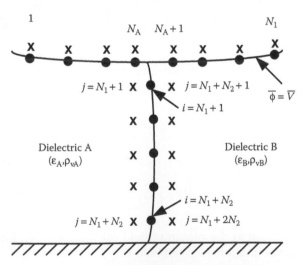

**FIGURE 16.10**
Multi-dielectric arrangement with volume resistivities.

$$\text{and} \sum_{j=1}^{N_1} P_{ij}\bar{Q}_j + \sum_{j=N_1+1}^{N_1+N_2} P_{ij}\bar{Q}_j = \bar{V} \quad \ldots i = N_A + 1, \ldots, N_1 \tag{16.48}$$

2. Potential continuity condition on the dielectric interface:

$$\bar{\phi}_A(i) = \bar{\phi}_B(i) \tag{16.49}$$

where:

The subscripts $A$ and $B$ denote dielectrics A and B, respectively

Equation 16.49 can be detailed explicitly as follows

$$\sum_{j=N_1+1}^{N_1+N_2} P_{ij}\bar{Q}_j - \sum_{j=N_1+N_2+1}^{N_1+N_2} P_{ij}\bar{Q}_j = 0 \quad \ldots i = N_1 + 1, \ldots, N_1 + N_2 \tag{16.50}$$

3. Continuity condition of $D_n$ on the dielectric interface:

$$\bar{D}_{nA}(i) - \bar{D}_{nB}(i) = \bar{\sigma}(i) \tag{16.51}$$

where:

$D_n$ and $\sigma$ represent normal component of electric flux density and surface charge density, respectively

Equation 16.51 can also be written as

$$\varepsilon_A \bar{E}_{nA}(i) - \varepsilon_B \bar{E}_{nB}(i) = \bar{\sigma}(i) \tag{16.52}$$

Now, the surface current density $\bar{J}(i)$ at any point $i$ on the dielectric interface due to volume resistance can be obtained as follows from Equation 16.53. For the case shown in Figure 16.11, $\bar{J}(i)$ is given by

$$\bar{J}(i) = \frac{\bar{E}_{nB}(i)}{\rho_{vB}} - \frac{\bar{E}_{nA}(i)}{\rho_{vA}} \tag{16.53}$$

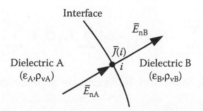

**FIGURE 16.11**
Determination of surface current density due to volume resistivities.

The surface charge density $\bar{\sigma}(i)$ at any point $i$ on the dielectric interface is given by

$$\bar{\sigma}(i) = \int \bar{J}(i)\,dt$$

Hence,

$$\bar{\sigma}(i) = \int \left[ \frac{\bar{E}_{nB}(i)}{\rho_{vB}} - \frac{\bar{E}_{nA}(i)}{\rho_{vA}} \right] dt \tag{16.54}$$

Now, for AC fields of angular frequency $\omega$,

$$\int \frac{\bar{E}_{nB}(i)}{\rho_{vB}}\,dt = \frac{\bar{E}_{nB}(i)}{j\omega\rho_{vB}}$$

Therefore, Equation 16.54 can be rewritten as

$$\bar{\sigma}(i) = \frac{\bar{E}_{nB}(i)}{j\omega\rho_{vB}} - \frac{\bar{E}_{nA}(i)}{j\omega\rho_{vA}} \tag{16.55}$$

Thus, Equations 16.52 and 16.55 lead to

$$\left( \varepsilon_A + \frac{1}{j\omega\rho_{vA}} \right)\bar{E}_{nA}(i) = \left( \varepsilon_B + \frac{1}{j\omega\rho_{vB}} \right)\bar{E}_{nB}(i) \tag{16.56}$$

$$\text{or,}\ \bar{\varepsilon}_A\bar{E}_{nA}(i) - \bar{\varepsilon}_B\bar{E}_{nB}(i) = 0 \tag{16.57}$$

where:

$$\bar{\varepsilon}_A = \left( \varepsilon_A + \frac{1}{j\omega\rho_{vA}} \right)$$

$$\bar{\varepsilon}_B = \left( \varepsilon_B + \frac{1}{j\omega\rho_{vB}} \right)$$

Equation 16.57 is given below in details

$$\bar{\varepsilon}_A\left( \sum_{j=1}^{N_1} F_{n,ij}\bar{Q}_j + \sum_{j=N_1+N_2+1}^{N_1+2N_2} F_{n,ij}\bar{Q}_j \right) - \bar{\varepsilon}_B\left( \sum_{j=1}^{N_1} F_{n,ij}\bar{Q}_j + \sum_{j=N_1+1}^{N_1+N_2} F_{n,ij}\bar{Q}_j \right) = 0 \tag{16.58}$$

Equations 16.47, 16.48, 16.50 and 16.58 can be represented in matrix form as given below.

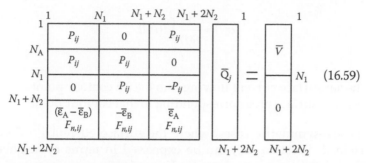

*Matrix 16.M.3 System of equations for capacitive-resistive field computation by CSM including volume resistance.*

It follows from Matrix 16.M.3, which is represented as Equation 16.59, that these are same as those for capacitive field, if the real values of the fictitious charges, permittivity, potential and field strength are replaced by their complex values.

### 16.6.2 Capacitive-Resistive Field Computation Including Surface Resistance

In fields including only surface resistance, true charges exist only on the boundary, that is, electrode and dielectric surfaces, and not inside the dielectric medium. As a result, the field distribution is always Laplacian inside each medium. This permits the application of CSM to capacitive-resistive field calculation, including surface resistance. The field distribution is obtained by superposing the effects of complex fictitious charges properly arranged inside the electrode and on both sides of the dielectric interface. The effect of true surface charges has to be incorporated into that of the complex fictitious charges.

The method can be better explained by considering the two-dielectric arrangement shown in Figure 16.12. The difference is that, in this case, the volume resistivities of the two dielectrics are considered to be infinite and a uniform surface resistivities $\rho_s$ is considered along the dielectric interface. The boundary conditions (1) and (2) as given by Equations 16.46 and 16.49, respectively, as well as the Equations 16.47, 16.48 and 16.50, in the case of volume resistance as given in sub-section 16.6.1, are also valid in the case of surface resistance. However, the expression of surface charge density $\sigma$ in continuity condition of $\bar{D}_n$, as given by Equation 16.51 in sub-section 16.6.1, has to be modified as follows.

In fields including surface resistance, the true surface charge density $\sigma(i)$ is expressed as given below.

$$\sigma(i) = \frac{1}{S(i)} \int_0^t I(i)\,dt \qquad (16.60)$$

where:
   $I(i)$ is the net surface current flowing into the $i$th contour point
   $S(i)$ is a small surface area corresponding to that contour point

For 2D or axi–symmetric cases, where the surface current $I(i)$ flows in a predetermined direction, $\sigma(i)$ can be expressed in terms of neighbouring potentials and resistances, as shown in Figure 16.13.

$$\sigma(i) = \frac{1}{S(i)} \int_0^t \left[ \frac{\phi(i-1)-\phi(i)}{R(i)} - \frac{\phi(i)-\phi(i+1)}{R(i+1)} \right] \qquad (16.61)$$

where:
   $R(i)$ and $R(i+1)$ are surface resistances corresponding to $i$th and $(i+1)$th contour points, respectively, as shown in Figure 16.13

The expressions for $R(i)$ and $S(i)$ are detailed below.

1. For 2D system (per unit length)

$$R(i) = \int_{i-1}^{i} \rho_s\,dl \qquad (16.62)$$

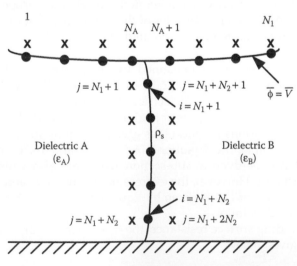

**FIGURE 16.12**
Multi-dielectric arrangement with surface resistivity.

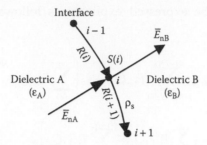

**FIGURE 16.13**
Determination of surface current density due to surface resistivity.

$$S(i) = \int\limits_{i-1}^{i+1} \frac{dl}{2} \tag{16.63}$$

2. For axi-symmetric system

$$R(i) = \int\limits_{i-1}^{i} \frac{\rho_s(i)}{2\pi r'} dl \tag{16.64}$$

$$S(i) = \int\limits_{i-1}^{i+1} \pi r' dl \tag{16.65}$$

where:

$r'$ is the $r$-coordinate of $dl$

$dl$ is a small length along the dielectric interface

For AC fields of angular frequency $\omega$, Equation 16.61 is modified as follows.

$$\overline{\sigma}(i) = \frac{1}{j\omega S(i)} \left[ \frac{\overline{\phi}(i-1) - \overline{\phi}(i)}{R(i)} - \frac{\overline{\phi}(i) - \overline{\phi}(i+1)}{R(i+1)} \right] \tag{16.66}$$

Hence, from Equations 16.52 and 16.66, it follows that

$$\varepsilon_A \overline{E}_{nA}(i) - \varepsilon_B \overline{E}_{nB}(i) = \frac{1}{j\omega S(i)} \left[ \frac{\overline{\phi}(i-1) - \overline{\phi}(i)}{R(i)} - \frac{\overline{\phi}(i) - \overline{\phi}(i+1)}{R(i+1)} \right] \tag{16.67}$$

Equation 16.67 can be expressed explicitly as follows for the system of Figure 16.12.

$$
\varepsilon_A \left( \sum_{j=1}^{N_1} F_{n,ij}\bar{Q}_j + \sum_{j=N_1+N_2+1}^{N_1+2N_2} F_{n,ij}\bar{Q}_j \right) - \varepsilon_B \left( \sum_{j=1}^{N_1} F_{n,ij}\bar{Q}_j + \sum_{j=N_1+1}^{N_1+N_2} F_{n,ij}\bar{Q}_j \right)
$$

$$
- \frac{1}{j\omega S(i)R(i)} \left( \sum_{j=1}^{N_1} P_{i-1,j}\bar{Q}_j + \sum_{j=N_1+1}^{N_1+N_2} P_{i-1,j}\bar{Q}_j \right)
$$

$$
- \frac{1}{j\omega S(i)R(i+1)} \left( \sum_{j=1}^{N_1} P_{i+1,j}\bar{Q}_j + \sum_{j=N_1+1}^{N_1+N_2} P_{i+1,j}\bar{Q}_j \right) \tag{16.68}
$$

$$
+ \frac{1}{j\omega S(i)} \left[ \frac{1}{R(i)} + \frac{1}{R(i+1)} \right] \left( \sum_{j=1}^{N_1} P_{i,j}\bar{Q}_j + \sum_{j=N_1+1}^{N_1+N_2} P_{i,j}\bar{Q}_j \right) = 0
$$

$$
\text{for } i = N_1+1,\ldots,N_1+N_2
$$

Now, the system of equations to be solved for unknown complex charges $\bar{Q}_j$, as given by Equations 16.47, 16.48, 16.50 and 16.68, can be expressed in matrix form, as shown in Matrix 16.M.4, which is represented as Equation 16.69.

Matrix 16.M.4 *System of equations for capacitive-resistive field computation by CSM including surface resistance.*

where:
$\bar{C}_1$ and $\bar{C}_2$ are two complex coefficients as given below:

$$
\bar{C}_1 = (\varepsilon_A - \varepsilon_B) F_{n,ij} - \frac{P_{i-1,j}}{j\omega S(i)R(i)} - \frac{P_{i+1,j}}{j\omega S(i)R(i+1)} + \frac{P_{ij}}{j\omega S(i)} \left[ \frac{1}{R(i)} + \frac{1}{R(i+1)} \right] \tag{16.70}
$$

$$
\bar{C}_2 = -\varepsilon_B F_{n,ij} - \frac{P_{i-1,j}}{j\omega S(i)R(i)} - \frac{P_{i+1,j}}{j\omega S(i)R(i+1)} + \frac{P_{ij}}{j\omega S(i)} \left[ \frac{1}{R(i)} + \frac{1}{R(i+1)} \right] \tag{16.71}
$$

From the matrix given in the sub-section 16.6.2 as well as the matrix given in sub-section 16.6.1, it is important to note that the equations contain not only complex charges but also complex coefficients.

## 16.7 Field Computation by CSM under Transient Voltage

Transient problems, where the applied voltage has an arbitrary waveform, are difficult to solve directly by CSM. For computing transient fields, there are two techniques. In the technique proposed by Singer [6], the transient voltages are decomposed into sinusoidal components by means of Fourier analysis or Fourier transformation. Field distribution due to individual sinusoidal components is calculated by CSM using complex fictitious charges. The complex AC responses for the needed frequencies are then weighted and summed up in order to get the time-dependent capacitive-resistive field distribution. In general, it is not necessary to calculate the field distribution for all the frequencies and it is sufficient to calculate the field distribution for some fixed reference frequencies. The results for the intermediate frequencies are then interpolated from the results for the reference frequencies.

In the other technique proposed by Takuma et al. [5] for transient field calculation, the integral of Equation 16.54 or 16.61 are approximated with a summation over a succession of sufficiently short time intervals to obtain the field distribution in relation to time discretely by CSM. Only real fictitious charges are used in this case for computation of transient capacitive-resistive field. This technique can be better explained in the following way.

In transient fields, where the voltage applied is $\phi = V(t)$, the field distribution is calculated by dividing the entire time span into short intervals $\Delta t$ and by converting the integral form of Equation 16.54 or 16.61 to the iterative summation. In this case, as the real fictitious charges are used, the Equations 16.47, 16.48 and 16.50 are also valid without the complex treatment, for the arrangement shown in Figure 16.10. However, the continuity condition of $\bar{D}_n$ has to be modified. The necessary modifications and the resulting equations for the transient field calculation including volume resistance or surface resistance are discussed below.

### 16.7.1 Transient Field Computation Including Volume Resistance

At the time instant $t_1 = \Delta t_1$, Equation 16.52 without the complex treatment can be written as

$$\varepsilon_A E_{nA}(i)_1 - \varepsilon_B E_{nB}(i)_1 = \sigma(i)_1 \tag{16.72}$$

where, the integral form of surface charge density, as given by Equation 16.54, is modified as follows:

$$\sigma(i)_1 = \Delta t_1 \left[ \frac{E_{nB}(i)_1}{\rho_B} - \frac{E_{nA}(i)_1}{\rho_A} \right] \tag{16.73}$$

where:
   The subscript 1 denotes time instant $t_1 = \Delta t_1$

Hence, Equation 16.72 can be written as follows:

$$\varepsilon_A E_{nA}(i)_1 - \varepsilon_B E_{nB}(i)_1 = \Delta t_1 \left[ \frac{E_{nB}(i)_1}{\rho_B} - \frac{E_{nA}(i)_1}{\rho_A} \right] \tag{16.74}$$

Then the real fictitious charges can be determined for the time instant $t_1 = \Delta t_1$ by solving a set of simultaneous linear equations constructed from Equation 16.74 along with Equations 16.47, 16.48 and 16.50 without the complex treatment.
   Then at the time instant $t_2 = \Delta t_1 + \Delta t_2$

$$\varepsilon_A E_{nA}(i)_2 - \varepsilon_B E_{nB}(i)_2 = \sigma(i)_2 \tag{16.75}$$

where:

$$\sigma(i)_2 = \sigma(i)_1 + \Delta t_2 \left[ \frac{E_{nB}(i)_2}{\rho_B} - \frac{E_{nA}(i)_2}{\rho_A} \right] \tag{16.76}$$

Because $\sigma(i)_1$ is known for the time instant $t_1 = \Delta t_1$, the real fictitious charges and the field distribution for the time instant $t_2 = \Delta t_1 + \Delta t_2$ can be obtained from Equations 16.75 and 16.76 along with Equations 16.47, 16.48 and 16.50 without the complex treatment. Thus, this iterative sequence gives the field distribution for $\phi = V(t)$ at any time instant.
   The equations to be solved for real fictitious charges at the $n$th time instant, $t_n = \Delta t_1 + \Delta t_2 + \cdots + \Delta t_n$ can be given in matrix form, as shown in Matrix 16.M.5, which is represented as Equation 16.77.

*Matrix 16.M.5 System of equations for transient field computation by CSM including volume resistance.*

where:

The subscripts $n$ and $n - 1$ denote $n$th and $(n - 1)$th time instants, respectively, and $K_1$, $K_2$ and $K_3$ are three real constants as detailed below

$$K_1 = \left( \varepsilon_A + \frac{\Delta t_n}{\rho_A} \right) - \left( \varepsilon_B + \frac{\Delta t_n}{\rho_B} \right), \quad K_2 = -\left( \varepsilon_B + \frac{\Delta t_n}{\rho_B} \right) \text{and } K_3 = \left( \varepsilon_A + \frac{\Delta t_n}{\rho_A} \right)$$

## 16.7.2 Transient Field Computation Including Surface Resistance

At the time instant $t_1 = \Delta t_1$, in the continuity condition of $D_n$ the expression for the surface charge density $\sigma(i)_1$ can be written as follows:

$$\sigma(i)_1 = \frac{\Delta t_1}{S(i)} \left[ \frac{\phi(i-1)_1 - \phi(i)_1}{R(i)} - \frac{\phi(i)_1 - \phi(i+1)_1}{R(i+1)} \right] \tag{16.78}$$

Equation 16.78 is a modified equation derived from the integral form, as given by Equation 16.61.

Thus, the continuity condition of $D_n$ can be written as follows:

$$\varepsilon_A E_{nA}(i)_1 - \varepsilon_B E_{nB}(i)_1 = \frac{\Delta t_1}{S(i)} \left[ \frac{\phi(i-1)_1 - \phi(i)_1}{R(i)} - \frac{\phi(i)_1 - \phi(i+1)_1}{R(i+1)} \right] \tag{16.79}$$

Hence, the real fictitious charges can be determined for the time instant $t_1 = \Delta t_1$ from Equation 16.79 along with Equations 16.47, 16.48 and 16.50 without the complex treatment.

Then at the time instant $t_2 = \Delta t_1 + \Delta t_2$

$$\varepsilon_A E_{nA}(i)_2 - \varepsilon_B E_{nB}(i)_2 = \sigma(i)_2$$

where:

$$\sigma(i)_2 = \sigma(i)_1 + \frac{\Delta t_2}{S(i)} \left[ \frac{\phi(i-1)_2 - \phi(i)_2}{R(i)} - \frac{\phi(i)_2 - \phi(i+1)_2}{R(i+1)} \right] \tag{16.80}$$

Using $\sigma(i)_1$ as obtained for the time instant $t_1 = \Delta t_1$ the real fictitious charges for the time instant $t_2 = \Delta t_1 + \Delta t_2$ can be obtained from Equation 16.80 along with Equations 16.47, 16.48 and 16.50 without the complex treatment. Hence, this iterative sequence gives the field distribution for $\phi = V(t)$ at any time instant.

The equations to be solved for real fictitious charges at the $n$th time instant, $t_n = \Delta t_1 + \Delta t_2 + \cdots + \Delta t_n$ can be given in matrix form, as shown in Matrix 16.M.6, which is represented as Equation 16.81.

$$
\begin{array}{c}
\quad\;\; 1 \qquad\quad N_1 \qquad N_1+N_2 \;\; N_1+2N_2 \qquad 1 \qquad\qquad 1 \\
\begin{array}{c}1\\N_A\\N_1\\N_1+N_2\\[2ex]N_1+2N_2\end{array}
\left[\begin{array}{ccc}
P_{ij} & 0 & P_{ij} \\
P_{ij} & P_{ij} & 0 \\
0 & P_{ij} & -P_{ij} \\
A_1 & A_2 & \dfrac{\varepsilon_A}{F_{n,ij}}
\end{array}\right]
\left[\,Q_{jn}\,\right]
=
\left[\begin{array}{c}
V_n \\
0 \\
\sigma_{n-1}
\end{array}\right]
\begin{array}{c}N_1\\N_1+N_2\\[2ex]N_1+2N_2\end{array}
\qquad (16.81)
\\[1ex]
\qqu\qquad\qquad N_1+2N_2 \qquad\qquad\qquad\qquad N_1+2N_2 \;\; N_1+2N_2
\end{array}
$$

*Matrix 16.M.6 System of equations for transient field computation by CSM including surface resistance.*

where:

$A_1$ and $A_2$ are two real coefficients as detailed below

$$
A_1 = (\varepsilon_A - \varepsilon_B)F_{n,ij} - \frac{\Delta t_n P_{i-1,j}}{S(i)R(i)} - \frac{\Delta t_n P_{i+1,j}}{S(i)R(i+1)} + \frac{\Delta t_n P_{ij}}{S(i)}\left[\frac{1}{R(i)} + \frac{1}{R(i+1)}\right] \qquad (16.82)
$$

$$
\text{and } A_2 = (-\varepsilon_B)F_{n,ij} - \frac{\Delta t_n P_{i-1,j}}{S(i)R(i)} - \frac{\Delta t_n P_{i+1,j}}{S(i)R(i+1)} + \frac{\Delta t_n P_{ij}}{S(i)}\left[\frac{1}{R(i)} + \frac{1}{R(i+1)}\right] \qquad (16.83)
$$

In general, the time interval $\Delta t$ for various time steps can have different values. However, they may be made equal for all the time steps for simplicity.

## 16.8 Accuracy Criteria

If the fictitious charges completely satisfy the boundary conditions, then these charges give the correct field distribution not only on the boundary but also everywhere outside it. But in the CSM, the fictitious charges are required to satisfy the boundary conditions only at a selected number of contour points. Again, the number of contour points is kept small in order to reduce the computer memory and computation time. Hence, it is essential to ensure that the simulation is accurate. To determine the simulation accuracy, the following criteria can be used.

1. The *potential error* on the electrode can be computed at a number of control points on the electrode surface between two contour points. Potential error is defined as the difference between the known potential of the electrode and the computed potential at the control point.

2. Compared to the potential error the *deviation angle* on the electrode surface is a more sensitive indicator of the simulation accuracy.

Deviation angle is defined as the angular deviation of the electric field intensity vector at the control point on the electrode surface from the direction of the normal to its surface.

Another very severe accuracy criterion is to check that the derivative of the potential gradient perpendicular to the electrode surface at the control point divided by the gradient itself is equal to the curvature at this point or not. This is especially applicable for the simulation of areas of the electrode with a small radius of curvature.

3. In multi-dielectric systems the *potential discrepancy* can be computed at a number of control points for each dielectric interface. Potential discrepancy is defined as the difference in the value of potential at the control point when computed from both the sides of the dielectric interface. Alternatively, the discrepancy in the tangential electric stress at the control points on the dielectric interface can also be computed. Another criterion for checking the simulation accuracy is to compute the discrepancy in the normal flux density at the control point on the dielectric interface.

For a good simulation all the above discrepancies should be small.

## 16.8.1 Factors Affecting Simulation Accuracy

The simulation accuracy in the CSM depends on the types and number of fictitious charges as well as locations of fictitious charges and contour points. In general, the simulation error can be reduced by increasing the number of charges. However, it has been found that increasing the number of fictitious charges beyond a certain limit does not necessarily improve the simulation accuracy. Generally, the *assignment factor* ($\lambda$) defined as the ratio of the distance between a contour point and the corresponding charge ($a2$) to the distance between two successive contour points ($a1$), as shown in Figure 16.14, considerably affects the simulation accuracy. Singer et al. [2] suggested that this factor should be between 1.0 and 2.0. Several others suggest a range of $0.7 < \lambda < 1.5$.

In a good simulation, potential error values as low as 0.001% are possible. However, for sharp corners and thin electrodes, such low values are difficult

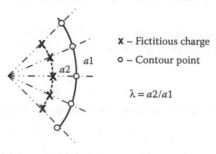

x – Fictitious charge

o – Contour point

$\lambda = a2/a1$

**FIGURE 16.14**
Definition of assignment factor.

to achieve. Because the electric field intensity error is an order of magnitude higher than the potential error, potential error values of about 0.1% are considered reasonable. For multi-dielectric systems, if the dielectric boundary has a complex shape, comparatively large potential discrepancy values of the order of 1% are usually acceptable.

Manufacturing tolerances of the conductors define the practical limit for the accuracy of the simulation of electrodes. In the same way, the accuracy of the determination of dielectric constants of the involved media puts the practical limit on the accuracy of the simulation of dielectrics.

### 16.8.2 Solution of System of Equations in CSM

Application of CSM for numerical field calculation involves solutions of linear systems of equations as explained in earlier sections. In the conventional CSM, for a single-dielectric case, the matrix of the linear system of equations to be solved is in general asymmetrical without a zero term, as detailed in Section 16.2. In such cases, the equations could be solved using the Gaussian elimination technique with or without partial or complete pivoting.

In multi-dielectric systems, the matrix of systems of equations to be solved is rather heterogeneous and is not symmetrical, as detailed in Section 16.3. Due to bad conditioning of the matrix, it is preferable to solve it by using a direct method, for example, Gaussian elimination technique, to avoid non-convergence problem, which may arise in the case of iterative methods. However, for complicated problems the size of the matrix becomes too large. In such cases, iterative methods such as Gauss–Seidel method or the successive over relaxation method with varying values of acceleration factor have also been found to be successful.

## 16.9 Other Development in CSM

### 16.9.1 Least Square Error CSM

In this method, compared to conventional CSM, boundary conditions are satisfied at a larger number of contour points than the number of charges, as shown in Figure 16.15.

Hence, in the case of least square error CSM (LSECSM), the matrix of system of equation is a rectangular one having $M$ rows and $N$ columns ($M > N$).

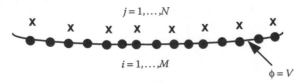

**FIGURE 16.15**
Fictitious charges and contour points for LSECSM.

$$[P]_{M \times N} [Q]_N = [V]_M \tag{16.84}$$

This equation is solved in the following way:

$$[P^t]_{N \times M} [P]_{M \times N} [Q]_N = [P^t]_{N \times M} [V]_M$$

$$\text{or, } [P']_{N \times N} [Q]_N = [V']_N \tag{16.85}$$

Equation 16.85 can be solved to find out the unknown fictitious charges. Anis et al. [7] most comprehensively presented this method. This method is expected to be more accurate than conventional CSM, but at the expense of more computation time. Again, the accuracy depends on the fitting ratio, that is, the ratio of number of contour points to the number of charges. This ratio should be kept between 1 and 2. This method is applicable to multi-dielectric systems too.

### 16.9.2 Optimized CSM

In conventional CSM and also in LSECSM, the positions of the charges are specified and their magnitudes are solved. In optimized CSM (OCSM), both magnitudes and locations of charges are determined by minimizing certain objective functions. Various versions of OCSM discussed in the literature differ in the choice of objective function and the optimization algorithm. Most authors have used the least square potential error as the objective function. Regarding the optimization techniques, constrained as well as unconstrained optimization has been used. Different algorithms such as the Fletcher method, Rosenbrock method and Pattern Search method can be used. OCSM are applicable to multi-dielectric systems also. Yalizis et al. [8] proposed OCSM in details.

For a fixed number of simulation charges, the optimized methods will produce more accurate results. However, such an increased accuracy will be obtained at the expense of more computation time as well as computer memory and will require more complex programming. Hence, it is recommended to be used in those problems where conventional CSM or LSECSM methods fail to produce adequate accuracy.

### 16.9.3 Region-Oriented CSM

Conventional CSM is also called *surface-oriented CSM* as discrete charges are used to simulate the electrode and dielectric surfaces. Conventional CSM suffers from difficulties associated with positioning the charges for complex geometries and thin electrodes. Region-oriented CSM (ROCSM) aims at removing these drawbacks and making CSM applicable for a wide variety of 2D and 3D problems in HV engineering. Blaszczyk et al. [9] proposed the ROCSM originally.

The basic concept of ROCSM is shown in Figure 16.16. A two-dielectric arrangement is divided into four regions. Each region is homogeneous with regard to its material properties, and consists of one linear dielectric (*R*1,

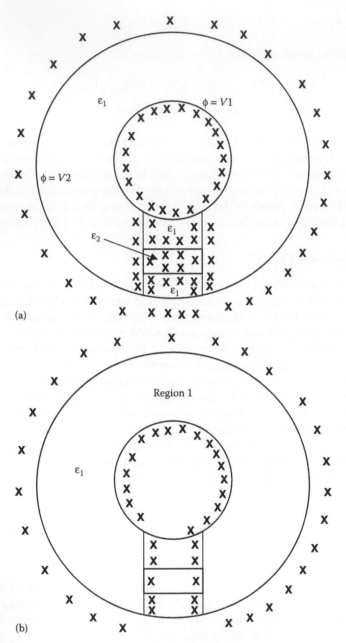

**FIGURE 16.16**
Concept of arrangement of charges in region-oriented CSM (ROCSM): (a) complete charge arrangement for ROCSM; (b) charge placement for Region 1; (c) charge placement for Region 2; (d) charge placement for Region 3 and (e) charge placement for Region 4.

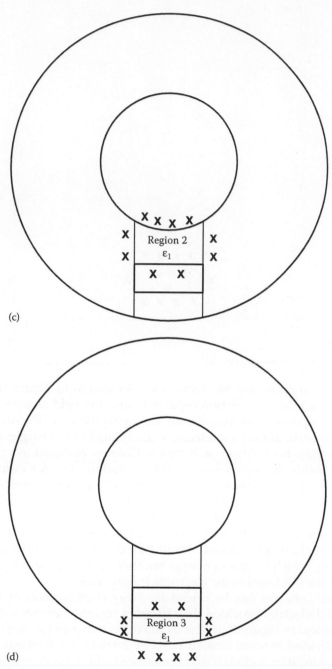

(c)

(d)

**FIGURE 16.16**
(Continued)

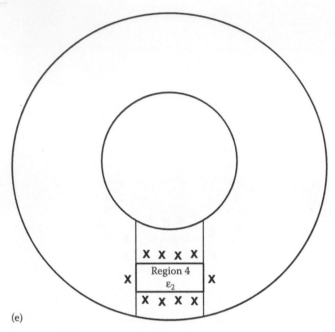

(e)

**FIGURE 16.16**
(Continued)

*R*2 and *R*3 contain $\varepsilon_1$ and *R*4 contains $\varepsilon_2$). As shown in Figure 16.16, a set of charges encloses each region separately and the field and the potential inside each region are calculated from the superposition of the surrounding charges. Interestingly, only a relatively small number of charges are necessary to calculate the fields in each region. Charges assigned to a region are not placed inside the region, but always at a certain distance away from its boundary. In this way, singularity problem can be avoided. Passing through an interface requires changing the set of charges used for field calculation.

An important advantage of the ROCSM is its ability to solve problems with thin conducting foils and thin dielectric layers. The conventional CSM requires that charges be placed within thin electrodes or dielectrics (*R*4 of Figure 16.16), which results in a large number of charges or is even impossible when the thickness of the electrode is very small.

In ROCSM, charges can be placed far away from surfaces of such electrodes and dielectrics provided that different regions have been defined on their two sides. In Figure 16.16, the large region between the two electrodes has been divided in some smaller parts by introducing fictitious boundaries. In this way, four regions have been created, although there are only two dielectric media in this example. Charges can then be placed easily for even smaller regions, as shown for Region 4 in Figure 16.16e.

## 16.10  Comparison of CSM with FEM

Both the FEM and CSM are extensively used for numerical calculation of electric field in high-voltage engineering.

In FEM, the entire ROI is subdivided into a large number of sub-regions and a set of coupled simultaneous linear equations, which minimize the electrostatic energy in the field region, are solved for unknown node potentials. On the other hand, in CSM, only the boundary surface, that is, the electrode surface and dielectric interface, is subdivided with fictitious charges, which are taken as unknowns. Therefore, it follows that the amount of time and effort needed for subdivision is greatly reduced in CSM. Moreover, the system of equations thus obtained by discretization is of smaller dimension in CSM.

FEM is useful for 2D and also 3D systems with or without symmetry and is advantageous for the calculation of fields where the boundaries have complicated shapes. However, for computing field distribution at a large distance from the HV electrodes by FEM, a large number of nodes and excessive computation time and computer memory space are required. Thus, FEM is more suited for problems where the space is bounded. On the contrary, the application of CSM is easy with high precision for field problems having infinitely extended unbounded region and for relatively simple boundary geometries but not so for fields with complex electrode configurations.

In FEM, exact field intensity at any point cannot be obtained. Instead average field intensity between two nodes is to be calculated from the known values of node potentials or numerical differentiation of the potential has to be done. But, in CSM, the electric field intensity can be obtained explicitly with the fictitious charges without resorting to numerical differentiation of the potential, which results in significant reduction in error. With proper positioning of the fictitious charges and the contour points and with the optimum number of fictitious charges, the potential and stress errors can be made less than 0.01% and 0.1%, respectively, in CSM. Though FEM is more suited for multiple dielectric problems, CSM can also be effectively employed for fields with many dielectrics.

A major disadvantage of CSM was that the electric field is difficult to calculate in systems having very thin electrodes because fictitious charges have to be placed within the electrodes. However, this disadvantage is obviated by the application of ROCSM in recent years. Further, CSM is usually, more accurate and less troublesome in computing Laplacian fields than FEM, but is difficult to use for non-Laplacian fields, for example, Poissonian fields. However, CSM with complex fictitious charges has been developed for calculating Poissonian field including volume and surface resistance providing very accurate results. Again, CSM is not suited for specific fields containing space charges where FEM can be employed very effectively. But, nowadays, suitable boundary conditions have been postulated for use in connection with CSM for computing spacer surface fields in compact gas-insulated substation (GIS) as modified by the charges accumulated on the spacer surface.

## 16.11 Hybrid Method Involving CSM and FEM

The most promising of the hybrid methods involving FEM and CSM is the so-called combination method, which has been independently proposed by Steinbigler [10] and Okubo et al. [11]. In general, it may be observed that CSM has more or less the opposite properties of FEM. Thus, attempts have been made to combine these two methods in a general-purpose high-precision method that takes the superior properties and excludes the inferior properties of these two methods. Higher precision in numerical field computation can be obtained if the field ROI is divided into parts to be analyzed by suitable methods. A field problem that could hardly be analyzed or that was analyzed approximately by only one method, may be analyzed with very good accuracy by applying an appropriate combination of different methods, for example, FEM and CSM.

In the FEM–CSM hybrid method, the entire space is divided into regions, which are to be analyzed separately by the CSM (C-domain) and by the FEM (F-domain). The boundary between the two regions is called the *combined surface*. Due to the properties of each method, the CSM is mainly used for open areas with infinite boundary and the FEM is used for finite enclosed space usually containing dielectric interfaces, conductive dielectrics, space charges and so on. It is to be noted that though CSM and FEM are two different methods, they result into similar linear system of equations. The coupling between C-domain and F-domain is based on the fact that the potential and the normal flux density must be continuous at the combined surface. Figure 16.17a and b show how the entire field region is divided into CSM-region and FEM-region in the application of hybrid method in a two-dielectric media and in space-charge-modified field computation, respectively.

Studies on combination method indicate that it offers advantages over the conventional CSM in 2D and 3D fields with axial symmetry for situation where space charges or conductive regions are present. Also, for the computation of 3D fields without axial symmetry, the advantages of combination method are significant.

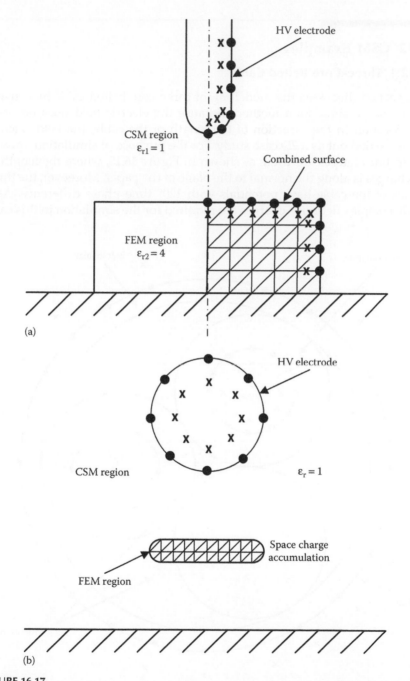

**FIGURE 16.17**
Separation of the field region into CSM region and FEM region in hybrid method: (a) application in a two-dielectric media and (b) application in space-charge modified field.

## 16.12 CSM Examples

### 16.12.1 Three-Core Belted Cable

This section discusses the modelling of three-core belted cable by conventional charge simulation method. Because the electric field does not vary with location in the direction of the length of the cable, the field calculation is carried out as a 2D case study. For the purpose of simulation, infinite length line charges are used, as shown in Figure 16.18, where the length of the charges is along the normal to the plane of the paper. Moreover, the three phases of the cable have potentials with 120° time phase difference. As a result, complex fictitious charges are required for the simulation in this case.

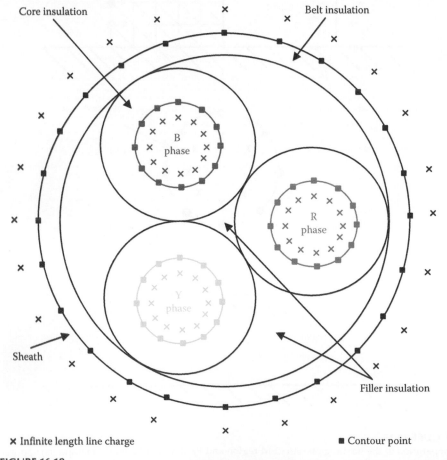

✕ Infinite length line charge          ■ Contour point

**FIGURE 16.18**
**(See colour insert.)** Simulation of a three-core belted cable by CSM using complex fictitious charges.

Although stranded conductors are used for cable core, the conductor surfaces are taken as smooth due to the presence of the conductive shield over the conductor strands. The insulating medium throughout the cross section of the cable may be taken as uniform, because the core, belt and filler insulations are made of either impregnated paper or impregnated jute fibre having similar relative permittivities between 3 and 4.

The sheath, which is at zero potential, could be handled in two different ways: (1) using fictitious charges placed beyond the sheath with appropriate contour points on the sheath surface, as shown in Figure 16.18. Because the general earth surface does not play any role in determining the field within the cable, the image charge with respect to the general earth surface, as discussed in Section 16.4.2, is not applicable in this case. Only the expression for the fictitious charge is to be taken into computation and (2) the sheath can also be incorporated in the simulation by taking the image of every fictitious charge in the manner discussed in Section 9.2.2. In such case, the number of fictitious charges required for simulation is less as no additional charge is needed for simulating the sheath surface and the obtained computational accuracy is also better.

## 16.12.2 Sphere Gap

Sphere-gap arrangements are extensively used in high-voltage systems for measurement as well as for switching purposes. Typically, a sphere-gap arrangement consists of two spheres with cylindrical shanks, as shown in Figure 16.19. Electric field computation in such arrangement is carried out as an axi-symmetric case study. Each cylindrical shank is simulated by one set of finite length line charges placed on the axis of symmetry. It is to be mentioned here that finite length line charges in an axi-symmetric arrangement should only be placed on the axis of symmetry. Because if a finite length line charge is placed in any place other than the axis of symmetry in an axi-symmetric system, then it no longer remains a line charge but becomes a cylinder/conical charge due to axial symmetry. Each sphere is simulated by one set of ring charges, although the tip of the sphere is simulated by one point charge, as shown in Figure 16.19. Considering air to be the dielectric between the two spheres, it is a case of electric field computation with only one dielectric medium.

## 16.12.3 Single-Core Cable Termination with Stress Cone

Insulated cables are used extensively for transmission and distribution of electrical power. Termination is the way in which the end of a cable is finished, so that the cable matches a power supply or another device/equipment. Such cable terminations are subjected to considerable electrical stresses during operation. A proper design of cable termination is essential in reducing the electric field distribution in and around cable termination. Figure 16.20 shows the conventional CSM simulation of a traditional single-core fluid-filled outdoor cable

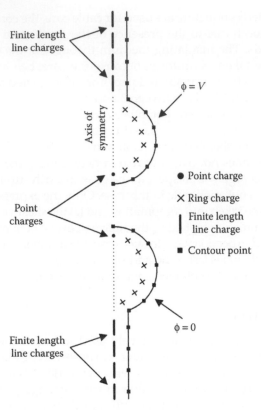

**FIGURE 16.19**
Simulation of a sphere-gap arrangement by CSM.

termination. Contrary to electric field computation within a cable, cable termination being of finite length needs to be analyzed as an axi-symmetric case study. Typically, such a cable termination comprises cylindrical cable core at live potential, earthed unit known as the deflector with semi-conducting covering, cable core insulation (e.g. cross-linked polyethylene or XLPE), insulating stress cone (e.g. silicone rubber) and the outer insulating cover (e.g. porcelain). The space within the outer insulating cover is filled with an insulating fluid. In Figure 16.20, the outer porcelain covering is taken to be cylindrical in shape for simplicity. It may be noted here that the shape of the outer covering plays insignificant role in determining the field stresses at the critical zones within the cable termination.

As shown in Figure 16.20, the cable core is simulated by one set of finite length line charges placed on the axis of symmetry and the earthed deflector is simulated by one set of ring charges. There are five dielectric–dielectric boundaries in this case, namely, XLPE–rubber, XLPE–oil, rubber–oil, oil–porcelain and porcelain–air boundaries. Each one of these five dielectric–dielectric boundaries is simulated by two sets of ring charges, such that the charges placed in a

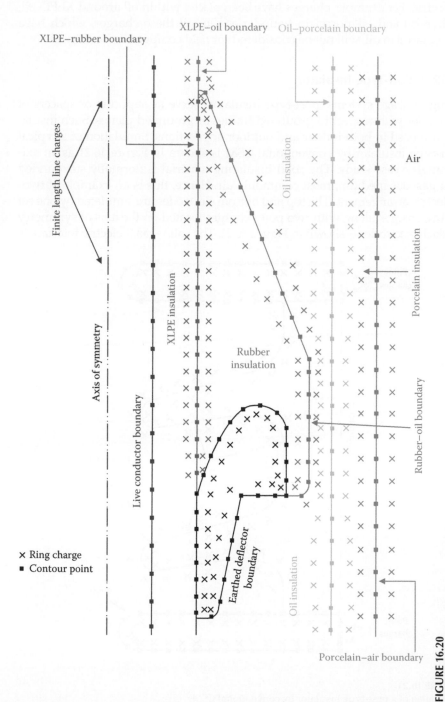

**FIGURE 16.20**

(See colour insert.) Simulation of fluid-filled outdoor cable termination by conventional CSM.

particular dielectric is not to be considered for field computation within that dielectric. For example, charges have been placed within oil around XLPE–oil, rubber–oil and oil–porcelain boundaries. But all these charges, which have been placed in oil, will not be considered for field computation within oil.

### 16.12.4 Post-Type Insulator

In high-voltage systems, post-type insulators serve as supports or spacers of electrodes with respect to grounded frames or grounded planes. Such insulators are used in both indoor and outdoor applications. Simulation of a typical post-type insulator by conventional CSM is shown in Figure 16.21 as an axisymmetric case study. The solid insulating material is normally surrounded by a gaseous medium, most commonly air. Hence, this is an example of two-dielectric arrangement. The top and bottom electrodes are simulated by one set of ring charges along with two point charges located on the axis of symmetry for each electrode, as shown in Figure 16.21. The solid–gas dielectric interface is

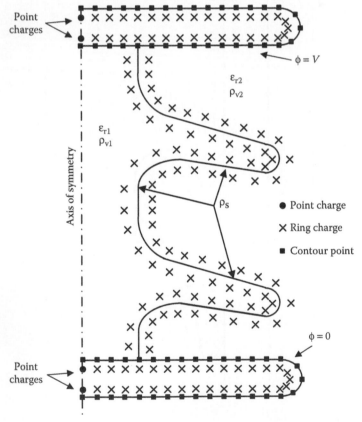

**FIGURE 16.21**
Simulation of a post-type insulator by conventional CSM.

simulated by two sets of ring charges, one set each for solid and gaseous media. The effect of volume as well as surface resistivities could be easily incorporated in the field computation, as discussed in Sections 16.6.1 and 16.6.2.

### 16.12.5 Asymmetric Sphere Gaps

Section 16.12.2 discussed symmetric sphere-gap arrangement, as shown in Figure 16.19. But if a third sphere along with its shank is introduced into the system, as shown in Figure 16.22, then it becomes an asymmetric configuration. Simulation of such a system is not possible with ring charges

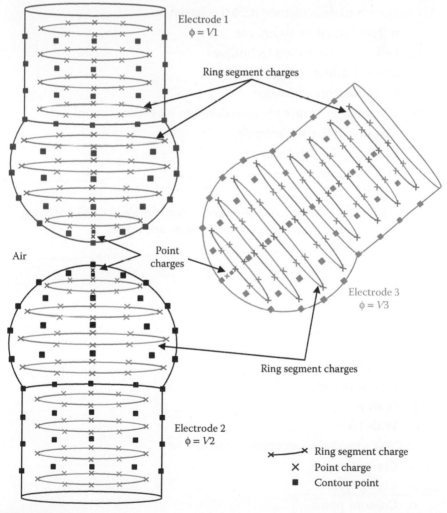

**FIGURE 16.22**
**(See colour insert.)** Simulation of an asymmetric sphere-gap arrangement by CSM.

as the charge density will not be uniform along the length of the ring charge. Consequently, such an arrangement needs to be computed as a 3D case study. As shown in Figure 16.22, ring segment charges are used for simulating the asymmetric sphere-gap arrangement. Only the tip of the three spheres is simulated by one point charge each. The expressions, as discussed in Section 16.4.6, are to be used in this case study.

## Objective Type Questions

1. Charge simulation method (CSM) is based on
   a. Integral equation technique
   b. Differential equation technique
   c. Monte Carlo technique
   d. Linear algebra technique
2. In CSM, image charges are considered to take into account
   a. Negative polarity electrode
   b. Positive polarity electrode
   c. Earth surface
   d. Both (a) and (c)
3. To simulate the conductor boundary in conventional CSM, for every fictitious charge how many contour points are taken?
   a. 1
   b. 2
   c. 3
   d. 0
4. For a point charge, electric field intensity at a distance $r$ from the charge varies
   a. Linearly with $r$
   b. Inversely with $r$
   c. With $r^2$
   d. With $1/r^2$
5. In CSM, simulation accuracy is checked at
   a. Control point
   b. Charge point
   c. Contour point
   d. None of the above

6. For the formation of coefficient matrix in CSM, fictitious charge
   a. Magnitudes are known but locations are unknown
   b. Magnitudes are unknown but locations are known
   c. Both magnitudes and locations are known
   d. Both magnitudes and locations are unknown
7. In CSM, control points are placed in between
   a. Two successive charge points
   b. A charge point and a contour point
   c. Two successive contour points
   d. None of the above
8. CSM is suitable for solving
   a. 2D Laplacian field
   b. 2D Poissonian field
   c. 3D Laplacian field
   d. Both (a) and (c)
9. CSM is suitable for simulating configurations
   a. Having thin electrodes
   b. Field including space charges
   c. Field including non-linear resistivity
   d. Field including isotropic media
10. In CSM, which one of the following is true?
    a. Entire field region is discretized
    b. Only conductor boundaries are discretized
    c. Only dielectric boundaries are discretized
    d. Both (b) and (c)
11. In CSM, singularity problem arises at the
    a. Control point
    b. Charge point
    c. Contour point
    d. Image charge point
12. Matrix representation of the system of equation in CSM gives rise to
    a. Sparse matrix
    b. Nearly full matrix
    c. Diagonal matrix
    d. Unit matrix

13. In CSM, if the number of charges is increased, then accuracy
    a. Initially increases then decreases
    b. Initially decreases then increases
    c. Initially increases then remains constant
    d. Initially decreases then remains constant

14. In CSM, fictitious charges are placed
    a. Within the field region of interest
    b. Outside the field region of interest
    c. On the boundaries
    d. Both (a) and (c)

15. For better simulation accuracy, the assignment factor in placing the fictitious charges should preferably lie between
    a. 1 and 2
    b. 0 and 1
    c. −1 and +1
    d. None of the above

16. On the conductor boundary, simulation accuracy is checked by calculating
    a. Potential error
    b. Potential discrepancy
    c. Deviation angle
    d. Both (a) and (c)

17. Use of fictitious charges having complex shapes results in
    a. Higher computation time and higher memory requirement
    b. Lower computation time and lower memory requirement
    c. Higher computation time and lower memory requirement
    d. Lower computation time and higher memory requirement

18. Ring charges of uniform charge density are suitable for simulating
    a. 2D configuration
    b. Axi-symmetric configuration
    c. 3D configuration
    d. Both (a) and (b)

19. In least square error CSM, the number of fictitious charges is
    a. Zero
    b. Equal to the number of contour points

c.  Less than the number of contour points

d.  More than the number of contour points

20. Optimized CSM results in

a.  Higher computation time and better accuracy

b.  Higher computation time and poor accuracy

c.  Lower computation time and better accuracy

d.  Lower computation time and poor accuracy

21. In conventional CSM, for every contour point on the dielectric–dielectric interface how many fictitious charges are taken?

a.  None

b.  1

c.  2

d.  3

22. To eliminate the problems of charge placement within thin electrodes, which one of the following methods is most suited?

a.  Least square error CSM

b.  Region-oriented CSM

c.  Optimized CSM

d.  Conventional CSM

23. For electric field calculation in a three-phase system with one dielectric medium, which one of the following is employed?

a.  Real fictitious charges and real permittivity

b.  Real fictitious charges and complex permittivity

c.  Complex fictitious charges and real permittivity

d.  Complex fictitious charges and complex permittivity

24. For resistive-capacitive field calculation by CSM, it is necessary that true charges

a.  Exist all over the field region

b.  Exist on boundaries only

c.  Exist on selected boundaries only

d.  Do not exist

25. For field calculation with volume resistance by CSM, which one of the following is employed?

a.  Real fictitious charges and real permittivity

b.  Real fictitious charges and complex permittivity

   c. Complex fictitious charges and real permittivity

   d. Complex fictitious charges and complex permittivity

26. For field calculation with surface resistance by CSM, which one of the following is employed?

   a. Real fictitious charges and real permittivity

   b. Real fictitious charges and complex permittivity

   c. Complex fictitious charges and real permittivity

   d. Complex fictitious charges and complex permittivity

27. Field calculation by CSM under transient voltage is carried out with the help of

   a. Fourier series

   b. Fourier transform

   c. Time-step discretization

   d. Both (a) and (c)

28. Field calculation by CSM under transient voltage with volume resistance with the help of time-step discretization is carried out employing

   a. Real fictitious charges

   b. Complex fictitious charges

   c. Complex permittivity

   d. None of the above

29. Field calculation by CSM under transient voltage with volume resistance with the help of Fourier series expansion is carried out employing

   a. Real fictitious charges

   b. Complex fictitious charges

   c. Complex permittivity

   d. Both (b) and (c)

30. Field calculation by CSM under transient voltage with surface resistance with the help of time-step discretization is carried out employing

   a. Real fictitious charges and real permittivity

   b. Real fictitious charges and complex permittivity

   c. Complex fictitious charges and real permittivity

   d. Complex fictitious charges and complex permittivity

**Answers:**   1) a; 2) c; 3) a; 4) d; 5) a; 6) b; 7) c; 8) d; 9) d; 10) d; 11) b; 12) b; 13) c; 14) b; 15) a; 16) d; 17) c; 18) b; 19) c; 20) a; 21) c; 22) b; 23) c; 24) b; 25) d; 26) c; 27) d; 28) a; 29) d; 30) a

## Bibliography

1. M.S. Abou-Seada and E. Nasser, 'Digital computer calculation of the potential and its gradient of a twin cylindrical conductor', *IEEE Transactions on Power Apparatus and Systems*, Vol. 88, pp. 1802–1814, 1969.
2. H. Singer, H. Steinbigler and P. Weiss, 'A charge simulation method for the calculation of high voltage fields', *IEEE Transactions on Power Apparatus and Systems*, Vol. 93, pp. 1660–1668, 1974.
3. D. Utmischi, 'Charge simulation method for three dimensional high voltage fields', *Proceedings of the 3rd ISH*, Milan, Italy, paper 11.01, August 28–31, IEEE North Italy Section Publication, 1979.
4. B. Bachmann, 'Models for computation of mixed fields with charge method', *Proceedings of the 3rd ISH*, Milan, Italy, paper 12.05, August 28–31, IEEE North Italy Section Publication, 1979.
5. T. Takuma, T. Kawamoto and H. Fujinami, 'Charge simulation method with complex fictitious charges for calculating capacitive-resistive fields', *IEEE Transactions on Power Apparatus and Systems*, Vol. 100, pp. 4665–4672, 1981.
6. H. Singer, 'Impulse stresses of conductive dielectrics', *Proceedings of the 4th ISH*, Athens, Greece, paper 11.02, September 5–9, National Technical University of Athens Publication, 1983.
7. H. Anis, A. Zeitoun, M. El-Ragheb and M. El-Desouki, 'Field calculations around non-standard electrodes using regression and their spherical equivalence', *IEEE Transactions on Power Apparatus and Systems*, Vol. 96, pp. 1721–1730, 1977.
8. A. Yalizis, E. Kuffel and P.H. Alexander, 'An optimized charge simulation method for the calculation of high voltage fields', *IEEE Transactions on Power Apparatus and Systems*, Vol. 97, pp. 2434–2440, 1978.
9. A. Blaszczyk and H. Steinbigler, 'Region-Oriented Charge Simulation', *IEEE Transactions on Magnetics*, Vol. 30, No. 5, pp. 2924–2927, 1994.
10. H. Steinbigler, 'Combined application of finite element method and charge simulation method for the computation of electric fields', *Proceedings of the 3rd ISH*, Milan, Italy, paper 11.11, August 28–31, IEEE North Italy Section Publication, 1979.
11. H. Okubo, M. Ikeda and M. Honda, 'Combination method for the electric field calculation', *Proceedings of the 3rd ISH*, Milan, Italy, paper 11.13, August 28–31, IEEE North Italy Section Publication, 1979.
12. M.S. Abou-Seada and E. Nasser, 'Digital computer calculation of the electric potential and field of a rod gap', *Proceedings of IEEE*, Vol. 56, No. 5, pp. 813–820, 1968.
13. M.P. Sarma and W. Janischewskyi, 'Electrostatic field of a system of parallel cylindrical conductors', *IEEE Transactions on Power Apparatus and Systems*, Vol. 88, pp. 1069–1079, 1969.
14. M. Khalid 'Computation of corona onset using the ring charge method', *Proceedings of IEEE*, Vol. 122, pp. 107–110, 1975.
15. H. Parekh, M.S. Selim and E. Nasser, 'Computation of electric field and potential around stranded conductor by analytical method and comparison with charge simulation method', *Proceedings of IEEE*, Vol. 122, pp. 547–550, 1975.
16. R. Malewski and P.S. Maruvada, 'Computer aided design of impulse voltage dividers', *IEEE Transactions on Power Apparatus and Systems*, Vol. 95, pp. 1267–1274, 1976.

17. A. Nosseir and A.A. Zaky, 'Application of the method of charge simulation of the study of electrical stresses in three-core cables', *IEEE Transactions on Electrical Insulation*, Vol. 12, pp. 262–266, 1977.

18. P.K. Mukherjee and C.K. Roy, 'Computations of fields in and around insulators by fictitious point charges', *IEEE Transactions on Electrical Insulation*, Vol. 13, pp. 24–31, 1978.

19. T. Takuma, T. Kouno and H. Matsuda, 'Field behavior near singular points in composite dielectric arrangements', *IEEE Transactions on Electrical Insulation*, Vol. 13, pp. 426–435, 1978.

20. A. Nosseir and A.A. Zaky, 'Representation of three-core cables using line charge simulation', *IEEE Transactions on Electrical Insulation*, Vol. 14, pp. 207–210, 1979.

21. M.D.R. Beasley, J.H. Pickles, G. d'Amico, L. Beretta, M. Fanelli, G. Giuseppetti, A. di Monaco, G. Gallet, J.P. Gregorie and M. Morin, 'A comparative study of three methods for computing electric fields', *Proceedings of IEEE*, Vol. 126, pp. 126–134, August 28–31, IEEE North Italy Section Publication, 1979.

22. H. Singer, 'Computation of optimized electrode geometries', *Proceedings of the 3rd ISH*, Milan, Italy, paper 11.06, August 28–31, IEEE North Italy Section Publication, 1979.

23. D. Metz, 'Optimization of high voltage fields', *Proceedings of the 3rd ISH*, Milan, Italy, paper 11.12, August 28–31, IEEE North Italy Section Publication, 1979.

24. S. Kato, 'An estimation method for the electric field error of a charge simulation method', *Proceedings of the 3rd ISH*, Milan, Italy, paper 11.09, August 28–31, IEEE North Italy Section Publication, 1979.

25. S. Sato, S. Menju, K. Aoyagi and M. Honda, 'Electric field calculation in 2 dimensional multiple dielectric by the use of elliptic cylinder charge', *Proceedings of the 3rd ISH*, Milan, Italy, paper 11.03, August 28–31, IEEE North Italy Section Publication, 1979.

26. T. Takuma and T. Kawamoto, 'Field calculation including surface resistance by charge simulation method', *Proceedings of the 3rd ISH*, Milan, Italy, paper 12.01, August 28–31, IEEE North Italy Section Publication, 1979.

27. Z. Haznadar, S. Milojkovic and I. Kamenica, 'Numerical field calculation of insulator chains for high voltage transmission lines', *Proceedings of the 3rd ISH*, Milan, Italy, paper 12.08, August 28–31, IEEE North Italy Section Publication, 1979.

28. T. Sakakibara, S. Sato, N. Kobayashi and S. Menju, *The Application of Charge Simulation Method to Three Dimensional Asymmetric Field with Two Dielectric Media*, Gaseous Dielectrics II, Pergamon Press, New York, pp. 312–321, 1980.

29. N.G. Trinh, 'Electrode design for testing in uniform field gaps', *IEEE Transactions on Power Apparatus and Systems*, Vol. 99, pp. 1235–1242, 1980.

30. S. Sato and S. Menju, 'Digital calculation of electric field by charge simulation method using axi-spheroidal charges', *Electrical Engineering in Japan*, Vol. 100, pp. 1–8, 1980.

31. T. Takashima, T. Nakae and R. Ishibashi, 'Calculation of complex fields in conducting media', *IEEE Transactions on Electrical Insulation*, Vol. 15, pp. 1–7, 1980.

32. M.J. Khan and P.H. Alexander, 'Charge simulation modeling of practical insulator geometries', *IEEE Transactions on Electrical Insulation*, Vol. 17, pp. 325–332, 1982.

33. K. Tsuruta and R. Terakado, 'Calculation of potential distribution and capacitance of coaxial system with an outer cylinder by the charge simulation method using plane sheet charges', *Electrical Engineering in Japan*, Vol. 102, pp. 1–6, 1982.

34. M.R. Iravani and M.R. Raghuveer, 'Accurate field solution in the entire inter-electrode space of a rod-plane gap using optimized charge simulation', *IEEE Transactions on Electrical Insulation*, Vol. 17, pp. 333–337, 1982.
35. S. Murashima, H. Kondo, M. Yokoi and H. Nieda, 'Relation between the error of the charge simulation method and the location of charges', *Electrical Engineering in Japan*, Vol. 102, pp. 1–9, 1982.
36. H. Okubo, M. Ikeda, M. Honda and T. Yanari, 'Electric field analysis by combination method', *IEEE Transactions on Power Apparatus and Systems*, Vol. 101, pp. 4039–4048, 1982.
37. A.S. Pillai, R. Hackam and P.H. Alexander, 'Influence of radius of curvature, contact angle and material of solid insulation on the electric field in vacuum (and gaseous) gaps', *IEEE Transactions on Electrical Insulation*, Vol. 18, pp. 11–22, 1983.
38. M.R. Iravani and M.R. Raghuveer, 'Numerical computation of potential distribution along a transmission line insulator chain', *IEEE Transactions on Electrical Insulation*, Vol. 18, pp. 167–170, 1983.
39. P.K. Mukherjee and C.K. Roy, 'Computation of electric field in a condenser bushing by using fictitious area charges', *Proceedings of the 4th ISH*, Athens, Greece, paper 12.09, September 5–9, National Technical University of Athens Publication, 1983.
40. M. Youssef, 'An accurate fitting-oriented charge simulation method for electric field calculation', *Proceedings of the 4th ISH*, Athens, Greece, paper 11.13, September 5–9, National Technical University of Athens Publication, 1983.
41. H. Anis and A. Mohsen, 'Application of the charge simulation method to time-varying voltages', *Proceedings of the 4th ISH*, Athens, Greece, paper 11.11, September 5–9, National Technical University of Athens Publication, 1983.
42. M.M.A. Salama and R. Hackam, 'Voltage and electric field distribution and discharge inception voltage in insulated conductors', *IEEE Transactions on Power Apparatus and Systems*, Vol. 103, pp. 3425–3433, 1984.
43. M.M.A. Salama, A. Nosseir, R. Hackam, A. Soliman and T. El-Sheikh, 'Methods of calculations of field stresses in a three core power cable', *IEEE Transactions on Power Apparatus and Systems*, Vol. 103, 1984, pp. 3424–3441.
44. A.S. Pillai and R. Hackam, 'Optimal electrode solid insulator geometry with accumulated surface charges', *IEEE Transactions on Electrical Insulation*, Vol. 19, pp. 321–331, 1984.
45. M.N. Horenstein, 'Computation of corona space charge, electric field and V-I characteristic using equipotential charge shells', *IEEE Transactions on Industry Applications*, Vol. 20, pp. 1607–1612, 1984.
46. N.H. Malik and A. Al-Arainy, 'Charge simulation modeling of three-core belted cables', *IEEE Transactions on Electrical Insulation*, Vol. 20, pp. 499–503, 1985.
47. A.A. Elmoursi and G.S.P. Castle, 'The analysis of corona quenching in cylindrical precipitators using charge simulation', *IEEE Transactions on Industry Applications*, Vol. 22, pp. 80–85, 1986.
48. M. Abdel-Salam and E.K. Stanek, 'Field optimization of high voltage insulators', *IEEE Transactions on Industry Applications*, Vol. 22, pp. 594–601, 1986.
49. A.A. Elmoursi and G.S.P. Castle, 'Modeling of corona characteristics in a wire-duct precipitator using the charge simulation technique', *IEEE Transactions on Industry Applications*, Vol. 23, pp. 95–102, 1987.
50. S. Sato, W.S. Zaengl and A. Knecht, 'A numerical analysis of accumulated surface charge of DC epoxy resin spacer', *IEEE Transactions on Electrical Insulation*, Vol. 22, pp. 333–340, 1987.

51. N.H. Malik and A. Al-Arainy, 'Electrical stress distribution in three core belted power cables', *IEEE Transactions on Power Delivery*, Vol. 2, pp. 589–595, 1987.

52. M. Abdel-Salam and E.K. Stanek, 'Optimizing field stress of high voltage insulators', *IEEE Transactions on Electrical Insulation*, Vol. 22, pp. 47–56, 1987.

53. P.L. Levin, J.F. Hoburg and Z.J. Cendes, 'Charge simulation and interactive graphics in a first course in applied electro-magnetics', *IEEE Transactions on Education*, Vol. 30, pp. 5–8, 1987.

54. T.H. Fawzi and Y. Safar, 'Boundary methods for the analysis and design of high voltage insulators', *Computer Methods in Applied Mechanics and Engineering*, Vol. 60, pp. 343–369, 1987.

55. N.H. Malik, A. Al-Arainy, A.M. Kailani and M.J. Khan, 'Discharge inception voltages due to voids in power cables', *IEEE Transactions on Electrical Insulation*, Vol. 22, pp. 787–793, 1987.

56. M. Abdel-Salam, H. Abdallah and S. Abdel-Sattar, 'Positive corona in point-plane gaps as influenced by wind', *IEEE Transactions on Electrical Insulation*, Vol. 22, pp. 775–786, 1987.

57. M.S. Rizk and R. Hackam, 'Performance improvement of insulators in a gas-insulated system', *IEEE Transactions on Electrical Insulation*, Vol. 22, pp. 439–446, 1987.

58. S. Chakravorti and P.K. Mukherjee, 'Efficient field calculation in three core belted cable by charge simulation using complex charges', *IEEE Transactions on Electrical Insulation*, Vol. 27, pp. 1208–1212, 1992.

59. S. Chakravorti and P.K. Mukherjee, 'Power frequency and impulse field calculation around a HV insulator with uniform or non-uniform surface pollution', *IEEE Transactions on Electrical Insulation*, Vol. 28, pp. 43–53, 1993.

60. M. Abdel-Salam and E.Z. Abdel-Aziz, 'New charge-simulation-based method for analysis of monopolar Poissonian fields', *Journal of Physics D: Applied Physics*, Vol. 27, pp. 807–817, 1994.

61. H. El-Kishky and R.S. Gorur, 'Electric potential and field computation along ac HV insulators', *IEEE Transactions on Dielectrics & Electrical Insulation*, Vol. 1, pp. 982–990, 1994.

62. S. Schmidt, 'Fast and precise computation of electrostatic fields with a charge simulation method using modern programming techniques', *IEEE Transactions on Magnetics*, Vol. 32, No. 3, pp. 1457–1460, 1996.

63. T. Takuma and T. Kawamoto, 'Numerical calculation of electric fields with a floating conductor', *IEEE Transactions on Dielectrics & Electrical Insulation*, Vol. 4, No. 2, pp. 177–181, 1997.

64. I.A. Metwally, 'Electrostatic and environmental analyses of high phase order transmission lines', *Electric Power Systems Research*, Vol. 61, pp. 149–159, 2002.

65. R. Nishimura, M. Nishihara, K. Nishimori and N. Ishihara, 'Automatic arrangement of fictitious charges and contour points in charge simulation method for two spherical electrodes', *Journal of Electrostatics*, Vol. 57, pp. 337–346, 2003.

66. R. Nishimura and K. Nishimori, 'Arrangement of fictitious charges and contour points in charge simulation method for electrodes with 3-D asymmetrical structure by immune algorithm', *Journal of Electrostatics*, Vol. 63, pp. 743–748, 2005.

67. H. Wei, Y. Fan, W. Jingang, Y. Hao, C. Minyou and Y. Degui, 'Inverse application of charge simulation method in detecting faulty ceramic insulators and processing influence from tower', *IEEE Transactions on Magnetics*, Vol. 42, pp. 723–726, 2006.

68. H.M. Ismail, 'Effect of oil pipelines existing in an HVTL corridor on the electric-field distribution', *IEEE Transactions on Power Delivery*, Vol. 22, pp. 2466–2472, 2007.
69. A. El-Zein, M. Talaat and M. El Bahy, 'A numerical model of electrical tree growth in solid insulation', *IEEE Transactions Dielectrics & Electrical Insulation*, Vol. 16, pp. 1724–1734, 2009.
70. A. Ranković and M.S. Savić, 'Generalized charge simulation method for the calculation of the electric field in high voltage substations', *Electrical Engineering*, Vol. 92, pp. 69–77, 2010.
71. A.M. Mahdy, 'Assessment of breakdown voltage of $SF_6/N_2$ gas mixtures under non-uniform field', *IEEE Transactions Dielectrics & Electrical Insulation*, Vol. 18, pp. 607–612, 2011.
72. H. Iwabuchi, T. Donen, A. Kumada and K. Hidaka, 'Electric field computation with improved charge simulation method considering volume and surface resistivities', *IEEE Transactions on Power and Energy*, Vol. 131, pp. 717–718, 2011.
73. R.M. Radwan, A.M. Mahdy, M. Abdel-Salam and M. Samy, 'Electric field mitigation under extra high voltage power lines', *IEEE Transactions Dielectrics & Electrical Insulation*, Vol. 20, pp. 54–62, 2013.

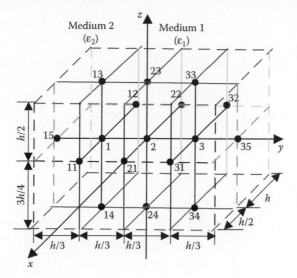

**FIGURE 14.12**
Nodal arrangement pertaining to Problem 14.5.

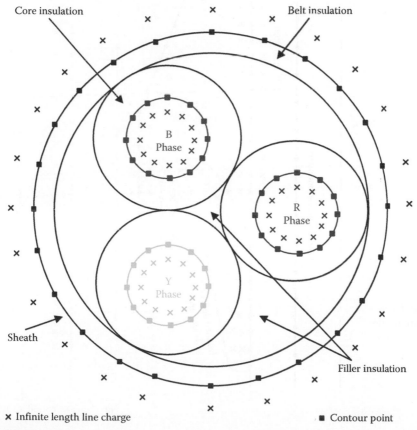

× Infinite length line charge      ■ Contour point

**FIGURE 16.18**
Simulation of a three-core belted cable by CSM using complex fictitious charges.

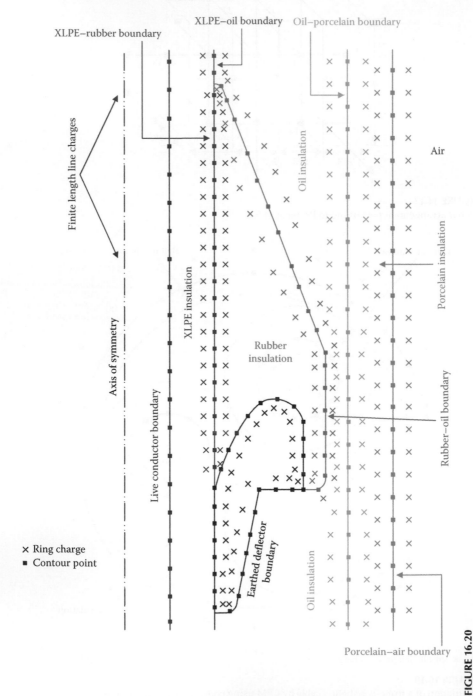

**FIGURE 16.20**
Simulation of fluid-filled outdoor cable termination by conventional CSM.

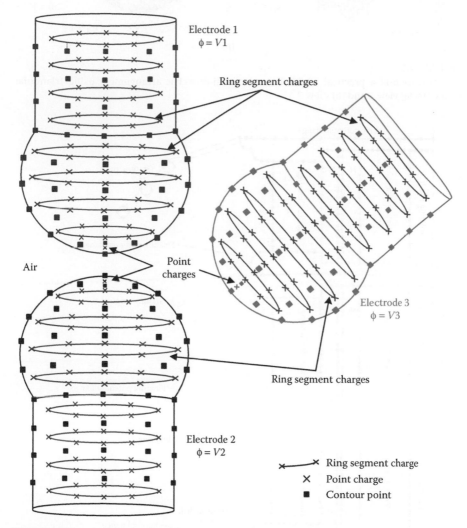

**FIGURE 16.22**
Simulation of an asymmetric sphere-gap arrangement by CSM.

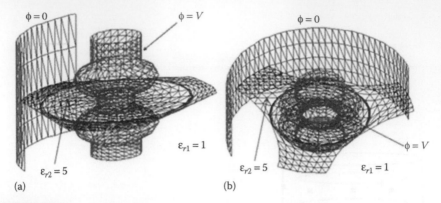

**FIGURE 17.5**
Discretization of a practical multi-electrode multi-dielectric arrangement using triangular elements: (a) view 1 and (b) view 2.

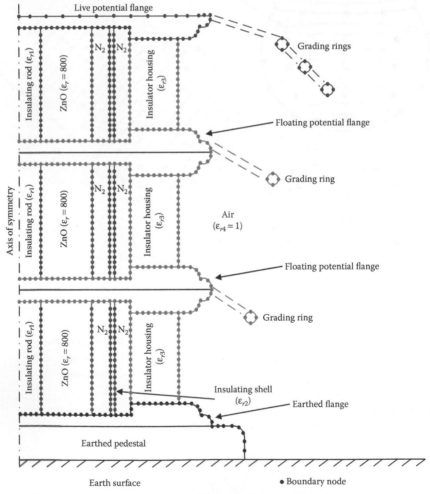

**FIGURE 17.11**
SCSM simulation of multiple unit assembly of metal oxide surge arrester.

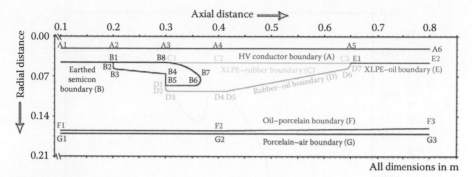

**FIGURE 18.2**
Geometry and boundaries of cable termination under study.

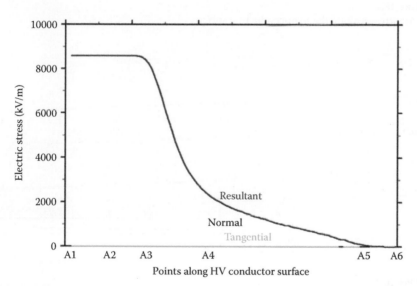

**FIGURE 18.3**
Electric stresses along the high-voltage conductor surface.

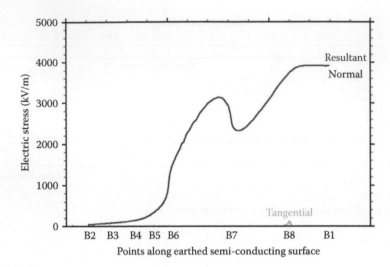

**FIGURE 18.4**
Electric stresses along the semi-conducting surface.

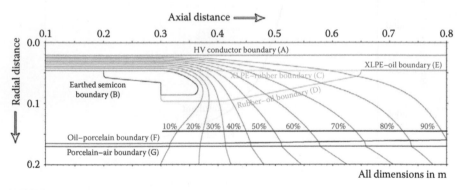

**FIGURE 18.7**
Equipotential lines for the cable termination.

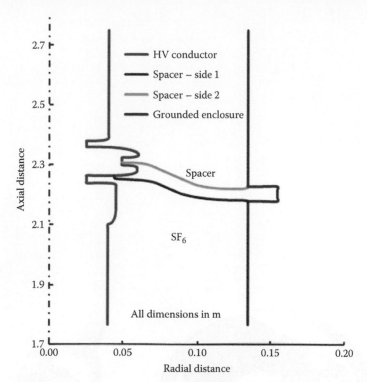

**FIGURE 18.20**
GIS configuration considered for electric field computation.

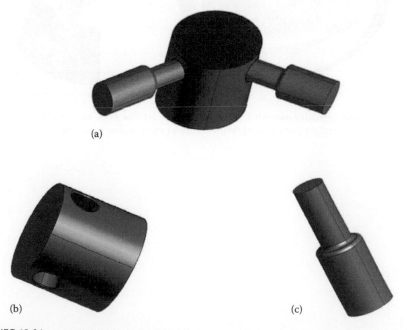

**FIGURE 19.21**
3D electrode–support insulator assembly for optimization: (a) electrode–insulator assembly; (b) live electrode and (c) support insulator.

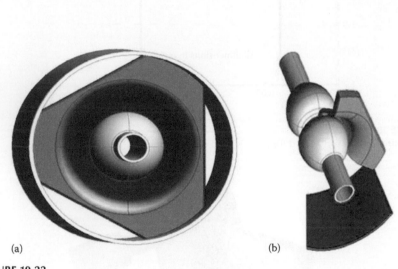

(a)                                                                    (b)

**FIGURE 19.22**
3D electrode–disc insulator assembly for optimization: (a) disc-type insulator with live and ground electrodes and (b) live electrode with sectional view of disc insulator.

# 17

# Numerical Computation of High-Voltage Field by Surface Charge Simulation Method

**ABSTRACT** Surface charge simulation method (SCSM) is aimed to improve the CSM technique, dealing with surface distributed charges instead of discrete ones, as adopted in the classical formulation. As it is very important in electric field analysis to simulate the boundaries accurately, the aim of SCSM is to divide the electrode and dielectric surfaces into suitable number of boundary elements having specific charge distribution. Over the last two and half decades, SCSM, which is also known as *indirect boundary element method*, has evolved into a powerful tool for electric field analysis in 2D, axi-symmetric as well as 3D systems with multiple dielectric media. SCSM fundamentals in such systems have been presented in details. Theoretical basis of implementing SCSM in systems having volume and surface resistances have now been firmly established and have also been discussed in this chapter.

## 17.1 Introduction

In charge simulation method (CSM), the discrete fictitious charges are so placed and their magnitudes are so determined that one of the equipotential will become identical to the electrode boundary of given potential and at the same time no singularity appears in the solution methodology. CSM is particularly attractive because of its simplicity and good accuracy in general, but suffers from large errors in the case of thin electrodes, which are often used to control and grade electric field near the HV terminals of power equipment, in condenser bushings and so on, where determination of electric field intensity is of paramount practical importance. It goes without saying that the true charges are distributed over the entire electrode boundary in a continuous manner. Therefore, the electric field could always be better computed if the charges are assumed to be continuously distributed over the boundary. With this in view, in SCSM, electric field is computed by assuming the charges to be present continuously on the boundary itself. As in the case of CSM, in SCSM, too, the potential and field functions are directly obtained from Coulomb's law and therefore fulfil Laplace's equation without any principal error. SCSM is an integral method wherein Fredholm's integral equations are used, in which the kernel function is the potential and field coefficients and the weight function is the charge density. Boundary conditions are given as follows: (1) Dirichlet formulation on electrode–dielectric boundary and (2) Neumann formulation on dielectric–dielectric boundary. The difficulties of numerical treatment of singularities and of numerical evaluation of complex integrals are overcome using suitable techniques in SCSM.

## 17.2 SCSM Formulation for Single-Dielectric Medium

For the calculation of electric fields, the distributed charges on the surface of the electrode are replaced by surface charge elements placed on the electrode boundary, as shown in Figure 17.1 for 2D or axi-symmetric configurations. For a closed-type boundary, as shown in Figure 17.1a, there are $N$ number of nodes or collocation points and for open-type boundary, as shown in Figure 17.1b, there are $N + 1$ number of nodes on the electrode boundary corresponding to $N$ number of surface charge elements.

Similar to CSM, SCSM also requires that at any one of these surface nodes, the potential resulting from superposition of effects all the surface charge elements is equal to the known electrode potential. The potential at any surface node due to a surface charge element is computed by integrating over the surface element considering a predetermined distribution of charge density along the surface element. Different types of charge density distribution may

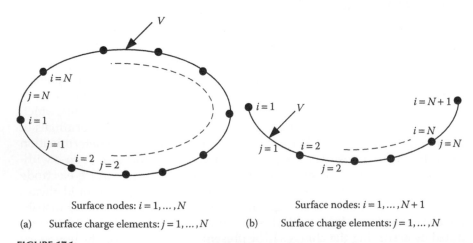

Surface nodes: $i = 1, \ldots, N$

(a)     Surface charge elements: $j = 1, \ldots, N$

Surface nodes: $i = 1, \ldots, N + 1$

(b)     Surface charge elements: $j = 1, \ldots, N$

**FIGURE 17.1**
Surface charge elements and nodes on electrode boundary: (a) for closed-type boundary and (b) for open-type boundary.

be considered. However, practical experience shows that linear charge density distribution gives accurate results for most of the practical cases. As a result, in this chapter, all subsequent discussions are based on linear distribution of charge density over a surface charge element. In SCSM, charge densities are considered at the surface nodes, the magnitudes of which are unknown and are to be solved. At any point on the $j$th surface charge element, the charge density $\sigma(\text{pelm})$ could be obtained from the charge densities of the nodes defining the surface element using linear charge density distribution. Then, at any surface node $i$, the potential due to the $j$th surface charge element will be given by

$$\phi_{ij} = \int_{elmj} P_{ij}\sigma(\text{pelm})d(\text{pelm}) \tag{17.1}$$

where:

$P_{ij}$ is the potential coefficient which depends on the type of surface charge element

$d(\text{pelm})$ depends on the system configuration

For 2D or axi-symmetric systems, $d(\text{pelm})$ is length and the integration is carried out over a line element and for 3D configurations $d(\text{pelm})$ is area and the integration is carried out over an area element. Similar to CSM, potential coefficients are evaluated analytically for different types of surface charge elements by solving Laplace's equation.

Electric field intensity components are obtained by differentiating Equation 17.1 as follows:

$$E_{x,ij} = \int_{elmj} -\frac{\partial P_{ij}}{\partial x}\sigma(\text{pelm})d(\text{pelm}) \tag{17.2a}$$

$$E_{y,ij} = \int_{elmj} -\frac{\partial P_{ij}}{\partial y}\sigma(\text{pelm})d(\text{pelm}) \tag{17.2b}$$

$$E_{z,ij} = \int_{elmj} -\frac{\partial P_{ij}}{\partial z}\sigma(\text{pelm})d(\text{pelm}) \tag{17.2c}$$

When Equation 17.1 is applied at $i$th node for all the $N$ number of surface charge elements, it gives the potential of node $i$, which according to Dirichlet's condition should be equal to the known electrode potential, that is,

$$\phi_i = \sum_{j=1}^{j=N} \phi_{ij} = V \tag{17.3}$$

When Equation 17.3 is applied to $N$ number of surface nodes, it leads to the following system of $N$ linear equations for $N$ unknown surface charge densities. In matrix form, it may be written as

$$
\begin{bmatrix}
\phi_{11} & \phi_{12} & \cdots & \phi_{1N} \\
\phi_{21} & \phi_{22} & \cdots & \phi_{2N} \\
& & \vdots & \\
\phi_{N1} & \phi_{N2} & \cdots & \phi_{NN}
\end{bmatrix} = [V] \tag{17.4}
$$

As in the case of CSM, in many cases, the effect of the ground plane is to be considered for electric field calculation. This plane can be taken into account by introducing the image of surface charge elements.

## 17.3 Surface Charge Elements in 2D and Axi-Symmetric Configurations

For the calculation of 2D and axi-symmetric fields, straight line and elliptical arc elements are most commonly used and are discussed here. A linear basis function is assumed for the description of charge density distribution along the element between the two nodes, which are the extremities of any given element.

In both the types of elements, as shown in Figure 17.2, a local coordinate $\xi$ is introduced for the purpose of integration over the element, the value of which varies from $-1$ at node $i$ to $+1$ at node $i + 1$.

### 17.3.1 Straight Line Element

With reference to Figure 17.2a, the location of any point lying on the element between node $i$ and node $i + 1$ is given by

For 2D system:

$$
x(\xi) = x_{i+1}\frac{1+\xi}{2} + x_i\frac{1-\xi}{2}
$$

$$
y(\xi) = y_{i+1}\frac{1+\xi}{2} + y_i\frac{1-\xi}{2}
\tag{17.5}
$$

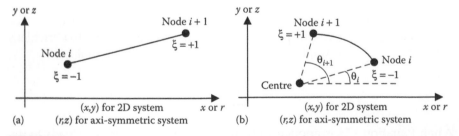

**FIGURE 17.2**
Surface charge elements in 2D and axi-symmetric configurations: (a) straight line element and (b) elliptic arc element.

For axi-symmetric system:

$$r(\xi) = r_{i+1}\frac{1+\xi}{2} + r_i\frac{1-\xi}{2}$$

$$z(\xi) = z_{i+1}\frac{1+\xi}{2} + z_i\frac{1-\xi}{2}$$

(17.6)

The charge density at any point on the element between node $i$ and node $i + 1$ is given by

$$\sigma(\xi) = \sigma_{i+1}\frac{1+\xi}{2} + \sigma_i\frac{1-\xi}{2}$$

(17.7)

For 2D system, the field is independent of the direction normal to the plane of the paper and hence the charge at any point on the element is equivalent to an infinitely long line charge, extending up to infinity in a direction normal to the plane of the paper. As a result, for computing the potential coefficient, the expression for infinitely long line charge given by Equation 16.20 in Chapter 16 could be used in SCSM. Electric field intensity coefficients could be computed using the Equations 16.21 and 16.22 given in Chapter 16.

For an axi-symmetric system, the z-axis is the axis of symmetry and hence the charge at any point on the element is equivalent to a ring charge, when the element is not lying on the z-axis. As a result, for computing the potential coefficient, the expression for ring charge given by Equation 16.26 in Chapter 16 could be used in SCSM. Electric field intensity coefficients could be computed using Equations 16.27 and 16.28 given in Chapter 16. When the element lies on the z-axis, then the elemental charge length is equivalent to a finite length line charge. In such cases, for computing the potential coefficient, the expression for finite length line charge given by Equation 16.23 in Chapter 16 could be used in SCSM. Electric field intensity coefficients could be computed using Equations 16.24 and 16.25 given in Chapter 16.

Here, it is to be noted that the expressions given in Chapter 16 incorporates the effect of the image charge. Hence, if the image charge is not to be taken into account, then the part of the expression that represents the effect of image charge need not be taken into computation.

### 17.3.2 Elliptic Arc Element

For the description of the elliptic arc, parametric representation of an ellipse is commonly used. Angle $\theta$ is the parameter that varies linearly from node $i$ and node $i + 1$ along the elliptic arc length.

With reference to Figure 17.2b, the parameter at any point on the element between node $i$ and node $i + 1$ is given by

$$\theta(\xi) = \theta_{i+1}\frac{1+\xi}{2} + \theta_i\frac{1-\xi}{2}$$

(17.8)

and the location of any point lying on the element between node $i$ and node $i + 1$ is given by

For 2D system:

$$x(\xi) = x_0 + A \cos \theta(\xi)$$

$$y(\xi) = y_0 + B \sin \theta(\xi)$$

(17.9)

For axi-symmetric system:

$$r(\xi) = r_0 + A \cos \theta(\xi)$$

$$z(\xi) = z_0 + B \sin \theta(\xi)$$

(17.10)

where:

$(x_0, y_0)$ or $(r_0, z_0)$ are the coordinates of the centre

$A$, $B$ are the semi-major and semi-minor axes of the elliptic arc element, respectively

The charge density at any point on the elliptic arc element between node $i$ and node $i + 1$ is also given by Equation 17.7. For both straight line and elliptic arc elements, the unit of charge density is Coulombs/m (C/m).

### 17.3.3 Contribution of Nodal Charge Densities to Coefficient Matrix

Consider four consecutive nodes defining three elements on any electrode boundary, as shown in Figure 17.3.

As per Equation 17.1, the potential of node $M$ due to element $M-1$ is given by

$$\phi_{M,M-1} = \int_{\xi=-1}^{\xi=+1} P_{M,M-1}\sigma(\xi)dl(\xi)$$

(17.11)

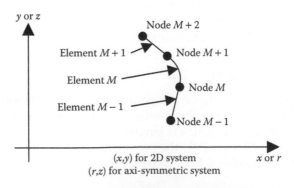

FIGURE 17.3

Pertaining to contribution of nodal charge densities to coefficient matrix.

where:
$\xi = -1$ at node $M-1$
$\xi = +1$ at node $M$ for element $M-1$

Using Equation 17.7, Equation 17.11 can be modified as follows:

$$
\begin{aligned}
\phi_{M,M-1} &= \int\limits_{\xi=-1}^{\xi=+1} P_{M,M-1}\left(\sigma_M \frac{1+\xi}{2} + \sigma_{M-1}\frac{1-\xi}{2}\right) dl(\xi) \\
&= \sigma_M \int\limits_{\xi=-1}^{\xi=+1} P_{M,M-1}\frac{1+\xi}{2} dl(\xi) + \sigma_{M-1}\int\limits_{\xi=-1}^{\xi=+1} P_{M,M-1}\frac{1-\xi}{2} dl(\xi) \\
&= \sigma_M C_{EM,M-1} + \sigma_{M-1}C_{SM,M-1}
\end{aligned}
\tag{17.12}
$$

Similarly, the potential of node $M$ due to element $M$ will be given by

$$
\begin{aligned}
\phi_{M,M} &= \int\limits_{\xi=-1}^{\xi=+1} P_{M,M}\left(\sigma_{M+1} \frac{1+\xi}{2} + \sigma_M \frac{1-\xi}{2}\right) dl(\xi) \\
&= \sigma_{M+1} \int\limits_{\xi=-1}^{\xi=+1} P_{M,M}\frac{1+\xi}{2} dl(\xi) + \sigma_M \int\limits_{\xi=-1}^{\xi=+1} P_{M,M}\frac{1-\xi}{2} dl(\xi) \\
&= \sigma_{M+1}C_{EM,M} + \sigma_M C_{SM,M}
\end{aligned}
\tag{17.13}
$$

Therefore, it may be seen that every nodal charge density will contribute twice towards the potential and also for electric field intensity of any node or point within the region of interest. This is because of the fact that every node belongs to two surface charge elements. In one element a given node will be the ending node, whereas in the next element, it will be the starting node. In Equations 17.12 and 17.13, the contribution of a nodal charge density is shown as $C_S$ when it is the starting node of an element and as $C_E$ when it is the ending node of another element. However, this is not valid for the starting node of the very first element and the ending node of the very last element of an open-type boundary, as shown in Figure 17.1b.

In light of Equations 17.12 and 17.13, the matrix of Equation 17.4 could be rewritten as follows for the closed-type boundary, as shown in Figure 17.1a.

$$
\begin{bmatrix}
(C_{S1,1}+C_{E1,N}) & (C_{S1,2}+C_{E1,1}) & \dots (C_{S1,i}+C_{E1,i-1}) & \dots (C_{S1,N}+C_{E1,1}) \\
(C_{S2,1}+C_{E2,N}) & (C_{S2,2}+C_{E2,1}) & \dots (C_{S2,i}+C_{E2,i-1}) & \dots (C_{S2,N}+C_{E2,1}) \\
& & \vdots & \\
(C_{Si,1}+C_{Ei,N}) & (C_{Si,2}+C_{Ei,1}) & \dots (C_{Si,i}+C_{Ei,i-1}) & \dots (C_{Si,N}+C_{Ei,1}) \\
& & \vdots & \\
(C_{SN,1}+C_{EN,N}) & (C_{SN,2}+C_{EN,1}) & \dots (C_{SN,i}+C_{EN,i-1}) & \dots (C_{SN,N}+C_{EN,1})
\end{bmatrix}
\begin{bmatrix}
\sigma_1 \\ \sigma_2 \\ \vdots \\ \sigma_i \\ \vdots \\ \sigma_N
\end{bmatrix}
=
\begin{bmatrix}
V \\ V \\ \vdots \\ V \\ \vdots \\ V
\end{bmatrix}
\tag{17.14}
$$

For open-type boundary, as shown in Figure 17.1b, Equation 17.14 will be modified as follows:

$$
\begin{bmatrix}
(C_{S1,1}) & (C_{S1,2}+C_{E1,1}) & \cdots & (C_{S1,i}+C_{E1,i-1}) & \cdots & (C_{E1,N-1}) \\
(C_{S2,1}) & (C_{S2,2}+C_{E2,1}) & \cdots & (C_{S2,i}+C_{E2,i-1}) & \cdots & (C_{E2,N-1}) \\
 & & \vdots & & & \\
(C_{Si,1}) & (C_{Si,2}+C_{Ei,1}) & \cdots & (C_{Si,i}+C_{Ei,i-1}) & \cdots & (C_{Ei,N-1}) \\
 & & \vdots & & & \\
(C_{SN,1}) & (C_{sN,2}+C_{EN,1}) & \cdots & (C_{SN,i}+C_{EN,i-1}) & \cdots & (C_{EN,N-1})
\end{bmatrix}
\begin{bmatrix}
\sigma_1 \\ \sigma_2 \\ \vdots \\ \sigma_i \\ \vdots \\ \sigma_N
\end{bmatrix}
=
\begin{bmatrix}
V \\ V \\ \vdots \\ V \\ \vdots \\ V
\end{bmatrix}
\tag{17.15}
$$

The difference between the matrices of Equations 17.14 and 17.15 is that in Equation 17.15 the first and the last columns have only one term, either the starting or the ending term, in the matrix elements. This is due to the fact that for an open-type boundary the starting node of the very first element belongs to only one element as the starting node and the ending node of the very last element belongs also to only one element as the ending node. One more point to be noted is that for an open-type boundary, if there are $N$ number of nodes, then there will be $N - 1$ number of elements instead of $N$. Accordingly the matrix elements of Equation 17.15 are different from those of Equation 17.14.

### 17.3.4 Method of Integration over a Surface Charge Element

In order to achieve good accuracy in SCSM, it is imperative that the integrations are carried out with high accuracy. It has been noted by the researchers that numerical integration by Gauss–Legendre Quadrature rule provides excellent accuracy. Hence, this integration rule is briefly discussed here.

Gauss–Legendre Quadrature rule approximates an integral as the sum of a finite number ($n$) of terms in the following way, by picking approximate values for $n$, $w_i$ and $\xi_i$

$$
\int_{-1}^{+1} f(\xi)\,d\xi \cong \sum_{i=1}^{n} w_i\, f(\xi_i)
\tag{17.16}
$$

where:
$\xi_i$ is known as Gauss point or abscissa
The values of $\xi_i$ are zeroes of the $n$th degree Legendre polynomial
$w_i$ is the weight of the function value at $i$th Gauss point or abscissa

Although this rule is originally defined for the interval $(-1,1)$, this is actually a universal function, because the limits of integration could be converted from any interval $(a,b)$ to the Gauss–Legendre interval $(-1,1)$ in the following manner

**TABLE 17.1**

High-Precision Abscissae and Weights of 16-Point Gauss–Legendre Quadrature Rule

| Abscissae ($\xi_i$) | Weights ($w_i$) |
|---|---|
| ±0.0950125098376374401853193 | 0.1894506104550684962853967 |
| ±0.2816035507792589132304605 | 0.1826034150449235888667637 |
| ±0.4580167776572273863424194 | 0.1691565193950025381893121 |
| ±0.6178762444026437484466718 | 0.1495959888165767320815017 |
| ±0.7554044083550030338951012 | 0.1246289712555338720524763 |
| ±0.8656312023878317438804679 | 0.0951585116824927848099251 |
| ±0.9445750230732325760779884 | 0.0622535239386478928628438 |
| ±0.9894009349916499325961542 | 0.0271524594117540948517806 |

$$\int_a^b f(\xi)d\xi \cong \frac{b-a}{2}\sum_{i=1}^n w_i\, f\left(\frac{b-a}{2}\xi_i + \frac{b+a}{2}\right) \tag{17.17}$$

In SCSM, 16-point or even 13-point Gauss–Legendre Quadrature rule gives excellent accuracy. Hence, abscissae and weights of 16-point Gauss–Legendre Quadrature rule are given in Table 17.1.

### 17.3.5 Electric Field Intensity Exactly on the Electrode Surface

It has already been discussed in Section 8.2.1 that the highest value of electric field intensity does not occur on the electrode surface, but occurs just off the electrode surface within the dielectric medium. Still it is relevant to compute the electric field intensity exactly on the surface. Here, it is to be noted that the electric field intensity is always directed along the normal to the surface at any point lying exactly on the electrode surface.

As in the case of electric potential given by Equation 17.12, the normal component of electric field intensity at node $M$ due to element $M$–1 can be computed as follows:

$$
\begin{aligned}
E_{NM,M-1} &= \int_{\xi=-1}^{\xi=+1} F_{NM,M-1}\left(\sigma_M \frac{1+\xi}{2} + \sigma_{M-1}\frac{1-\xi}{2}\right)dl(\xi) \\
&= \sigma_M \int_{\xi=-1}^{\xi=+1} F_{NM,M-1}\frac{1+\xi}{2}dl(\xi) + \sigma_{M-1}\int_{\xi=-1}^{\xi=+1} F_{NM,M-1}\frac{1-\xi}{2}dl(\xi) \\
&= \sigma_M A_{EM,M-1} + \sigma_{M-1}A_{SM,M-1}
\end{aligned}
\tag{17.18}
$$

where:
$F_N$ is normal field intensity coefficient

Similarly, the normal component of electric field intensity at node $M$ due to element $M$ is given by

$$E_{NM,M} = \int\limits_{\xi=-1}^{\xi=+1} F_{NM,M}\left(\sigma_{M+1}\frac{1+\xi}{2} + \sigma_M\frac{1-\xi}{2}\right)dl(\xi)$$

$$= \sigma_{M+1}\int\limits_{\xi=-1}^{\xi=+1} F_{NM,M}\frac{1+\xi}{2}dl(\xi) + \sigma_M\int\limits_{\xi=-1}^{\xi=+1} F_{NM,M}\frac{1-\xi}{2}dl(\xi) \tag{17.19}$$

$$= \sigma_{M+1} A_{EM,M} + \sigma_M A_{SM,M}$$

For any node $i$ the normal component of electric field intensity due to all the elements on the electrode boundary could be obtained from Equations 17.14, 17.18 and 17.19 for closed-type boundary as follows:

$$E_{Ni} = (A_{Si,1} + A_{Ei,N})\sigma_1 + (A_{Si,2} + A_{Ei,1})\sigma_2 + \cdots + (A_{Si,i} + A_{Ei,i-1})\sigma_i$$

$$+ \cdots + (A_{Si,N} + A_{Ei,1})\sigma_N \tag{17.20}$$

and from Equations 17.15, 17.18 and 17.19 for open-type boundary as follows:

$$E_{Ni} = A_{Si,1}\sigma_1 + (A_{Si,2} + A_{Ei,1})\sigma_2 + \cdots + (A_{Si,i} + A_{Ei,i-1})\sigma_i + \cdots + A_{Ei,N-1}\sigma_N \tag{17.21}$$

Because node $i$ has a charge density $\sigma_i$, there will be problem of singularity in evaluating the coefficient terms associated to $\sigma_i$. Therefore, the coefficients associated to $\sigma_i$ are computed leaving aside a small part around node $i$. However, it may be mentioned here that in Gauss–Legendre Quadrature rule, the integrand is approximated at the selected abscissae, which do not coincide with the nodes lying at the extremities of the element. Hence, the integrand is not computed at the nodes and so singularity does not appear. But, to be precise and correct in computation, it is necessary to assume that the values obtained from Equations 17.20 and 17.21 are without the small part around the node under consideration. Then the charge on the small part left aside around node $i$ is considered as a point charge of magnitude $\sigma_i$ and the field intensity, which is normal to the electrode surface, due to this point charge is given by $[\sigma_i/(2\varepsilon_r\varepsilon_0)]$. This additional normal electric field intensity has to be added to the value obtained either from Equation 17.20 or 17.21 to get the value exactly on the node.

## 17.4 SCSM Formulation for Multi-Dielectric Media

In the case of multi-dielectric configuration, free charges are present on the electrode boundaries and bound charges are present on the dielectric–dielectric boundaries, which appear due to dielectric polarization. In ideal condition, there should not be any free charge present on a dielectric–dielectric boundary. But in many practical cases, due to volume as well as surface conduction, partial discharge and so on, there could be free charges present on dielectric–dielectric boundary. In the case of Laplacian field, the free charge density on dielectric boundary is zero, while for Poissonian field the effect of the free charges present on the dielectric boundary needs to be taken into consideration.

In SCSM, Neumann condition for normal component of electric flux density is satisfied on dielectric–dielectric boundary as mentioned below:

$$\varepsilon_1 E_{n1} - \varepsilon_2 E_{n2} = \sigma_s \tag{17.22}$$

where:

$\sigma_s$ is free charge density on dielectric–dielectric boundary, which is finite for Poissonian field and is zero for Laplacian field

Consider a multi-electrode and multi-dielectric arrangement, as shown in Figure 17.4. For this arrangement, all the boundaries are discretized into suitable number of surface charge elements and the field at any point within the region of interest is due to the superimposed effect of all such surface charge elements.

Let there be *ELB* number of electrode boundaries and *DLB* number of dielectric–dielectric boundaries. At any node *i* on the dielectric boundary, the normal component of electric field intensity is computed as

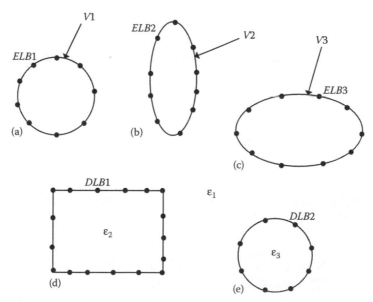

**FIGURE 17.4**
Schematic diagram of a multi-electrode and multi-dielectric arrangement. (a) ELB1, electrode boundary 1; (b) ELB2, electrode boundary 2; (c) ELB3, electrode boundary 3; (d) DLB1, dielectric boundary 1; (e) DLB2, dielectric boundary 2.

$$E_{n,i}^- = \sum_{e=1}^{ELB} E_{n,i} + \sum_{d=1(d\neq DLBi)}^{DLB} E_{n,i} + \sum_{d=(DLBi-AB)} E_{n,i} \qquad (17.23)$$

where:

e and d denote electrode and dielectric boundaries, respectively
$DLBi$ is the dielectric boundary on which the node $i$ is located
$AB$ is a small part around node $i$

The left-hand side of Equation 17.23 is denoted as $E_{n,i}^-$ because the small part $AB$ around node $i$ is not considered for the computation. The terms on the right-hand side of Equation 17.23 are computed using either Equation 17.20 or 17.21.

Normal electric field intensity in free space due to the small part $AB$ around node $i$ can be written as follows, with orientation to side 1 of the dielectric–dielectric boundary:

$$E_{1n,i}^{AB} = -E_{2n,i}^{AB} = \frac{\sigma_i}{2\varepsilon_0} \qquad (17.24)$$

Thus, the normal component of electric field intensity at the node $i$ on both sides of the dielectric–dielectric boundary can be written as

$$E_{1n,i} = E_{n,i}^- + \frac{\sigma_i}{2\varepsilon_0} = \sum_{e=1}^{ELB} E_{n,i} + \sum_{d=1(d\neq DLBi)}^{DLB} E_{n,i} + \sum_{d=(DLBi-AB)} E_{n,i} + \frac{\sigma_i}{2\varepsilon_0} \qquad (17.25)$$

and

$$E_{2n,i} = E_{n,i}^- - \frac{\sigma_i}{2\varepsilon_0} = \sum_{e=1}^{ELB} E_{n,i} + \sum_{d=1(d\neq DLBi)}^{DLB} E_{n,i} + \sum_{d=(DLBi-AB)} E_{n,i} - \frac{\sigma_i}{2\varepsilon_0} \qquad (17.26)$$

Combining Equations 17.22, 17.25 and 17.26, the boundary condition can thus be written as

$$\varepsilon_1 \left( \sum_{e=1}^{ELB} E_{n,i} + \sum_{d=1(d\neq DLBi)}^{DLB} E_{n,i} + \sum_{d=(DLBi-AB)} E_{n,i} + \frac{\sigma_i}{2\varepsilon_0} \right)$$

$$- \varepsilon_2 \left( \sum_{e=1}^{ELB} E_{n,i} + \sum_{d=1(d\neq DLBi)}^{DLB} E_{n,i} + \sum_{d=(DLBi-AB)} E_{n,i} - \frac{\sigma_i}{2\varepsilon_0} \right) = \sigma_{s,i} \qquad (17.27)$$

## 17.5 SCSM Formulation in 3D System

In contrast to finite element method (FEM), in SCSM only the boundary surfaces are approximated by finite number of boundary elements. The boundary elements are mainly of two types: (1) triangles and (2) rectangles. Depending on how precisely these boundary elements need to be defined, shape functions of different orders are used. In most of the practical cases, boundary elements are either of first order or of second order, which can be defined analytically by natural coordinate system. Thus, the problem of solving integral equations can be looked upon as defining a set of functions of equivalent surface charge densities defined on each boundary element.

Figure 17.5a and b show typical discretization of a multi-electrode multi-dielectric arrangement used in a gas-insulated system. The discretization is done by triangular elements. One of the main reasons for using triangular elements is that any irregular polygon can be looked upon as a union of several triangles. The simplicity of the approximating functions used with triangular elements lies in its two basic properties, namely, (1) these functions always ensure the continuity of the desired potential along all the boundaries between triangles provided only that continuity is imposed at the vertices of the triangle and (2) the approximations are independent of the global coordinates of the triangles.

In order to obtain the numerical solution of integral equations, it is required to introduce basis functions of suitable order for approximating the surface charge density ($\sigma$) on each surface. Similar to the case of FEM, it is convenient to express the basis function in local coordinate system. In most of the cases, the boundary elements are approximated by first order polynomials, that is, using linear approximation. It follows that the number of nodes on each of the triangular boundary elements will be three. In other words, the surface charge densities are assumed at the vertices of each triangular boundary

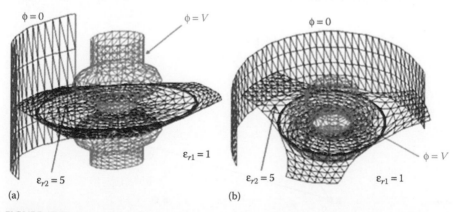

**FIGURE 17.5**
**(See colour insert.)** Discretization of a practical multi-electrode multi-dielectric arrangement using triangular elements: (a) view 1 and (b) view 2.

element. Thus, the surface charge density function over a boundary element can be written in terms of the shape functions as

$$\sigma_{ele} = L_1\sigma_1 + L_2\sigma_2 + L_3\sigma_3 \tag{17.28}$$

where:
$$L_1 + L_2 + L_3 = 0$$

The shape functions and natural coordinates for linear triangular elements have been discussed in detail in Sections 15.4.4 and 15.4.5. Other types of surface elements have also been discussed in Section 15.4.6. The relationships described in the above-mentioned sections of Chapter 15 can be used here.

With reference to Figure 17.6, the contribution of a particular linear boundary element $M$ to the potential at any arbitrary point $G$ is given by

$$\phi_{G,M} = \frac{1}{4\pi\varepsilon_0} \int_0^1 \int_0^{1-L1} \frac{L_1\sigma_1 + L_2\sigma_2 + (1-L_1-L_2)\sigma_3}{r_{GM}(L_1,L_2)} \left| J(L_1,L_2) \right| dL_1 dL_2 \tag{17.29}$$

where:
$J(L_1,L_2)$ is the Jacobian matrix of the transformation for shape functions

When expressed in natural $(\xi - \eta)$ coordinates, Equation 17.29 is modified to

$$\phi_{G,M} = \frac{1}{4\pi\varepsilon_0} \int_0^1 \int_0^{1-\xi} \frac{\xi\sigma_1 + \eta\sigma_2 + (1-\xi-\eta)\sigma_3}{r_{GM}(\xi,\eta)} \left| J(\xi,\eta) \right| d\xi d\eta \tag{17.30}$$

where:
$J(\xi,\eta)$ is the Jacobian matrix of the transformation for natural coordinates and is given in Section 15.4.5

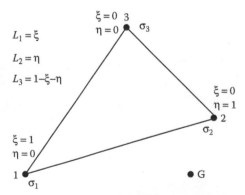

**FIGURE 17.6**
Linear triangular boundary element.

Equation 17.30 shows that potential coefficient has three terms as given below.

$$P_1 = \frac{1}{4\pi\varepsilon_0} \int_0^1 \int_0^{1-\xi} \frac{\xi\sigma_1}{r_{GM}(\xi,\eta)} |J(\xi,\eta)| d\xi d\eta$$

$$P_2 = \frac{1}{4\pi\varepsilon_0} \int_0^1 \int_0^{1-\xi} \frac{\eta\sigma_2}{r_{GM}(\xi,\eta)} |J(\xi,\eta)| d\xi d\eta \qquad (17.31)$$

$$P_3 = \frac{1}{4\pi\varepsilon_0} \int_0^1 \int_0^{1-\xi} \frac{(1-\xi-\eta)\sigma_3}{r_{GM}(\xi,\eta)} |J(\xi,\eta)| d\xi d\eta$$

Electric field intensity coefficients could be obtained in a similar manner.

---

## 17.6 Capacitive-Resistive Field Computation by SCSM

SCSM can be employed to compute capacitive-resistive field accurately in 2D, axi-symmetric and 3D systems including volume and surface resistances. The methodology for 2D and axi-symmetric systems are similar and are described in the next sub-section, whereas the 3D formulations are given in the subsequent sub-section.

### 17.6.1 Capacitive-Resistive Field Computation in 2D and Axi-Symmetric Systems

For capacitive-resistive field computation Dirichlet's condition on the electrode boundary remains unaltered, but $\sigma_s$ of Equation 17.22 need to be determined from the general condition of current density vector at node $i$ to include volume and surface resistances in the following manner.

$$J_{1n,i} - J_{2n,i} + \Delta J_{s,i} + \frac{\partial \sigma_{s,i}}{\partial t} = 0 \qquad (17.32)$$

With reference to Figure 15.11 of Chapter 15 and for isotropic media, Equation 17.32 can be rewritten as

$$\frac{E_{1n,i}}{\rho_{v1}} - \frac{E_{2n,i}}{\rho_{v2}} + \Delta J_{s,i} + \frac{\partial \sigma_{s,i}}{\partial t} = 0 \qquad (17.33)$$

where:
$\rho_{vx}$ is volume resistivity of dielectric $x$, $x = 1,2$

For sinusoidal fields, incorporating complex notations Equation 17.33 can be written as

$$\frac{\bar{E}_{1n,i}}{\rho_{v1}} - \frac{\bar{E}_{2n,i}}{\rho_{v2}} + \Delta\bar{J}_{s,i} + j\omega\bar{\sigma}_{s,i} = 0 \qquad (17.34)$$

With reference to Figure 15.13, surface current density term in Equation 17.34 can be written as

$$\Delta\bar{J}_{s,i} = \frac{1}{S(i)}\left[\bar{\phi}_i\left(\frac{1}{R(i)} + \frac{1}{R(i+1)}\right) - \frac{\bar{\phi}_{i+1}}{R(i+1)} - \frac{\bar{\phi}_{i-1}}{R(i)}\right] \qquad (17.35)$$

where:
  $R(i)$ and $R(i+1)$ are surface resistances corresponding to $i$th and $(i+1)$th boundary node, respectively
  $S(i)$ is a small surface area corresponding to $i$th boundary node

The expressions for $R(i)$ and $S(i)$ are given for a 2D system in Equations 15.62 and 15.63, respectively, and for an axi-symmetric system in Equations 15.64 and 15.65, respectively, in Chapter 15.

Finally, Equations 17.27, 17.34 and 17.35 can be combined together in the following form.

$$\frac{2(\bar{\varepsilon}_1 - \bar{\varepsilon}_2)}{(\bar{\varepsilon}_1 + \bar{\varepsilon}_2)}\bar{E}_{n,i} - \frac{2j}{\omega S(i)(\bar{\varepsilon}_1 + \bar{\varepsilon}_2)}\left[\bar{\phi}_i\left(\frac{1}{R(i)} + \frac{1}{R(i+1)}\right) - \frac{\bar{\phi}_{i+1}}{R(i+1)} - \frac{\bar{\phi}_{i-1}}{R(i)}\right] + \frac{\bar{\sigma}_i}{2\varepsilon_0} = 0 \quad (17.36)$$

where:
  $\bar{\varepsilon}_x = \varepsilon_0\varepsilon_{rx} - [j/(\omega\rho_{vx})]$, $x = 1,2$ and $\bar{E}_{n,i}$ are given in details in Equation 17.23 without complex notations

### 17.6.2 Capacitive-Resistive Field Computation in 3D System

Incorporating volume resistivity into a 3D capacitive-resistive field computation is similar to a 2D or an axi-symmetric system, because the current density due to volume resistance is caused by the normal component of the electric field intensity at a surface node, as given in Equation 17.33. Hence, volume resistivity could be incorporated by considering the permittivity of the dielectric medium as a complex quantity and expressing it as $\bar{\varepsilon}_x = \varepsilon_0\varepsilon_{rx} - [j/(\omega\rho_{vx})]$, where, $\varepsilon_x$ is the permittivity of medium $x$ and $\rho_{vx}$ is its volume resistivity.

However, incorporation of surface resistance into 3D capacitive-resistive field computation requires special treatment as discussed below. In this treatment, hexagonal discretization using triangular elements has been considered. As shown in Figure 17.7, the target node 0 is connected to six other nodes, namely, 1, 2, 3, 4, 5 and 6. The potentials of these nodes are $V_0$, $V_1$, $V_2$, $V_3$, $V_4$, $V_5$ and $V_6$, respectively, as shown in Figure 17.7. $R_1$, $R_2$, $R_3$, $R_4$, $R_5$

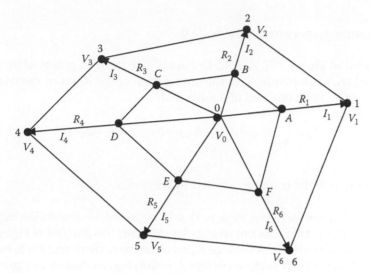

**FIGURE 17.7**
Incorporation of surface resistance into SCSM 3D formulation.

and $R_6$ are the resistances of the branches 01, 02, 03, 04, 05 and 06, respectively. Surface currents flowing through the branches 01, 02, 03, 04, 05 and 06 are $I_1$, $I_2$, $I_3$, $I_4$, $I_5$ and $I_6$, respectively. The directions of all these currents are assumed to be from node 0 to node $k$, $k = 1$, 6.

Then, application of Kirchhoff's current law at node 0 gives

$$\sum_{i=1}^{6} \frac{V_0 - V_i}{R_i} = 0 \tag{17.37}$$

For AC field, charge could be computed in the following way.

$$q_s(t) = \int i_s(t)dt = \int I_{sm} e^{j\omega t}\, dt = \frac{i_s(t)}{j\omega} \tag{17.38}$$

Removing $t$ and bringing in complex notation and combining Equations 17.37 and 17.38, surface charge can be written as

$$\overline{q_s} = \frac{1}{j\omega} \sum_{i=1}^{6} \frac{\overline{V}_0 - \overline{V}_i}{R_i} \tag{17.39}$$

Then, the surface charge density around node 0 over the surface can be obtained as follows:

$$\overline{\sigma_s} = \frac{1}{j\omega S_0} \sum_{i=1}^{6} \frac{\overline{V}_0 - \overline{V}_i}{R_i} \tag{17.40}$$

where:

$S_0$ is a surface area around the node 0

As depicted in Figure 17.7, $A$, $B$, $C$, $D$, $E$ and $F$ are the mid points of 01, 02, 03, 04, 05 and 06, respectively. $S_0$ is the summation of the areas of the triangles 0AB, 0BC, 0CD, 0DE, 0EF and 0FA such that

$$S_0 = \frac{1}{4}\left(\Delta_{0AB} + \Delta_{0BC} + \Delta_{0CD} + \Delta_{0DE} + \Delta_{0EF} + \Delta_{0FA}\right) \tag{17.41}$$

where:

$\Delta$ is the area of the triangle denoted by the suffix

Surface resistance between two nodes is obtained in the following way. Figure 17.8 shows two adjacent triangular elements 056 and 061 of Figure 17.7. Consider that the surface resistance $R_6$ between the nodes 0 and 6 is to be computed, between which surface current $I_6$ is flowing, as shown in Figure 17.7. Now the side 06 belongs to both the triangular elements 056 and 061. As shown in Figure 17.8, OG and OH are the medians of triangular elements 056 and 061, respectively. Thus,

$$\Delta_{0G6} = \frac{1}{2}\Delta_{056} \text{ and } \Delta_{06H} = \frac{1}{2}\Delta_{061} \tag{17.42}$$

Let the length of 06 be $L_{06}$ and the equivalent lengths that are normal to the direction of flow of $I_6$ are as shown by the hatched lines in the triangles 0G6 and 06H in Figure 17.8. Let the height of the triangles 0G6 and 06H be $L_{h1}$ and $L_{h2}$, respectively. Considering $L_{06}$ as the base of the triangles 0G6 and 06H, $L_{h1}$ and $L_{h2}$ can be obtained as

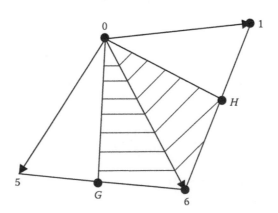

**FIGURE 17.8**
Computation of surface resistance on the boundary.

$$\Delta_{0G6} = \frac{1}{2} L_{h1} L_{06} \text{ and } \Delta_{06H} = \frac{1}{2} L_{h2} L_{06} \tag{17.43}$$

Combining Equations 17.42 and 17.43,

$$L_{h1} = 2\frac{\Delta_{0G6}}{L_{06}} = \frac{\Delta_{056}}{L_{06}} \text{ and } L_{h2} = 2\frac{\Delta_{06H}}{L_{06}} = \frac{\Delta_{061}}{L_{06}} \tag{17.44}$$

Because the surface resistance associated to the side 06 is $R_6$,

$$R_6 = \left( \rho_{s,06} \frac{L_{06}}{L_{h1}} \right) \Big\| \left( \rho_{s,06} \frac{L_{06}}{L_{h2}} \right) = \rho_{s,06} \frac{L_{06}}{L_{h1} + L_{h2}} \tag{17.45}$$

Combining Equations 17.44 and 17.45

$$R_6 = \rho_{s,06} \frac{L_{06}^2}{\Delta_{056} + \Delta_{061}} \tag{17.46}$$

where:
$\rho_{s,06}$ is the surface resistivity associated to side 06

In the same way, the surface resistance associated to all the sides, as shown in Figure 17.7 can be written as follows:

$$R_1 = \rho_{s,01} \frac{L_{01}^2}{\Delta_{061} + \Delta_{012}}, R_2 = \rho_{s,02} \frac{L_{02}^2}{\Delta_{012} + \Delta_{023}}, R_3 = \rho_{s,03} \frac{L_{03}^2}{\Delta_{023} + \Delta_{034}}$$

$$R_4 = \rho_{s,04} \frac{L_{04}^2}{\Delta_{034} + \Delta_{045}} \text{ and } R_5 = \rho_{s,05} \frac{L_{05}^2}{\Delta_{045} + \Delta_{056}} \tag{17.47}$$

From Equations 17.46 and 17.47 it is obvious that only half the area of the triangles to which the resistance path belongs should be considered while approximating the equivalent length perpendicular to the current path. Otherwise, every triangular element would have been considered twice while computing the resistance of the adjacent sides.

Then the Neumann condition at node 0 on dielectric–dielectric boundary is satisfied as follows:

$$\varepsilon_1 \overline{E}_{n1,0} - \varepsilon_2 \overline{E}_{n2,0} = \overline{\sigma}_{s,0} \tag{17.48}$$

where:
$\overline{\sigma}_{s,0}$ is obtained from Equation 17.40 and is given below in expanded form

$$\overline{\sigma}_{s,0} = \frac{1}{j\omega S_0} \left[ \begin{array}{l} \overline{V}_0 \left( \dfrac{1}{R_1} + \dfrac{1}{R_2} + \dfrac{1}{R_3} + \dfrac{1}{R_4} + \dfrac{1}{R_5} + \dfrac{1}{R_6} \right) \\ - \left( \dfrac{\overline{V}_1}{R_1} + \dfrac{\overline{V}_2}{R_2} + \dfrac{\overline{V}_3}{R_3} + \dfrac{\overline{V}_4}{R_4} + \dfrac{\overline{V}_5}{R_5} + \dfrac{\overline{V}_6}{R_6} \right) \end{array} \right] \tag{17.49}$$

where:

$\overline{V}_k$ is the potential of the $k$th node

$R_k$ is the resistance of the side connecting node 0 and the $k$th node. $R_k$ is given by Equation 17.46 or 17.47

$S_0$ is the area associated with node 0 as given by Equation 17.41

In order to compute capacitive-resistive fields incorporating both volume and surface resistances, Equation 17.48 is modified by bringing in complex permittivity.

$$\overline{\varepsilon}_1 \overline{E}_{n1,0} - \overline{\varepsilon}_2 \overline{E}_{n2,0} = \overline{\sigma}_{s,0} \tag{17.50}$$

where:

$\overline{\varepsilon}_x = \varepsilon_0 \varepsilon_{rx} - [j / (\omega \rho_{vx})]$, $x = 1,2$ and $\rho_{vx}$ is volume resistivity of medium $x$

$\overline{\sigma}_{s,0}$ is computed according to Equation 17.49

## 17.7 SCSM Examples

### 17.7.1 Cylinder Supported on Wedge

In many practical cases, electrodes are supported by insulating wedges. Figure 17.9 shows a metallic cylindrical electrode supported by solid insulating wedge. The point where the electrode touches the solid insulating material is called *triple junction*, as electrode, solid insulating material and gaseous dielectric (air in this case) meet at this point. The dimension of air gap between the electrode and the solid insulation depends on the angle $\alpha$. Ideally the angle $\alpha$ should be made 90°. But it is not always possible to make it so. Hence, it is important to study the effect of the angle $\alpha$ on the field distribution particularly around the triple junction, where field concentration takes place. In this example, the metallic electrode boundary and the solid–air dielectric interfaces on both the sides of the electrode are discretized by line and arc elements as this is a 2D case study. The elements are located between two successive nodes as marked on Figure 17.9. The sizes of the elements need to be adjusted according to the field concentration. The element lengths are small near the triple junction while such lengths could be higher on those sections of the solid–air dielectric

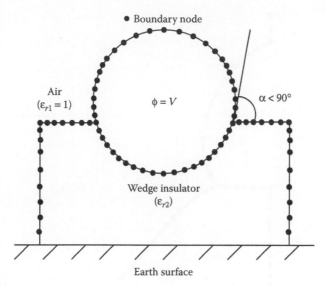

**FIGURE 17.9**
SCSM simulation of cylindrical electrode supported by solid insulating wedge.

interface which are away from the electrode or triple junction. The effect of the earth surface is taken into account by considering images of surface charge elements.

### 17.7.2 Conical Insulator in Gas-Insulated System

Conical spacers are used in gas insulated substation (GIS) for providing insulating support to the live central conductor. The shape of the conductor at the junction of the electrode and solid spacer at both live and earth ends are suitably designed to reduce field concentration at these critical locations. Consequently, such electrode spacer configuration needs to be analyzed carefully from electrical field distribution viewpoint as the dimensions of GIS are relatively small and hence the margin of error is also small. Figure 17.10 shows a typical conical spacer with central and outer conductor arrangement used in GIS. It is an axi-symmetric configuration and is simulated by line and arc elements lying between two successive nodes as marked on Figure 17.10. There are four boundaries in this case that need to be discretized by boundary elements, for example, live conductor boundary, earthed conductor boundary and two solid insulation-$SF_6$ gas interfaces. The element lengths are to be adjusted according to the nature of field distribution. The elements near the spacer are smaller in size, while the elements on the sections of the electrodes away from the spacer are relatively larger. To avoid any edge effect affecting the results near the spacer, the electrode boundaries at the top and bottom of Figure 17.10 are to

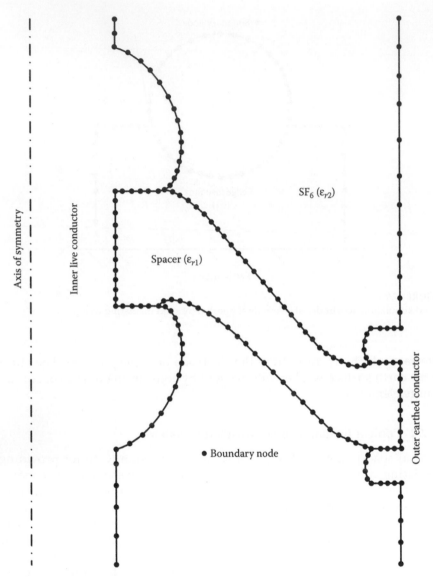

**FIGURE 17.10**
SCSM simulation of conical insulator used in GIS.

be considered extended up to a considerable distance from the spacer. But for simplicity it is not shown in Figure 17.10. In high-voltage DC (HVDC) GIS, conical spacers are often coated with a lossy dielectric for reducing problems due to charge accumulation. Such cases could also be studied by considering suitable surface resistivity along the two surfaces of the spacer.

### 17.7.3 Metal Oxide Surge Arrester

For high-voltage–high-energy cases, the assembly of multiple units of metal oxide surge arresters are commonly used in power system. As shown in Figure 17.11, the metal oxide elements are held together by a central insulating rod and are placed within an insulating shell. Then this arrangement is housed in a protective outer cover very often made of porcelain. Nowadays, polymeric insulation is also used for outer covering. To provide an inert atmosphere within the insulator housing, $N_2$ gas is put instead

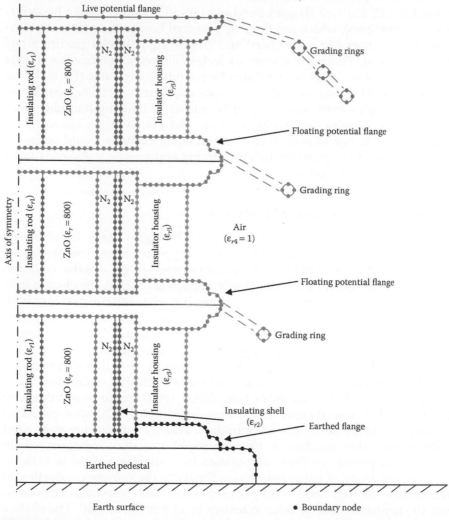

**FIGURE 17.11**
**(See colour insert.)** SCSM simulation of multiple unit assembly of metal oxide surge arrester.

of air. To linearize field distribution as much as possible field grading rings are used at the top live flange in all the applications. Sometimes grading rings are also used at the intermediate connecting flanges, as shown in Figure 17.11. The intermediate metallic flanges and the associated grading rings are floating potential electrodes and need to be simulated accordingly. The grading rings are closed-type boundaries while the other boundaries are open-type boundaries in this axi-symmetric example. The metal oxide elements have a dielectric constant of around 800 and very high resistivity before the start of conduction. The following are conductor boundaries that need to be simulated by boundary elements: (1) live flange boundary; (2) earthed flange boundary; (3) intermediate flange boundaries (two numbers), which are floating potential boundaries; (4) live potential grading rings (three numbers) and (5) floating potential grading rings (two numbers). There are several dielectric–dielectric boundaries in this example that are also to be simulated by boundary elements: (1) insulating rod–metal oxide boundary; (2) metal oxide–$N_2$ boundary; (3) $N_2$–insulating shell boundaries (two numbers); (4) $N_2$–insulator housing boundary and (5) insulator housing–air boundary. Line and arc elements are used for simulation. Change of resistance of metal oxide elements could be incorporated in field computation by considering appropriate value of volume resistivity of metal oxide elements.

### 17.7.4 Condenser Bushing of Transformer

Condenser bushings are very often used in high-voltage transformers (Figure 17.12). In this type of bushings floating potential grading electrodes are embedded in solid insulation (typically paper) to get a better field distribution. Consequently, it is practically important to study the electric field distribution in such bushing configuration. It is an axi-symmetric case with multiple electrode as well as dielectric boundaries. There are three types of electrode boundaries: (1) live central conductor boundary, (2) earthed metallic tank boundary and (3) floating potential grading electrode boundaries. It is to be noted here that these grading electrodes are very thin in dimension and SCSM is particularly useful in simulating such thin electrodes. There are following dielectric–dielectric boundaries, which are to be simulated by boundary elements: (1) paper–transformer oil boundary; (2) paper–porcelain boundary and (3) porcelain–air boundary. When the paper becomes aged with service, then conduction through paper insulation increases, which can be incorporated in field computation by assuming a suitable value of volume resistivity of paper insulation. Similarly, the volume conduction in transformer oil could also be duly considered in field computation by choosing the required value of volume resistivity of transformer oil. The surface pollution of porcelain outer cover could be taken into account by considering appropriate value of surface resistivity along porcelain–air boundary.

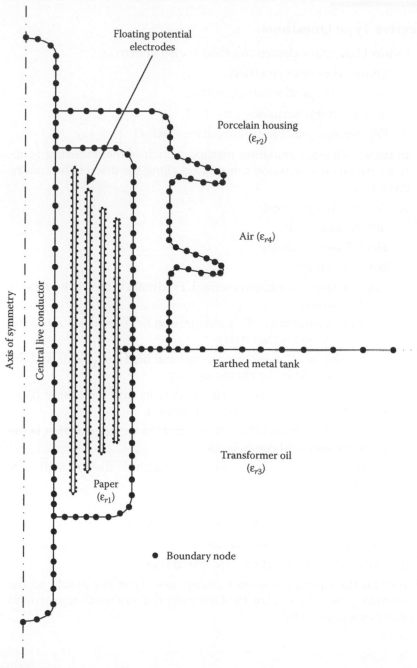

**FIGURE 17.12**
SCSM simulation of condenser bushing used in high-voltage transformers.

## Objective Type Questions

1. Indirect boundary element method is also known as
   a. Charge simulation method
   b. Volume charge simulation method
   c. Surface charge simulation method
   d. Region-oriented charge simulation method

2. In surface charge simulation method, which of the following functions are taken as unknown functions along the discrete boundary elements?
   a. Surface charge density
   b. Electric potential
   c. Electric field intensity
   d. Both (b) and (c)

3. In surface charge simulation method, Fredholm's integral equations are used, in which
   a. The kernel function is the potential and field coefficients and the weight function is the charge
   b. The kernel function is the potential and field coefficients and the weight function is the charge density
   c. The kernel function is the charge density and the weight function is the potential and field coefficients
   d. The kernel function is the charge and the weight function is the potential and field coefficients

4. In surface charge simulation method, which of the following are discretized?
   a. Entire volume of the regions of interest
   b. All the boundaries of the regions of interest
   c. Only the electrode boundaries
   d. Only the dielectric–dielectric boundaries

5. If $\sigma(i)$ is the equivalent surface charge density at any point $i$ on the boundary, then the electric field intensity due to a small area around that point is given by
   a. $\sigma_i \varepsilon_0$
   b. $\sigma_i/\varepsilon_0$
   c. $\sigma_i 2\varepsilon_0$
   d. $\sigma_i/(2\varepsilon_0)$

6. For the calculation of 2D and axi-symmetric fields using surface charge simulation method, most commonly used elements are
   a. Straight line and triangular elements
   b. Elliptic arc and triangular elements
   c. Straight line and elliptic arc elements
   d. Straight line and circular arc elements

7. For the calculation of 3D fields using surface charge simulation method, most commonly used element is
   a. Curvilinear triangle
   b. Curvilinear rectangle
   c. Curvilinear square
   d. Irregular polygon

8. In surface charge simulation method, for an open-type boundary if the number of surface charge elements is $N$, then the number of surface nodes is
   a. $N - 2$
   b. $N - 1$
   c. $N$
   d. $N + 1$

9. In surface charge simulation method, for a closed-type boundary if the number of surface charge elements is $N$, then the number of surface nodes is
   a. $N - 2$
   b. $N - 1$
   c. $N$
   d. $N + 1$

10. In surface charge simulation method, for the computation of electric potential in a 2D or an axi-symmetric system, the integration is carried out over a
    a. Line
    b. Triangle
    c. Quadrilateral
    d. Volume

11. For a 2D field computation by surface charge simulation method, which one of the following charge configurations could be used?
    a. Point charge
    b. Infinite length line charge

    c.  Finite length line charge

    d.  Ring charge

12. For an axi-symmetric field computation by surface charge simulation method, which one of the following charge configurations could be used?

    a.  Point charge

    b.  Finite length line charge

    c.  Ring charge

    d.  All of the above

13. For 2D and axi-symmetric field computation by surface charge simulation method using linear basis function, charge density of unknown magnitude is considered on the discrete boundary elements

    a.  At the mid-point of the element

    b.  At the extremities of the element

    c.  Near but not at the mid-point of the element

    d.  Near but not at the extremities of the element

14. In surface charge simulation method, which one of the following integration rules gives excellent accuracy?

    a.  Trapezoidal rule

    b.  Simpson's rule

    c.  Romberg's rule

    d.  Gauss–Legendre Quadrature rule

15. In surface charge simulation method, the simplicity of the approximating functions used with triangular elements lies in which of the following basic properties

    a.  These functions always ensure the continuity of the desired potential along all the boundaries between triangles provided only that continuity is imposed at the vertices of the triangle

    b.  The approximations are independent of the global coordinates of the triangles

    c.  The approximations are independent of the local coordinates of the triangles

    d.  Both (a) and (b)

16. Surface charge simulation method formulations give rise to a coefficient matrix that by property is a

    a.  Sparse matrix

    b.  Nearly full matrix

    c.  Diagonal matrix

    d.  Unit matrix

17. At any node $i$ on the dielectric boundary, let the normal component of electric field intensity be $E_{n,i}^-$, when the small area $AB$ around node $i$ is not considered in the computation. Then which one of the following is incorporated to compute the normal component of electric field intensity at node $i$

   a. $\sigma_i/(2\varepsilon_0)$

   b. $\sigma_i 2\varepsilon_0$

   c. $\sigma_i/\varepsilon_0$

   d. $\sigma_i \varepsilon_0$

18. The basic formulations of surface charge simulation method for capacitive field can be used for capacitive-resistive field, when

   a. Only real quantities are used

   b. Complex potentials are used

   c. Complex charge densities are used

   d. Both (b) and (c)

19. In surface charge simulation method, volume resistivity could be incorporated by expressing the permittivity of the dielectric medium as

   a. $\bar{\varepsilon}_x = \omega\rho_{vx} - [j/(\varepsilon_0\varepsilon_{rx})]$

   b. $\bar{\varepsilon}_x = \omega\rho_{vx} - j\varepsilon_0\varepsilon_{rx}$

   c. $\bar{\varepsilon}_x = \varepsilon_0\varepsilon_{rx} - [j/(\omega\rho_{vx})]$

   d. $\bar{\varepsilon}_x = \varepsilon_0\varepsilon_{rx} - j\omega\rho_{vx}$

20. By considering the general condition of current density vector at any node which one of the following could be included in field computation by surface charge simulation method?

   a. Volume resistance

   b. Surface resistance

   c. Both (a) and (b)

   d. Space charges

**Answers:**   1) c; 2) a; 3) b; 4) b; 5) d; 6) c; 7) a; 8) d; 9) c; 10) a; 11) b; 12) d; 13) b; 14) d; 15) d; 16) b; 17) a; 18) d; 19) c; 20) c

## Bibliography

1. S. Sato and W.S. Zaengl, 'Effective 3-dimensional electric field calculation by surface charge simulation method', *IEEE Proceedings A*, Vol. 133, No. 2, pp. 77–83, 1986.
2. T. Kouno, 'Computer calculation of electric fields', *IEEE Transactions on Electrical Insulation*, Vol. 21, No. 6, pp. 869–872, 1986.

3. H. Tsuboi and T. Misaki, 'The optimum design of electrode and insulator contours by nonlinear programming using the surface charge simulation method', _IEEE Transactions on Magnetics_, Vol. 24, No. 1, pp. 35–38, 1988.

4. B.L. Qin, G.Z. Chen, Q. Qiao and J.N. Sheng, 'Efficient computation of electric field in high voltage equipment', _IEEE Transactions on Magnetics_, Vol. 26, No. 2, pp. 387–390, 1990.

5. H. Steinbigler, D. Haller and A. Wolf, 'Comparative analysis of methods for computing 2-D and 3-D electric fields', _IEEE Transactions on Electrical Insulation_, Vol. 26, No. 3, pp. 529–536, 1991.

6. D. Beatovic, P.L. Levin, S. Sadovic and R. Hutnak, 'A Galerkin formulation of the boundary element for two-dimensional and axi-symmetric problems in electrostatics', _IEEE Transactions on Electrical Insulation_, Vol. 27, No. 1, pp. 135–143, 1992.

7. Z. Andjelic, B. Krstajic, S. Milojkovic, A. Blaszczyk, H. Steinbigler and M. Wohlmuth, _Integral Methods for the Calculation of Electric Fields_, Scientific Series of the International Bureau, Vol. 10, Forschungszentrum Jülich GmbH, Germany, 1992.

8. J. Zhou, G.D. Theophilus and K.D. Srivastava, 'Pre-breakdown field calculations for charged spacers in compressed gases under impulse voltages', _Proceedings of the IEEE ISEI_, Baltimore, MD, pp. 273–278, July 3–8, IEEE Publication, 1992.

9. P.L. Levin, A.J. Hansen, D. Beatovic, H. Gan and J.H. Petrangelo, 'A unified boundary-element finite-element package', _IEEE Transactions on Electrical Insulation_, Vol. 28, No. 2, pp. 161–167, 1993.

10. F. Gutfleisch, H. Singer, K. Foerger and J.A. Gomollon, 'Calculation of HV fields by means of the boundary element method', _IEEE Transactions on Power Delivery_, Vol. 9, No. 2, pp. 743–749, 1994.

11. G. Mader and F.H. Uhlmann, 'Computation of 3-D capacitances of transmission line discontinuities with the boundary element method', _Second International Conference on Computation in Electromagnetics_, London, pp. 8–11, April 12–14, IEE (UK) Publication, 1994.

12. J. Weifang, W. Huiming and E. Kuffel, 'Application of the modified surface charge simulation method for solving axial symmetric electrostatic problems with floating electrodes', _Proceedings of the IEEE ICPADM_, Brisbane, Australia, Vol. 1, pp. 28–30, July 3–8, IEEE Publication, 1994.

13. A. Vishnevsky and A. Lapovok, 'Conservative methods in boundary-element calculations of static fields', _IEEE Proceedings – Science, Measurement and Technology_, Vol. 142, No. 2, pp. 151–156, 1995.

14. R. Schneider, P.L. Levin and M. Spasojevic, 'Multiscale compression of BEM equations for electrostatic systems', _IEEE Transactions on Dielectrics & Electrical Insulation_, Vol. 3, No. 4, pp. 482–493, 1996.

15. E. Cardelli and L. Faina, 'Open-boundary, single-dielectric charge simulation method with the use of surface simulating charges', _IEEE Transactions on Magnetics_, Vol. 33, No. 2, pp. 1192–1195, 1997.

16. S. Chakravorti and H. Steinbigler, 'Electric field calculation including surface and volume resistivities by boundary element method', _Journal of the Institution of Engineers (India): Electrical Engineering Division_, Vol. 78, No. 1, pp. 17–23, 1997.

17. N. de Kock, M. Mendik, Z. Andjelic and A. Blaszczyk, 'Application of the 3D boundary element method in the design of EHV GIS components', _IEEE Electrical Insulation Magazine_, Vol. 14, No. 3, pp. 17–22, 1998.

18. S. Chakravorti and H. Steinbigler, 'Capacitive-resistive field calculation in and around HV bushings by boundary element method', *IEEE Transactions on Dielectrics & Electrical Insulation*, Vol. 5, No. 2, pp. 237–244, 1998.
19. A. Ahmed, H. Singer and P.K. Mukherjee, 'A numerical model using surface charges for the calculation of electric fields and leakage currents on polluted insulator surfaces', *Proceedings of the IEEE CEIDP*, Atlanta, GA, Vol. 1, pp. 116–119, April 30–May 3, IEEE Publication, 1998.
20. B.Y. Lee, S.H. Myung, J.K. Park, S.W. Min and E.S. Kim, 'The use of rational B-spline surface to improve the shape control for three-dimensional insulation design and its application to design of shield ring', *IEEE Transactions on Power Delivery*, Vol. 13, No. 3, pp. 962–968, 1998.
21. P.K. Mukherjee, A. Ahmed and H. Singer, 'Electric field distortion caused by asymmetric pollution on insulator surfaces', *IEEE Transactions on Dielectrics & Electrical Insulation*, Vol. 6, No. 2, pp. 175–180, 1999.
22. A. Ahmed and H. Singer, 'New modelling of the boundary between wet and dry zones on the surface of polluted high-voltage insulators', *Proceedings of the 11th ISH Symposium*, London, Vol. 2, pp. 35–38, 1999.
23. S. Chakravorti and H. Steinbigler, 'Boundary element studies on insulator shape and electric field around HV insulators with or without pollution', *IEEE Transactions on Dielectrics & Electrical Insulation*, Vol. 7, No. 2, pp. 169–176, 2000.
24. A. Ahmed and H. Singer, 'Effect of thermal capacities of ceramic insulators on their electrical and thermal analysis under contaminated surface conditions', *Proceedings of the IEEE CEIDP*, Victoria, BC, Canada, Vol. 1, pp. 238–241, October 15–18, IEEE Publication, 2000.
25. S. Hamada and T. Takuma, 'Electrostatic field calculation of multi-particle systems with and without contact points by the surface charge simulation method', *Proceedings of the 7th IEEE ICPADM*, Nagoya, Japan, Vol. 1, pp. 27–32, June 1–4, IEEE Publication, 2003.
26. J.A. Gomollon and R. Palau, 'Steady state 3-D-field calculations in three-phase systems with surface charge method', *IEEE Transactions on Power Delivery*, Vol. 20, No. 2, pp. 919–924, 2005.
27. B. Zhang, S. Han, J. He, R. Zeng and P. Zhu, 'Numerical analysis of electric-field distribution around composite insulator and head of transmission tower', *IEEE Transactions on Power Delivery*, Vol. 21, No. 2, pp. 959–965, 2006.
28. B. Zhang, J. He, X. Cui, S. Han and J. Zou, 'Electric field calculation for HV insulators on the head of transmission tower by coupling CSM with BEM', *IEEE Transactions on Magnetics*, Vol. 42, No. 4, pp. 543–546, 2006.
29. W. Ling, Z. Rong, B. Zhang, G. Shanqiang, C. Lian and H. Jinliang, 'Electric-field calculation for hot-line working on the head of transmission tower', *12th Biennial IEEE Conference on Electromagnetic Field Computation*, Miami, FL, p. 194, April 30–May 3, IEEE Publication, 2006.
30. B. Zhang, J. Zhao, R. Zeng and J. He, 'Numerical analysis of DC current distribution in AC power system near HVDC system', *IEEE Transactions on Power Delivery*, Vol. 23, No. 2, pp. 960–965, 2008.
31. S. Hamada and T. Kobayashi, 'Analysis of electric field induced by ELF magnetic field utilizing fast-multipole surface-charge simulation method for voxel data', *Electrical Engineering in Japan*, Vol. 165, No. 4, pp. 1–10, 2008.

32. S. Hamada, M. Kitano and T. Kobayashi, 'Accuracy estimation of induced electric field calculated by fast-multipole surface-charge-simulation method for voxel data', *IEEE Transactions on Fundamentals and Materials*, Vol. 128, No. 4, pp. 223–234, 2008.

33. C. Zhuang, R. Zeng, B. Zhang, P. Zhu, L. Wang and J. He, 'The optimization of entering route for live working on 750 kV transmission towers by space electric-field analysis', *IEEE Transactions on Power Delivery*, Vol. 25, No. 2, pp. 987–994, 2010.

34. J.A. Gomollon, E. Santome and R. Palau, 'Calculation of 3D-capacitances with surface charge method by means of the basis set of solutions', *IEEE Transactions on Dielectrics & Electrical Insulation*, Vol. 17, No. 1, pp. 240–246, 2010.

35. B.S. Reddy and U. Kumar, 'Simulation of potential and electric field profiles for transmission line insulators', *International Journal of Modelling and Simulation*, Vol. 30, No. 4, pp. 490–498, 2010.

36. B.S. Reddy and U. Kumar, 'Investigations on the corona performance of a ceramic disc insulators integrated with field reduction electrode', *IEEE Industry Applications Society Annual Meeting*, Houston, TX, pp. 1–8, October 3–7, IEEE Publication, 2010.

37. L. Xiaoming, W. Qi, C. Yundong and X. Qiumin, 'Analyses of electric field for vacuum circuit breaker using response surface-charge simulation method', *Proceedings of the Asia-Pacific Power and Energy Engineering Conference*, Chengdu, China, pp. 1–4, March 28–31, IEEE Publication, 2010.

38. S.S. Chowdhury, A. Lahiri and S. Chakravorti, 'Surface resistance modified electric field computation in asymmetric configuration using surface charge simulation method: A new approach', *IEEE Transactions on Dielectrics & Electrical Insulation*, Vol. 19, No. 3, pp. 1068–1075, 2012.

# 18

# Numerical Computation of Electric Field in High-Voltage System – Case Studies

**ABSTRACT**  Several components, which are widely used in high-voltage systems, are critical in nature from the viewpoint of failure due to electrical discharges. Two common examples of such components are bushings and cable terminations. Insulators that are used in large numbers in all high-voltage systems are prone to failure due to surface pollution or surface wetting. These unwanted failures could be prevented by appropriate design only if the electric field distribution is estimated accurately through numerical field computations. However, results of numerical field computations need to be validated using benchmark models for which analytical solutions are available. This chapter presents a few such benchmark models and reports results of electric field analysis in bushings, cable terminations as well as insulators under various operating conditions.

## 18.1 Introduction

Numerical computation of electric field for high-voltage system components and devices are carried out to determine the adequacy of insulation to reliably withstand electric stresses that may arise in the system. Some of these components are critical to the system. For example, if the bushing of a transformer fails, then it leads to terminal short circuit and causes very high fault levels. Again bushings are of different types depending on the system voltage level. Thus, field distribution within bushings is to be determined accurately, so that proper insulation could be provided to prevent any unwanted failure. Similarly, cable terminations, where the cable is connected to another equipment, are critical components that are prone to failure due to excessive field concentration. Stress diverters of suitable design are used in such cable terminations, the shapes of which are normally finalized through extensive electric field computations. Insulators that are able to withstand electric stresses satisfactorily under dry condition may not be able to withstand electric stresses under wet or polluted conditions due to change in the field distribution. In order to ensure safe operation of insulators under all environmental conditions, it is necessary to find electric field distribution in and around insulators in all such conditions. However, there is a need for validating the results obtained from numerical field analysis. This is typically done by considering benchmark models for which analytical solutions are available. Taking the analytical results as reference, the accuracy of numerical results is determined through comparison of analytical and numerical results.

## 18.2 Benchmark Models for Validation

Two benchmark models are commonly used for validation of numerical codes for purely capacitive field analysis as detailed below: (1) for two-dimensional system – cylinder in uniform external field and (2) for axi-symmetric system – sphere in uniform external field. For capacitive-resistive field in axi-symmetric system, a dielectric sphere in uniform external field is used as a benchmark model.

### 18.2.1 Cylinder in Uniform External Field

Analytical solutions for capacitive field distribution in the case of a cylinder in uniform external field have been discussed in detail in Section 10.3. For validation of numerical output, results for a conducting cylinder as well as a dielectric cylinder in uniform external field are used. In the case of a conducting cylinder in uniform field, numerical results for electric potential could be compared to that obtained from Equation 10.54, while numerically computed values of electric field intensity components could be compared to the corresponding values obtained from Equations 10.55 and 10.56. For a dielectric cylinder in uniform external field, Equations 10.65 and 10.66 are used for validating results for electric potential and the results for electric field intensity components are validated using the partial derivatives of Equations 10.65 and 10.66.

### 18.2.2 Sphere in Uniform External Field

Electric field in and around a sphere in uniform external field has been solved analytically in Section 10.2 in the absence of resistivity of dielectric material, that is, in the case of purely capacitive field. These analytical solutions are useful for validating numerical field computation results in axi-symmetric systems. For a conducting sphere in uniform external field, Equations 10.22 through 10.24 are used for validating results for electric potential and electric field intensity components, respectively. For a dielectric sphere in uniform external field, Equations 10.33 and 10.34 are used for validating results for electric potential within and outside the dielectric sphere, respectively.

Results for electric field intensity components within and outside the dielectric sphere could be validated using the partial derivatives of Equations 10.33 and 10.34, respectively.

### 18.2.3 Dielectric Sphere Coated with a Thin Conducting Layer in Uniform External Field

Results for capacitive-resistive field in axi-symmetric system could be validated using the benchmark model comprising a dielectric sphere coated with a thin conducting layer in uniform sinusoidal external field [1]. The configuration is schematically shown in Figure 18.1. Uniform sinusoidal field is represented as $e_u = E_{um} \sin \omega t$. In this configuration, for the dielectric sphere as well as the surrounding dielectric medium conductivity is zero, that is, $\sigma_1 = \sigma_3 = 0$. For the thin conducting layer on the dielectric sphere, relative permittivity is unity and conductivity ($\sigma_2$) is variable. Radius of the dielectric sphere is $r$ and the thickness of conducting layer is $t$.

Analytical solution for maximum electric field intensity at the point $P$, as shown in Figure 18.1, is given by

$$E_{max} = E_{um}\left[\frac{3\bar{\varepsilon}_2(2\bar{\varepsilon}_2 + \bar{\varepsilon}_1)t_r^3 + 6\bar{\varepsilon}_2(\bar{\varepsilon}_1 - \bar{\varepsilon}_2)}{(2\bar{\varepsilon}_3 + \bar{\varepsilon}_2)(2\bar{\varepsilon}_2 + \bar{\varepsilon}_1)t_r^3 - 2(\bar{\varepsilon}_3 - \bar{\varepsilon}_2)(\bar{\varepsilon}_1 - \bar{\varepsilon}_2)}\right] \tag{18.1}$$

where:

$$\bar{\varepsilon}_x = \sigma_x + j\omega\varepsilon_0\varepsilon_{rx} \text{ for } x = 1,2,3$$

$$t_r = \left(\frac{t}{r} + 1\right)$$

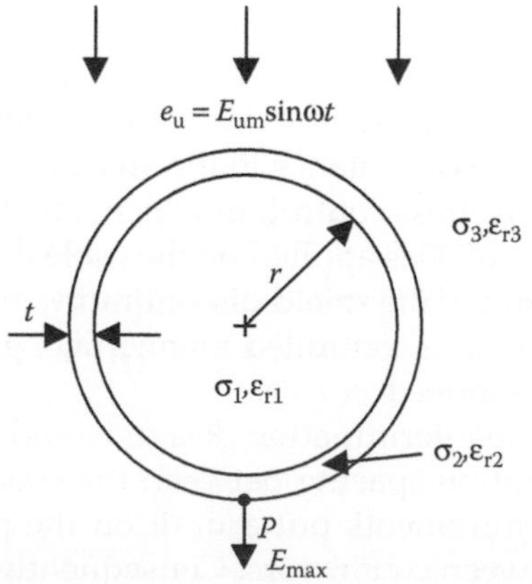

**FIGURE 18.1**
Dielectric sphere coated with a thin conducting layer in uniform sinusoidal external field.

## 18.3  Electric Field Distribution in the Cable Termination

Terminations are required in the case of connecting the cable to another line or a busbar or to equipments such as a transformer or a switchgear. A high-voltage cable termination typically provides (1) adequate electric stress control for the cable insulation shield terminus; (2) comprehensive external leakage insulation between the high-voltage conductor(s) and grounded ends and (3) proper sealing to prevent the entry of the external environmental elements, for example, moisture and corrosive chemicals, into the cable and also to maintain the pressure, if any, within the cable system. The requirements of an AC cable termination are detailed in IEEE Std 48-1975.

In a co-axial shielded cable, the electric field does not vary along the cable axis and there is variation in electric field only in the radial direction. In terminating a shielded cable, it is necessary to remove the cable insulation shield up to a certain distance from the exposed conductor, depending on the voltage level of the cable and properties of dielectric media in use. The removal of the insulation shield results in a discontinuity in the axial geometry of the cable. As a result the electric field is no longer invariant along the cable axis, but exhibits variations along all three Cartesian axes, which can be suitably modelled by axi-symmetric representation.

Electric flux lines emanating from the conductor converge in the vicinity of the shield discontinuity at the end of the shield causing high electric stresses in this area. Such high stresses may cause premature failure of the cable. But more importantly partial discharges will occur continuously at the site of high-field concentration, which will shorten the life of the cable significantly. Hence, cable terminations must provide stress control to reduce the stresses occurring near the end of the shield. Two commonly used methods for stress control in cable terminations are (1) geometric stress control, in which a stress cone is used, which optimizes the geometry at the terminating discontinuity in order to reduce the stresses at that location. The stress cones use the geometrical solution by controlling the capacitance in the area of the insulation shield terminus and (2) capacitive stress control, in which a material of high dielectric permittivity of the order of 30 is applied on the cable dielectric at the termination. Located near the end of the shield discontinuity, the material changes the electric field distribution in a controlled manner along the entire area where the shielding has been removed.

The problem of the cable-termination design would have been simple had there been no constraint on spacing between the conductors. But for many installations, space requirements put a limit on the practical distance that can be maintained between conductors. Consequently, the determination of electric field distribution in and around the cable termination is of practical

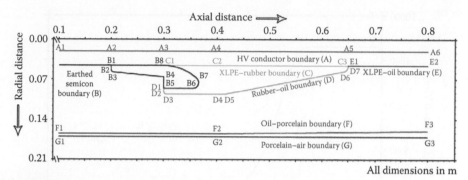

**FIGURE 18.2**
**(See colour insert.)** Geometry and boundaries of cable termination under study.

importance. This section presents the results of electric field analysis of a typical cable termination used in 220 kV system.

The cable termination for a single-core cable with a porcelain outer cover, as shown in Figure 18.2, is modelled as an axi-symmetric system, where the radial distances are taken from the axis of symmetry. Relative dielectric permittivities that are considered in modelling the cable termination are as follows: (1) cross-linked polyethylene (XLPE) – 2.3, (2) insulation grade rubber – 2.8, (3) oil – 2.2 and (4) porcelain – 6.0.

The various boundaries that are considered in this study are shown in Figure 18.2. Following are the abbreviations that are used in this case: (1) high-voltage (HV) conductor surface – boundary A, (2) earthed semi-conducting surface – boundary B, (3) XLPE–rubber interface – boundary C, (4) rubber–oil interface – boundary D, (5) XLPE–oil interface – boundary E, (6) oil–porcelain interface – boundary F and (7) porcelain–air interface – boundary G. For simplicity, the outer surface of the porcelain cover is taken as cylindrical.

Figure 18.3 shows the plot of normal, tangential and the resultant stresses along the HV conductor boundary. Electric stresses are plotted from A1 to A6 with reference to Figure 18.2. As it is a conductor boundary, the tangential stresses are negligible and hence the normal and resultant stresses are same. The maximum stress occurs between A1 to A3 where the semi-conducting surface is co-axial with the XLPE insulation.

Figure 18.4 shows the plot of normal, tangential and the resultant stresses along the earthed semi-conducting surface. Electric stresses are plotted from B1 to B8 with reference to Figure 18.2. Maximum electric stress occurs between B8 to B1 where the semi-conducting boundary is co-axial with the XLPE insulation. Tangential stresses are negligible except at B8 where very acute angle is formed with XLPE–rubber boundary. Theoretically, such tangential stresses should not occur near B8 as it is a conducting boundary. These stresses arise due to error in numerical simulation of boundary involving very acute angle.

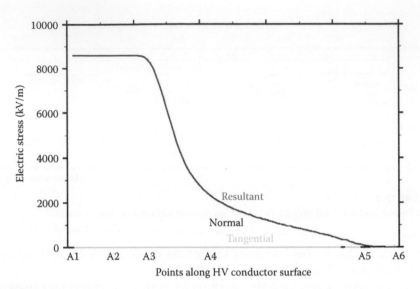

**FIGURE 18.3**
**(See colour insert.)** Electric stresses along the high-voltage conductor surface.

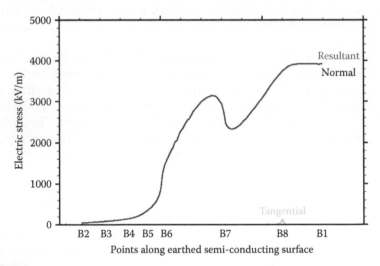

**FIGURE 18.4**
**(See colour insert.)** Electric stresses along the semi-conducting surface.

Figure 18.5 depicts the plot of normal, tangential and the resultant stresses on the rubber side along the XLPE–rubber interface. Electric stresses are plotted from C1 to C3 with reference to Figure 18.2. Maximum resultant stresses occur at the zone near C1, where the rubber insulation meets the semiconducting surface making very acute angle. Tangential stresses are significant around location C2.

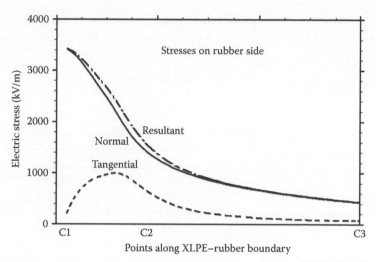

**FIGURE 18.5**
Electric stresses on the rubber side along the XLPE-rubber interface.

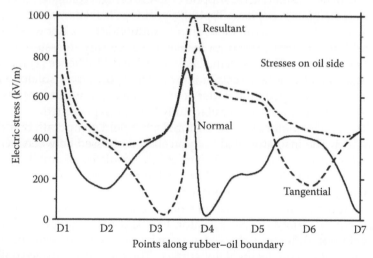

**FIGURE 18.6**
Electric stresses on the oil side along the rubber–oil interface.

The plot of normal, tangential and resultant stresses along the rubber–oil interface is presented in Figure 18.6. Electric stresses are plotted from D1 to D7 with reference to Figure 18.2. Maximum stresses occur at the zone near D4 where the rubber starts converging towards the XLPE surface.

Figure 18.7 shows the equipotential lines for the cable termination shown in Figure 18.2.

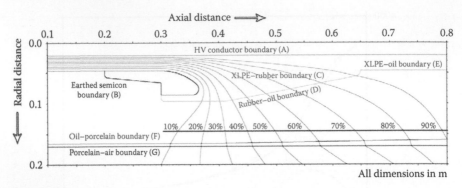

**FIGURE 18.7**
**(See colour insert.)** Equipotential lines for the cable termination.

## 18.4 Electric Field Distribution around a Post-Type Insulator

Post-type insulators usually act as support or spacer of high-voltage electrode with respect to earthed frame or plane. Typically, post-type insulators are surrounded by gaseous dielectric such as air or sulphur hexafluoride ($SF_6$). Knowledge of the electric field distribution around an insulator is necessary to assure reliability in operation of high-voltage system. Chakravorti and Mukherjee [2] reported a detailed study on electric field distribution around a post-type insulator without as well as with surface pollution under power frequency as well as impulse voltages. In this section, results of a similar study have been presented.

The post-type insulator considered for electric field computation is shown in Figure 18.8. The insulator made of porcelain is stressed between two electrodes and is surrounded by air. The electrode–insulator assembly is an axisymmetric configuration, having two dielectric media, namely, porcelain ($\varepsilon_r = 6$) and air. Electric field computations have been carried out for uniform surface pollution for which a constant value of surface resistivity is considered along the entire surface of the insulator. For field computation with partial surface pollution different values of surface resistivity have been considered at different locations on the insulator surface. The severity of surface pollution depends on the type of pollution, for example, marine pollution, industrial pollution, and dryness of the pollution. Surface resistivity of pollution layer decreases drastically as the pollution layer is wetted. As a result field computations are carried out over a wide range of surface resistivity from $10^{15}$ to $10^4$ $\Omega$ in order to simulate varying degrees of surface pollution severity.

### 18.4.1 Effect of Uniform Surface Pollution

Results of field computations under power frequency (50 Hz) voltage show that the field distribution is capacitive in nature for $\rho_s \geq 10^{11}$ $\Omega$ and it is

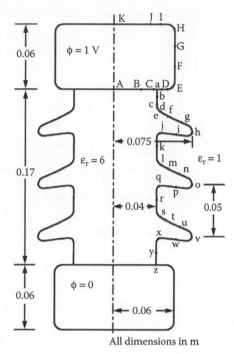

**FIGURE 18.8**
Post-type insulator stressed between two electrodes.

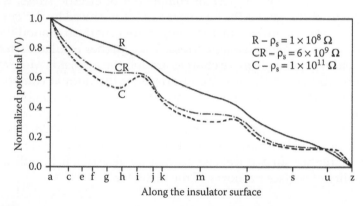

**FIGURE 18.9**
Potential distribution along the insulator surface under power frequency voltage with uniform surface pollution.

resistive in nature for $\rho_s \leq 10^8\ \Omega$. In between the field is capacitive-resistive in nature. Figure 18.9 presents the potential distribution along the insulator surface showing the change in field nature from capacitive to resistive with the change in surface resistivity. Figure 18.10 shows the variation of resultant stress along the insulator surface for capacitive as well as resistive fields.

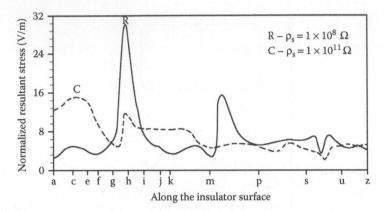

**FIGURE 18.10**
Resultant stress distribution along the insulator surface under power frequency voltage for capacitive and resistive fields.

The stresses are reported for air side of the porcelain–air interface from the viewpoint of the surface flashover, which occurs in air. It may be seen that the highest resultant stress in the case of resistive field is nearly two times higher than that in the case of capacitive field and occurs near the tip of the topmost shed of the insulator.

Charges that may be present around the insulator surface are drawn on the insulator surface by the normal component of electric stress, whereas the movement of charges on the insulator surface is caused by the tangential component of electric stress. Further charges are generated around the insulator surface due to high resultant stress. Thus, higher resultant stresses in resistive field will generate more charges around the insulator. More charges will then be drawn on the insulator surface by the normal component of electric stress, which under the influence of the tangential component will move along the insulator surface. As a result, the surface leakage current will increase in the case of resistive field compared to capacitive field. It is known that the onset of surface flashover is related to a critical surface leakage current. Hence, higher surface leakage current for resistive field will increase the possibility of surface flashover of the insulator.

### 18.4.2 Effect of Partial Surface Pollution

Figures 18.11 and 18.12 present the potential and resultant stress distribution, respectively, along the insulator surface for uniform as well as partial surface pollution. In the case of partial surface pollution, the section of the insulator surface from a to h, as shown in Figure 18.8, is considered to be polluted and hence the surface resistivity in this section is taken to be $10^8\ \Omega$ corresponding to resistive field, whereas the rest of the insulator surface is considered to be pollution free for which surface resistivity is taken to be

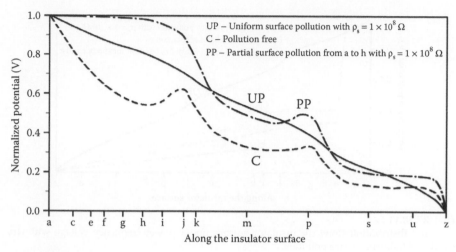

**FIGURE 18.11**

Potential distribution along the insulator surface under power frequency voltage with uniform and partial surface pollution vis-a-vis no surface pollution.

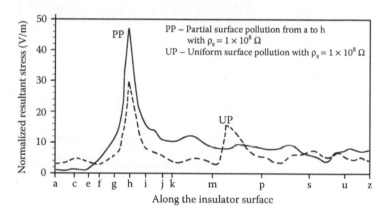

**FIGURE 18.12**

Resultant stress distribution along the insulator surface under power frequency voltage with uniform and partial surface pollution.

$10^{11}$ Ω, corresponding to capacitive field. This case of partial surface pollution is found to be most onerous, as tangential stress is more than twice and the resultant and normal stresses are about 1.6 times higher than the corresponding values for resistive field. These results clearly indicate the danger posed by partial surface pollution.

### 18.4.3 Effect of Dry Band

Researchers have recognized that failure of insulators very often is caused by dry bands in the pollution layer on the insulator surface. The dry band can be simulated by a zone of high resistivity in the uniformly polluted surface

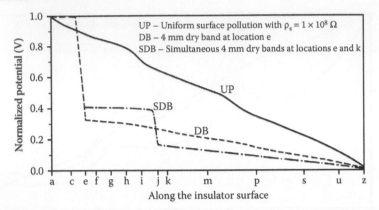

**FIGURE 18.13**
Potential distribution along the insulator surface under power frequency voltage with dry band and uniform surface pollution.

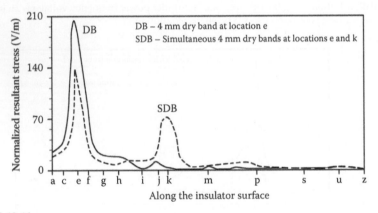

**FIGURE 18.14**
Resultant stress distribution along the insulator surface under power frequency voltage with dry band.

having a low surface resistivity. The width of the dry band can be varied by varying the length of the zone having high resistivity.

Figures 18.13 and 18.14 present the potential and resultant stress distribution, respectively, along the insulator surface for dry band as well as uniform surface pollution. As a thumb rule, many researchers assume that the entire potential difference between the electrodes appears across the dry band. But from the results of field computation it is possible to determine exactly how much potential appears across the dry band and what will be the stresses due to the dry band. These results are very helpful in the studies related to growth of dry band on the insulator surface and it may be stated that the growth of dry band depends on the width as well as the location of the dry band on the insulator surface, as the potential across the dry band depends on these two factors.

Figures 18.13 and 18.14 also show the effect of two simultaneously occurring dry bands on potential and resultant stresses, respectively. It may be seen that if multiple dry bands occur at the same time on the insulator surface, then the resultant stresses are reduced as compared to the occurrence of a single dry band.

### 18.4.4 Impulse Field Distribution

Electric field computations could be extended to study the field distribution under both lightning and switching impulses. The standard waveshape of lightning impulse voltage is 1.2/50 μs and that for switching impulse is 250/2500 μs. Because the effective frequency of impulse voltage is much higher than power frequency, the value of surface resistivity for which the field will be capacitive or resistive under impulse voltage will be much lower than that for power frequency. Further, the effective frequency of lightning impulse is higher than that for switching impulse and hence the transition of field from capacitive to resistive occurs at lower values of surface resistivity for lightning impulse compared to those for switching impulse.

Results of electric field computations show that for switching impulse the field is capacitive for $\rho_s \geq 10^9 \, \Omega$ and resistive for $\rho_s \leq 10^6 \, \Omega$, while for lightning impulse the field is capacitive for $\rho_s \geq 10^7 \, \Omega$ and resistive for $\rho_s \leq 10^4 \, \Omega$.

## 18.5 Electric Field Distribution in a Condenser Bushing

Bushings are used when a high-voltage conductor is to pass through a barrier having a different potential, for example, an earthed metal tank or an earthed wall. As a result bushings are integral parts of high-voltage equipment such as transformers and shunt reactors. For high-voltage equipment, condenser bushings fitted with cylindrical floating potential electrodes are preferred in practice, as the field nature in a condenser bushing is much less non-linear than the traditional porcelain bushing. The critical region of a bushing from the viewpoint of electrical stress is the zone where the distance between the high-voltage conductor and the barrier is minimum. If the electric field intensity in the critical zone exceeds the breakdown strength of the dielectric used, then discharge will start in this zone. Such discharges may grow and lead to complete breakdown of the bushing. Electric field distribution in and around an outdoor bushing is governed by bushing geometry, permittivity and volume as well as surface resistivities of dielectric media. Out of these factors, surface resisitivity of the outer cover of the bushing plays a major role in determining the field distribution, as it is the most variable parameter affected by atmospheric conditions. Researchers have identified that the deposition of pollution on the outer cover and wetting of pollution layer by fog or condensation are major causes of outdoor bushing failure. Breakdown of bushing is equivalent to terminal short circuit of the equipment such as transformer and results in very high fault levels causing severe damage to the equipment in particular and the system is general. Considering all these practical aspects, Chakravorti and Steinbigler [3] reported detailed results of capacitive-resistive field computation in and around a condenser bushing considering volume and surface resistivities. Results of a similar study are presented in this section.

Figure 18.15 shows the condenser bushing configuration for which capacitive-resistive field computations have been carried out. It comprises a central conductor that is cylindrical in shape, three cylindrical floating potential electrodes embedded in paper, a porcelain outer cover and an earthed metal tank filled with transformer oil. Thus, the configuration under study is taken to be an axi-symmetric one having four dielectric media, namely, paper ($\varepsilon_r = 5$), transformer oil ($\varepsilon_r = 2.2$), porcelain ($\varepsilon_r = 6$) and air ($\varepsilon_r = 1$). The potential of the central conductor is considered to be sinusoidal of frequency 50 Hz. For the bushing shown in Figure 18.15 there are three dielectric–dielectric boundaries: (1) paper–oil interface – boundary A, (2) paper–porcelain

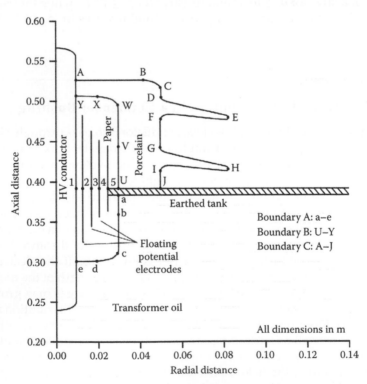

**FIGURE 18.15**
Condenser bushing arrangement considered for field computation.

interface – boundary B and (3) porcelain–air interface – boundary C. Surface resistivity of three boundaries are considered as follows: (1) for boundary A, $\rho_{sA}$ is taken to be always uniform; (2) for boundary B, $\rho_{sB}$ is taken to be always infinite and (3) for boundary C, $\rho_{sC}$ is taken to be both uniform as well as non-uniform to take into account uneven surface pollution. Considering aging-related degradation of paper and oil, volume resistivities of paper ($\rho_{vp}$) and oil ($\rho_{vo}$) have been taken to be high for new condition and low for aged condition, whereas the volume resistivity of porcelain is taken to be always infinite considering very little degradation.

Results of field computations for the following seven cases have been noted to be significant and are presented below: (1) Case 1 – purely capacitive field with infinite surface and volume resistivities; (2) Case 2 – uniform $\rho_{sC} = 10^7\ \Omega$ and $\rho_{sA} = \rho_{vp} = \rho_{vo} = \infty$; (3) Case 3 – $\rho_{vp} = 10^6\ \Omega.m$ and $\rho_{sA} = \rho_{sC} = \rho_{vo} = \infty$; (4) Case 4 – $\rho_{vp} = \rho_{vo} = 10^6\ \Omega.m$, uniform $\rho_{sA} = 10^6\ \Omega$ and $\rho_{sC} = \infty$; (5) Case 5 – $\rho_{sC} = 10^7\ \Omega$ between A and E while $\rho_{sC} = \infty$ on rest of boundary C, as shown in Figure 18.15, and $\rho_{sA} = \rho_{vp} = \rho_{vo} = \infty$; (6) Case 6 – as in Case 5 with $\rho_{vp} = 10^6\ \Omega.m$ and $\rho_{sA} = \rho_{vo} = \infty$ and (7) Case 7 – $\rho_{vo} = 10^6\ \Omega.m$ and $\rho_{sA} = \rho_{sC} = \rho_{vp} = \infty$.

Figure 18.16 shows the variation of resultant field intensity along the line 1-2-3-4-5 depicted in Figure 18.15, where 2, 3 and 4 correspond to the locations of the three floating potential electrodes from left to right. This zone is the critical region within the bushing. From Figure 18.16 it may be seen that electric stress near HV conductor and earthed tank are highest for Case 6 and Case 3, respectively, and are lowest for Case 3 and Case 6, respectively. Figure 18.16 also shows that partial pollution of boundary C along with a lower volume resistivity of paper insulation (Case 6) causes highest electric stresses in the critical zone.

Figure 18.17 shows electric stress distribution along the paper–oil boundary (boundary A) for different cases, where the stresses are determined for

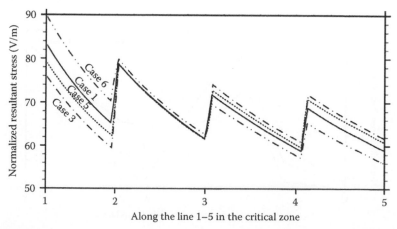

**FIGURE 18.16**
Electric stress distribution in the critical zone of the bushing.

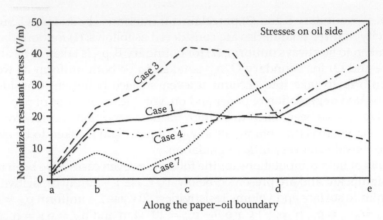

**FIGURE 18.17**
Electric stress distribution along the paper–oil boundary.

the oil side of the interface. It may be noted that in all the cases where the volume resistivities of paper and oil have been taken to be low electric stresses are higher. Moreover, when the volume resistivities are taken to be low, then a lower value of $\rho_{sA}$ does not affect electric stresses much. For a low value of $\rho_{vp}$ (Case 3), the maximum stress occurs at location c on paper–oil boundary, while the maximum stress occurs at location e when volume resistivity of oil ($\rho_{vo}$) is taken to be low (Case 7).

Electric stresses on the paper side of the paper–porcelain boundary (boundary B) are shown in Figure 18.18. It shows that for purely capacitive field (Case 1) electric stresses near the point V on boundary B are high, although the highest stress occurs at the point Y. However, for the other cases the stresses at the location Y are relatively lower. From Figure 18.18 it is also clear

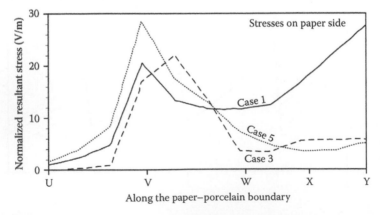

**FIGURE 18.18**
Electric stress distribution along the paper–porcelain boundary.

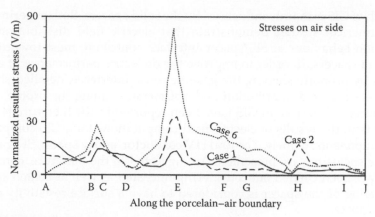

**FIGURE 18.19**
Electric stress distribution along the porcelain–air boundary.

that partial pollution of boundary C (Case 5) increases the stresses near the location V significantly. Results of computations show that the stresses along boundary B are affected much more by a lower value of $\rho_{vp}$ than uniform or non-uniform $\rho_{sC}$.

Electric stresses as computed on the air side of porcelain–air boundary (boundary C) are shown in Figure 18.19. It may be seen that for purely capacitive field (Case 1) electric stresses are comparatively lower. The difference between the maximum and minimum stresses on boundary C is higher for uniform surface pollution (Case 2) than capacitive field. For uniform surface pollution the maximum stress occurs at location E on boundary C. This result is similar to that obtained for uniform surface pollution of porcelain post insulator, as shown in Figure 18.10. Partial surface pollution increases the stresses to a great extent and for Case 6, the maximum stress is quite high at location E. Detailed results of computations revealed that a lower value of $\rho_{vp}$ increases the stresses on boundary C to some extent even when the outer surface is uniformly or partially polluted.

## 18.6 Electric Field Distribution around a Gas-Insulated Substation Spacer

Fully encapsulated and compact $SF_6$ gas-insulated substation (GIS) is considered an integral part of modern-day power installations. Major advantages of GIS are high reliability and compactness achieved by compressed $SF_6$ gas insulation. At the same time, the downsizing of GIS leads to high electric field intensities. Basic insulation components of GIS are $SF_6$ gas and solid spacers. In this respect, it has been recognized that the breakdown strength of GIS is mainly influenced by the solid spacers. Both experimental and numerical studies demonstrate that electric field distribution and insulation behaviour at $SF_6$/spacer interface contribute most towards the failure of spacers. In order to improve the dielectric performance of epoxy spacers by properly shaping the gas–dielectric interfaces, detailed studies on the electric field distribution and optimization along the profile of the gas–dielectric interface in GIS have been reported in the literature [4,5]. In this section, the results of electric field computations for a GIS comprising four components, namely, a central HV conductor, epoxy spacer, $SF_6$ gas and metallic enclosure, are presented. As suggested in the literature, computations have been carried out considering uncoated as well as coated spacer. The coating of the spacer is considered to have a surface resistivity of the order of $10^7\ \Omega$.

Figure 18.20 shows the axi-symmetric configuration of the GIS for which electric field computations have been carried out. The dimensions of the GIS are commensurate to voltage rating of 110 kV. Relative permittivity of epoxy spacer has been taken to be 5 and that of $SF_6$ as 1.005. The HV conductor is inserted into the solid spacer from the viewpoint of field reduction at the conductor–spacer–gas triple junction. The condition of solid spacer touching

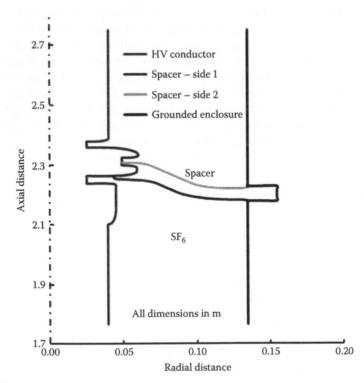

**FIGURE 18.20**
**(See colour insert.)** GIS configuration considered for electric field computation.

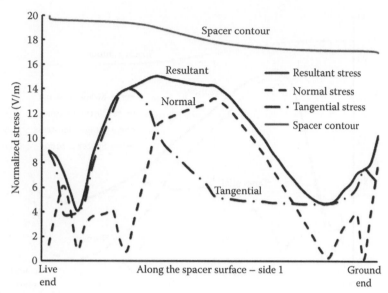

**FIGURE 18.21**
Electric field intensity along spacer surface on side 1 for capacitive field.

the conductor at right angle has also been maintained. All results are presented in a normalized format.

For uncoated spacer, resultant electric field intensity as well as the two components, namely, normal and tangential, along the spacer surface on side 1 has been presented in Figure 18.21. For this case, surface resistivity of the epoxy spacer is assumed to be infinite, which corresponds to capacitive field distribution. Similar distribution of electric field intensities along the spacer surface on side 2 have been presented in Figure 18.22. Both Figures 18.21 and 18.22 show that the stresses have been reduced near the HV conductor due to metal insert electrode design. On the other hand, such electrode design has shifted the maximum stresses to occur somewhere in the mid region of the spacer surface on both side 1 and side 2.

For coated spacer, resultant, normal and tangential field intensities along the spacer surface on side 1 have been presented in Figure 18.23. For this case, surface resistivity of the epoxy spacer is assumed to be $10^7\ \Omega$, which corresponds to resistive field distribution. Similar distribution of electric field intensities along the spacer surface on side 2 have been presented in Figure 18.24. Figure 18.23 shows that a resistive coating on side 1 of the spacer has unfavourable effects on electric stresses. It increases the field stresses near the HV conductor compared to the case of capacitive field. Although the values of maximum field intensities for resistive field are higher than capacitive field, the increment in the maximum value is not much. The field stresses near the grounded enclosure are lower for resistive field compared to capacitive field.

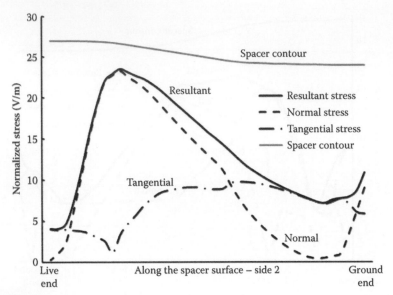

**FIGURE 18.22**
Electric field intensity along spacer surface on side 2 for capacitive field.

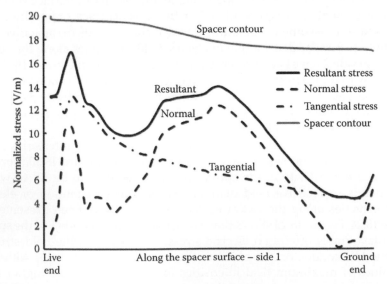

**FIGURE 18.23**
Electric field intensity along spacer surface on side 1 for resistive field.

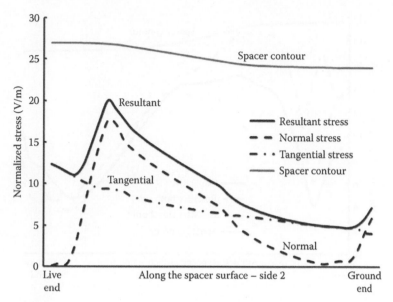

**FIGURE 18.24**
Electric field intensity along spacer surface on side 2 for resistive field.

Figure 18.24, on the other hand, shows that a resistive coating has favourable effect on field intensities on side 2 of the spacer as the maximum field stresses are reduced and the field stresses near the HV conductor are lower compared to the stresses in the mid-region of spacer surface on side 2.

Thus studies have also been carried out to determine the effects of resistive coating on any one side of the spacer. Figure 18.25 presents a comparison of resultant field intensity along the spacer surface on side 1 for three different cases: (1) uncoated spacer (capacitive field), (2) spacer coated on both sides by resistive layer of $10^7\,\Omega$ (resistive field) and (3) spacer coated on side 2 by resistive layer of $10^7\,\Omega$. It may be seen from Figure 18.25 that coating only side 2 with a resistive layer increases the field intensity on side 1 compared to both capacitive and resistive fields. It should also be mentioned here that if only side 1 is coated with a resistive layer and side 2 is left uncoated, then the field intensities on side 1 of spacer surface are similar to the case when both sides are coated.

Comparison of resultant field intensity along the spacer surface on side 2 has been depicted in Figure 18.26 for three different cases: (1) uncoated spacer (capacitive field), (2) spacer coated on both sides by resistive layer of $10^7\,\Omega$ (resistive field) and (3) spacer coated on side 1 by resistive layer of $10^7\,\Omega$. Figure 18.26 shows that field stresses for coating of side 1 only are almost the same as capacitive field. Hence, from Figures 18.25 and 18.26, it may be stated

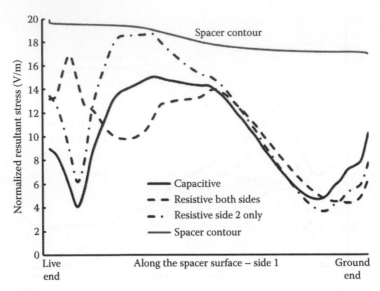

**FIGURE 18.25**
Comparison of resultant field intensity along spacer surface on side 1.

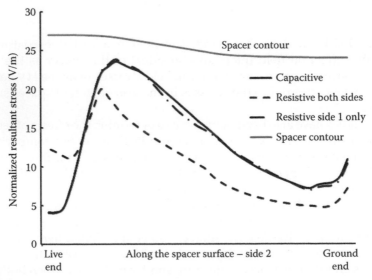

**FIGURE 18.26**
Comparison of resultant field intensity along spacer surface on side 2.

that if a resistive coating is to be used for field reduction, then both sides of the spacer should be coated with resistive layer.

In the case of high-voltage DC (HVDC) GIS, an additional problem arises in the form surface charging of spacer surface with time. The normal component of field intensity brings charges from gas insulation on to the spacer surface, which will get trapped on the surface if the surface resistivity is very high. Accumulation of surface charges reduces the normal component of field intensity, so that less and less charges are drawn to the surface. With passage of time, when the normal component of field intensity becomes nearly zero, then no more charges are drawn to the spacer surface and the surface may be stated to be fully charged in the absence of any surface resistivity [6]. Such accumulated surface charges change the field distribution around the spacer. In such cases, a resistive coating on the spacer surface is useful, as it helps in draining out the charges from the spacer surface [7,8].

Field computation for fully charged surface could be carried out with the boundary condition that $E_{nor} = 0$ on the spacer surface [9], where $E_{nor}$ is the normal component of electric field intensity. Such field computation has been carried out for the spacer configuration under study. Figure 18.27 shows the comparison of resultant stresses on side 1 of spacer surface for three different cases: (1) capacitive field, (2) resistive field and (3) fully charged side-1 of surface. Figure 18.27 shows that the field distribution for fully charged surface is quite different from capacitive field, as the maximum field intensity is shifted close to the grounded enclosure. However, the maximum stress is not increased compared to capacitive field. The maximum stress is highest for resistive field, as shown in Figure 18.27.

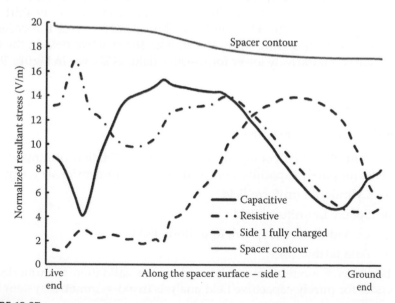

**FIGURE 18.27**
Comparison of resultant field intensity along spacer surface on side 1 for fully charged surface with capacitive and resistive fields.

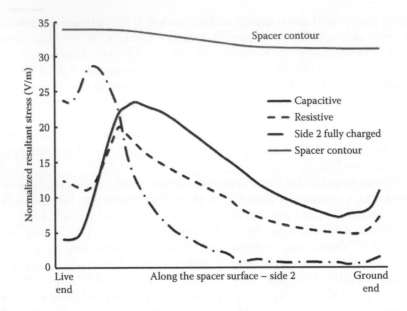

**FIGURE 18.28**
Comparison of resultant field intensity along the spacer surface on side 2 for fully charged surface with capacitive and resistive fields.

Results of a similar study for side 2 of the spacer are shown in Figure 18.28. Figure 18.28 shows that full charging of side-2 of surface not only shifts the maximum stress near the HV conductor but also increases the value of the maximum stress significantly. The stresses for resistive field in the case of side 2 are comparatively lower for resistive field, as shown in Figure 18.28.

## Objective Type Questions

1. Benchmark model commonly used for the validation of numerical codes for purely capacitive field analysis in two-dimensional system is
   a. Cylinder in uniform field
   b. Sphere in uniform field
   c. Coated dielectric sphere in uniform field
   d. Both (a) and (b)
2. Benchmark model commonly used for the validation of numerical codes for purely capacitive field analysis in axi-symmetric system is
   a. Cylinder in uniform field
   b. Sphere in uniform field

   c. Coated dielectric sphere in uniform field

   d. Both (b) and (c)

3. Benchmark model commonly used for the validation of numerical codes for capacitive-resistive field analysis in axi-symmetric system is

   a. Cylinder in uniform field

   b. Sphere in uniform field

   c. Coated dielectric sphere in uniform field

   d. Both (b) and (c)

4. A high-voltage cable termination typically provides

   a. Adequate electric stress control

   b. Comprehensive external leakage insulation

   c. Proper sealing

   d. All the above

5. Electric field distribution within a high-voltage cable termination is typically

   a. Two-dimensional in nature

   b. Three-dimensional in nature

   c. Axi-symmetric in nature

   d. None of the above

6. Commonly used method for stress control in a high-voltage cable termination is

   a. Geometric stress control

   b. Guard ring stress control

   c. Capacitive stress control

   d. Both (a) and (c)

7. Error in the numerical simulation of boundary occurs when a dielectric boundary meets a conductor boundary at

   a. Very acute angle

   b. An angle nearly equal to 90°

   c. 90°

   d. None of the above

8. For a numerical field computation, a post-type insulator is an example of

   a. Two-dimensional case study

   b. Three-dimensional case study

   c. Axi-symmetric case study

   d. None of the above

9. In the case of a post-type insulator, if the surface resistivity increases from a low value to a high value, then the power–frequency field distribution changes from
   a. Resistive to capacitive–resistive to capacitive
   b. Resistive to capacitive to capacitive–resistive
   c. Capacitive to capacitive–resistive to resistive
   d. Capacitive to resistive to capacitive–resistive

10. Free charges that are present around an insulator surface are drawn on the insulator surface by the
    a. Resultant surface field intensity
    b. Normal component of surface field intensity
    c. Tangential component of surface field intensity
    d. All the above

11. Movement of free charges on an insulator surface is caused by the
    a. Resultant surface field intensity
    b. Normal component of surface field intensity
    c. Tangential component of surface field intensity
    d. All the above

12. In the case of an outdoor porcelain insulator, which one of the following gives rise to highest field intensity on the outer surface?
    a. No surface pollution
    b. Uniform surface pollution
    c. Partial surface pollution over a large area
    d. Formation of dry band on the outer surface

13. The range of surface resistivity over the field distribution for an outdoor porcelain insulator changes from capacitive to resistive is lowest for
    a. DC field
    b. Power frequency field
    c. Lightning impulse field
    d. Switching impulse field

14. If a high-voltage conductor is to pass through an earthed barrier, then which one of the following is used?
    a. Bushing
    b. Pin-type insulator
    c. Post-type insulator
    d. Disc-type insulator

15. Accumulation of charges on the surface of an insulator
    a. Increases normal component of surface field intensity
    b. Decreases normal component of surface field intensity
    c. Increases tangential component of surface field intensity
    d. Decreases tangential component of surface field intensity
16. In the case of a high-voltage DC gas-insulated substation, if the surface of a spacer is fully charged, then which one of the following becomes zero?
    a. Resultant component of surface field intensity
    b. Normal component of surface field intensity
    c. Tangential component of surface field intensity
    d. All the above

**Answers:**  1) a; 2) b; 3) c; 4) d; 5) c; 6) d; 7) a; 8) c; 9) a; 10) b; 11) c; 12) d; 13) c; 14) a; 15) b; 16) b

---

# Bibliography

1. A. Blaszczyk, 'Computation of Quasi-static electric fields with region-oriented charge simulation', *IEEE Transactions on Magnetics*, Vol. 32, No. 3, pp. 828–831, 1996.
2. S. Chakravorti and P.K. Mukherjee, 'Power frequency and impulse field calculation around a HV insulator with uniform or non-uniform surface pollution', *IEEE Transactions on Electrical Insulation*, Vol. 28, No. 1, pp. 43–53, 1993.
3. S. Chakravorti and H. Steinbigler, 'Capacitive-resistive field calculation in and around HV bushings by boundary element method', *IEEE Transactions on Dielectrics & Electrical Insulation*, Vol. 5, No. 2, pp. 237–244, 1998.
4. N.G. Trinh, F.A.M. Rizk and C. Vincent, 'Electrostatic field optimization of the profile of epoxy spacers for compressed SF6 insulated cable', *IEEE Transactions on Power Apparatus and Systems*, Vol. 99, pp. 2164–2174, 1980.
5. K. Itaka, T. Hara, T. Misaki and H. Tsuboi, 'Improved structure avoiding local field intensification on spacers in $SF_6$ gas', *IEEE Transactions on Power Apparatus and Systems*, Vol. 102, pp. 250–255, 1983.
6. T. Nitta and K. Nakanishi, 'Charge accumulation on insulating spacers for HVDC GIS', *IEEE Transactions on Electrical Insulation*, Vol. 26, pp. 418–427, 1991.
7. F. Messerer, W. Boeck, H. Steinbigler and S. Chakravorti, *Enhanced Field Calculation for HVDC GIS*, Gaseous Dielectrics IX, Springer, New York/London, pp. 473–483, 2001.
8. F. Messerer and W. Boeck, 'High resistance surface coating of solid insulating components for HDVC metal enclosed equipment', *Proceedings of the 11th ISH*, London, Vol. 4, pp. 63–66, August 23–27, IEE (UK) Publication, 1999.
9. S. Chakravorti, 'Modified E-field analysis around spacers in $SF_6$ GIS under DC voltages', *Proceedings of the 11th ISH*, London, Vol. 2, pp. 91–94, August 23–27, IEE (UK) Publication, 1999.

# 19

## Electric Field Optimization

**ABSTRACT** The electricity supply system nowadays is characterized by two aspects: (1) high voltages at both transmission and distribution levels and (2) distribution of electric power in densely populated areas. Electric field distribution and the design of insulation system is affected by both these aspects. The higher voltage levels cause higher stresses in the insulation system, whereas the supply of power to populated regions demands compactness of the power equipment, which in turn increases the stresses within the insulation system of the equipment. Critical domains of high-voltage arrangement, which will have high field concentrations, cannot be designed by simple bodies such as spheres, rings or cylinders, which are easier to manufacture. Optimization methods are applied to high-voltage system so as to design electrodes as well as insulator contours in such a way that prescribed field distributions on defined surfaces in the critical domain are obtained. Consequently, such methodologies help to use the field space optimally within high-voltage system. This chapter presents a review of the works done in the area of optimization of high-voltage field and then discusses a few specific application examples that involve classical optimization techniques as well as techniques based on artificial neural network (ANN) and evolutionary algorithms.

## 19.1 Introduction

An important objective in insulating system design is to obtain electrodes and insulator contours, which will withstand all the electrical stresses that may appear during normal operation as well as during occasional transient events. For safe and reliable operation of any high-voltage equipment, the maximum values of electric field stresses in the insulating system must be lower, or at the most equal, to the allowable values. From this viewpoint, the optimal insulating system element contours are those on which the uniformly distributed field stresses or the desired distribution of field stresses remain within the allowable limit.

In the classical approach to insulating system design, system elements with simple geometrical forms such as sphere, ring, cylinder and so on were used and electric field strength was controlled by varying the distances between the system elements. This approach although works in practice but results in non-uniform electric field distribution, leading to suboptimal utilization of the capability of insulation system. The present-day need for better space utilization in the case of high-voltage supply system demands application of electrode and insulator contour optimization in insulating system design. The contours obtained through optimization processes have either uniformly distributed field intensity or a desired distribution of field intensity, which is achievable because of their not so simple geometric shapes. From another viewpoint, an efficient approach to electrode and insulator design is to have optimized profile(s) of electrode and/or insulator such that the dimensions are minimum for a given voltage rating, thus minimizing the space needed for and also the cost of installation.

The criterion most commonly used for the optimization of electrode contour is the minimization of the normal component of the electric field intensity on the surface of the electrodes, as it increases the discharge initiation voltage. On the other hand, the minimization of the tangential component of the electric field intensity along the surface is the preferred criterion for insulator contour optimization, as it increases the onset voltage of surface flashover. However, the minimization of the resultant field intensity on the insulator surface is also used as criterion for insulator contour optimization.

## 19.2 Review of Published Works

From a mathematical viewpoint, as early as in 1915, it was reported by Spielrein [1] that the change in the field on the surface of a given electrode can be obtained by a change in the curvature of the contour as given below:

$$\frac{(\partial^2 \phi / \partial n^2)}{(\partial \phi / \partial n)} = \frac{(\partial E / \partial n)}{E} = -\kappa \tag{19.1}$$

where:
$\kappa$ is the total curvature of the electrode surface at any arbitrary point

Equation 19.1 shows the relationship between the electric field intensity, decrease in electric field intensity due to a displacement normal to the contour and the total curvature of the contour at any arbitrary point on the contour. Thus, the change of curvature necessary to achieve the desired change in field can be deduced from Equation 19.1. This change in curvature ($\Delta\kappa$) at every contour point, in accordance with the deviation of electric field intensity ($\Delta E$) from the desired field strength, can be calculated as follows:

$$\Delta\kappa \approx \frac{(\partial E / \partial n)}{E^2} \Delta E \tag{19.2}$$

In order to maintain the proportional relationship between the deviation of electric field intensity and the required change in curvature, small differences are to be assumed. The optimization procedure for electrodes must alter the contour in such a way that at any contour point the electric field intensity equals the desired value of field intensity, which is below the value required for discharge initiation with a safety margin.

Determination of the right shape for the boundary part of a plane sparking gap based on the theory of conformal mapping was reported by Rogowski [2] in 1923. Later in 1950, Félici [3] reported the use of conformal mapping for designing capacitor plate to have an electric field of constant strength.

### 19.2.1 Conventional Contour Correction Techniques for Electrode and Insulator Optimization

In 1975, Singer and Grafoner [4] published a methodology for optimizing electrodes and insulators designed for practical use. In their method, domains with constant field intensity were obtained for axi-symmetric arrangements by displacing contour points successively. Some more conventional methods of electric field optimization were reported in References [5–7].

In 1982, Misaki et al. [8] reported a method to get the optimum design of epoxy pole spacers used in sulphur hexafluoride ($SF_6$) gas-insulated cables. In their method an improved surface charge method using curved surface elements was employed for the computation of three-dimensional (3D) electric field distribution and the results of electric field computation were used for the optimization of insulator design. The optimum insulator design was performed automatically by correcting the insulator contour using a simple vector computation. In this method, the insulator contour was corrected in such a way that both tangential field intensity and normal field intensity became uniform along the new contour. If the insulator contour moves in the direction of the normal vector, then the tangential field intensity decreases, and if the insulator contour moves in the direction opposite to that of the normal vector, then the normal field intensity decreases on the insulator surface. Therefore, the direction in which correction of the insulator contour had to be carried out was decided by taking note of the normal stress on the insulator surface. The insulator contour was corrected in proportion to the normal Maxwell's stress, which was obtained from both tangential and normal field intensities. A method for 3D electrode contour optimization was reported by Misaki et al. [9] in 1983, in which the electric field intensity and force on the electrode surface were computed, and then the electrode contour was moved in the opposite direction of the force and the contour displacement was done in proportion to the magnitude of the force.

In 1983, Grönewald [10] described an algorithm based on the computer-aided design (CAD) concept for optimizing electrode contour to obtain a given field distribution on the surface. They proposed a two-step procedure that strictly separated electric field calculation and geometrical contour corrections. At any point on the contour, the changes in curvature depending on the desired field were described with the help of relationships developed between field quantity and geometry. The new contour was found by a simultaneous displacement of all discretizing points on the contour.

Stih [11] in his paper described an iterative procedure for designing optimally stressed insulating system. The described procedure combined the approach of varying the distances between system elements and optimizing their contours. Axially symmetric electric fields were computed by an integral equation technique introducing cubic spline expansion for charge density distribution function. In an extension of the contour correction techniques reported in References [4,8,9], an additional step was introduced in this work in the form of smooth approximation of the corrected contour by circular arcs, which enabled relatively simple geometrical description of the optimal contours.

Abdel-Salam and Stanek [12,13] reported a method for optimizing the field stress on high-voltage insulators through the modification of their profile to obtain a uniform distribution of the tangential field along the insulator surface. An algorithm based on a modified charge simulation technique was developed for calculating the tangential field component along the insulator surface. After computation of the tangential field distribution for the non-optimized profile, the insulator profile was enlarged or reduced on going up or down along its axis. If the field distribution increased, when moving along the axis of the insulator, the profile radius was increased in that direction and if the field distribution decreased, then the profile radius was decreased. Exponential mathematical expressions, which defined a smooth enlargement/reduction of the profile radius, were used in this work.

Mosch et al. [14] published a method of optimization of large electrodes of ultra high voltage (UHV) testing equipment designed on the basis of inception of streamer and leader discharges. The stochastic nature of these two discharge processes leads to the well-known area effects, and also to time effects. Hence, the proposed model was related to the volume–time concept. The design procedure started with choosing the principle of field control and the main geometric parameters of the electrode. Then the electric field was calculated numerically and from the results of electric field computation the inception voltages $V_{iS}$, for streamer discharge, and $V_{iL}$, for leader discharge, were determined. The calculated values were compared to the maximum output voltage. If the calculated values were less, then the geometry of the electrode was improved and the cycle of computation was repeated till an optimum design was achieved. A similar study was reported by Kato et al. in Reference [15]. In this paper, they described a technique for the optimization of high-voltage electrode contour to make the electrical insulation performance highest by introducing the volume–time characteristics into the optimization procedure. Optimization technique was also investigated from the viewpoints of computation efficiency and accuracy.

The basic principle of electrode optimization method suggested by Liu and Sheng [16] was that the electrode was divided into two parts: the fixed part and the part to be optimized. The fixed part was simulated by the simulation charges and the second part by the optimization charges, in which the magnitudes and coordinates were assigned first. Normally, the magnitudes of the optimization charges were taken as optimization variables. In this method, the coordinates instead of the magnitudes of optimization charges were taken as the optimization variables. The reported results of optimization examples indicated that the use of the proposed method yielded good results. Liu et al. [17] also described an automatic procedure for the optimization of axi-symmetric electrodes using boundary element method. The criterion employed for the optimization of the shape of the electrodes was the minimization of the maximum field intensity on the surface of the electrodes. The proposed method was applicable in 3D fields too. Judge and Lopez-Roldan in Reference [18] discussed an approach to design components for high-voltage engineering based on the boundary element method in two and three dimensions. Optimization of the components was performed using an iterative method of refining a design and re-analyzing without having to resort to repeated prototype building and testing. Studies on the optimization of electrode contours using circular contour elements [19] and optimization of multi-electrode system [20] have also been reported in the literature.

Däumling and Singer [21] described a new algorithm based on the CAD concept for the optimization of insulator contours to get a given field distribution along the surface. Three optimization procedures with different targets were investigated: (1) uniform tangential field strength, (2) uniform resultant field strength and (3) uniform electrostatic pressure along the insulator contour. It was found that achieving a low tangential field strength component was not sufficient, and it was necessary to reduce the maximum value of the resultant field strength as far as possible, especially in the case of high air humidity.

Caminhas et al. [22] determined the optimum profile and location of shielding electrode used in high-voltage equipment with the aim to obtain a linear voltage distribution along the axis of symmetry. It was shown that the combined usage of the charge simulation method and the Broyden–Fletcher–Goldfarb–Shannon optimization technique was both efficient and simple. Very good attempt was made to review various optimization methods applicable to the specific case.

### 19.2.2 Optimization of High-Voltage System Elements

Optimization of field stresses in high-voltage bushings have been reported in References [23,24] and special techniques for electrode and insulator contour optimization, which were claimed to be efficient, have been reported in References [25–27]. Efficient techniques for electric field optimization using personal computers were reported in References [28,29]. Studies on

3D electric field optimization in high-voltage equipment was reported by Trinitis [30,31]. In Reference [31] optimal electric field strength distribution for 3D problems was achieved by utilizing a parametric CAD modelling system coupled to a 3D electric field calculation program. These two components were then linked to a numerical optimization algorithm. The package comprising the above-mentioned three components was then able to automatically optimize arbitrary 3D field problems in high-voltage engineering. Optimization of high-voltage insulators in 3D field configurations was reported in Reference [32].

High gradient insulators (HGI) consist of a periodic array of insulator and metal rings, which have been found to be more resilient to vacuum surface breakdown than homogeneous insulators of the same length. Studies based on calculations and experiments were reported by Leopold et al. [33] to understand the effect of geometry on the performance of well conditioned, flat surface HGI assemblies.

### 19.2.3 Soft-Computing Techniques for Electrode and Insulator Optimization

With the advent of artificial intelligence, it was realized that instead of iterative calculation of electric field and contour modification, electric field optimization could be done based on artificial neural network (ANN). Once the ANN is trained, optimum solution satisfying desired specification can be found without iterative calculation, resulting in high-speed method. ANN with supervised learning by error back-propagation method as used for high-voltage electrode optimization was reported by Chakravorti and Mukherjee [34]. It was found that the trained ANN can give results with mean absolute error (MAE) of about 1% in comparison with analytically obtained results. For electrode contour optimization, the results of electric field calculations for some pre-determined contours of an axi-symmetric electrode arrangement were used for training the ANN. Then the trained ANN was used to give the optimized electrode contour to obtain a desired field strength distribution on the electrode surface. Similar study on application of ANN in the design of toroidal electrodes was reported by Bhattacharya et al. [35]. Optimization of HV electrode systems by ANN with resilient propagation was carried out by Mukherjee et al. [36]. Two axi-symmetric examples were optimized: (1) the first one was the termination of a single-phase gas-insulated substation (GIS) bus and (2) the second one was a shield ring in a three-phase transformer. Charge simulation method was used for preparation of the training sets as well as for checking the test outputs from the ANN. The procedure of the electric field optimization method based on ANN as reported by Okubo et al. [37] comprised two ANNs. As the electric field strength on a high-voltage electrode surface has empirical relationship with the curvature and the gap length, one ANN, called 'NN1', was trained to learn this relationship. Another ANN, called 'NN2', was trained to learn the relationship between a set of curvature, gap length and electrode contour. Once the learning of the two ANNs were completed, the curvature, gap length and electrode contour for the optimized electrode configuration were obtained as output by giving the target electric field distribution as input into the trained ANN system.

In Reference [38], ANN-based studies have been presented for the optimization of insulator contour in multiple dielectric systems, where the degree of field nonlinearity is more than that for single-dielectric configurations. ANN-aided contour optimization of axi-symmetric insulators in multi-dielectric arrangements had been carried out to obtain not only a uniform but also a complex electric stress distribution along the insulator surface. Multilayer feed-forward networks with error-back propagation as well as resilient propagation learning algorithms were employed.

As each ANN application requires a certain amount of training to achieve the desired accuracy, Chatterjee et al. [39] in their paper discussed the development of a self-organizing fuzzy inference system for designing optimized electrode contours. The process of optimization was accelerated using a fuzzy inference system that eliminated training time. Reliability of the proposed methodology was further improved by implementing an algorithm for automatic generation of the fuzzy rule base from input–output data sets.

Alotto et al. [40] gave an overview of several stochastic optimization strategies, namely, evolution strategies, genetic algorithms (GA) and simulated annealing (SA), as applied to electromagnetic optimization problems. The application of evolution strategy for optimal design of high-voltage bushing electrode in transformer was investigated in Reference [41]. Several regular circular arcs and straight lines were used to get the optimal shape of electrode. The aim of optimization was to obtain lower maximum field intensity and well-distributed electric stresses on the surface of electrode, greater reliability during operation and easy manufacturability in large-scale production.

Lahiri and Chakravorti [42] carried out the optimization of contours of 3D electrode–spacer arrangements used in gas-insulated transmission line (GIL) by ANN-aided GA. Input–output data set used for training the ANN was prepared by means of electric field calculations using surface charge simulation method varying several design parameters of each of the two 3D arrangements considered in their work. The trained ANN was coupled to a GA loop. At each exploratory move of the GA loop a new set of values of the design parameters were set and the corresponding maximum resultant field intensity along the insulator surface was obtained from the trained ANN instead of running a comprehensive field computation routine. This maximum resultant field intensity was returned to the GA loop as the value of the cost function to decide on the fact whether the loop has converged or not. In this way, a time-saving optimization technique had been developed to obtain an optimized field distribution along the insulator surfaces. In another paper, Lahiri and Chakravorti [43] used ANN-aided SA algorithm for the optimization of stress distribution on and around 3D electrode–spacer arrangements. By coupling the trained neural net with the annealing algorithm, the execution speed of the optimization routine was significantly increased to evaluate the optimum values for the design parameters of the electrode–spacer arrangements, because there is no need for the cost function calculation via the entire process for electric field calculation at every move of the optimization algorithm, which makes the optimization routine very fast.

Banerjee et al. [44] in their paper used support vector machine (SVM) for the optimization of electric field along the support insulators used in high-voltage systems. The SVM designed for insulator contour optimization was first trained with the results obtained from electric field computations for some predetermined contours of the arrangements. Then the trained SVM was used to provide the optimized insulator contour in such a way that the desired stress distribution was obtained on the insulator surface.

### 19.2.4 Optimization of Switchgear Elements

Kitak et al. [45] described an algorithm for the design of medium-voltage switchgear insulation elements using numerical calculations on the basis of finite element method in connection with evolutionary optimization methods. Differential evolution and evolution strategy algorithms were used for optimization. The task of both optimization algorithms was to find an adequate capacitance of the voltage divider and the optimal distribution of electric field strength. The highlight of the work was simultaneous use of parametric representation in geometry, a novel mesh generator, numerical computation with finite element method (FEM) and a genetic optimization algorithm. The proposed methodology thus represented a generalized method of optimization for various objective functions of switchgear elements.

For accurate insulation design of the vacuum interrupter, an optimization technique was developed by Kato et al. [46] to improve the electrical insulation performance. Several design variables necessary for optimizing the electrode contour of the main contactor and centre shield in vacuum interrupter were evaluated. The electrode area effect on vacuum breakdown process as well as the electric field distribution was considered in the optimization process. Electrode contour optimization of arc quenching chamber of extra-high-voltage $SF_6$ circuit breaker had been carried out by Liu et al. [47] to obtain not only a uniform field but also a dynamic distribution of electric field strength along different contact surfaces. The variable interval GA was used to perform multivariable global optimization of the electrode contour of a single-break 550 kV $SF_6$ arc-quenching chamber.

Functionally graded material (FGM) spacer, the permittivity of which changes gradually, exhibits considerable reduction in the maximum electric field when compared to a conventional spacer with uniform permittivity. However, it is difficult to realize a gradual permittivity variation in the FGM spacer in real-life product processing due to its complicated shape. Thus, optimization processes were used by Ju et al. [48,49] to modify the shape of both the electrode and the FGM spacer on a commercial gas-insulated switchgear configuration to increase the possibility of real FGM insulator manufacturing. Modification of the spacer configuration was performed with the design of experiments. For a reliable and effective design process, the full factorial design method and the response surface methodology were employed in the development of an effective computational approach.

### 19.2.5 Optimization of Bushing Elements

Although it is entirely possible that the use of uniform field profile electrodes could achieve the goal of reducing the electric field in critical areas, the manufacturers of bushings often deemed that this was not an economically feasible solution. Hence, Monga et al. [50] used a basic bushing configuration as the starting point and the goal of the work was to design optimum grading hardware for the larger bushings. In this paper, the authors illustrated the use of electric field computation based on boundary element method to optimize the design of gas filled high-voltage composite bushings. The optimized design used both internal and external elements for electric stress grading at critical parts of the bushing.

Hesamzadeh et al. [51] proposed a methodology for finding the optimum electrical design of high-voltage condenser bushings using an improved GA. In this paper, the authors determined the optimal values of bushing design parameters to achieve well-distributed electric stress with the lowest possible maximum value and also a constant voltage drop between different layers of concentric conductive foils, which are isolated from each other, subject to practical and technological constraints. The proposed method was applied for optimal design of a 145 kV oil impregnated paper bushing and the performance of optimally designed bushing were satisfactorily verified under IEC 60137 tests.

### 19.2.6 User-Friendly Optimization Environment

Precise simulation and geometric optimization of the electric field distribution on electrodes and insulators are key aspects in the design and optimization process of high-voltage apparatus. Because these simulations and optimizations are computation intensive, an engineer working in industry demands a user-friendly working environment requiring as little knowledge as possible with regard to the computer specific aspects. From the user's viewpoint, a user-friendly design and optimization environment should (1) require minimum interaction and knowledge about the simulation system and (2) provide a solution as quickly as possible. Trinitis [52] achieved the first requirement by a distributed design and optimization system based on a model-driven architecture and the second requirement was achieved by both parallelization of the simulation process and acceleration of the simulation process on each of the parallel computing nodes. By optimizing the simulation software with regard to the hardware it was running on, computation time for an overall optimization run was kept at an acceptable level.

## 19.3 Field Optimization Using Contour Correction Techniques

Conventionally, the profile of electrodes or insulators were optimized using contour correction techniques to obtain pre-determined field values along the optimized contour. Two such examples are discussed in the following sub-sections.

### 19.3.1 Insulator Contour Optimization by Simultaneous Displacement

Considering two successive contour points on the surface of the insulator the tangential field intensity may be given by

$$E_t = -\frac{\Delta\phi}{\Delta l} \approx \frac{\phi_1 - \phi_2}{\Delta l} \tag{19.3}$$

where:
  $\phi_1$ and $\phi_2$ are the potentials of the two successive contour points 1 and 2, respectively
  $\Delta l$ is the distance between the contour points 1 and 2

If the actual value of the tangential field intensity does not match with the desired value, then either the potential difference between the points or the distance between the points need to be altered. Accordingly, there are two methods of displacement of contour points: (1) displacement keeping $\Delta\phi$ constant and (2) displacement keeping $\Delta l$ constant.

#### 19.3.1.1 Contour Correction Keeping Potential Difference Constant

Singer and Grafoner [4] presented a technique for contour correction by displacing the contour points keeping $\Delta\phi$ constant. As shown in Figure 19.1, the contour point 2 is moved along the equipotential corresponding to $\phi_2$ either to the right or to the left of the insulator boundary. In this way, $\Delta l$ is either lengthened or shortened to modify the tangential field intensity. If the actual value of tangential field intensity ($E_{ta}$) is higher than the desired value ($E_{td}$), then point 2 is moved towards right, otherwise it is moved towards left. Then the actual value of distance between the contour points ($\Delta l_a$) and the desired value ($\Delta l_d$) are related as follows:

$$\Delta l_d = \frac{E_{ta}}{E_{td}} \Delta l_a \tag{19.4}$$

From Equation 19.4 the displacement $\Delta s$ of point 2 can be obtained. The side of the insulator contour to which point 2 is to be displaced is determined by the value of $E_{ta}$ with respect to that of $E_{td}$.

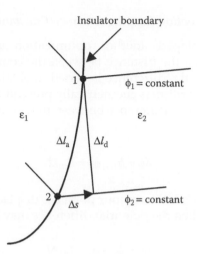

**FIGURE 19.1**
Displacement of contour points keeping Δφ constant.

Based on the above-mentioned procedure, a method of simultaneous displacement of contour points was proposed by Grönewald [10]. In this proposed methodology, the distance between the contour points after displacement ($\Delta l_d$) was calculated as a function of the actual distance ($\Delta l_a$) and the displacement $\Delta s$ such that

$$\Delta l_d \approx \Delta l_a + k\Delta s \tag{19.5}$$

where:
$k$ is a proportionality constant

For simultaneous displacement of all the contour points, using linearized matrix notation, Equation 19.5 can be written as

$$[\Delta L_d] = [\Delta L_a] + [K][\Delta S] \tag{19.6}$$

Combining Equations 19.4 with 19.6, it may be written that

$$[K][\Delta S] = [\Delta L_a]^T \left( \frac{E_{ta}}{E_{td}} - 1 \right) \tag{19.7}$$

where:
$[\Delta L_a]^T$ is the transposed vector of $[\Delta L_a]$

However, Grönewald reported that the algorithm based on the displacement of contour points along the corresponding equipotential lines is not applicable in general, because of many geometric restrictions.

### 19.3.1.2 Contour Correction Keeping Distance Constant

Grönewald [10] developed another optimization algorithm based on Equation 19.3, in which the distance between the contour points was kept constant and the potential difference is varied to obtain the desired electric field intensity. This algorithm is geometrically presented in Figure 19.2. From Equation 19.3 the potential difference between the contour points 1 and 2 can be expressed as

$$\Delta\phi = \phi_1 - \phi_2 = E_{ta}\Delta l \tag{19.8}$$

As shown in Figure 19.2, if the contour point 2 is displaced by $\Delta s$ keeping the distance $\Delta l$ constant, then the potential difference may be written as

$$\Delta\phi' = \phi_1 - \phi_2' = E_{td}\Delta l \tag{19.9}$$

If the point 1 is kept fixed, then the potential difference is given by

$$\Delta\phi_2 = \phi_2' - \phi_2 \tag{19.10}$$

Then the potential of the displaced point 2' may be expressed with the help of Equations 19.8 through 19.10 as follows:

$$\phi_2' = \phi_2 + \Delta\phi_2 = \phi_1 - E_{td}\Delta l = \phi_1 - E_{ta}\Delta l + \Delta\phi_2 \tag{19.11}$$

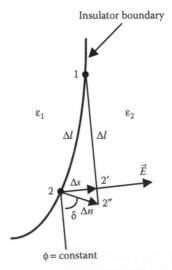

**FIGURE 19.2**
Displacement of contour points keeping $\Delta l$ constant.

From Equation 19.11, it is evident that

$$\Delta\phi_2 = (E_{ta} - E_{td})\Delta l \tag{19.12}$$

If the field in the vicinity of the contour point 2 is considered to be uniform, then the potential of the displaced point 2' can be written as follows:

$$\phi_2' = \phi_2 - E\Delta s \tag{19.13}$$

Combining Equations 19.12 with 19.13, the expression for $\Delta s$ can be obtained as follows:

$$\Delta s = \frac{E_{td} - E_{ta}}{E}\Delta l \tag{19.14}$$

Considering $\Delta s$ to be small compared to $\Delta l$ and the length $\Delta n$ to be normal to the contour,

$$\Delta n \approx \frac{\Delta s}{\sin\delta} = \frac{(E_{td} - E_{ta})}{E\sin\delta}\Delta l \tag{19.15}$$

The direction of the displacement of the contour point, that is, towards right or left of the point 2, with reference to Figure 19.2, is defined by the sign of the term $(E_{td} - E_{ta})$.

The contour optimization algorithm begins with a starting configuration for which electric field computation is carried out, commonly with the help of numerical method. The contour correction is carried out in the second step. Thus, the entire optimization routine is an iterative process in which successive numerical field computation and contour correction were carried out till the desired field distribution is obtained.

Grönewald [10] reported a study on contour optimization of a disc-shaped insulator, which is schematically shown in Figure 19.3. The contour was represented by a specified number of contour points. The displacements of the contour points $\Delta n_i$ were taken to be normal to the insulator contour as shown in Figure 19.2. Then for all the contour points, in matrix form it may be written that

$$[K][\Delta N] + [E_{ta} - E_{td}] = [0] \tag{19.16}$$

where:
$K$ is the matrix of displacement coefficients

The optimized contour was obtained by solving Equation 19.16 for the values of $\Delta n_i$. For this reason this method was called *method of simultaneous displacement*

Figure 19.4 shows the optimized contour for one constant value of tangential field intensity along the insulator surface and Figure 19.5 depicts the tangential field distribution along the starting contour vis-à-vis the tangential field distribution along the optimized contour [10].

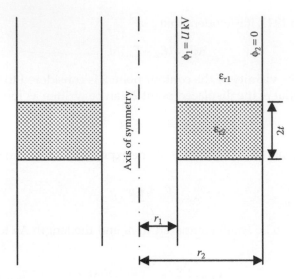

**FIGURE 19.3**
Schematic representation of the disc insulator considered for optimization study. (Data from Grönewald, H., *IEEE Proc. C,* 130, 4, 201–205, 1983.)

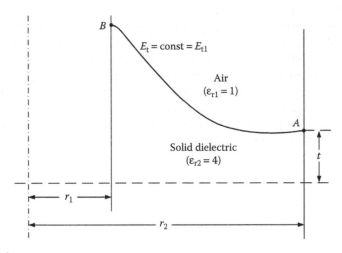

**FIGURE 19.4**
Optimized contour for a constant tangential field intensity along insulator surface.

## 19.3.2 Electrode and Insulator Contour Correction with Approximation of Corrected Contour

Stih [11] proposed a procedure for optimization of electrode and insulation system combining the approach of varying the distances between contour points and then approximating the corrected contour by circular arcs. The procedure was applied for optimizing axi-symmetric configurations. In the

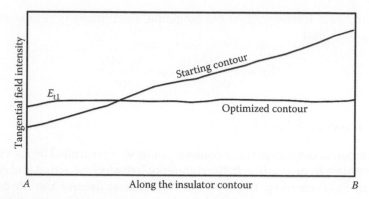

**FIGURE 19.5**
Tangential field intensity distribution along the starting and optimized insulator contours.

proposed approach, the target was to attain uniform distribution of normal field intensity on the electrode surface and uniform distribution of tangential field intensity on the insulator surface. The contour correction principle, as presented in Reference [11] is described below.

As shown in Figure 19.6a if the point 1 moves to 1' in the normal direction while the potential $V_A$ of the electrode is kept constant, then the normal field intensity $E_A$ will increase to $E_{A'}$, such that

$$E_A = \frac{V_A - V_B}{l_{12}} \text{ and } E_{A'} = \frac{V_A - V_B}{l_{1'2}}$$ (19.17)

where:
$E_{A'} > E_A$ as $l_{1'2} < l_{12}$

On the other hand, if the point $P$ on the insulator surface moves to the $P'$, as shown in Figure 19.6b, along the equipotential line, that is, normal to

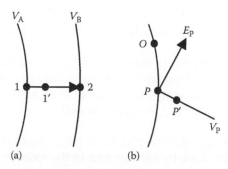

**FIGURE 19.6**
Principle of contour correction: (a) for electrode and (b) for insulator.

the electric field intensity, then the tangential field intensity decreases from $E_{tP}$ to $E_{tP'}$ such that

$$E_{tP} = \frac{V_O - V_P}{l_{OP}} \text{ and } E_{tP'} = \frac{V_O - V_P}{l_{OP'}} \tag{19.18}$$

where:
$E_{OP'} < E_{OP}$ as $l_{OP'} > l_{OP}$

The quantum of displacement of contour points was controlled by moving the contour points in proportion to the difference between the actual and desired values of field intensity. Let $\vec{l}_{Ai}$ be the vector that defines the $i$th contour point on the actual contour, while $\vec{l}_{ci}$ be the vector that defines the $i$th contour point on the corrected contour. Then these two vectors were related as

$$\vec{l}_{ci} = \vec{l}_{Ai}\left(1 + \frac{\Delta E_i}{E_i}\hat{u}_{ni}\right) \tag{19.19}$$

where:
$\hat{u}_{ni}$ is the unit vector normal to the electrode or to the electric field intensity at the $i$th contour point on the surface of electrode or insulator, as the case may be

In Equation 19.19,

$$\Delta E_i = E_{di} - E_{Ai} \qquad \text{for} \quad \Delta E_{mx} < f_1 E_{di}$$

$$\text{and} \quad \Delta E_i = \frac{E_{di} - E_{Ai}}{E_{mx}} f_1 \qquad \text{for} \quad \Delta E_{mx} > f_1 E_{di} \tag{19.20}$$

where:

$$\Delta E_{mx} = \max\left\{|E_{di} - E_{Ai}|\right\} \quad i = 1, \dots, N$$

where:
$E_{di}$ is the desired value of electric field intensity
$E_{Ai}$ is the value of electric field intensity on the actual contour at the $i$th contour point
$N$ is the number of contour points for which correction is carried out
$f_1$ is a factor that limits the quantum of displacement of the contour points, as Equations 19.17 and 19.18 are valid only for small displacements. Typically the value of $f_1$ was chosen as 0.1

In the next step of the optimization procedure [11], smooth approximation of the corrected contour was done by circular arcs in the following way. For smooth approximation of a particular set of points $H = \left[P_i(x_i, y_i)\right]$, $i = 1, \dots, N$ by a circular arc, the conditions that are to be satisfied are as follows: (1) the centre of

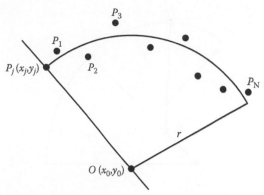

**FIGURE 19.7**
Approximation of a set of points by circular arc.

the circle should be on a given line $y = m_h x + c_h$, (2) the circle passes through the given point $P_j(x_j,y_j)$ and (3) the circle is the best approximation of the given set of points. With reference to Figure 19.7, it may be written that

$$y_0 = m_h x_0 + c_h \tag{19.21}$$

$$\text{and } (x_j - x_0)^2 + (y_j - y_0)^2 = r^2 \tag{19.22}$$

where:
  $r$ is the radius
  $(x_0,y_0)$ are the coordinates of the centre of the circular arc

Considering the conditions given by Equations 19.21 and 19.22, for best approximation the minimum is to be found of the function $G$, which is given below

$$G = \sum_{i=1}^{N} \left( \sqrt{(x_i - x_0)^2 + (y_i - y_0)^2} - r \right)^2 \tag{19.23}$$

For this purpose, the minimum of the function $F$ is to be determined, where $F$ is given as follows:

$$F = G + \alpha(y_0 - m_h x_0 - c_h) + \beta \left[ (x_j - x_0)^2 + (y_j - y_0)^2 - r^2 \right] \tag{19.24}$$

The conditions to be satisfied for minimum $F$ are

$$\frac{\partial F}{\partial x_0} = 0, \frac{\partial F}{\partial y_0} = 0, \frac{\partial F}{\partial r} = 0, \frac{\partial F}{\partial \alpha} = 0, \frac{\partial F}{\partial \beta} = 0 \tag{19.25}$$

From these conditions a system of equations for the unknown variables $x_0$, $y_0$, $r$, $\alpha$ and $\beta$ was obtained, which in turn was reduced to a non-linear equation for $x_0$ that was solved numerically.

Further, the smooth approximation of complex electrode and insulator contours by several circular arcs is based on the fact that two circular arcs are

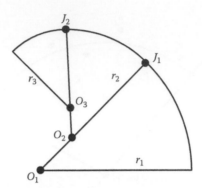

**FIGURE 19.8**
Smooth joining of circular arcs.

joined smoothly at a point $J1$ or $J2$ if the centres of the two circles lie on the line passing through the point $J1$ or $J2$, as shown in Figure 19.8.

The optimization process was started by choosing an initial contour from not only electrical considerations but also considering mechanical, thermal and other technical constraints. The electric field distribution was computed numerically using integral equation technique. The distances between the contour points were changed using the following rule: if the maximum field intensity is lower than the allowable value then the distances are decreased and vice versa. The stopping criteria of the optimizations steps were

$$\frac{E_{max}}{E_{min}} < f_u$$

(19.26)

$$\text{and} \quad E_{allow} > E_{max} > f_e E_{allow}$$

where:

$E_{max}$, $E_{min}$ and $E_{allow}$ are maximum, minimum and allowable values of electric field intensities, respectively
$f_u$ is the field uniformity factor
$f_e$ is the chosen limit of maximum field intensity with respect to allowable field intensity. Typical range of values of $f_u$ and $f_e$ were $1 < f_u < 1.1$ and $0.9 < f_e < 1$, respectively

Stih [11] applied the methodology for optimization of the shielding electrode of a transformer winding. The initial geometry of the system is shown in Figure 19.9, where the shielding electrode is chosen as semicircle. The optimized final geometry of the shielding electrode is shown in Figure 19.10, which was obtained for $f_u = 1.06$ and $f_e = 0.99$ [11].

### 19.3.3 Parametric Optimization of Insulator Profile

Abdel-Salam and Stanek [12] described a method to obtain uniform tangential field intensity along high-voltage insulator surface by modifying the profile.

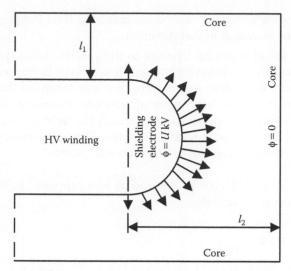

**FIGURE 19.9**
Initial geometry of shielding electrode with field distribution.

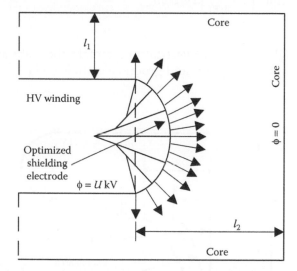

**FIGURE 19.10**
Final optimized geometry of shielding electrode with field distribution.

In this algorithm actual value of tangential field intensity was used to correct the mathematical expression of the insulator contour to achieve the final optimized profile. In order to search for a profile for which tangential field intensity is uniform along the profile, the following procedure was adopted:

1. Tangential field intensity values were computed along the non-optimized profile numerically. If the tangential field intensity increases

while moving along the axis of the insulator, then the profile radius needs to be increased in that direction.

2. In order to get a smooth change in the profile radius, pre-defined mathematical expression for the profile was used in this work. While formulating the profile expression, care was taken, so that the contact angle at the insulator–electrode junction remained 90°. In other words, the profile expression must satisfy the right angle contact criterion for the insulator with electrodes. This is necessary to avoid excessive field concentration at the contact points.

As shown in Figure 19.11, the profile radius was proposed to have an exponential variation as given below.

$$r = R_0 \quad \text{for} \quad 0 \le z \le \frac{xh}{100}$$
$$r = R_0 + DR\left[e^{0.693[z-(xh/100)][100/(100-2x)h]} - 1\right] \quad \text{for} \quad \frac{xh}{100} < z \le \left(1 - \frac{x}{100}\right)h \qquad (19.27)$$
$$r = R_0 + DR \quad \text{for} \quad \left(1 - \frac{x}{100}\right)h < z \le h$$

where:
$x\%$ of the height $h$ was kept vertical to maintain the right angle contact criterion at both the live and ground electrode ends

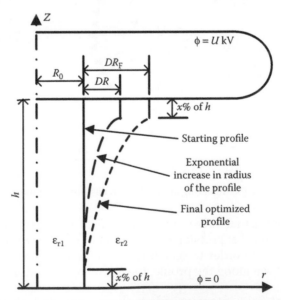

**FIGURE 19.11**
Profile optimization by exponential variation of profile radius.

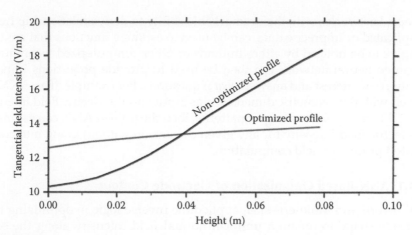

**FIGURE 19.12**
Distribution of tangential field intensity along non-optimized as well as optimized profiles.

Typically $x$ was chosen to be 5%. $DR$ was the largest value of enlargement in profile radius through exponential increase along the $z$-direction from the ground end to live end. As a result, the radius of the insulator at the ground end was $R_0$ and that at the live end was $R_0 + DR$.

$N$ number of contour points were chosen along the insulator profile and the tangential field intensities were computed at these points by numerical field computation. Then the value of $DR$ was changed iteratively to the final value $DR_F$ to achieve an acceptable degree of uniformity in tangential field intensity along the optimized insulator profile, which is depicted in Figure 19.11. Figure 19.12 shows the tangential field distribution for non-optimized as well as for the optimized profiles [12].

## 19.4 ANN-Based Optimization of Electrode and Insulator Contours

ANNs offer a completely different approach to problem solving. An ANN is a behavioural model built through learning from a number of examples of the behaviour. It transforms the given data pertaining to a problem into a model or predictor, and then applies this model to the present data to obtain an estimate. ANN has two fundamental features: (1) the ability to represent any function, linear or not, simple or complicated. ANNs are what mathematicians call *universal approximators* and (2) the ability to learn from representative examples so that model building is automatic. The advantages of ANNs are (1) one goes directly from factual data to the model without any manual work, without tainting the result with oversimplification or pre-conceived ideas and (2) there is no need to postulate a model or to amend it.

Neural networks, with their remarkable ability to derive relationship from complicated or imprecise data, can be used to estimate functions that are too complex to be noticed by either humans or other computerized techniques. A trained neural network can then be used to provide projections in new situations of interest and answer *what if* questions. For example, if an ANN is trained with the geometric dimensions as input and the electric field intensities at different points as output patterns, then the trained ANN can estimate the electric field intensity for any given dimension without going into complicated process of field computation.

### 19.4.1 ANN-Based Optimization of Electrode Contour

Chakravorti and Mukherjee [34] applied the inverse logic in optimizing the electrode shape to obtain a uniform normal field intensity along the end profile of a parallel disc electrode arrangement. In this approach, the ANN was trained with the electric field intensities as input patterns and the geometric dimensions as the output patterns. Then the trained ANN was used to predict the geometric dimensions of the electrode to get a uniform field intensity along the electrode end profile. Chakravorti and Mukherjee [34] optimized the end profile of an axi-symmetric electrode arrangement in the form of parallel discs using the above-mentioned inverse logic. For the purpose of training the ANN, input–output patterns need to be generated. These training patterns were generated by numerical field computations carried out for pre-defined electrode shapes for which the end profile was taken to be circular, as shown in Figure 19.13.

For the purpose of generating the training data, radius of parallel discs and the separation distance between the discs were kept constant at $r_d$ and

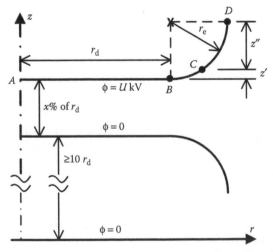

**FIGURE 19.13**
Parallel disc electrode configuration with circular end profile.

30% of $r_d$, respectively. $N_C$ numbers of different contours were obtained by varying the radius of end profile ($r_e$) in steps. For each of these $N_C$ contours, electric field intensities were calculated at ($N_P + 2$) number of points on the electrode surface, out of which two were on the parallel surface and the rest $N_P$ were on the end profile. The electric field intensities at the ($N_P + 2$) points in each contour were then given as input pattern vector to the ANN, while the coordinates of the $N_P$ points on the end profile were given as output pattern vector. The $N_P$ points on the end profile were taken at fixed $z$-coordinates for all the circular end profiles having pre-defined radii. Hence, only the $r$-coordinates of the $N_P$ points were given as output pattern vector. The ANN is thus made up of ($N_P + 2$) and $N_P$ neurons in the input and output layers, respectively, and there were $N_C$ number of input–output patterns. For preparing the training set, electric field computations were carried out considering the potential of the live electrode as 0.1 kV, where 100 V represented the percent potential difference between the two electrodes.

For electrode contour optimization, for every set of input pattern the output pattern was known. Hence, for this problem, ANN with supervised learning was implemented involving multilayer feed-forward network with error-back propagation [34]. After completion of the training process, an optimized end profile was determined as a test case with the desired field intensity value of 390 V/m in the critical domain between $B$ and $C$ on the end profile, as shown in Figure 19.13. For this optimized end profile, as shown in Figure 19.14, electric field computations were carried out to determine the electric field intensities along the end profile to find out the deviation of the test results from the

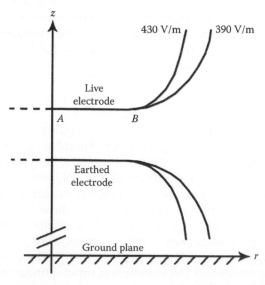

**FIGURE 19.14**
Optimized end profiles for two values of desired uniform field intensities in the critical domain.

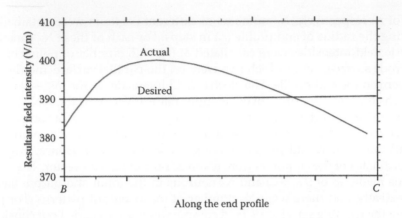

**FIGURE 19.15**
Actual field intensity distribution vis-à-vis the desired uniform field intensity in the critical domain of the end profile.

desired value of 390 V/m. MAE in field results along profiles obtained from ANN as output with respect to the desired values was found to be ≈3% [34]. The variation of actual electric field intensity distribution for a desired uniform field intensity of 390 V/m is shown in Figure 19.15 [34].

### 19.4.2 ANN-Based Optimization of Insulator Contour

Using an approach similar to that discussed in Section 19.4.1, Bhattacharya et al. [38] reported ANN-based contour optimization of axi-symmetric insulators in multi-dielectric arrangements to obtain not only uniform but also complex electric field intensity distributions along the insulator surface. For the purpose of generating the training data, electric field computations were carried for conical support insulators having linear contour, as shown in Figure 19.16.

In electric field computations $h$, $h_1$ and $h_2$ were kept constant. Typically $h_1$ and $h_2$ were taken as 10% of $h$. $N_R$ numbers of different values of $R_2$ were taken when $R_1$ was kept constant. Similarly, $N_R$ numbers of different values of $R_1$ were taken when $R_2$ was kept constant. Therefore, there were $2N_R$ numbers of contours for generating the training data. Tangential stresses were calculated at $N_P$ numbers of points on all the $2N_R$ numbers of pre-defined insulator surfaces. For each of the $N_P$ points, the z-coordinate was kept fixed on all the pre-defined contours. For every one of these $2N_R$ contours, $N_P$ tangential field intensities were given to the ANN as input pattern vector and the corresponding r-coordinates of the $N_P$ points were given as output pattern vector during training. Therefore, the ANN had $N_P$ neurons each for input and output layers and $2N_R$ numbers of input–output training patterns. Multi-layer feed-forward network with error-back propagation as well as resilient propagation were employed for supervised learning of ANN.

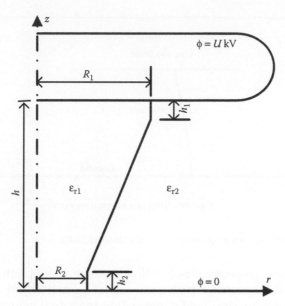

**FIGURE 19.16**
Conical support insulator having linear contour.

After completion of successful training, a normalized uniform tangential field intensity of 0.1 V/cm was desired along the insulator surface, as shown in Figure 19.17. $N_P$ values of uniform tangential field intensities at the $N_P$ specified points were fed to the trained ANN as inputs and the corresponding $r$-coordinates of the $N_P$ points were obtained as outputs from the trained ANN. These $r$-coordinates, as obtained from the trained ANN as outputs,

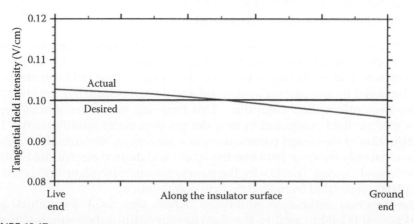

**FIGURE 19.17**
Desired and actual distributions of tangential field intensity along the optimized insulator contour.

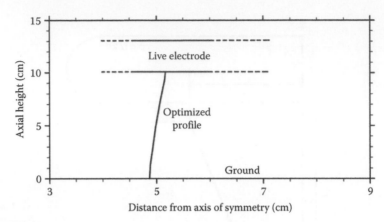

**FIGURE 19.18**
Optimized insulator contour as obtained from the trained ANN.

along with their respective $z$-coordinates were then plotted to get the optimized insulator profile, which is shown in Figure 19.18 [38]. Subsequently, with the help of numerical field computation the actual values of tangential field intensities at these $N_P$ points on the optimized insulator contour were determined. MAE between the actual and desired tangential field intensities was found to be $\approx 2\%$ [38]. The actual and desired tangential field intensity distributions along the optimized insulator profile of Figure 19.18 are shown in Figure 19.17.

## 19.5  ANN-Aided Optimization of 3D Electrode–Insulator Assembly

A simplified flow chart, as shown in Figure 19.19, explains the iterative structure of conventional procedure for the optimization of electrode or insulator contours. It is a two-step procedure, in which, electric field distributions are computed by any particular method and design parameters are changed using any optimization algorithm. This two-step procedure strictly separates electric field computation and design parameter modifications. The modification of the design parameters does not require absolute electric field values, but only the error between the actual and desired electric field values in each iterative step. That is why the parameter updation algorithm could be completely separated from electric field computation.

The function estimation ability of ANN was used by Lahiri and Chakravorti [42,43] to modify the flowchart of optimization process shown in Figure 19.19. It has been mentioned in Section 19.4 that if an ANN is trained with the geometric dimensions as input and the electric field intensities

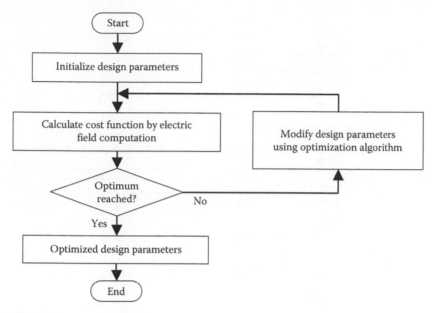

**FIGURE 19.19**
Flowchart of conventional optimization process.

as output patterns, then the trained ANN can estimate the electric field intensity for any set of given dimensions without going into detailed electric field computation. Because of decoupling of electric field computation and parameter updation routines in the optimization process, it was possible to replace the electric field computation routine by a trained ANN to get an estimate of the electric field distribution for the values of design parameters in each iterative step. The modified flowchart is shown in Figure 19.20.

Major saving in computation time was achieved in References [42,43] by coupling a trained ANN in the optimization loop. At each move of the optimization loop, the evaluation of the cost function through electric field computation routine would require, say, $T$ units of time. Now supposing the optimization algorithm requires $N$ number of moves to converge to the optimum value of the cost function, the total time required for the entire optimization process would be $N^*(T + t)$ units of time, where $t$ is the execution time of each step of the optimization algorithm. Typically, for any numerical field computation method, the value of $T$ will be of the order of several minutes for real-life 3D configurations. But if the cost function is estimated by a trained ANN, then it requires only a fraction of a second to do so, instead of $T$ minutes. However, in this proposed method, the time consuming part is the training of ANN plus the time required for electric field computation routine to generate the desired number of training sets, which requires several minutes. But, it is to be noted here that the training set is required to be generated only once and the ANN is required to be trained once, too. Hence,

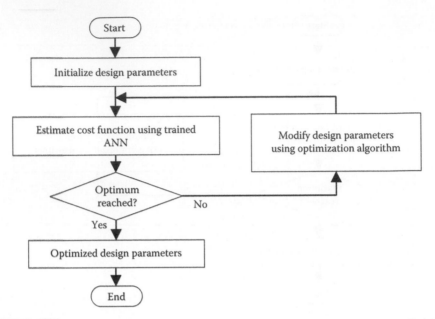

**FIGURE 19.20**
Flowchart of ANN-aided optimization process.

any optimization algorithm coupled to a trained ANN offers considerable saving in time for 3D electric field optimization. Another major advantage of the proposed method was that the upper and lower bounds of the design parameters could be easily modified by varying the range of training data for the ANN.

Lahiri and Chakravorti [42] optimized the 3D electrode–support insulator assembly shown in Figure 19.21 using the above-mentioned methodology employing GA. This configuration is often used in gas-insulated transmission lines (GIL). In this arrangement, the most critical parts of the profiles were (1) the shape of the hole in the electrode through which the support insulator was to be inserted as shown in Figure 19.21b, and (2) the curved profile of the insulator in the middle, as shown in Figure 19.21c. As a result the design parameters chosen for optimization were related to the shape of these critical parts.

In another work, Lahiri and Chakravorti [43] optimized 3D electrode-disc insulator assembly used in GIL shown in Figure 19.22 employing SA for optimization. In this arrangement, the most critical parts of the profiles were the curvatures of the live electrode as well as the disc-type insulator as shown in Figure 19.22b. Consequently, the design parameters chosen for optimization were related to these critical curvatures.

For both these optimization problems, sensitivity studies were carried out first to determine which design parameters of the electrode–insulator assembly are to be modified to get a minimum value of field intensity within

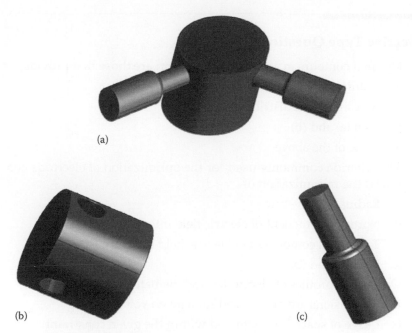

**FIGURE 19.21**
**(See colour insert.)** 3D electrode–support insulator assembly for optimization: (a) electrode–insulator assembly; (b) live electrode and (c) support insulator.

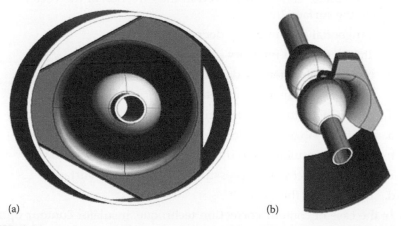

**FIGURE 19.22**
**(See colour insert.)** 3D electrode–disc insulator assembly for optimization: (a) disc-type insulator with live and ground electrodes and (b) live electrode with sectional view of disc insulator.

the GIL. The most sensitive design parameters were initialized and then modified using the flowchart given in Figure 19.20. The cost function for optimization was maximum value of the electric field intensity, which was minimized through optimization routines such as GA or SA.

## Objective Type Questions

1. Optimal contours in an insulating system are those that provide
   a. A uniform field distribution
   b. A desired field distribution
   c. Both (a) and (b)
   d. None of the above

2. The criterion commonly used for the optimization of electrode contour is the minimization of
   a. Radius of curvature
   b. Normal component of electric field intensity
   c. Tangential component of electric field intensity
   d. Both (a) and (b)

3. Optimized profiles of electrodes and insulators are such that
   a. Dimensions are minimized for a given voltage rating
   b. Usage of space is maximized within the given constraint
   c. Cost of installation is minimized
   d. All the above

4. If the insulator contour is moved in the direction of the vector normal to the surface, then
   a. Tangential field intensity decreases
   b. Tangential field intensity increases
   c. Normal field intensity decreases
   d. Normal field intensity increases

5. Target for insulator contour optimization is achieving
   a. Uniform tangential field intensity on the insulator surface
   b. Uniform resultant field intensity on the insulator surface
   c. Uniform electrostatic pressure on the insulator surface
   d. Both (a) and (b)

6. In the case of contour correction technique, insulator contour optimization is done by keeping
   a. Potential difference between two successive contour points constant
   b. Field intensity difference between two successive contour points constant

   c. Distance between two successive contour points constant

   d. Both (a) and (c)

7. In the case of contour correction technique, the quantum of displacement of contour points is controlled by moving the contour points in proportion to the difference between the actual and desired values of

   a. Electric potential

   b. Electric field intensity

   c. Electric flux density

   d. Both (a) and (b)

8. In the field optimization algorithms, the modification of the design parameters requires

   a. Absolute values of electric potential

   b. Absolute values of electric field intensity

   c. Error between the actual and desired values of electric potential

   d. Error between the actual and desired values of the electric field intensity

9. In ANN-aided electric field optimization, it is possible to replace the electric field computation routine by a trained ANN to get an estimate of the electric field distribution because of

   a. Decoupling of electric field computation and parameter updation routines

   b. Coupling of electric field computation and parameter updation routines

   c. Decoupling parameter initialization and parameter updation routines

   d. Both (b) and (c)

10. In ANN-aided electric field optimization, the upper and lower bounds of the design parameters could be easily modified by varying the range of

   a. Training data of the ANN

   b. Testing data of the ANN

   c. Both (a) and (b)

   d. None of the above

**Answers:** 1) c; 2) b; 3) d; 4) a; 5) d; 6) d; 7) b; 8) d; 9) a; 10) a

## References

1. J. Spielrein, 'Geometrisches zur elektrischen Festigkeitsrechnung', *Archiv für Elektrotechnik*, Vol. 4, pp. 78–85, 1915.
2. W. Rogowski, 'Die elektrische Festigkeit am Rande des Platten-Kondensators', *Archiv für Elektrotechnik*, Vol. 12, pp. 1–14, 1923.
3. N. Félici, 'Les surfaces á champ électrique constant', *Revue Générale de l'électricité*, Vol. 59, pp. 479–501, 1950.
4. H. Singer and P. Grafoner, 'Optimization of electrode and insulator contours', *Proceedings of the 2nd ISH*, Zurich, Switzerland, pp. 111–116, September 9–13, 1975.
5. H. Singer, 'Computation of optimized electrode geometries', *Proceedings of the 3rd ISH*, Milan, Italy, Paper No. 11.06, August 28–31, IEEE North Italy Section Publication, 1979.
6. D. Metz, 'Optimization of high voltage fields', *Proceedings of the 3rd ISH*, Milan, Italy, Vol. 1, Paper No. 11.12, August 28–31, IEEE North Italy Section Publication, 1979.
7. H. Okubo, T. Amemiya and M. Honda, 'Borda's profile and electric field optimization by using charge simulation method', *Proceedings of the 3rd ISH*, Milan, Italy, August 28–31, IEEE North Italy Section Publication, Paper No. 11.16, 1979.
8. T. Misaki, H. Tsuboi, K. Itaka and T. Hara, 'Computation of three-dimensional electric field problems and its application to optimum insulator design', *IEEE Transactions on Power Apparatus and Systems*, Vol. PAS-101, pp. 627–634, 1982.
9. T. Misaki, H. Tsuboi, K. Itaka and T. Hara, 'Optimization of three-dimensional electrode contour based on surface charge method and its application to insulation design', *IEEE Power Engineering Review*, Vol. 82, p. 44, 1983.
10. H. Grönewald, 'Field optimisation of high-voltage electrodes', *IEEE Proceedings C – Generation, Transmission and Distribution*, Vol. 130, No. 4, pp. 201–205, 1983.
11. Z. Stih, 'High voltage insulating system design by application of electrode and insulator contour optimization', *IEEE Transactions on Electrical Insulation*, Vol. 21, No. 4, pp. 579–584, 1986.
12. M. Abdel-Salam and E.K. Stanek, 'Field optimization of high-voltage insulators', *IEEE Transactions on Industry Applications*, Vol. 22, No. 4, pp. 594–601, 1986.
13. M. Abdel-Salam and E.K. Stanek, 'Optimizing field stress on HV insulators', *IEEE Transactions on Electrical Insulation*, Vol. 22, No. 1, pp. 47–56, 1987.
14. W. Mosch, M. Dietrich, E. Lemke and W. Hauschild, 'Statistical design of large UHV electrodes in air', *IEEE Proceedings A – Physical Science, Measurement and Instrumentation, Management and Education – Reviews*, Vol. 133, No. 8, pp. 547–551, 1986.
15. K. Kato, X. Han and H. Okubo, 'Optimization technique of high voltage electrode contour considering V-t characteristics based on volume-time theory', *Proceedings of the 11th ISH*, London, Vol. 2, pp. 123–126, August 23–27, IEE (UK) Publication, 1999.
16. J. Liu and J. Sheng, 'The optimization of the high voltage axi-symmetrical electrode contour', *IEEE Transactions on Magnetics*, Vol. 24, No. 1, pp. 39–42, 1988.
17. J. Liu, E.M. Freeman, X. Yang and J. Sheng, 'Optimization of electrode shape using the boundary element method', *IEEE Transactions on Magnetics*, Vol. 26, No. 5, pp. 2184–2186, 1990.

18. T.N. Judge and J. Lopez-Roldan, 'Optimization of high voltage equipment design using boundary element method based electromagnetic analysis tools', *Proceedings of the 11th ISH*, London, Vol. 2, pp. 140–143, August 23–27, IEE (UK) Publication, 1999.

19. J.D. Welly, 'Optimization of electrode contours in high voltage equipment using circular contour elements', *Proceedings of the 5th ISH*, Braunschweig, Germany, Paper No. 31.03, August 24–28, 1987.

20. D. Huimin et al., 'Optimization of high voltage fields of multi-electrode systems', *Proceedings of the 6th ISH*, New Orleans, LA, Paper No. 40.03, August 28–September 1, 1989.

21. H.H. Däumling and H. Singer, 'Investigations on field optimization of insulator geometries', *IEEE Transactions on Power Delivery*, Vol. 4, No. 1, pp. 787–793, 1989.

22. W.M. Caminhas, R.R. Saldanha and G.R. Mateus, 'Optimization methods used for determining the geometry of shielding electrodes', *IEEE Transactions on Magnetics*, Vol. 26, No. 2, pp. 642–645, 1990.

23. Z. Fang, J. Jicun and Z. Ziyu, 'Optimal design of HV transformer bushing', *3rd International Conference on Properties and Applications of Dielectric Materials*, Tokyo, Japan, pp. 434–437, July 8–12, IEEE Publication, 1991.

24. M. Abdel-Salam and A. Mufti, 'Optimizing field stress on high voltage bushings', *IEEE International Symposium on Electrical Insulation*, Pittsburg, PA, pp. 225–228, June 5–8, IEEE Publication, 1994.

25. E.S. Kim et al., 'Electric field optimization using NURB curve and surface', *Transactions of IEEJ*, Vol. 113-B, No. 10, pp. 1081–1087, 1993.

26. K. Kato et al., 'A highly efficient method for determination of electric field optimum contour on high voltage electrode', *Proceedings of the 9th ISH*, Graz, Austria, Paper No. 8358, August 28–September 1, 1995.

27. J.A.G. Gacia et al., 'Contour optimization of high-voltage insulators by means of smoothing cubic splines', *Proceedings of the 9th ISH*, Graz, Austria, Paper No. 8343, August 28–September 1, 1995.

28. H. Okubo et al., 'The development of electric field optimization technique using personal computer', *Proceedings of the 8th ISH*, Yokohama, Japan, Paper No. 11.04, August 23–27, 1993.

29. K. Kato, M. Hikita, N. Hayakawa, Y. Kito and H. Okubo, 'Development of personal-computer-based high efficient technique for electric field optimization', *European Transactions on Electrical Power Engineering*, Vol. 5, pp. 401–407, 1995.

30. C. Trinitis et al., 'Accelerated 3-D optimization of high voltage apparatus', *Proceedings of the 9th ISH*, Graz, Austria, Paper No. 8867, August 28–September 1, 1995.

31. C. Trinitis, 'Field optimization of three dimensional high voltage equipment', *Proceedings of the 11th ISH*, London, Vol. 2, pp. 75–78, August 23–27, IEE (UK) Publication, 1999.

32. J.A. Gomollon, G. Gonzalez-Filgueira and E. Santome, 'Optimization of high voltage insulators within three dimensional field distributions', *Proceedings of the 6th IEEE ICPADM*, Xian, China, Vol. 2, pp. 629–632, June 21–26, IEEE Publication, 2000.

33. J.G. Leopold, U. Dai, Y. Finkelstein and E. Weissman, 'Optimizing the performance of flat-surface, high-gradient vacuum insulators', *IEEE Transactions on Dielectrics & Electrical Insulation*, Vol. 12, No. 3, pp. 530–536, 2005.

34. S. Chakravorti and P.K. Mukherjee, 'Application of artificial neural networks for optimization of electrode contour', *IEEE Transactions on Dielectrics & Electrical Insulation*, Vol. 1, No. 2, pp. 254–264, 1994.
35. K. Bhattacharya, S. Chakravorti and P.K. Mukherjee, 'An application of artificial neural network in the design of toroidal electrodes', *Journal of the IE(I), Pt. CP*, Vol. 76, pp. 14–20, 1995.
36. P.K. Mukherjee, C. Trinitis and H. Steinbigler, 'Optimization of HV electrode systems by neural networks using a new learning method', *IEEE Transactions on Dielectrics & Electrical Insulation*, Vol. 3, No. 6, pp. 737–742, 1996.
37. H. Okubo, T. Otsuka, K. Kato, N. Hayakawa and M. Hikita, 'Electric field optimization of high voltage electrode based on neural network', *IEEE Transactions on Power Systems*, Vol. 12, No. 4, pp. 1413–1418, 1997.
38. K. Bhattacharya, S. Chakravorti and P.K. Mukherjee, 'Insulator contour optimization by artificial neural network', *IEEE Transactions on Dielectrics & Electrical Insulation*, Vol. 8, No. 2, pp. 157–161, 2001.
39. A. Chatterjee, A. Rakshit and P.K. Mukherjee, 'A self-organizing fuzzy inference system for electric field optimization of HV electrode systems', *IEEE Transactions on Dielectrics & Electrical Insulation*, Vol. 8, No. 6, pp. 995–1002, 2001.
40. P.G. Alotto, B. Brandstätter, E. Cela, G. Fürntratt, C. Magele, G. Molinari, M. Nervi, K. Preis, M. Repetto and H.R. Richter, 'Stochastic algorithms in electromagnetic optimization', *IEEE Transactions on Magnetics*, Vol. 34, No. 5, pp. 3674–3684, 1998.
41. Z. Guoqiang, Z. Yuanlu and C. Xiang, 'Optimal design of high voltage bushing electrode in transformer with evolution strategy', *IEEE Transactions on Magnetics*, Vol. 15, pp. 1690–1693, 1999.
42. A. Lahiri and S. Chakravorti, 'Electrode-spacer contour optimization by ANN aided genetic algorithm', *IEEE Transactions on Dielectrics & Electrical Insulation*, Vol. 11, No. 6, pp. 964–975, 2004.
43. A. Lahiri and S. Chakravorti, 'A novel approach based on simulated annealing coupled to artificial neural network for 3D electric field optimization', *IEEE Transactions on Power Delivery*, Vol. 20, No. 3, pp. 2144–2152, 2005.
44. S. Banerjee, A. Lahiri and K. Bhattacharya, 'Optimization of support insulators used in HV systems using support vector machine', *IEEE Transactions on Dielectrics & Electrical Insulation*, Vol. 14, No. 2, pp. 360–367, 2007.
45. P. Kitak, J. Pihler, I. Ticar, A. Stermecki, C. Magele, O. Bíró and K. Preis, 'Use of an optimization algorithm in designing medium-voltage switchgear insulation elements', *IEEE Transactions on Magnetics*, Vol. 42, No. 4, pp. 1347–1350, 2006.
46. K. Kato, S. Kaneko, S. Okabe and H. Okubo, 'Optimization technique for electrical insulation design of vacuum interrupters', *IEEE Transactions on Dielectrics & Electrical Insulation*, Vol. 15, No. 5, pp. 1456–1463, 2008.
47. X. Liu, Y. Cao, F. Wen and E. Wang, 'Optimization of extra-high-voltage $SF_6$ circuit breaker based on improved genetic algorithm', *IEEE Transactions on Magnetics*, Vol. 44, No. 6, pp. 1138–1141, 2008.
48. H.J. Ju, K.C. Ko and S.K. Choi, 'Optimal design of a permittivity graded spacer configuration in a gas insulated switchgear', *Journal of Korean Physical Society*, Vol. 55, No. 5, p. 1803, 2009.
49. H. Ju, B. Kim and K. Ko, 'Optimal design of an elliptically graded permittivity spacer configuration in gas insulated switchgear', *IEEE Transactions on Dielectrics & Electrical Insulation*, Vol. 18, No. 4, pp. 1268–1273, 2011.

50. S. Monga, R.S. Gorur, P. Hansen and W. Massey, 'Design optimization of high voltage bushing using electric field computations', *IEEE Transactions on Dielectrics & Electrical Insulation*, Vol. 13, No. 6, pp. 1217–1224, 2006.
51. M.R. Hesamzadeh, N. Hosseinzadeh and P. Wolfs, 'An advanced optimal approach for high voltage AC bushing design', *IEEE Transactions on Dielectrics & Electrical Insulation*, Vol. 15, No. 2, pp. 461–466, 2008.
52. C. Trinitis, 'Automatic high voltage apparatus optimization: Making it more engineer-friendly', *International Symposium on Parallel Computing in Electrical Engineering*, Bialystok, Poland, September 13–17, IEEE Publication, 2006.

# Index

*Note:* Locators followed by "*f*" and "*t*" denote figures and tables in the text

Printed and bound by CPI Group (UK) Ltd, Croydon, CR0 4YY

24/10/2024

01778302-0017